AF777833

TiO_2 Nanotube Arrays

Craig A. Grimes • Gopal K. Mor

TiO_2 Nanotube Arrays

Synthesis, Properties, and Applications

Springer

Craig A. Grimes
Pennsylvania State University
Electrical Engineering
Department
University Park, PA 16802
USA
cgrimes@engr.psu.edu

Gopal K. Mor
Pennsylvania State University
Materials Research Institute
University Park, PA 16802
USA
gkm1@ psu.edu

ISBN 978-1-4419-0067-8 e-ISBN 978-1-4419-0068-5
DOI 10.1007/978-1-4419-0068-5
Springer Dordrecht Heidelberg London New York

Library of Congress Control Number: 2009930642

© Springer Science + Business Media, LLC 2009
All rights reserved. This work may not be translated or copied in whole or in part without the written permission of the publisher (Springer Science+Business Media, LLC, 233 Spring Street, New York, NY 10013, USA), except for brief excerpts in connection with reviews or scholarly analysis. Use in connection with any form of information storage and retrieval, electronic adaptation, computer software, or by similar or dissimilar methodology now known or hereafter developed is forbidden.
The use in this publication of trade names, trademarks, service marks, and similar terms, even if they are not identified as such, is not to be taken as an expression of opinion as to whether or not they are subject to proprietary rights.

Printed on acid-free paper

Springer is part of Springer Science+Business Media (www.springer.com)

Preface

As the title of this book implies, we examine the fabrication, properties, and different applications of highly ordered vertically oriented TiO_2 nanotube arrays made by Ti anodization. Since the first report of the highly ordered TiO_2 nanotube arrays by Gong et al. (J Mater Res 16:3331–3334, 2001), the field has grown so rapidly as to make it difficult to summarize the scope of all related work. Built upon a suite of synthesis chemistries that enable significant differences in nanotube morphologies, the TiO_2 nanotube arrays have demonstrated remarkable properties that have successfully given rise to a large number of diverse applications that include, but are certainly not limited to, super capacitors, solar cells, drug eluting surfaces, hydrogen sensors, biofiltration membranes, and photoelectrochemical cells for the solar generation of hydrogen. It appears, for the moment, impossible to completely canvas the field as it moves more rapidly than one can report on it. Be that as it may we have tried to at least touch on key aspects of the field, reviewing the contributions made by various research groups to the fabrication, and application of the nanotube arrays, and apologize to those authors of significant work we may have overlooked.

The book is intended for researchers in the field, which could encompass undergraduate students, graduate students and beyond. It is meant as both an introductory work as well as a reference to the field. While the book could be read cover to cover by those having such energy and interest, the individual chapters are self-contained per their specific topic. Chapter 1 considers nanotube array synthesis techniques. Chapter 2 summarizes the structural, elemental, optical, mechanical, and electrical properties of the TiO_2 nanotube arrays. Chapter 3 examines use of TiO_2 nanotube arrays as hydrogen gas sensors. Chapter 4 considers use of the material architecture for photoelectrochemical water splitting. Chapter 5 considers application of TiO_2 nanotube arrays to dye-sensitized solar cells and polymer-based solid state heterojunction solar cells. Chapter 6 discusses the use of TiO_2 nanotubes in biological applications, including their use as a base for biosensors, drug eluting surfaces, osseointegration, biotemplating, enhanced blood clotting, and as biofiltration membranes.

We wish to thank Emeritus Professor Dale M. Grimes for helpful comments towards improving the manuscript, the anonymous reviewers of the book who put forth many helpful suggestions, and thank the many authors who generously granted us permission to use their figures in this book, with a particular nod to Dr. Radim Beránek, Department of Chemistry & Pharmacy, University of Erlangen-Nürnberg, who was willing to redraw some of his figures for use herein.

University Park, PA

Craig A. Grimes
Gopal K. Mor

Contents

Introduction

Highly ordered, vertically oriented TiO_2 nanotube-arrays fabricated by potentiostatic anodization of titanium constitute a material architecture that offers a large internal surface area without a concomitant decrease in geometric and structural order. The precisely oriented nature of the nanotube arrays, see Figs. 1 and 2, makes them excellent electron percolation pathways for vectorial charge transfer between interfaces. TiO_2 nanotube arrays have been found to possess outstanding charge transport and carrier lifetime properties enabling a variety of advanced applications, including their use in sensors [1–9], dye sensitized solar cells [10–14], hydrogen generation by water photoelectrolysis [15–22], photocatalytic reduction of CO_2 under outdoor sunlight [23], and supercapacitors [24]. Furthermore, the nanotube arrays have demonstrated great utility in biomedical related applications including biosensors, molecular filtration, drug delivery, and tissue engineering [25–31]. In this introduction we briefly consider some aspects of material synthesis and, as a means to motivate reading of this book, touch-on some of the interesting applications of this material architecture.

The application of TiO_2 nanotube arrays to hydrogen gas sensing serves to illustrate their remarkable, and readily measurable, properties. At room temperature, in response to 1,000 ppm hydrogen TiO_2 nanotube arrays $\approx 1\ \mu m$ in length demonstrate an unprecedented change in electrical resistance of 8.7 orders of magnitude, or 50,000,000,000% [6, 7]; this is the largest known sensitivity of any material, to any gas, at any temperature. In their use as hydrogen sensors, the TiO_2 nanotube arrays possess such excellent photocatalytic properties that they are able to self-clean from even extreme contamination with exposure to ambient UV light, see Fig. 3, regaining their gas sensing properties [8, 9].

The self-organized architecture allows for the precise design and control of the geometrical features, allowing one to achieve a material with specific light absorption and propagation characteristics [11, 37–39]. The nanotube array geometry appears to be an, if not the, ideal architecture for water photoelectrolysis where

Fig. 1 (*Left*): FESEM cross-sectional image of titanium oxide nanotube array sample reported by Gong and co-workers in 2001 [32], achieved by anodization of a Ti foil sample in 0.5%wt HF solution at 20 V for 20 min. Use of the HF aqueous electrolyte limited nanotube array length to less than a micron. (*Right*): Illustrative FESEM image of a TiO_2 nanotube array, mechanically fractured, made by anodization of Ti foil in an aqueous KF containing electrolyte, reported by Cai and co-workers in 2005 [33]. The pore size is approximately 30 nm, wall thickness 13 nm. Use pH controlled KF or NaF aqueous baths extended nanotube array lengths to several microns enabling new applications

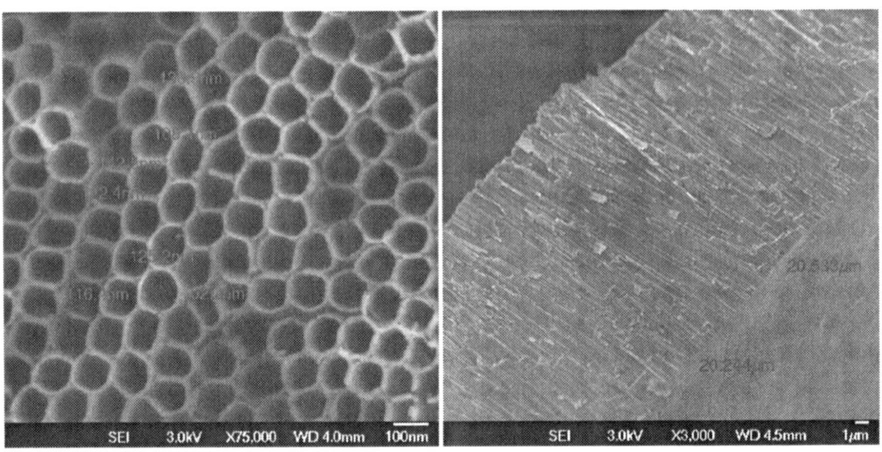

Fig. 2 Illustrative top and cross-sectional FESEM images of titania nanotube arrays synthesized by anodization of Ti films in an ethylene glycol electrolyte [20, 34–36]

under UV-spectrum illumination (320–400 nm) nanotube arrays $\approx$30 μm long demonstrate a photoconversion conversion efficiency of 16.25% [10, 20]. Considering such remarkable properties, it appears that TiO_2 nanotube arrays are more than worthy of careful consideration by the scientific community.

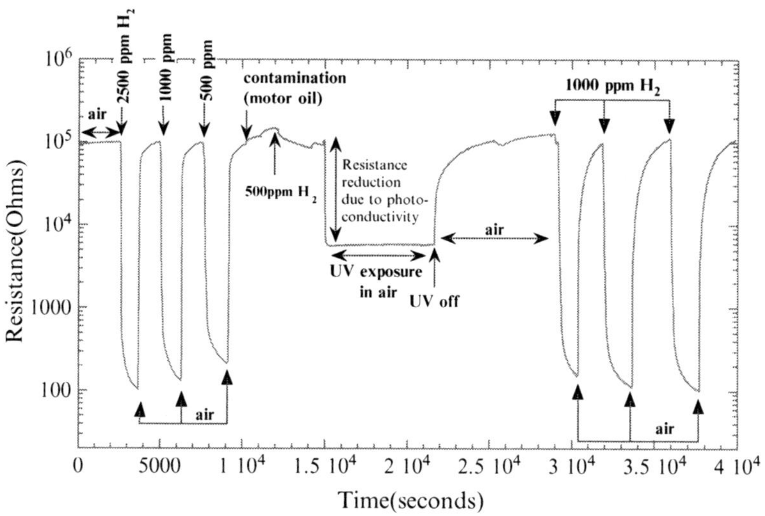

Fig. 3 Time variation of resistance change before and after self-cleaning from a layer of motor oil, that completely extinguished the gas sensing capabilities of the sensor, with UV exposure. The semi-log plot shows that the sensor regains the original hydrogen sensitivity [8]

Gong and co-workers [32] reported first generation titania nanotube arrays of up to 500 nm length by electrochemical oxidation of titanium in a HF aqueous electrolyte, see Fig. 1. Recently, changing the cathode material from Pt to Fe Allam et al. reported nanotube arrays to 2.5 μm using aqueous based electrolytes [40]. In the second synthesis generation Cai and co-workers [34, 35] adjusted the pH of both KF and NaF aqueous electrolytes to reduce the chemical dissolution of the oxide, increasing the nanotube length to several microns with a growth rate of approximately 0.25 μm/h. The third synthesis generation of titania nanotube arrays, initially reported by Paulose and co-workers [20], used organic electrolytes such as ethylene glycol (EG), di-ethylene glycol, formamide (FA), *N*-methylformamide (NMF), and dimethyl sulfoxide (DMSO) in combination with HF, KF, NaF, NH_4F, Bu_4NF, or $BnMe_3NF$ to provide fluoride ions to achieve nanotube arrays of up to 1,000 μm in length with growth rates of up to 15 μm/h [34–36, 40–43]. Figure 4 shows illustrative FESEM images of nanotube arrays made using an ethylene glycol electrolyte containing 0.25 wt% NH_4F at an anodization potential of 60 V. Nanotube arrays greater than several tens of microns in length are mechanically robust enough that they can be made in the form of self-standing membranes [36, 44]. The ability to fabricate nanotube array membranes with long range order between the nanotubes, manifested in hexagonal close packing, offers an exciting potential for application of the material as a photonic waveguide, photocatalytic nanotemplate, or filter. Allam et al. reported a fourth synthesis generation, fabrication of nanotube arrays by Ti anodization using non-fluorine bearing species, notably HCl [45, 46] and HCl in combination with H_2O_2 [47].

Fig. 4 Illustrative FESEM cross-sectional, bottom, and top images of a nanotube array sample grown at 60 V in an ethylene glycol electrolyte containing 0.25 wt% NH_4F [20, 34–36]

It is now well established that the properties of the nanotube arrays are dependent upon their specific architecture, including length, wall thickness, wall roughness, pore diameter, and tube-to-tube spacing. The geometrical features of the nanotube arrays are controlled by a variety of parameters that are not specific to one geometrical feature, including anodization potential, electrolyte composition and properties thereof (conductivity, viscosity), as well as anodization time and temperature. Virtually identical tubes can be obtained in dissimilar electrolytes by the control of different anodization variables. Control of fabrication parameters [48–50] has enabled variation of pore size from 12 to 350 nm, outer diameters from 48 to 256 nm, wall thicknesses from 5 nm to, while still a discernable wall 34 nm (a nanoporous topology can obtained in TiO_2 like that of alumina where there are no definable walls between the pores), and tube-to-tube spacing from adjacent to microns. Anodization of the starting metal films has now been used to achieve self-organized nanotube arrays of alumina [51], zirconia [52, 53], tantala [54, 55], and recently iron oxide [56].

Most commonly, the nanotube arrays have been made from a titanium thick film or foil, in which case the resulting nanotubes rest upon an underlying Ti substrate as separated by a barrier layer. This barrier layer, after thermal annealing for crystallization, is rutile, with the nanotube arrays themselves being anatase. The nanotube arrays have also been fabricated from titanium thin films sputtered onto a variety of substrates [5, 57], of particular interest glass substrates coated with transparent conductive layers, e.g., fluorine doped tin oxide (FTO), for use in dye sensitized or heterojunction solar cells. The major fabrication steps needed to achieve transparent nanotube array films, made by anodization of titanium films sputter-deposited onto FTO coated glass substrates are illustrated in Fig. 5. The films are anodized to a point at which the remaining Ti film is so thin as to become discontinuous, at which time the sample is removed from the anodization bath and rinsed to remove the fluorine ions. The initially amorphous nanotube-array film is then oxygen annealed at $\approx$450°C, whereupon the residual Ti islands oxidize and the sample becomes transparent. Figure 5 shows visual images of: (a) a deposited Ti film, (b) the film upon anodization, and (c) the film upon subsequent anneal. Also shown in Fig. 5 are FESEM images of the resulting nanotube array films.

The formation of nanotube arrays in a fluoride containing electrolyte is the result of three simultaneously occurring processes [59]: (1) field assisted oxidation of Ti metal to form titanium dioxide, (2) field assisted dissolution of Ti metal ions in the electrolyte, and (3) chemical dissolution of Ti and TiO_2 due to etching by fluoride ions. In general, it appears that the key to successfully achieving very long titania nanotube arrays using polar organic electrolytes is to minimize water content to a few percent [20]. With organic electrolytes the donation of oxygen is more difficult in comparison to water and results in a reduced tendency to form oxide, while the reduction in water content allows for thinner or lower quality barrier layers through which ionic transport may be enhanced. Shankar and co-workers [42] demonstrated how the electrolyte cation strongly influences both the nanotube growth rate and resulting nanotube length, with the length and aspect ratio of the nanotubes

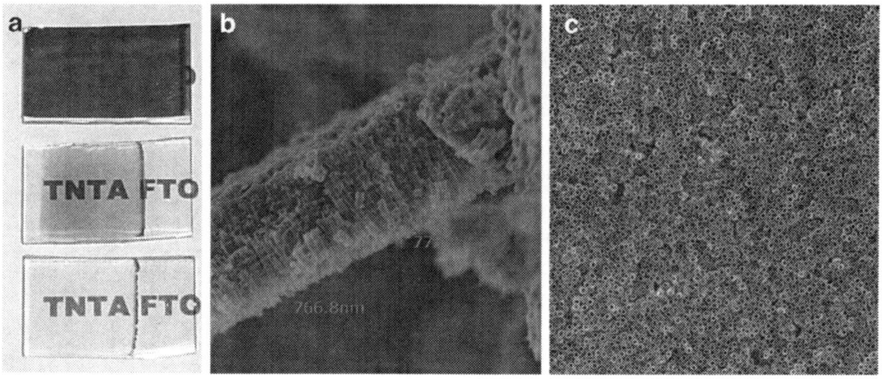

Fig. 5 (*Left*): Key stages in fabrication of a transparent TiO_2 nanotube-array film (illustrated on FTO coated glass): (**a**) Deposition of a Ti film of desired thickness; (**b**) anodization of resulting film, and (**c**) heat treatment to crystallize tubes and oxidize remaining metallic islands. (*Center, right*): FESEM images of transparent nanotube array films on FTO coated glass synthesized using a DMSO electrolyte [58]

increasing with increasing cation size. Under similar conditions, electrolytes containing a tetrabutylammonium cation resulted in the longest nanotubes (~94 μm), while the shortest nanotubes (~3 μm) were obtained when H^+ ions were the sole cationic species in the anodization electrolyte. The difference in growth characteristics was attributed to the inhibitory effect of the quarternary ammonium ions that restrict the thickness of the interfacial (barrier) oxide layer; a thinner interfacial oxide layer facilitates ionic transport, thus, enhancing nanotube growth. Tables 1 and 2 are illustrative of some of the inherent nanotube array fabrication variables. It is important to note that a tube is not necessarily a tube; nanotube array formation in aqueous electrolytes, and under certain conditions using organic electrolytes, is purely a subtractive process: one starts with a titanium film of thickness X and achieves a nanotube array having a thickness less than X. In contrast, in some polar organic electrolytes, one can begin with a titanium film of thickness Y and achieve a nanotube array of thickness greater than Y with material removed from the pores going into wall formation. It appears that a comparative study on the electrical properties of the TiO_2 nanotubes as a function of their fabrication conditions has yet to be reported, and would be a most interesting and valuable contribution to the field.

As initially fabricated, the nanotube arrays are amorphous, crystallizing with a high temperature anneal [60]. Higher temperature anneals improve the crystallinity of the anatase nanotube walls. For nanotube arrays made from, and resting upon, an underlying titanium layer higher annealing temperatures increase the thickness of the barrier layer and the rutile content in the barrier layer. There are clues that indicate control of the anodization process may possibly enable one to directly achieve crystalline nanotube arrays [41]. We note the work of Allam et al. [61] that

Table 1 Effect of anodization potential and molar F^- concentration on maximum nanotube array length, in microns, that can be obtained in a NH_4F and formamide bath. The time to maximum length is a function of the F^- concentration as well as anodization voltage, with the nanotubes reaching this maximum sooner at lower potentials [49]

F^- Concentration (M)	20 V	35 V
0.11	24	44
0.14	28	58
0.22	32	72
0.27	38	76
0.37	42	78

Table 2 Effect of NH_4F concentration and H_2O concentration in ethylene glycol electrolyte on the nanotube array length (μm); 60 V anodization potential, 17 h duration [49]

Volume% H_20	0.25 wt% NH_4F	0.3 wt% NH_4F	0.50 wt% NH_4F
1	139	156	45
2	177	220	107
3	80	160	142

directly synthesized crystalline nanotube arrays a few microns in length by anodization of chemically oxidized Ti films.

TiO_2 nanotube array films are readily converted into sensors by sputter deposition of platinum electrodes onto their top surface, to which electrical connections can be made by wire-bonding. Figure 6 shows the typical response of a 1 μm long, 30 nm pore size TiO_2 nanotube array sample, crystallized by oxygen anneal at 480°C, on switching the ambient atmosphere between air and 1,000 ppm hydrogen in nitrogen (a standard test of a hydrogen sensor) at room temperature [6]. On exposure to hydrogen, a reduction in resistance of 8.7 orders of magnitude is observed, with no measurement hysteresis and a minimal baseline low-level drift over approximately 12 h. In response to 20 ppm hydrogen, a variation in electrical resistance of 7.2 orders of magnitude is seen. With drift compensation [2], the low-level linear sensitivity indicates the ability to detect hydrogen at 200 ppt [7]. Longer nanotube arrays show less sensitivity with significantly longer response/recovery times [62]. The hydrogen sensors have been successfully used in clinical trials as an indicator of neonatal necrotizing enterocolitis, measuring transcutaneous hydrogen concentrations of a few ppm from an un-heated skin temperature (≈37°C) sensor [7]. Within the auspices of a hydrogen economy, the described sensors would be of great utility in monitoring hydrogen concentrations, including the presence of unwanted hydrogen leaks.

TiO_2 nanotube arrays are essentially all surface and no bulk. High hydrogen sensitivity in the absence of oxygen indicates that direct chemisorption of hydrogen on the nanotubes is the dominant mechanism leading to the tremendous resistance reduction in the nanotubes [6, 7], while the presence of oxygen facilitates the removal of the chemisorbed hydrogen. Hydrogen activation appears to occur on the walls of the undoped nanotubes at highly active surface states. The dissociated hydrogen species form OH groups with the surface oxygen accompanied by electron transfer to the titania conduction band and formation of an electron rich

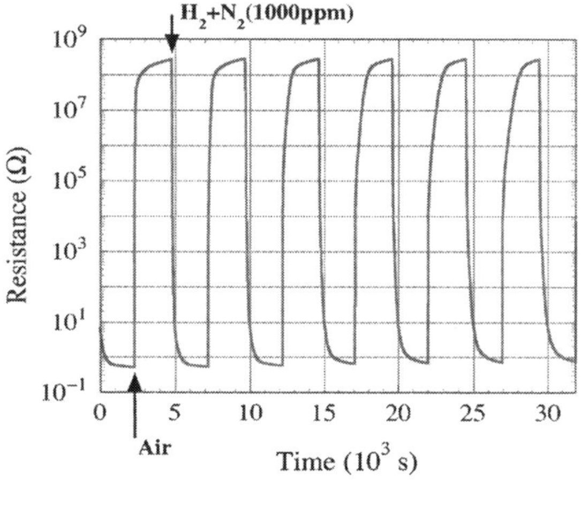

Fig. 6 Room temperature resistance variation of TiO_2 nanotube-array sample, 1 μm length, 30 nm pore size, 13 nm wall thickness, crystallized by a 480°C anneal. The sample is alternately exposed to air and 1,000 ppm hydrogen in nitrogen showing a variation in electrical resistance of 8.7 orders of magnitude (50,000,000,000%) [6, 7]

region within the nanotube walls. The adsorption of oxygen in air takes place on either side of the nanotube walls creating an electron depletion region. The width of the space charge layer L is given by $L = L_D\,[2eV/\mathrm{k}T]^{1/2}$, where $L_D = [\varepsilon_0 \varepsilon kT/2e^2 N_D]^{1/2}$ is the Debye length, eV is the barrier height, kT is the thermal energy, and N_D the ionized donor density. In metal oxides, the space charge layer extends to a few tens of nanometers [63]. If the nanotube wall half-thickness $t/2$ is significantly greater than the width of the space charge region, as shown in Fig. 7a, oxygen removal by hydrogen and subsequent hydrogen chemisorption will have little effect on device resistance, hence, high sensitivity cannot be expected. In contrast, when $t/2$ is comparable to or less than the space charge region, the shift in the electrical resistance on exposure to hydrogen can be very high (Fig. 7b, c), with a flat-band condition existing when the wall thickness is less than the width of the space charge region. The nanotube array samples showing the highest sensitivity had a wall thickness of ≈13 nm, corresponding to the geometry of Fig. 7b. The inter-wall connecting points also appear to play a significant role in enabling the ultra-high hydrogen sensitivity. The oxygen adsorption and its removal by hydrogen atoms as well as chemisorption of hydrogen at these constricted points, Fig. 7d, regulate the current passing from nanotube to nanotube. Extension of the material architecture to other metal oxides should enable dramatically improved broad-spectrum gas sensing materials.

Figure 8a shows the measured photocurrent density of nanotube array samples, ≈20 μm in length with an outer diameter of 205 nm, fabricated in an electrolyte of

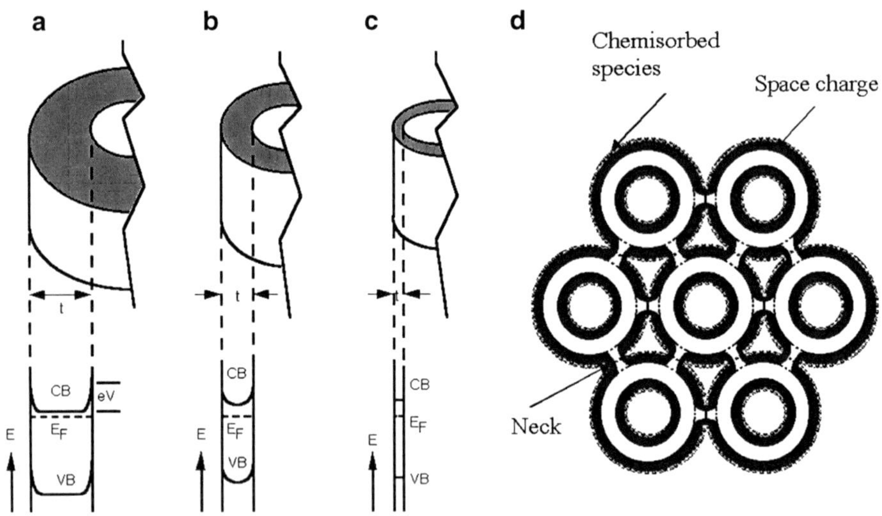

Fig. 7 The influence of nanotube wall thickness on band bending due to oxygen chemisorption: (**a**) When nanotube wall half-thickness ($t/2$) is much greater than the space charge layer, (**b**) when comparable, and (**c**) when $t/2$ is less than the width of the space charge region. (**d**) Schematic illustration of nanotubes, top view, and the tube-to-tube connecting points corresponding to case shown in (**a**)

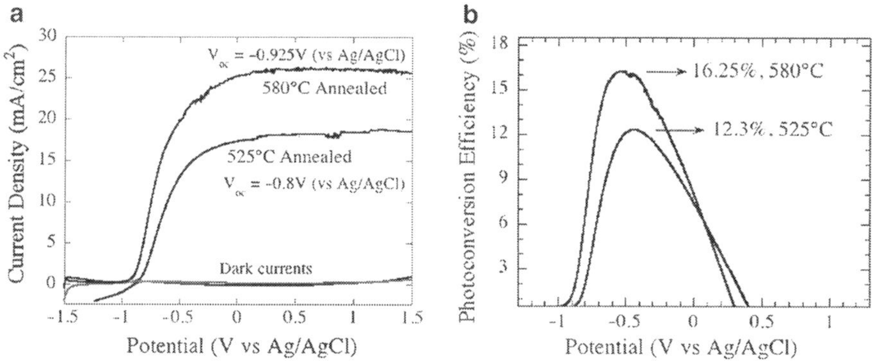

Fig. 8 (**a**) Photocurrent density and (**b**) corresponding photoconversion efficiency of nanotube array samples fabricated in an electrolyte of 1.2 g of NH_4F in a solution of 5 ml de-ionized water+ 95 ml formamide at 35 V. The samples were annealed at indicated temperatures for 1 h in oxygen prior to measurement. The resulting nanotube array samples were 30 μm in length, with an outer diameter of 205 nm [10, 20]. The samples were measured under 320 to 400 nm 100 mW/cm^2 illumination in a 1 M KOH solution

1.2 g of NH_4F in a solution of 5 ml de-ionized water +95 ml formamide at 35 V, measured at room temperature under 320 to 400 nm 100 mW/cm^2 illumination in a 1 M KOH solution [10, 20]. The samples were annealed at 525°C and 580°C for 1 h in oxygen prior to measurement. Figure 8b shows the corresponding photoconversion efficiency of the two samples; the 580°C annealed sample demonstrated a sustained hydrogen evolution rate via photoelectrolysis of approximately 106 mL/ W h (energy time normalized). When the nanotube length exceeds several microns, the great majority of the charge carriers are photogenerated in the nanotube walls [37]; hence, nanotube photoelectrochemical properties are significantly improved by increased crystallinity of the nanotube walls. In the temperature regime 500–620°C, a trend is observed whereby higher photoconversion efficiencies are obtained for nanotubes annealed at the higher temperatures. However the annealing temperature cannot be increased indefinitely since oxide growth below the nanotube layer during a high temperature anneal eventually distorts and finally destroys the nanotube array layer [60].

A key aspect of Fig. 8 is that results were obtained under UV illumination, which makes up but a small fraction of the incident solar spectrum energy. Hence, nanotube arrays of lower bandgap materials, including α-Fe_2O_3 [48, 56, 64], Cu–Ti–O [16] as well as Fe–Ti–O ternary oxides [15], are a focus of ongoing research with an aim toward achieving materials capable of efficiently responding to visible spectrum light while maintaining the outstanding charge transport properties demonstrated by the TiO_2 nanotube arrays.

As a further example of the remarkable photocatalytic properties of the nanotube arrays, Varghese and co-workers [23] recently reported efficient solar conversion of carbon dioxide and water vapor to methane and other hydrocarbons using titania nanotube arrays, with a wall thickness low enough to facilitate effective carrier transfer to the adsorbing species, surface-loaded with nano-dimensional islands of co-catalysts platinum and/or copper (subsequently oxidized to copper oxide). All experiments were conducted in outdoor sunlight at University Park, PA. Intermediate reaction products, hydrogen and carbon monoxide, are also detected with their relative concentrations underlying hydrocarbon production rates and dependent upon the nature of the co-catalysts on the nanotube array surface. Using outdoor global AM 1.5 sunlight, 100 mW/cm^2, a hydrocarbon production rate of 111 ppm cm^{-2} h^{-1}, or $\approx$160 μl/g h, is obtained when the nanotube array samples are surface decorated with both Cu and Pt nanoparticles. This rate of CO_2 to hydrocarbon production obtained under outdoor sunlight is at least 20 times higher than previously published reports, which were conducted under laboratory conditions using UV illumination [23]. Taking into account both hydrogen and hydrocarbons, the production rate was 273 ppm cm^{-2} h^{-1}. Figure 9 depicts one exciting prospect offered by such remarkable photocatalytic properties, a high-rate flow through photocatalytic membrane for CO_2 reduction.

In photocatalysis and photoelectrolysis application of the nanotube arrays, there are of course an optimal length and geometric area where the absorption of the incident light is balanced by recombination of the photogenerated electron-hole pairs. Too short a nanotube array and the light is not fully absorbed [37]; too long a nanotube array, that is a length greater than that needed for full light absorption, and the photoconversion efficiency suffers from recombination of the photogenerated electron-hole pairs. A larger geometric surface area entails the availability of a larger number of active sites for chemical reactions to occur and allows photogenerated holes to access a large number of solution ions. Greater nanotube length enables higher absorption of incident photons and results in a larger surface area for fixed nanotube pore size and outer diameter. The maximum photocurrent is a

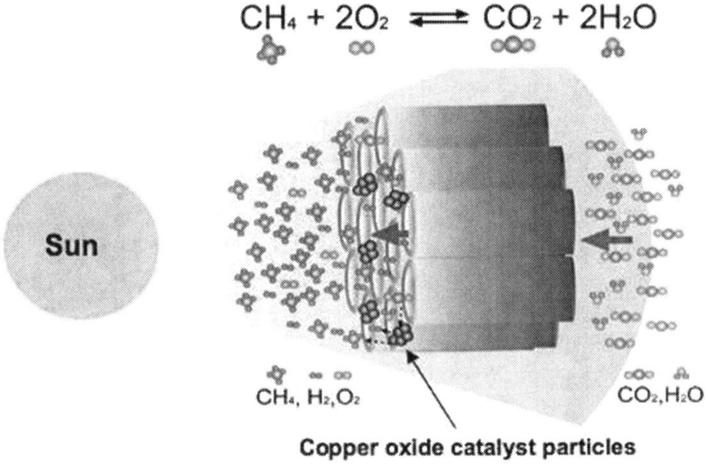

Fig. 9 Depiction of co-catalyst loaded flow-through nanotube array membrane for high rate photocatalytic conversion of CO_2 and water vapor into hydrocarbon fuels [23]

measure of the generation, and subsequent collection of charge carriers and correlates well with the surface area of the nanotubes [10]. In contrast, the photoconversion efficiency also depends on the separation processes of the photogenerated electron-hole pairs, that is its recombination characteristics, which for the nanotube arrays are a complex function of the temperature and duration of crystallization, the wall-thickness of the nanotubes, barrier-layer thickness, and the incorporation (if any) of dopants from the electrolyte during anodization.

While back-side illuminated dye sensitized solar cells can be readily constructed using nanotube arrays easily fabricated from thick-film foil Ti samples [65, 66] the geometry is not optimal, with the photoconversion efficiency of such devices significantly decreased due to unwanted reflection of light from the counter electrode and unwanted light absorption by the electrolyte. Gratzel and co-workers report [67] an absolute difference in dye-sensitized solar cell photoconversion efficiency between back-side illumination (through the cathode), and front-side illumination (through the photoanode) of 4%. Hence, it is of significant interest that Shankar and co-workers [10] reported a back-side illuminated dye sensitized solar cell comprised of a 20 μm long nanotube array, on a Ti foil substrate, with 6.89% photoconversion efficiency under (NREL calibrated) AM 1.5 illumination (N719 dye, $J_{sc}=12.72$ mA/cm^2, $V_{oc}=0.82$ V, fill factor=0.663). Frank and co-workers [39] used intensity-modulated photocurrent and photovoltage spectroscopy of nanotube array films on Ti substrates to determine that the electron recombination time constants of nanotube array films were an order of magnitude greater than those of nanoparticle films of equal thickness, with calculations indicating that the charge-collection efficiency of the nanotube array films is 25% larger than that of nanoparticulate films, implying that nanotube arrays can be made much thicker than the nanoparticulate films for a given recombination loss allowing for a higher light-harvesting efficiency [39].

For use in a heterojunction solar cell, an optically transparent nanotube array film, see Fig. 5, serving as the photoanode as seen in Fig. 10 [57, 58], is desired to achieve maximum photoconversion efficiency [67]. Transparent nanotube array films up to 32 μm in length have recently been reported by Varghese and co-workers [58], and used in the fabrication of dye sensitized solar cells of 7.8% efficiency.

All said, highly ordered TiO_2 nanotube arrays demonstrate remarkable behaviors that warrant further study, underlying the large number of researchers now working in this field as well as the motivation for this book. Chapter 1 considers nanotube array synthesis techniques detailing the different synthesis generations. Chapter 2 summarizes the structural, elemental, optical, mechanical, and electrical properties of the TiO_2 nanotube arrays. Chapter 3 examines the use of TiO_2 nanotube arrays as hydrogen gas sensors. Chapter 4 considers the use of the material architecture for photoelectrochemical water splitting [22, 68], including efforts at bandgap engineering through doping, by the combination with low bandgap materials, and fabrication of ternary oxide nanotube arrays. Chapter 5 considers the application of TiO_2 nanotube arrays to dye sensitized solar cells (DSCs) and polymer based solid state bulk heterojunction solar cells. Chapter 6 discusses the use of TiO_2 nanotubes

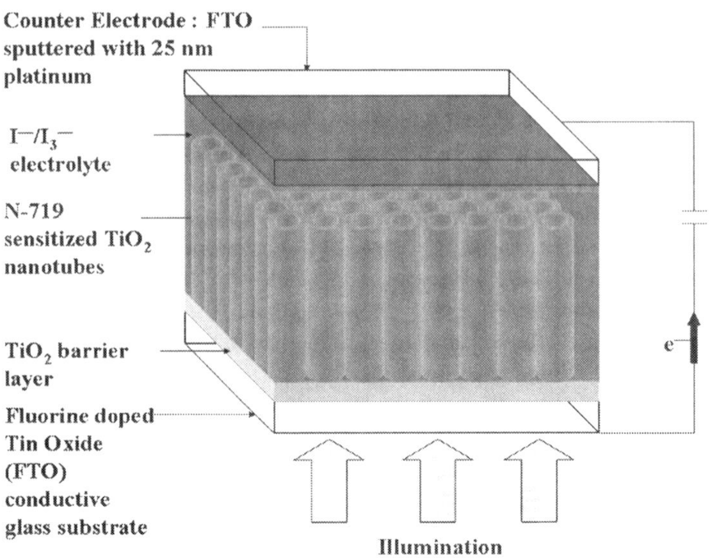

Fig. 10 Integration of transparent nanotube array architecture into (front-side illuminated) dye-sensitized solar cells

in biological applications, including their use as a base for biosensors, drug eluting surfaces, osseointegration, biotemplating, enhanced blood clotting, and as biofiltration membranes.

References

1. Varghese OK, Gong D, Paulose M, Ong KG, Grimes CA (2003) Hydrogen sensing using titania nanotubes. Sens Actuators B 93:338–344
2. Varghese OK, Mor GK, Grimes CA, Paulose M, Mukherjee N (2004) A titania nanotube-array room-temperature sensor for selective detection of hydrogen at low concentrations. J Nanosci Nanotechnol 4:733–737
3. Varghese OK, Gong D, Paulose M, Ong KG, Dickey EC, Grimes CA (2003) Extreme changes in the electrical resistance of titania nanotubes with hydrogen exposure. Adv Mater 15:624–662
4. Mor GK, Varghese OK, Paulose M, Ong KG, Grimes CA (2006) Fabrication of hydrogen sensors with transparent titanium oxide nanotube-array thin films as sensing elements. Thin Solid Films 496:42–48
5. Varghese OK, Grimes CA (2003) Metal oxide nanoarchitectures for environmental sensing. J Nanosci Nanotechnol 3:277–293
6. Paulose M, Varghese OK, Mor GK, Grimes CA, Ong KG (2006) Unprecedented ultra-high hydrogen gas sensitivity in undoped titania nanotubes. Nanotechnol 17:398–402
7. Varghese OK, Yang X, Kendig J, Paulose M, Zeng K, Palmer C, Ong KG, Grimes CA (2006) A transcutaneous hydrogen sensor: from design to application. Sensor Lett 4:120–128
8. Mor GK, Varghese OK, Paulose M, Grimes CA (2003) A self-cleaning room temperature titania-nanotube hydrogen gas sensor. Sens Lett 1:42–46

9. Mor GK, Carvalho MA, Varghese OK, Pishko MV, Grimes CA (2004) A room temperature TiO_2 nanotube hydrogen sensor able to self-clean photoactively from environmental contamination. J Mater Res 19:628–634
10. Shankar K, Mor GK, Prakasam HE, Yoriya S, Paulose M, Varghese OK, Grimes CA (2007) Highly-ordered TiO_2 nanotube arrays up to 220 μm in length: use in water photoelectrolysis and dye-sensitized solar cells. Nanotechnol 18:065707
11. Mor GK, Shankar K, Paulose M, Varghese OK, Grimes CA (2006) Use of highly-ordered TiO_2 nanotube arrays in dye-sensitized solar cells. Nano Lett 6:215–218
12. Mor GK, Shankar K, Paulose M, Varghese OK, Grimes CA (2007) High efficiency double heterojunction polymer photovoltaic cells using highly ordered TiO_2 nanotube arrays. Appl Phys Lett 91:152111
13. Shankar K, Mor GK, Paulose M, Varghese OK, Grimes CA (2008) Effect of device geometry on the performance of TiO_2 nanotube array-organic semiconductor double heterojunction solar cells. J Non-Cryst Solids 354:2767–2771
14. Shankar K, Mor GK, Prakasam HE, Varghese OK, Grimes CA (2007) Self-assembled hybrid polymer-TiO_2 nanotube array heterojunction solar cells. Langmuir 23:12445–12449
15. Mor GK, Prakasam HE, Varghese OK, Shankar K, Grimes CA (2007) Vertically oriented Ti–Fe–O nanotube array films: toward a useful material architecture for solar spectrum water photoelectrolysis. Nano Lett 7:2356–2364
16. Mor GK, Varghese OK, Wilke RHT, Sharma S, Shankar K, Latempa TJ, Choi KS, Grimes CA (2008) p-Type Cu–Ti–O nanotube arrays and their use in self-biased heterojunction photoelectrochemical diodes for hydrogen generation. Nano Lett 8:1906–1911
17. Chen SG, Paulose M, Ruan C, Mor GK, Varghese OK, Grimes CA (2006) Electrochemically synthesized CdS nanoparticle-modified TiO_2 nanotube-array photoelectrodes: preparation, characterization, and application to photoelectrochemical cells. J Photochem Photobiol A 177:177–184
18. Seabold JA, Shankar K, Wilke RHT, Paulose M, Varghese OK, Grimes CA, Choi KS (2008) Photoelectrochemical properties of heterojunction CdTe/TiO_2 electrodes constructed using highly ordered TiO_2 nanotube arrays. Chem Mater 20:5266–5273
19. Varghese OK, Paulose M, Shankar K, Mor GK, Grimes CA (2005) Water-photolysis properties of micron-length highly-ordered titania nanotube-arrays. J Nanosci Nanotechnol 5:1158–1165
20. Paulose M, Shankar K, Yoriya S, Prakasam HE, Varghese OK, Mor GK, Latempa TA, Fitzgerald A, Grimes CA (2006) Anodic growth of highly ordered TiO_2 nanotube arrays to 134 μm in length. J Phys Chem B 110:16179–16184
21. Mor GK, Shankar K, Paulose M, Varghese OK, Grimes CA (2005) Enhanced photocleavage of water using titania nanotube arrays. Nano Lett 5:191–195
22. Grimes CA, Varghese OK, Ranjan S (2008) Light, water, hydrogen: the solar generation of hydrogen by water photoelectrolysis. Springer, New York
23. Varghese OK, Paulose M, LaTempa TJ, Grimes CA (2009) High-rate solar photocatalytic conversion of CO_2 and water vapor to hydrocarbon fuels. Nano Lett 9:731–737
24. Fabregat-Santiago F, Barea EM, Bisquert J, Mor GK, Shankar K, Grimes CA (2008) High carrier density and capacitance in TiO_2 nanotube arrays induced by electrochemical doping. J Am Chem Soc 130:11312–11316
25. Popat KC, Eltgroth M, LaTempa TJ, Grimes CA, Desai TA (2007) Titania nanotubes: a novel platform for drug-eluting coatings for medical implants? Small 11:1878–1881
26. Popat KC, Leoni L, Grimes CA, Desai TA (2007) Influence of engineered titania nanotubular surfaces on bone cells. Biomaterials 28:3188–3197
27. Peng L, Mendelsohn AD, LaTempa TJ, Yoriya S, Grimes CA, Desai TA (2009) Long-term small molecule and protein elution from TiO_2 nanotubes. Nano Lett 9:1932–1936
28. Liu S, Chen A (2005) Coadsorption of horseradish peroxidase with thionine on TiO_2 nanotubes for biosensing. Langmuir 21:8409–8413
29. Xie Y, Zhoua L, Huang H (2007) Bioelectrocatalytic application of titania nanotube array for molecule detection. Biosens Bioelectron 22:2812–2818

30. Roy SC, Paulose M, Grimes CA (2007) The effect of TiO_2 nanotubes in the enhancement of blood clotting for the control of hemorrhage. Biomater 28:4667–4672
31. Popat KC, Eltgroth M, LaTempa TJ, Grimes CA, Desai TA (2007) Decreased Staphylococcus epidermis adhesion and increased osteoblast functionality on antibiotic-loaded titania nanotubes. Biomater 28:4880–4888
32. Gong D, Grimes CA, Varghese OK, Hu W, Singh RS, Chen Z, Dickey EC (2001) Titanium oxide nanotube arrays prepared by anodic oxidation. J Mater Res 16:3331–3334
33. Cai Q, Paulose M, Varghese OK, Grimes CA (2005) The effect of electrolyte composition on the fabrication of self-organized titanium oxide nanotube arrays by anodic oxidation. J Mater Res 20:230–236
34. Prakasam HE, Shankar K, Paulose M, Grimes CA (2007) A new benchmark for TiO_2 nanotube array growth by anodization. J Phys Chem C 111:7235–7241
35. Albu SP, Ghicov A, Macak JM, Schmuki P (2007) 250 μm long anodic TiO_2 nanotubes with hexagonal self-ordering. Physica Status Solidi 1:R65–R67
36. Paulose M, Prakasam HE, Varghese OK, Peng L, Popat KC, Mor GK, Desai TA, Grimes CA (2007) TiO_2 nanotube arrays of 1000 μm in length by anodization of titanium foil: phenol red diffusion. J Phys Chem C 111:14992–14997
37. Ong KG, Varghese OK, Mor GK, Grimes CA (2005) Numerical simulation of light propagation through highly-ordered titania nanotube arrays: dimension optimization for improved photoabsorption. J Nanosci Nanotech 5:1801–1808
38. Ong KG, Varghese OK, Mor GK, Shankar K, Grimes CA (2007) Application of finite difference time domain to dye-sensitized solar cells: the effect of nanotube-array negative electrode dimensions on light absorption. Solar Energy Mater & Solar Cells 91:250–257
39. Zhu K, Neale NR, Miedaner A, Frank AJ (2006) Enhanced charge-collection efficiencies and light scattering in dye-sensitized solar cells using oriented TiO_2 nanotubes arrays. Nano Lett 7:69–74
40. Allam NK, Grimes CA (2008) Effect of cathode material on the morphology and photoelectrochemical properties of vertically oriented TiO_2 nanotube arrays. Sol Energy Mater sol cells 92:1468–1475
41. Yoriya S, Mor GK, Sharma S, Grimes CA (2008) Synthesis of ordered arrays of discrete, partially crystalline titania nanotubes by Ti anodization using diethylene glycol electrolytes. J Mater Chem 18:3332–3336
42. Shankar K, Mor GK, Fitzgerald A, Grimes CA (2007) Cation effect on the electrochemical formation of very high aspect ratio TiO_2 nanotube arrays in formamide–water mixtures. J Phys Chem C 111:21–26
43. Yoriya S, Paulose M, Varghese OK, Mor GK, Grimes CA (2007) Fabrication of vertically oriented TiO_2 nanotube arrays using dimethyl sulfoxide electrolytes. J Phys Chem C 111:13770–13776
44. Paulose M, Peng L, Popat KC, Varghese OK, LaTempa TJ, Bao N, Desai TA, Grimes CA (2008) Fabrication of mechanically robust, large area, polycrystalline nanotubular/porous TiO_2 membranes. J Membrane Sci 319:199–205
45. Allam NK, Grimes CA (2007) Formation of vertically oriented TiO_2 nanotube arrays using a fluoride free HCl aqueous electrolyte. J Phys Chem C 111:13028–13032
46. Chen X, Schriver M, Suen T, Mao SS (2007) Fabrication of 10 nm diameter TiO_2 nanotube arrays by titanium anodization. Thin Solid Films 515:8511–8514
47. Allam NK, Shankar K, Grimes CA (2008) Photoelectrochemical and water photoelectrolysis properties of ordered TiO_2 nanotubes fabricated by Ti anodization in fluoride-free HCl electrolytes. J Mat Chem 18:2341–2348
48. Shankar K, Basham JI, Allam NK, Varghese OK, Mor GK, Feng X, Paulose M, Seabold JA, Choi KS, Grimes CA (2009). A review of recent advances in the use of TiO_2 nanotube and nanowire arrays for oxidative photoelectrochemistry. J. Phys Chem C 113:6327–6359 (in press)
49. Grimes CA (2007) Synthesis and application of highly-ordered arrays of TiO_2 Nanotubes. J Mater Chem 17:1451–1457

50. Mor GK, Varghese OK, Paulose M, Shankar K, Grimes CA (2006) A review on highly ordered, vertically oriented TiO_2 nanotube arrays: fabrication, material properties, and solar energy applications. Sol Energy Mater Sol Cells 90:2011–2075
51. Woo Lee W, Roland Scholz R, Ulrich Go U (2008) A continuous process for structurally well-defined Al_2O_3 nanotubes based on pulse anodization of aluminum. Nano Lett 8:2155–2160
52. Lee WJ, Smyrl WH (2005) Zirconium oxide nanotubes synthesized via direct electrochemical anodization. Electrochem Solid-State Lett 8:B7–B9
53. Shin Y, Lee S (2009) A freestanding membrane of highly ordered anodic ZrO2 nanotube arrays. Nanotechnology 20: Art. No. 105301
54. Allam NK, Feng X, Grimes CA (2008) Self-assembled fabrication of vertically oriented Ta_2O_5 nanotube arrays, and membranes thereof, by one-step tantalum anodization. Chem Mater 20:6477–6481
55. Ruckh T, Porter JR, Allam NK, Feng X, Grimes CA, Popat KC (2009) Nanostructured tantala as a template for enhanced osseointegration. Nanotechnology 20:045102
56. LaTempa TJ, Grimes CA (2009) Hematite nanotube arrays by potentiostatic anodization of Ti Foil: photoelectrochemical properties. J Phys Chem C 113:mmm-mmm (in press)
57. Mor GK, Varghese OK, Paulose M, Grimes CA (2005) Transparent highly-ordered TiO_2 nanotube-arrays via anodization of titanium thin films. Adv Funct Mater 15:1291–1296
58. Maggie Paulose, Oomman K. Varghese and Craig A. Grimes. Transparent films of vertically aligned long titania nanotube arrays on TCO glass for highly efficient dye sensitized solar cells. In Press, Nat Nanotechnol
59. Mor GK, Varghese OK, Paulose M, Mukherjee N, Grimes CA (2003) Fabrication of tapered, conical-shaped titania nanotubes. J Mater Res 18:2588–2593
60. Varghese OK, Gong D, Paulose M, Ong KG, Grimes CA, Dickey EC (2003) Crystallization and high-temperature structural stability of titanium oxide nanotube arrays. J Mater Res 18:156–165
61. Allam NK, Shankar K, Grimes CA (2008) A general method for the anodic formation of crystalline metal oxide nanotube arrays without the use of thermal annealing. Adv Mater 20:3942–3946
62. Yoriya S, Prakasam HE, Varghese OK, Shankar K, Paulose M, Mor GK, Latempa TJ, Grimes CA (2006) Initial studies on the hydrogen gas sensing properties of highly-ordered high aspect ratio TiO_2 nanotube-arrays 20 μm to 222 μm in length. Sensor Lett 4:334–339
63. Morimoto T, Nagao M, Tokuda F (1969) Relation between amounts of chemisorbed and physisorbed water on metal oxides. J Phys Chem 73:243–248
64. Prakasam HE, Paulose M, Varghese OK, Mor GK, Grimes CA (2006) Synthesis and photoelectrochemical properties of nanoporous iron (III) oxide by potentiostatic anodization. Nanotechnology 17:4285–4291
65. Paulose M, Shankar K, Varghese OK, Mor GK, Hardin B, Grimes CA (2006) Backside illuminated dye-sensitized solar cells based on titania nanotube array electrodes. Nanotechnology 17:1446–1448
66. Paulose M, Shankar K, Varghese OK, Mor GK, Grimes CA (2006) Application of highly-ordered TiO_2 nanotube-arrays in dye-sensitized solar cells. J Physics D 39:2498–2503
67. Ito S, Ha NC, Rothenberger G, Liska P, Comte P, Zakeeruddin SM, Pechy P, Nazeeruddin MK, Gratzel M (2006) High-efficiency (7.2%) flexible dye-sensitized solar cells with Ti-metal substrate for nanocrystalline-TiO_2 photoanode. Chem Commun 38:4004–4006
68. Fujishima A, Honda K (1972) Electrochemical photolysis of water at a semiconductor electrode. Nature 238:37–38

Chapter 1
Fabrication of TiO_2 Nanotube Arrays by Electrochemical Anodization: Four Synthesis Generations

1.1 Introduction

One-dimensional (1-D) nanowire and nanotube systems with high surface-to-volume ratios have been found to possess significant, useful, and unique properties. The synthesis of highly ordered 1-D materials using localized chemical dissolution with controlled, field-assisted oxidation and dissolution reactions is particularly noteworthy for it permits achievement of a precisely ordered, nanoscale self-assembly. Comparative studies show that ordered arrays of TiO_2 outperform colloidal TiO_2 for photocatalytic applications [1–5], sensing [6–9], photoelectrolysis [10–16], polymer-based bulk heterojunction photovoltaics [17–19], dye-sensitized solar cells [20–26], biofluids filtration, drug delivery and other biomedical applications [27–30]. Initial investigations indicate that they also may be useful for energy storage devices such as Li-ion batteries and supercapacitors; these applications of TiO_2 nanotube arrays are among those discussed in the subsequent chapters.

TiO_2 nanotubes and arrays thereof have been produced by a number of methods. These include: using a template of nanoporous alumina [31–34], sol-gel transcription processes using organo-gelator templates [35, 36], seeded growth mechanisms [37], and hydrothermal techniques [38–40]. None of these methods, however, offer superior control over the nanotube dimensions than does the anodization of titanium in a fluoride-based electrolyte [41–48].

In 1999, Zwilling and co-workers achieved self-organized porous TiO_2 by anodizing a Ti-based alloy in an acidic, fluoride-based electrolyte [49, 50]. In 2001, Gong and co-workers fabricated self-organized, highly uniform TiO_2 nanotube arrays by anodizing Ti in an aqueous dilute HF electrolyte [41, 42]. Maximum nanotube lengths in this first synthesis generation were approximately 500 nm. In subsequent work, the second-generation, the nanotube array length was increased to approximately 7 μm by proper control of the anodization electrolyte pH thereby reducing the chemical dissolution of TiO_2 during anodization [43, 46]; the pH should be high but remain acidic. In later work, the third-generation, TiO_2 nanotube arrays with lengths of up to approximately 1000 μm were achieved using

C.A. Grimes and G.K. Mor, *TiO_2 Nanotube Arrays*,

DOI 10.1007/978-1-4419-0068-5_1, © Springer Science+Business Media, LLC 2009

a non-aqueous, polar organic electrolyte such as formamide, dimethylsulfoxide, ethylene glycol or diethylene glycol [44, 51–54]. We suggest the fourth synthesis generation is that of non-fluoride-based anodization chemistries.

In this chapter, we review the contribution of different research groups to the fabrication of the titania nanotube arrays, and we discuss a mechanistic growth model for nanotube array formation.

1.1.1 The Electrochemical Anodization Process

Anodization is an electrolytic process that creates a protective or decorative oxide film over a metallic surface. Anodization typically increases both the thickness and density of the oxide layer that forms on any metal surface exposed to the earth's atmosphere. To accomplish it, the conducting piece undergoing anodization is connected to the positive terminal of a dc power supply and placed in an electrolytic bath where it serves as the anode. The cathode is commonly a plate or rod of platinum, although materials such as carbon are sometimes used. When power is applied electrons are forced from the electrolyte to the positive anode. The process leaves surface metal atoms exposed to oxygen ions within the electrolyte. The atoms react and become an *in situ* integral part of the oxide layer. The electrons travel through the power source and return to the cathode where, if an appropriate electrolyte pH is present, they react with hydrogen ions and the combination bubbles off as hydrogen gas.

Since the metal oxide partially dissolves in any electrolyte, it is necessary to use only those electrolytes for which the oxide forms more rapidly than it dissolves. The electrolyte composition is also the primary determinant of whether the oxide film is porous or if it forms a barrier layer. Oxide barrier layers grow in those neutral or slightly alkaline solutions in which titanium dioxide is largely insoluble. Porous oxide layers grow in acidic electrolytes with fluoride or chloride ions in which oxide forms and then rapidly dissolves. The acid cations also affect the resulting nanotube array structures.

A depiction of an electrochemical anodization cell is shown in Fig. 1.1. With a titanium anode and a platinum cathode immersed in an aqueous electrolyte of dilute

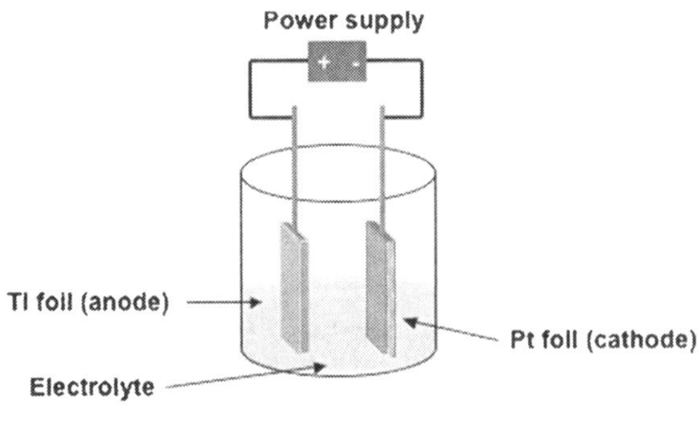

Fig. 1.1 Depiction of a electrochemical cell in which the Ti samples are anodized. Fabrication variables include temperature, voltage, pH and electrolyte composition

acid to which a small dc voltage is applied the surface layer is sufficiently resistive to prevent current flow. Increasing the applied voltage produces no additional current flow until a threshold level is reached where the electric field intensity within the barrier is sufficient to force oxygen ions to diffuse across it, producing an ionic current. These oxygen ions react with the metal and increase the thickness and/or density of the oxide barrier. This process of high-field ionic conduction is central to anodization. Of course, the same process liberates hydrogen gas from the cathode. Since the electric resistance of the layer increases in proportion to its thickness and since the rate of oxide growth is proportional to the current density, the thinner portions of the layer carry more current than the thicker ones. Hence, a thin section grows faster than a thick one, creating an even more uniform layer. As the layer thickens the applied voltage necessary to maintain a constant current increases. The process continues until, for each bath composition and temperature, a maximum applied voltage is reached above which other, non-desired reactions become manifest, for example: oxygen evolution, solute oxidation, or sparking due to electron avalanche through the oxide.

Fabrication of TiO_2 nanotube arrays via anodic oxidation of Ti was first reported in 2001 [41]. Later studies focused on precise control and extension of the nanotube morphology [42], length and pore size [43], and wall thickness [55]. Illustrative TiO_2 nanotube array morphologies, as grown by Ti anodization, are shown in Fig. 1.2; these vertically oriented nanotubes are firmly attached to the substrate with an open top and a bottom closed by a barrier layer of metal oxide. The use of different electrolytic solvents permits control of the architecture from well-separated, stand-alone nanotubes to densely packed arrays. A typical nanotube's wall thickness ranges from 5 to 30 nm, the pore size from 20 to 350 nm, and the length from 0.2 to 1000 μm. The aspect ratio, defined as the nanotube length-to-outer diameter ratio, can be controlled from about 10 to approximately 20,000 by selection of appropriate anodization variables.

1.2 Nanotube Array Synthesis Using Aqueous Electrolytes: The First Generation

1.2.1 HF-Based Electrolytes

Anodization of titanium foil (note, the titanium sample does not have to be planar) were conducted in a two-electrode electrochemical cell at a constant-applied voltage and with a platinum foil cathode. Anodization experiments were conducted at room temperature with magnetic stirring of electrolyte. Stirring reduces the thickness of the double layer at the metal/electrolyte interface and ensures uniform local current density and temperature over the surface of the Ti electrode [56]. Foils were anodized in 0.5 wt% HF aqueous solution at anodizing voltages of: 3, 5, 10 and 20 V [41, 57–61]. At low anodization voltages, the film morphology is similar

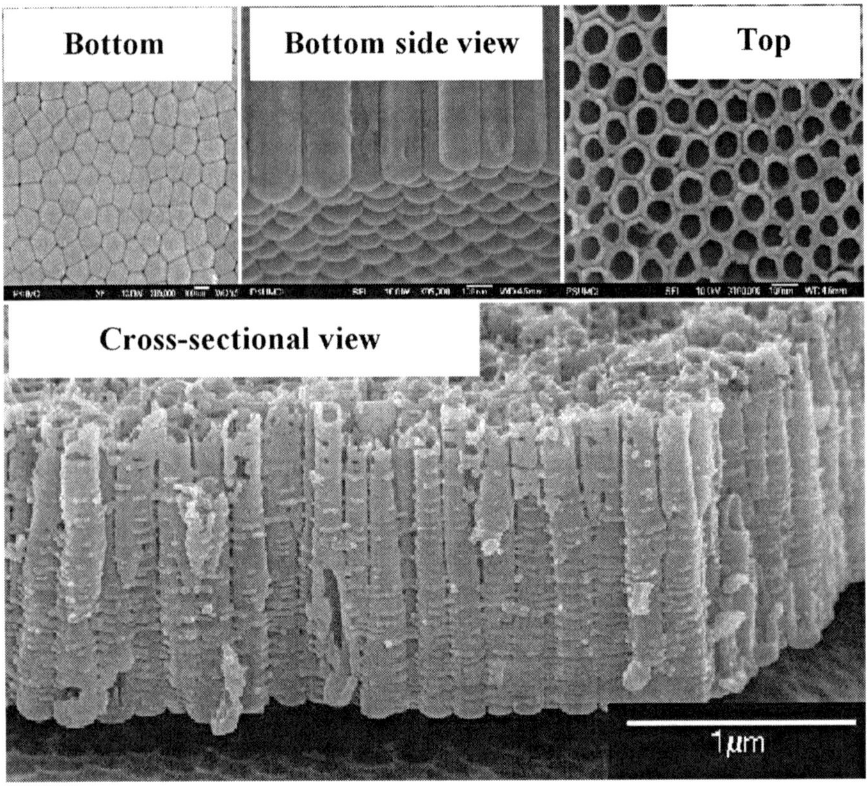

Fig. 1.2 TiO_2 nanotube array morphologies achieved by Ti anodization. The upper images are the architecture resulting from Ti anodization in an ethylene glycol bath; the lower image is the architecture resulting from Ti anodization in an aqueous KF bath

to that of porous (sponge-like) alumina [62], with a typical pore sizes from 15 to 30 nm. At somewhat higher applied voltages, the surface becomes nodular in nature. Increasing the voltage to 10 V removes the particulate appearance and replaces it with discrete, hollow, cylindrical tube-like features. Nanotube samples prepared in this manner but using anodization voltages of 10, 14 and 20 V have, respectively, inner diameters of 22, 53 and 76 nm; wall thickness of 13, 17 and 27 nm; and lengths of 200, 260 and 400 nm. After anodizing the samples for 45 min, a uniform nanotube array grows atop the titanium foil together with an electrically insulating barrier layer between the nanotubes and the underlying (unanodized) titanium foil. The nanotube structure is lost at anodizing voltages greater than 23 V and replaced by a sponge-like, randomly porous structure. Adding acetic acid to the 0.5% HF electrolyte in a 1:7 ratio results in more mechanically robust nanotubes without changing their shape or size [7, 63].

To examine the efficiency of nanotube formation we determined the fraction of the titanium converted into soluble products and into nanotubes [57]. A titanium foil was weighed before and after anodization in a 0.5% by weight HF solution and held at 20 V for 2 h; the average weight loss was 8.3×10^{-4} g/cm^2. Estimating a tube length of 250 nm, pore size of 100 nm, and a wall thickness of 10 nm we calculated the fraction of the area occupied by nanotubes to be 0.23. The TiO_2 nanotube array volume is 5.75×10^{-6} cm^3 per each cm^2 of foil surface area, giving 2.208×10^{-5} g of TiO_2 (density 3.84 g/cm^3) formed in each cm^2 surface area. Since oxygen increases the mass of the barrier layer, if TiO_2 formation was the only process inherent to anodization a mass gain would be observed, not a mass loss. Analysis of the used anodization solution via Inductively Coupled Plasma Mass Spectrometry (ICMS) revealed that most of the mass loss is the loss of titanium as it is dissolved into the solution; only about 2% of the titanium was converted into nanotubes and the underlying barrier layer.

1.2.2 *Tapered Conical Shape Nanotubes*

A constant anodization voltage produces straight nanotubes. Sweeping the anodization voltage produces tapered, conical shaped nanotubes. In 0.5% hydrofluoric acid solution, pH$\approx$1.0, and with an applied voltage between 10 and 23 V, nanotubes of well-defined shape are obtained; the length and diameter are proportional to the applied voltage. The anodization voltage was ramped up from 10 to 23 V at rates from 0.43 to 2.6 V/min to achieve conical nanotubes with a continuously increasing pore size from top to bottom [42]. Two sets of samples were prepared: in Set-1 the voltage increased linearly from 10 to 23 V and then held constant at 23 V for a total anodization time of 40 min, in Set-2 the voltage was held constant at 10 V for 20 min and then ramped up at 0.5 V/min, in one case, and 1.0 V/min in another. The voltage was then held constant at 23 V for 2 min (a total anodization time of 35 min for the 1.0 V/min ramp, and 47 min for the 0.5 V/min ramp); Fig. 1.3 shows field emission scanning electron microscope (FESEM) images of the resulting

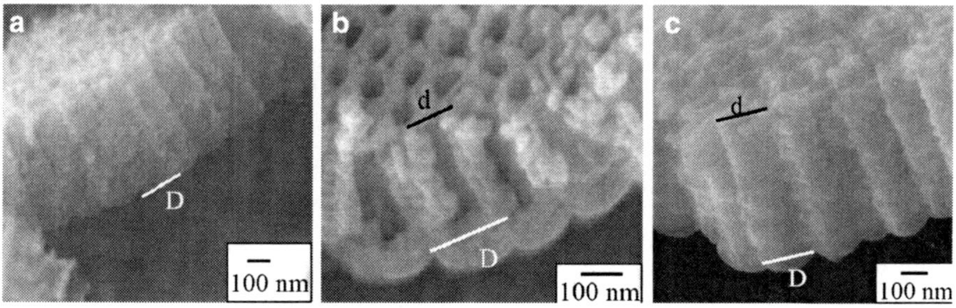

Fig. 1.3 FESEM cross-sectional views of tapered nanotubes obtained: (**a**) by ramping the anodization voltage from 10 to 23 V over a 30 min period, 0.43 V/min, then holding the voltage at 23 V for 10 min, (**b**) by initially anodizing the sample at 10 V for 20 min then increasing the voltage at 1.0 V/min to 23 V then held constant at 23 V for 2 min, (**c**) straight nanotubes obtained by applying a constant 23 V for 45 min. Here, *d* denotes diameter of the conical apex, and *D* diameter of conical base [42]

tapered nanotubes. Figure 1.3a shows the nanotube arrays resulting from a 0.43 V/min anodization voltage ramp of 30 min followed by a constant 23 V anodization for 10 min. Figure 1.3b shows the nanotubes fabricated by anodizing the sample for 20 min at 10 V followed by ramping the voltage at the rate of 1.0 V/min, and finally holding the voltage at 23 V for 2 min. For comparison, Fig. 1.3c shows nanotubes prepared using a constant 23 V anodization. In all cases the average outer-diameter of the base of the tube is about 166 nm, a value nearly equal to that of a tube fabricated using a constant 23 V anodization voltage. It is difficult to determine the inner-diameter of the bottom of the tube bottom that requires tube cleavage. The average inner diameters of the tapered end of the tubes in Set-1 are approximately 70, 80, 85, and 100 nm, respectively, for sweep rates of 0.43, 0.65, 0.87 and 2.6 V/min. The average inner diameters of the tubes from Set-2 are approximately 36 and 42 nm, respectively, for the 0.5 and 1.0 V/min ramps. A sweep-rate greater than 1.0 V/min leads to the collapse of the nanotubes. Tapered nanotubes did not result from sweeping the anodization voltage from 23 to 10 V followed by a constant 10 V anodization for a total of 40 min anodization time; the fabricated tubes were straight with a constant 22 nm inner diameter and 200 nm long. These dimensions equal those achieved using a constant 10 V anodization.

1.2.3 Wall Thickness Variation

Chemical dissolution and electrochemical etching are critical factors in the growth of nanotube arrays via anodic oxidation of titanium. Varying the electrolyte bath temperature changes the rate of both etching processes [55]. Nanotube arrays were grown with a constant 10 V anode potential and a titanium foil anode in an electrolyte of acetic acid plus 0.5% HF mixed in 1:7 ratio and kept at each of four different electrolyte bath temperatures: 5°C, 25°C, 35°C and 50°C. Figure 1.4

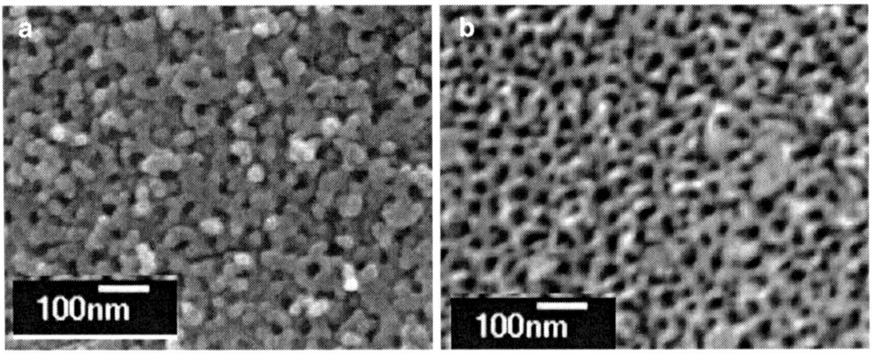

Fig. 1.4 FESEM images of 10 V nanotube arrays anodized at: (**a**) 5°C showing an average wall thickness of 34 nm, and (**b**) 50°C showing an average wall thickness of 9 nm. The pore size is ≈22 nm for all samples [55]

shows FESEM images of TiO_2 nanotubes fabricated by anodization at 10 V at (a) 5°C and (b) 50°C. The pore diameter of 22 nm is essentially the same for the 10 V anodized TiO_2 nanotube arrays fabricated at the different temperatures, but the wall thickness changes by an approximate factor of four and the tube-length by an approximate factor of two. With decreasing anodization temperature the wall thickness increases from 9 nm at 50°C to 34 nm at 5°C and the voids in both the interpore areas and tubes become more interconnected; the discrete tube-like structure approaches a nanoporous structure in appearance. Also, with decreasing anodization bath temperature, the length of the nanotubes increases from 120 nm at 50°C to 224 nm at 5°C. Table 1.1 shows the variation in 10 V wall thicknesses and tube length as a function of anodization temperature. FESEM images of TiO_2 nanotubes fabricated by anodization at 20 V and (a) 5°C and (b) 25°C. The inner pore diameter of 76 nm shows that the nanotube wall thickness increases from 17 nm at 25°C to 27 nm at 5°C, confirming the trend of increasing nanotube wall-thickness with lower anodization temperature [55].

1.2.4 Using HNO_3/HF

During anodization of titanium foil in a 2.5% HNO_3 plus 1% HF water solution an initial application of a 20 V anodization potential resulted in high current densities in which dielectric breakdown prevented formation of an oxide coating, both with or without addition of boric acid. The applied anodic potential was next ramped from 0 to 20 V at a rate of 0.1 V/s and then held constant at 20 V for 4 h [44]. Figure 1.5 shows the current density for electrolytes with and without boric acid as a function of anodization time after the potential reached 20 V. In an HNO_3–HF electrolyte without boric acid, the current density rapidly decreases with formation of the barrier layer, then slightly increases with formation of a partially porous structure, and then remains relatively constant. In contrast, for boric acid (0.5M) containing HNO_3–HF electrolyte the current densities decrease relatively slowly to a minimum and then slowly increase until reaching a plateau after approximately 110 min.

The surface morphology of a nanotube-array sample anodized in an electrolyte containing 2.5% HNO_3 and 1% HF at 20 V for 4 h showed a uniform, clean, regular structure with the nanotubes having an average pore size of about 100 nm and a wall thickness of about 20 nm [44]. The length of nanotubes was found to be about

Table 1.1 Average wall-thickness and tube-length of 10 V TiO_2 nanotube arrays anodized in HF aqueous electrolyte at different bath temperatures [55]

Anodization temperature	Wall thickness(nm)	Tube-length (nm)
5°C	34	224
25°C	24	176
35°C	13.5	156
50°C	9	120

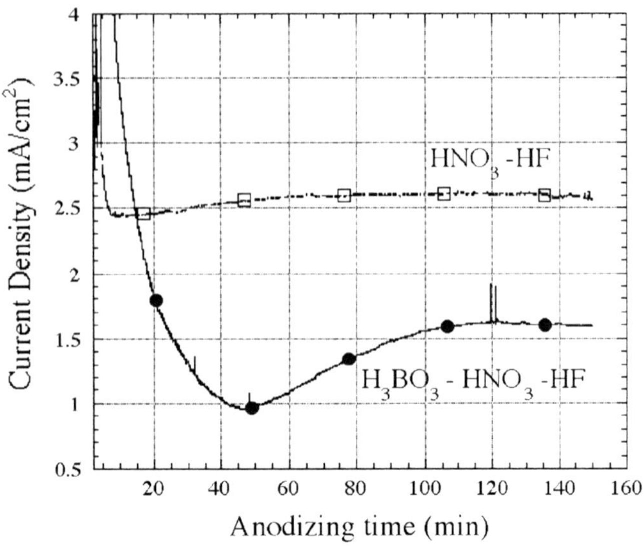

Fig. 1.5 Time variation of titanium electrode etching current density for a nitric acid–hydrofluoric acid electrolyte, and a boric acid/nitric acid/hydrofluoric acid electrolyte [44]

400 nm. The TiO_2 nanotube-array anodized in an electrolyte of 0.5M H_3BO_3–2.5% HNO_3–1%HF at 20 V for 4 h was found to have a precipitate layer ≈400 nm thick. The precipitate layer could be removed by washing with a dilute HF solution. In these samples, there is a pore irregularity with sizes ranging from 10 to 120 nm. The average wall thickness is 20 nm and the length is about 560 nm.

1.2.5 *Using H_2SO_4/HF*

Titanium foils were anodized in 1M H_2SO_4 electrolytes containing HF (0.15 wt%); nanotube arrays 500 nm long were formed [64, 65] using a potential ramp of 50 mV/s from the open-circuit potential to 20 V, followed by a constant 20 V anodization for 2 h. Macak and co-workers used the same recipe to obtain nanotubes by anodizing 500 nm thick Ti film ion-beam sputter deposited on a silicon wafer [66].

1.2.6 *Using $H_2Cr_2O_7$/HF*

Zwilling and co-workers anodized Ti and Ti–6Al–4V alloy in an electrolyte containing 0.5 mol/l chromic acid and 0.095 mol/l HF [49, 50]. Anodization was conducted using a static potential at room temperature in a two-electrode cell with 4 cm electrode separation. The procedure was to increase the potential from 0 V up to the final voltage of 5 or 10 V by five equal steps spaced 1 min apart. The sample was then held at the final voltage for a time between 1 and 55 min. The resulting films were described as a self-organized porous layer with a thin oxide under layer. Pore size increased with voltage. At all potentials the pores were found to be smaller on the Ti substrate than on the TA6V alloy substrate. A typical porous structure obtained in Ti-6Al-4V is shown in Fig. 1.6. A maximum film thickness of 155 ± 20 nm was determined, with pore sizes of 75 ± 20 nm for the Ti-6Al-4V substrate when anodized at 10 V for 20 min. In the growth and pore formation process, it was suggested that a poison-antidote competition controlled the growth – dissolution similar to one suggested for the random porous structure seen for Li in $SiOCl_2$ [67].

1.2.7 Using CH_3COOH/NH_4F, H_2SO_4/NH_4F

Formation of TiO_2 nanotubes was observed by electrochemical anodization of 99.6% impurity free 0.1 mm thick Ti foil samples in nearly anhydrous CH_3COOH, 0.2% water, containing 0.5% NH_4F at potentials of 10 to 120 V [68]. The applied potential was continuously increased from the open-circuit potential to several fixed values at several different scan rates and then held at a fixed potential for several hours. FESEM images of the TiO_2 nanotubes formed at 30 V in CH_3COOH containing 0.5 wt% NH_4F showed the pore diameter to be approximately 10 nm

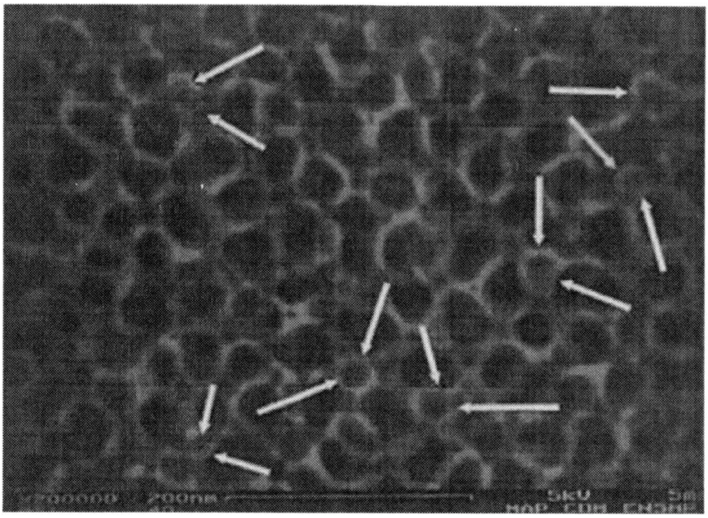

Fig. 1.6 High-resolution scanning electron micrograph of anodized TA6V in chromic acid/HF solution and held at 10 C for 20 min. The arrows point to possible protrusions ([50]; Copyright Wiley-VCH Verlag GmbH & Co. KGaA. Reproduced with permission)

and tube length about 200 nm. The tube diameter increases slightly from top to bottom of the porous structure; pore widening is presumed to be caused by changes in the local chemistry within the pores while it was growing. At potentials higher than 50 V the layer exhibited some "morphology burst," where a local morphology different that the rest of the surface was obtained with the "burst" area resembling a coral reef. This burst morphology was quite different from layers formed at potentials lower than 50 V. The fraction of surface area showing burst-locations increased with increasing potential up to 120 V, while outside the burst areas a uniform surface coverage with tubes 20 nm diameter was obtained similar to those formed at lower applied potentials. Using other electrolytes, 40 nm diameter tubes were reported in phosphate-based electrolyte [70] and 100 nm diameter tubes were reported sulfate-based electrolytes [45], indicating again that electrolyte characteristics can control details of the tube geometry. Using 0.24 mol/l H_2SO_4 in 0.5 wt% NH_4F, a well-ordered TiO_2 nanotube array with an inner diameter of about 120 nm and length of about 300 nm was obtained at an anodization voltage of 20 V [71].

1.2.8 *Using H_3PO_4/HF, H_3PO_4/NH_4F*

Ti anodization at constant 10 V for 2 h in 1M H_3PO_4 with fluoride ion concentrations between 0.1–0.4 wt%, yielded self-organized nanotube structures with 50 nm pore diameters. Optimized conditions for self-organized pore formation occurred at 0.3 wt% HF in 1M H_3PO_4, with both tube length and pore diameter potential dependent. From 1 to 25 V, measured with respect to the reference Ag/AgCl electrode, self-organized porous nanotubes with diameters ranging from 15 to 120 nm and length from 20 nm to 1.1 μm were achieved.

Using an electrolyte with pH of 4.5 and consisting of [72] 1M H_3PO_4+1M NaOH+0.5 wt%, anodizing Ti foil at 20 V for 2 h, and with an initial voltage ramp rate of 0.5 V/s from open circuit potential to 20 V, nanotube arrays were achieved with a diameter of 100 ± 10 nm, length 1 ± 0.1 μm, and tube wall thickness of 10 ± 2 nm [70]. Using the same electrolyte, Premchand et al. obtained self-organized nanotube arrays of TiO_2 by anodizing sputter-deposited titanium films 1.9 μm thick on p-type Si(100) substrates [73]. An average tube length of 560 nm, diameter of about 80 nm, wall thicknesses of about 20 nm were achieved. Yu and co-workers fabricated nanoporous/nanotube array films by anodizing sputter-deposited Ti films in 0.1–2 wt% HF between 3 and 20 V [74].

1.2.8.1 Effect of Different Cathode Metals

Overvoltage is the excess potential required for the discharge of an ion at an electrode over and above the equilibrium potential of the electrode. In metal anodization, both anode and cathode selectively influence the rates of certain reaction steps and hence overvoltage. For example, hydrogen evolution reaction (HER) rates differ by several orders of magnitude among possible cathodes [75, 76]. Thus, the overpotential influences the morphology, dimensions as well as growth rate of the nanotubes. Although platinum is commonly used as the cathode material for anodization due to its high catalytic activity and hence low overpotential losses in addition to its high stability, it may not provide the desired nanotube morphology. Therefore, to investigate the most suitable electrode system for Ti anodization, as well as to develop an alternative to Pt based cathodes, Allam et al. selected a variety of cathode materials and investigated their effects on the morphology of the resulting TiO_2 nanotubes, using both aqueous and ethylene glycol electrolytes that contained NH_4F and H_3PO_4 [77]. Anodization was performed using a two-electrode cell with titanium foil as the working anode and foil samples of Pt, Pd, Ni, Fe, Co, Cu, Ta, W, Sn, Al, and a Carbon rod as the cathode. Anodization were carried out for 10 h at a constant applied voltage of 20 V at room temperature (about 22°C) in either: (A) 0.2M NH_4F+0.1M H_3PO_4 aqueous electrolyte, or (B) ethylene glycol (EG) containing 0.2M NH_4F+0.1M H_3PO_4+2% H_2O electrolyte. Table 1.2 lists the cathode materials tested, grouped according to their position in the periodic table, with the average diameters and lengths of the nanotube arrays fabricated using these cathode materials. The tube dimensions were compiled after detailed examination of the resulting morphologies using FESEM.

Ni and Pd, both Pt-group metals, produced architectures similar to those obtained using Pt. Within the transition metals, Fe, Co and Cu cathodes produced well-aligned nanotubes of lengths, respectively: 2.5, 1.9 and 1.27 μm. Pore diameters of Fe – 9 nm, Co – 135 nm, and Cu – 81 nm were not correlated with length. The pore size with Pt was 105 nm; Ta produced the largest pore size of 140 nm. Nanotube arrays produced using a W cathode were the shortest. Of the non-transition metals, although arrays fabricated using C and Sn showed comparable

Table 1.2 The cathode materials tested, grouped according to their position in the periodic table, with the average diameters and lengths of the nanotube arrays fabricated using these cathode materials, 10 h anodization duration, in both aqueous and EG electrolytes [77]

Group	Cathode material	Fabricated nanotubes in aqueous electrolytes		Fabricated nanotubes in EG electrolytes	
		Average diameter ±5–7 (nm)	Average length ±10 (nm)	Average diameter ±5 (nm)	Average length ±10 (nm)
Pt-group Elements	Ni	143	1200	65	1510
	Pd	134	1435	61	2500
	Pt	105	1520	65	1725
Non-Pt transition Elements	Fe	99	2470	68	2000
	Co	135	1900	143	2100
	Cu	81	1265	83	1130
	Ta	140	1175	70	1300
	W	91	690	114	2400
Non-transition Elements	C	143	1300	81	1220
	Al	96	570	85	1600
	Sn	147	1220	90	1060

dimensions arrays produced using a C cathode were more uniformly aligned. Al cathodes produced the shortest nanotubes of all: 0.57 μm and with relatively poor uniformity.

FESEM images of Ti foil samples anodized in EG electrolytes containing 0.2M NH_4F+0.1M H_3PO_4 at 20 V using Ni and Pt cathodes produced TiO_2 nanotube array surfaces covered with precipitate debris. In contrast, nanotube arrays fabricated using a Pd cathode had no such debris. Cross-sectional images show the Pd fabricated nanotube arrays were more highly aligned than those fabricated using either Pt or Ni. FESEM micrographs of nanotube arrays synthesized using a Fe cathode showed no surface debris. Although cross-sectional views of the nanotubes fabricated using Co, Cu, Ta and W cathodes showed well-organized nanotube arrays, top view micrographs showed that the tubes were clogged with debris. Nanotube arrays fabricated using C, Sn and Al cathodes showed the formation of well-aligned structures with very little debris.

The results confirm that the nature of the cathode material plays a significant role in the appearance of surface precipitate. The overpotential of the cathode is a critical factor that affects the dissolution kinetics of the Ti anode and, in turn, the activity of the electrolyte and morphology of the architectures. Different cathode materials lead to different morphologies that appear to result from overvoltages within the electrolyte.

1.3 Nanotube Array Synthesis Using Buffered Electrolytes: The Second Generation

Cai and co-workers first fabricated nanotube arrays several microns long using either KF or NaF electrolytes of variable pH [43, 96]. A summary is presented in Table 1.3. They obtained the desired pH by adding NaOH, sulfuric acid (pH 1–2), sodium hydrogen sulfate, or citric acid (pH 2.5–7.5) to the electrolyte before adding KF. The F^- concentration was held fixed at 0.1 mol/L. In the 0.1 mol/L F^- and 1 mol/L H_2SO_4 medium, the potential window for nanotube formation is between 10 and 25 V (Samples 01 to 08); outside this potential range no nanotubes were formed. Sample 01 was produced with an applied voltage of 5 V; the electrochemical etch rate was slow and only a few pits appear on the sample surface. Sample 08 was produced with an applied voltage of 30 V; the electrochemical etch is fast enough to prevent nanotube formation and only a highly disturbed porous structure was obtained [43]. The nanotube pore size was found to be proportional to the potential applied (Samples 02 to 05) and independent of the anodization time (Samples 04 and 07) and electrolyte concentration (Samples 04 and 08). Increasing the potential from 10 to 25 V increased the diameter of the resulting nanotubes from 40 to 110 nm. No significant difference was observed in the pore size for anodization times of 1 h (Sample 04) and 6.5 h (Sample 07) or for electrolyte concentrations 1 mol/L H_2SO_4 (Sample 04) and 2 mol/L H_2SO_4 (Sample 08).

Table 1.3 Electrolyte pH and composition, anodization conditions, and size of the resulting nanotubes [43]

No.	Elecrolyte[a,d,e]				pH[b]	V	t	D	L	Q[c]
	F^-	SO_4^{2-}	PO_4^{3-}	Cit		(V)	(h)	(nm)	(μm)	
01	0.1	1.0	–	–	<1	5	1	10±2	–	No NT
02	0.1	1.0	–	–	<1	10	1	40±5	0.28±0.02	NT
03	0.1	1.0	–	–	<1	15	1	80±9	–	NT
04	0.1	1.0	–	–	<1	20	1	100±11	0.48±0.03	NT
05	0.1	1.0	–	–	<1	25	1	110±12	0.56±0.04	NT
06	0.1	1.0	–	–	<1	30	1	–	–	No NT
07	0.1	1.0	–	–	<1	20	6.5	100±11	0.43±0.03	NT
08	0.1	2.0	–	–	<1	20	1	100±11	0.45±0.03	NT
09	0.1	1.0	–	0.2	1.3	10	20	30±5	0.32±0.03	NT
10	0.1	1.0	–	0.2	2.8	10	20	30±5	0.59±0.05	NT
11	0.1	1.0	–	0.2	2.8	15	20	50±5	1.00±0.05	NT
12	0.1	1.0	–	0.2	2.8	25	20	115±10	1.50±0.04	NT
13	0.1	1.0	–	0.2	3.8	10	20	30±5	0.80±0.06	NT
14	0.1	1.0	–	0.2	3.8	10	60	30±5	1.80±0.06	NT
15	0.1	1.0	–	0.2	3.8	10	90	30±5	2.30±0.08	NT
16	0.1	1.0	–	0.2	4.5	10	20	30±5	1.05±0.04	NT
17	0.1	1.0	–	0.2	4.5	25	20	115±5	4.40±0.10	NT
18	0.1	1.0	–	0.2	5.0	10	20	30±5	1.40±0.06	NT
19	0.1	1.0	–	0.2	5.0	25	20	115±5	6.00±0.40	NT
19	0.1	1.0	0.1	0.2	6.4	10	24	–	–	No NT
20	–	2.0	–	–	<1	10	24	–	–	No NT

[a]Electrolyte components are in mol/L

[b]pH < 1 represents a 1.0 or 2.0 mol/L H_2SO_4 medium

[c]Quality Q of resulting nanotubes. NT: nanotubes uniformly across substrate. No NT: no nanotubes or partly developed nanotube/porous structures

[d]*Cit* citrate; *t* time; *D* inner diameter of nanotube; *L* length of nanotube

[e]SO_4^{2-} is from addition of H_2SO_4 or $NaHSO_4$; PO_4^{3-} is addition of potassium hydrogen phosphate $K_2HP_3O_4$; Cit denotes citric acid from its salt, $HO(CO_2Na)(CH_2CO_2Na)_2 \bullet 2H_2O$

Electrolyte pH affects both the behavior of the electrochemical etch and the chemical dissolution due to the hydrolysis of titanium ions. With increasing pH the hydrolysis content increases and it, in turn, slows the rate of chemical dissolution. As shown in Fig. 1.7 (Samples 10 and 13 for 10 V and 12 and 17 for 25 V) and Table 1.3, longer nanotubes are formed in higher pH solution. With a potential of 25 V and with pH increasing from strong acidity (Sample 05, pH < 1) to weak acidity (Sample 17, pH 4.5), nanotube length increased from 0.56 to 4.4 μm; for 10 V equivalent sample lengths increased from 0.28 μm (Sample 2, pH < 1) to 1.4 μm (Sample 18, pH 5.0). At a specific pH value, the length increases with applied potential (Samples 10–12 and 16–17). When the potential increased from 10 to 25 V the length increased from 0.59 to 1.5 μm for pH = 2.8 and from 1.05 to 4.4 μm for pH = 4.5. Also, at a specific pH, the pore size of the nanotubes increases with anodization potential; this is shown in the inset of Fig. 1.7 (Samples 10, 11, and 12). However, at a specific applied voltage, the pore size was independent of pH.

Samples 04 and 07 illustrate that increasing the anodization time does not increase the nanotube length in highly acidic electrolytes, i.e., pH < 1. Samples 13

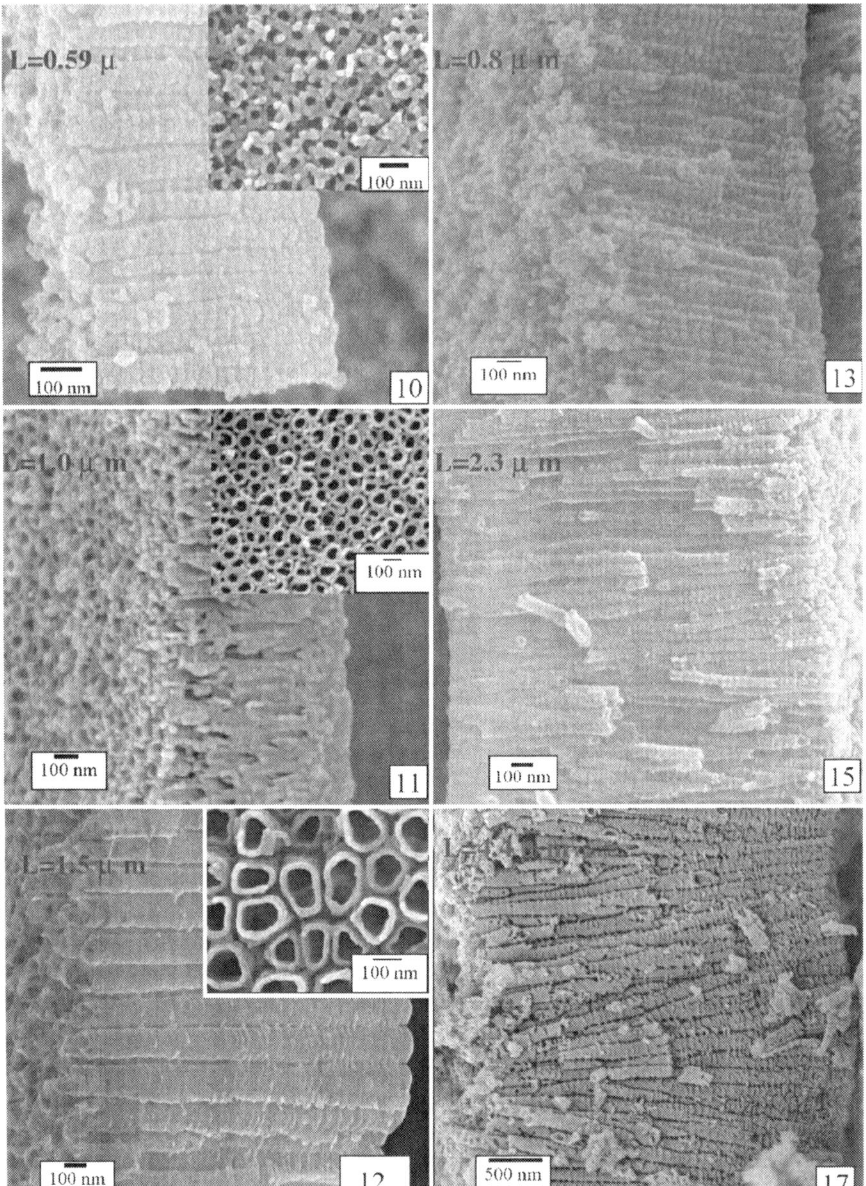

Fig. 1.7 Lateral view of the nanotubes formed in different pH solutions (pH>1). The anodization conditions for each sample are listed in Table 1.3. Inset to Samples 10 and 12 show variation of pore size with anodization potential for a pH 2.8 electrolyte [43]

to 15 show that nanotube length is time-dependent in weakly acidic electrolytes, see Fig. 1.7; increasing the anodization time from 23 to 90 h increases the length from 0.8 to 2.3 μm. At increased pH values the hydrolysis content increases and that, in turn, produces a significant amount of hydrous titanic oxide as a precipitate on the nanotube surface. The best pH values for formation of longer nanotubes appeared between three and five; lower pH values produced shorter but clean nanotubes and higher pH values produce longer tubes with unwanted precipitates. Alkaline solutions are not favorable for self-organized nanotube formation. In the case of Sample 19, see Table 1.3, the anodization was done in pH 6.4 with 0.1 mol/L PO_4^{3-} with no nanotubes produced, rather a dense layer of hydrous TiO_2 salts; in the presence of phosphate ions a more acidic electrolyte is required to avoid titanic phosphate precipitate. No nanotube array formation has been achieved without F^-, even in 2 mol/L H_2SO_4 solutions, see sample 20.

1.3.1 Step-by-Step Procedure: Solution Preparation, Mixing and pH Adjustment

With anodic oxidation of titanium in a KF bath, titanium foil 0.25 mm thick 99.7% purity from Aldrich is used as the anode, and platinum as the cathode. Anodization performed at 25 V in a KF bath pH of 6.0 for 17 h results in 6 μm long, amorphous titania nanotube array film.

1.3.2 Solution Set Preparation

S1 Using sodium hydrogen sulfate monohydrate (Reagent grade from Alfa Aesar, Formula weight – 138.07), make 2.0M solution (to make 500 ml solution, add 138.07 g of $NaHSO_4$ in D.I. water).

S2 Using sodium citrate tribasic dihydrate (ACS reagent ≥99% from Sigma Aldrich, Formula weight – 294.1), make 1.0M solution (to make 250 ml solution, add 73.625 g of sodium citrate in D.I. water).

S3 Using sodium hydroxide (anhydrous, 97% from Alfa Aesar), make a saturated solution adjusting solution pH.

S4 Using potassium fluoride (ACS reagent from Alfa Aesar, Formula weight – 58.1), make 2.0M solution (to make 250 ml solution, add 29.05 g of KF in D.I. water).

We mix 1.0M sodium hydrogen sulfate (*S1*), 0.2M sodium citrate (*S2*) and 0.1M potassium fluoride (*S4*) solution to prepare KF buffer solution. The steps for making 100 ml KF buffer solution of pH 5.0 are as follow:

Step 1 Take 50 ml solution from *S1* and mix with 20 ml solution *S2*.

Step 2 Measure pH of the solution (it will be less than 1). Add *S3* solution until pH becomes 4.5. (Note: Adding NaOH to acid solution results in a warm solution; let it stand normalized to room temperature for 2 h).

Step 3 Once solution temperature equates to room temperature, add *S3* solution to make pH close to 4.85. Add D.I. water to make the total volume 90 ml.

Step 4 Add a few drops of *S3* solution to make the pH 5.0. (Note: at this stage, even a single drop of saturated NaOH solution can make a great change in pH).

Step 5 Add D.I water to make volume 95 ml.

Step 6 Add 5 ml of *S4* solution, bringing total volume to 100 ml. Mix the solution in a Nalgene (or plastic) beaker.

When titanium foil is anodized in a KF electrolyte, the solution sometimes wicks (travels) up the sample and touches the clips holding the Ti foil. This problem can be avoided by making an electrical contact to Ti foil with a copper wire, typically using a conducting epoxy, and covering the rest of the area with non-conducting epoxy. There should be 2–4 cm separation between the anode and cathode during anodization. The size of the foil sample exposed to the bath should be chosen to avoid electrolyte heating and its resulting affect upon the growth and morphology of nanotube arrays. When anodization is completed leave the applied voltage on the sample, but remove the sample from the bath and drop it into D.I. water, thoroughly washing it. A white precipitate on the top surface will occur if the pH of the bath is $\geq$6.0; the precipitate can be removed by applying ultrasonic agitation to the sample in D.I. water for 10–15 s, followed by a subsequent wash and nitrogen dry.

To achieve high-aspect-ratio growth of TiO_2 nanotubes, Macak and co-workers sought to adjust the dissolution rate of TiO_2 by localized acidification at the pore bottom while a protective environment was maintained along the pore walls and at the pore mouth [45, 69, 97, 98]. Localized acidification at the pore bottom follows from the oxidation and hydrolysis of elemental titanium. Using a buffered neutral solution as electrolyte and choosing the proper anodic current value, acid can be created at the pore bottom while higher pH values are established at the pore mouth as a result of migration and diffusion effects. A voltage-sweep was applied to achieve a steady current and to establish the desired pH profile. For this purpose, Ti foil 0.1 mm thick with 99.6% purity was anodized in a 1.0M $(NH_4)_2SO_4$ electrolyte containing 0.5 wt% NH_4F, using a potential sweep from open-circuit potential to 20 V with a sweep rate 0.1 Vs^{-1}. The resulting self-organized nanoporous TiO_2 had a thickness of about 2.5 μm and pore diameters in range between 90 and 110 nm. Higher sweep rates resulted in wider pores [97].

1.3.3 Anodization with Constant Current Density

Anodization of Ti foil using a constant current density of 0.5–1 mA/cm^2 in 1.0M $(NH_4)_2SO_4$+0.5 wt% NH_4F electrolyte resulted in a tubular structure [99]. Samples anodized at 1 mA/cm^2 resulted in a 950 nm thick tubular layer, with a tube diameter ranging from 60 to 90 nm. It was found that for current densities lower than 1 mA/cm^2 the cell voltage rapidly increases and then plateaus. For currents densities ranging from 1 to 2 mA/cm^2, after a slight delay voltage oscillations begin that result in layered structures. For an anodic current density of 1 mA/cm^2, voltage oscillations begin at about 90 min. Damped oscillations were observed in all experiments with anodic current densities between 2 and 4 mA/cm^2. For higher current densities, the delay time decreases strongly and often is less than 5 min for anodization currents of 4 mA/cm^2. Ti samples anodized at 0.5 mA/cm^2 are porous and have a layer thickness of 500+100 nm, with the layer consisting of individual tubes with a tube diameter of 40+10 nm. Samples anodized at 0.75 mA/cm^2 show a 900 nm thick tubular layer with tube diameters of 60+10 nm. For anodization at 1 or 1.1 mA/cm^2, where no potential oscillations were detected, the thickness and the tube diameter are, respectively, 950 nm and 75+15 nm. It is evident from the potential vs. time curves that the "steady" potential amplitude before oscillations occur depends only weakly on the current, i.e. a value of about 11 V is observed for 0.75, 1, 1.1 mA/cm^2. In all cases nanotube array films about 900 nm thick with 60 nm pore diameters are obtained. Results suggest that, indeed, to a large extent potential determines the tube diameter.

The potential vs. time behavior was observed in constant current density anodization experiments on Ti in 1.0M $(NH_4)_2SO_4$+0.5 wt% NH_4F electrolyte [99]. A voltage maximum in the first few minutes is ascribed to the growth of a compact layer. After the initial peak the voltage decreases, indicating the onset of the first breakdown events due to the presence of F^- ions. At the breakdown sites the pores nucleate and deepen; the longer electrolyte path causes the potential necessary to drive the same current to slightly increase. In the oscillation-free cases, a steady state situation is eventually attained where the chemical dissolution and growth rates balance and the layer thickness remains unchanged.

The difference in electrolyte pH, which in turn determines oxide dissolution rate, leads to significant variation in tube length and pore diameter. Anodizing the titanium foil at a constant potential of 20 V after a potential ramp of 0.1 V/s in 1M Na_2SO_4 electrolyte with 0.5 wt% NaF for about 6 h resulted in a highly ordered porous TiO_2 layer of a maximum 2.4 μm thickness and typical pore diameters of 100 nm [100]. However, in 1M $(NH_4)H_2PO_4$ containing 0.5 wt% NH_4F TiO_2 nanotubes formed at 20 V had a maximum length of 4 μm with a pore diameter of 100 nm after 40 h of anodization. The nanotube length and pore diameter are, 500 and 50 nm, respectively, for 1M H_3PO_4+0.5 wt% HF, and 4 μm and 100 nm in 1M, respectively, for $(NH_4)H_2PO_4$+0.5 wt% NH_4F [69, 70, 101, 102]. DC magnetron sputtering of titanium films, less than a micron in thickness, on silicon substrates at a substrate temperature during deposition of 500°C were shown to form nanotube arrays upon anodization in 1M Na_2SO_4+0.2M citric acid with the addition of 0.4 wt % NaF in the potential range from 10 to 20 V [103, 104]. At 20 V, the length and average tube outer diameter of the arrays were, respectively, 458 and 100 nm.

Efforts were made to study different cations in neutral sulfate-containing electrolytes and compare their influence on the pore morphology and thickness of nanoporous oxide films grown on titanium [105]. In each electrolyte, both the anionic fluoride and sulfate ions are coupled to identical cations of ammonium, cesium, and potassium or sodium. A three-electrode configuration was used for anodic oxidation at 20 V. Top views of samples formed in 0.2M Cs_2SO_4+0.5 wt% CsF show a non-uniform structure with some small patches of pores between large non-porous areas. Samples fabricated in 0.2M K_2SO_4+0.5 wt% KF showed somewhat organized pores covering almost all the entire area (~99%) were coated with a precipitated whisker-like layer of insoluble salts. The porous structure formed by anodization in 1M Na_2SO_4+0.5 wt% NaF and 1M $(NH_4)_2SO_4$+0.5 wt% NH_4F solutions resulted in far more homogenous tubular structures; after an anodization of several hours, the nanotube layers were uniform and regular, with little precipitate. Cross-sectional images of the self-organized nanotubes formed in Cs^+ and K^+ electrolytes revealed that the nanotubes were covered with a precipitate layer; the layer interfered with the continuity and uniformity of the ion flow. Nanotube array lengths of about 500 nm were achieved [105]. When anodizing in Na^+- or NH_4^+-electrolytes the resulting structure consisted of self-organized, vertically standing, and well-adhered nanotube array layers on the Ti substrate [105]. For anodization times of 2 h or more there was little indication of debris or precipitates. The maximum nanotube length achieved in sulfate-based electrolytes was 2.4 μm with addition of NaF [47], and 2.5 μm with the addition of NH_4F [45].

1.4 Synthesis of Nanotube Arrays Using Polar Organic Electrolytes: The Third Generation

1.4.1 Using Formamide and Dimethyl formamide electrolyte

Figure 1.8 shows nanotubes nearly 70 μm long grown in a formamide (FA) based electrolyte, 48 h anodization at a constant potential of 35 V [51]. The average outer diameter of the nanotubes is approximately 180 nm, resulting in an aspect ratio of ≈390; the average and wall-thickness is about 24 nm. The key to successfully growing long nanotube arrays is to keep the water content of the anodization bath below 5%. With organic electrolytes, compared with water, the reduced availability of oxygen reduces oxide formation [106]. At the same time reduced water content decreases the chemical dissolution of the oxide in fluorine-containing electrolytes thus aiding formation of longer nanotubes. Illustrative formamide electrolyte compositions include formamide (99%) and/or N-methylformamide (NMF; 99%) solutions containing 1–5 wt% of deionized water and 0.3–0.6 wt% NH_4F (98%) [51, 53, 107]. Nanotubes with a smaller pore diameter but approximately 10% longer were obtained using a NMF electrolyte. Lower anodization potentials generally result in shorter nanotubes with smaller diameters. The increase in nanotube length with anodization voltage is attributed to the increased driving force for ionic transport through the barrier layer at the bottom of the pore, in turn resulting in faster movement of the Ti/TiO_2 interface into the Ti metal. As seen for the aqueous and buffered aqueous electrolytes, pore diameter increases with anodization voltage.

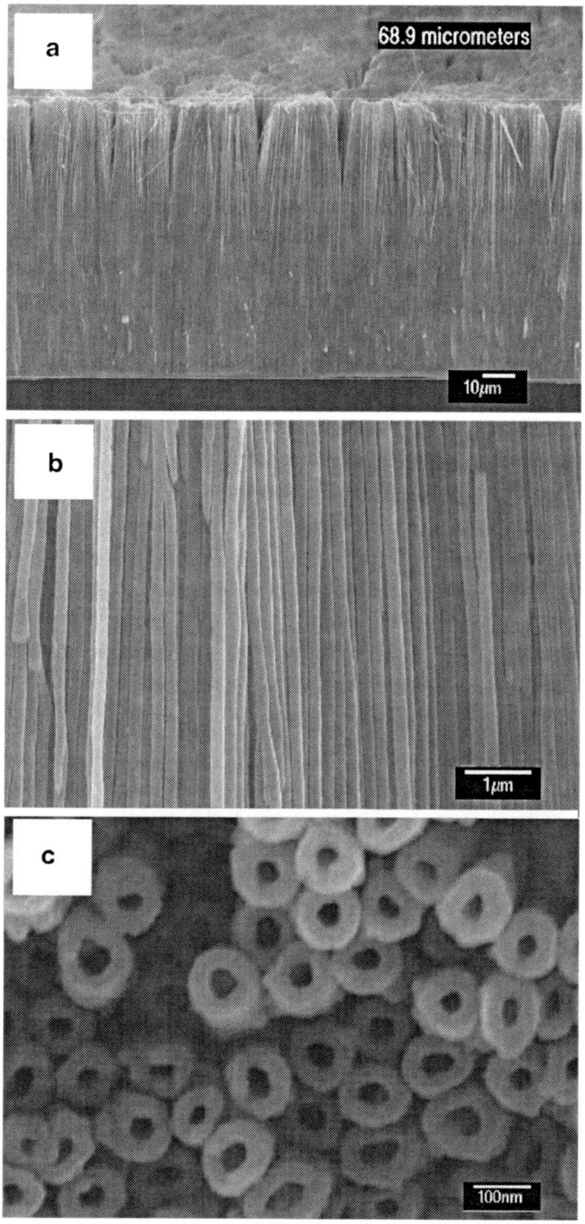

Fig. 1.8 FESEM images of TiO_2 nanotubes grown in FA based electrolyte at 35 V for 48 h showing: (**a**) cross-section at lower magnification, (**b**) cross-section at high magnification, and (**c**) top surface image [51]

The range of applied anodization potentials over which nanotube arrays are obtained in FA/NMF based electrolytes, 10 to 50 V, is significantly larger than aqueous KF or NaF electrolytes, 10 to 29 V. With FA/NMF based electrolytes anodization at 60 V is unstable with sharp fluctuations in the current; anodization at voltages below 10 V was not attempted. The anodization duration is also important since lengths increase with time up to a maximum length, beyond which they shorten, presumably due to dissolution-induced cleaving of the tubes. For example, with a 35 V anodization potential and duration of 48 h the average length was 70 μm, after 88 h it was 30 μm and after 164 h it was only 10 μm. The time to maximum length is a function of the anodization voltage, and the lower the voltage the sooner the maximum is reached.

Shankar and co-workers [53] used formamide-water mixtures containing fluoride ions to study the effect of five different cationic species on TiO_2 nanotube array formation. Cation choice is a significant factor both in nanotube growth rate and its length. Under similar conditions, electrolytes containing the cation tetrabutylammonium (from tetrabutylammonium fluoride Bu_4NF) produced the longest nanotubes, about 94 μm. The shortest nanotubes, about 3 μm, grew when H^+ ions were the only cationic species in the electrolyte. This length difference is attributed to the inhibitory effect of the quaternary ammonium ions that restrict the thickness of the barrier oxide layer; a thinner interfacial oxide layer facilitates ionic transport and thus enhances nanotube growth. The aspect ratio of the resulting nanotubes is also voltage dependent with the highest aspect ratio, about 700, obtained at an anodizing voltage of 20 V in an electrolyte containing tetrabutylammonium ions. Table 1.4

Table 1.4 Effect of cation type, cation concentration, and anodization duration on the morphological features of TiO_2 nanotube arrays, at a fixed potential. In every case the electrolytes contain 5% water in formamide and anodization was performed at room temperature, 22°C. The anodization duration is the approximate time required to obtain maximum nanotube length [53]

Cation	Molar (M) cation concentration	Anodization duration (h)	Outer diameter (nm)	Wall thickness (nm)	Nanotube length (μm)
Anodization voltage 35 V					
H^+	0.14	101	256	21	5.6
H^+	0.27	48	214	20	7.3
NH_4^+	0.14	88	208	17	29.2
NH_4^+	0.27	30	159	15	37.4
Na^+	Saturated solution (~0.04)	48	48	18	9.6
Bu_4N^+	0.27	48	190	22	68.9
Anodization voltage 20 V					
H^+	0.27	48	99	22	2.9
NH_4^+	0.14	55	90	19	14.4
NH_4^+	0.27	24	90	17	19.6
Bu_4N^+	0.27	34	90	16	35.2
Anodization voltage 15 V					
NH_4^+	0.14	110	81	29	8.2
Bu_4N^+	0.27	46	80	15	20.0
$BnMe_3^+$	0.27	42	70	18	2082

summarizes the effect of different cations and cation concentrations on the nanotube morphology for formamide-based electrolytes containing 5% water at different anodization potentials. A trend of decreasing wall-thickness at higher fluoride ion concentrations is attributed to increased chemical etching by higher F^- concentrations. At larger anodization voltages, the driving force for ionic transport through the barrier layer at the bottom of the pore is greater and results in faster movement of the Ti/TiO_2 interface into the Ti metal. This enhanced pore deepening resulted in greater nanotube lengths obtained at larger anodization voltages. Figure 1.9 shows, respectively, the real time constant potential anodization behavior of Ti anodized at 20 V in 95% formamide+5% water solutions containing 0.27M of NH_4F, $(C_4H_7)_4NH_4F$, HF and $C_6H_6(CH_3)_3NH_4F$. Figure 1.10 shows the real time anodization behavior of Ti foil anodized at 20 V in electrolytes containing an identical concentration of NH_4F (0.27M) but with different amounts of water, ranging from 100% water (no formamide) to 2.5% water+97.5% formamide. For a given electrolyte we observe an increase in the nanotube diameter as the anodization voltage is increased [41, 108]. At anodization potentials from 10 V to about 20 V the nanotube length increases faster than nanotube diameter, i.e. increasing voltage results in an increased aspect ratio. At potentials larger than 20 V the increase in length was less dramatic and generally matched by the increasing pore diameter. Consequently,

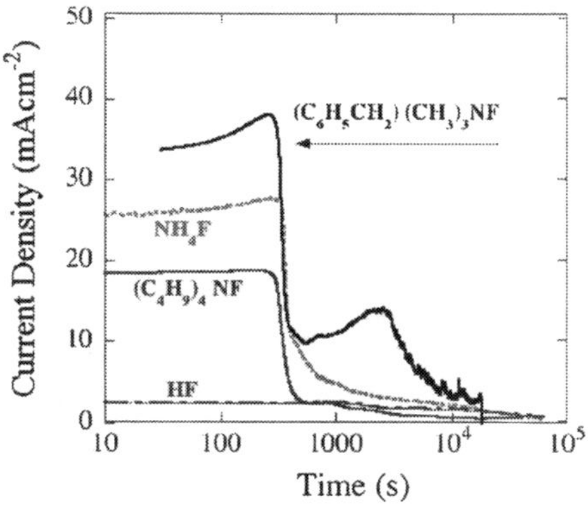

Fig. 1.9 Anodization current-time behavior for Ti foil at 20 V in formamide solution containing 5% H_2O and 0.27M of fluoride ion bearing compounds with four different cationic species: H^+, NH_4^+, $[C_4H_9]_4N^+$, and $[C_6H_5CH_2][CH_3]_3N^+$ [53]

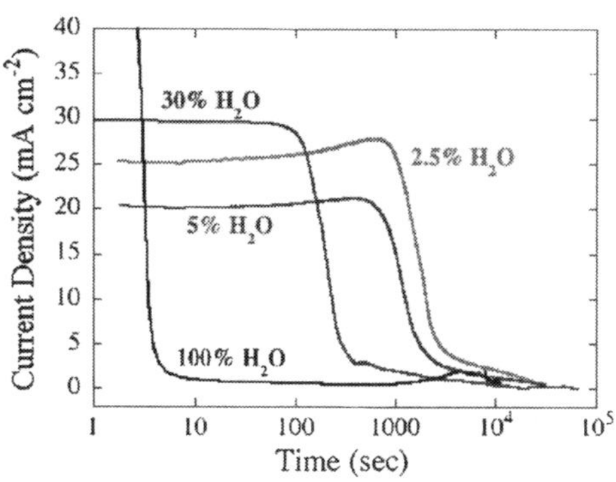

Fig. 1.10 The anodization current-time behavior of a Ti foil (99.8% pure) anodized at 20 V in an electrolyte containing 0.27M NH_4F in formamide of variable water content [53]

for the electrolytes investigated the highest aspect ratio nanotubes are obtained at a potential of 20 V.

Maximum nanotube length was obtained with a shorter anodization period, implying a faster growth rate. We attribute this behavior to: (a) a thinner interfacial oxide layer promotes faster migration of the oxide-metal interface into the metal, (b) a decrease in the rate of chemical etching, and (c) the higher conductivity and smaller solvation shell of the Bu_4N^+ ion that enables a greater amount of charge to enter. Quaternary ammonium salts effectively inhibit the acid corrosion of metals in the presence of halide ions [109]. The inhibitory effect of quaternary ammonium salts is explained by structuring the solvent in the diffuse part of the electrical double layer and by decreasing the mobility of hydronium ions in the near-electrode layer [110]. The inhibiting action of organic compounds containing polar groups such as nitrogen and sulfur is also based on the adsorption ability of the molecules where the resulting adsorbed film protects the metal from corrosion.

1.4.2 Dimethyl Sulfoxide Electrolytes

In 2005, Ruan and co-workers reported synthesis of titania nanotube arrays by Ti anodization in an electrolyte comprised of a fluorinated dimethyl sulfoxide (DMSO) and ethanol mixture [111], with the addition of ethanol limiting nanotube array length to $\approx 2.5\,\mu m$. Nanotube arrays of extended length from DMSO electrolytes were reported by Paulose and co-workers in 2006 [51]. Ti foils were anodized at different DC voltages in a two-electrode electrochemical cell using an electrolyte containing DMSO (99.6%) and hydrofluoric acid (HF, 48% aqueous solution). The anodization voltages were varied from 10 to 70 V, HF concentration varied from 1 to 6%, and anodization duration from 20 to 90 h [51, 112]. To effectively template the surface, some samples were pre-anodized in a 0.5% HF aqueous solution at 20 V before anodizing in a 2.0% HF-DMSO solution at 40 V. Electrode spacing was maintained at 1 cm. All experiments were conducted at room temperature, approximately 22°C. Voltage affects pore size, wall thickness, and length of the nanotubes, as shown in Table 1.5. After anodizing at 40 V the wall thickness was in the range of 8–17 nm, the variation was because of fluctuations in dissolution of the titanium metal. An average wall thickness of about 15 nm increased to approximately 30 nm with crystallization; as a result the inner diameter of the annealed nanotubes

Table 1.5 Effect of anodization voltage on the morphology of the resulting TiO_2 nanotube arrays, for a 70 h 2% HF-DMSO anodization [112]

	40 V anodized sample	60 V anodized sample
Pore size (nm)	~120	~150
Bottom size (nm)	~250	~400
Wall thickness (nm)	~15 (8–17 nm variation)	~50
Length (μm)	45 (after sonication)	93
Calculated geometrical surface area	~1800	~3200

reduced from about 120 to 105 nm. The nanotube array packing appears to be very nearly an ideal hexagonal close packed arrangement. Figure 1.11 shows FESEM images of samples prepared in the 2.0% HF-DMSO solution and anodized at 40 V (images a,c,e) and 60 V (images b,d,f) for 70 h [51, 112]. The surface of the as-anodized nanotube samples are covered with unwanted debris, precipitate from solution that can be removed by ultrasonic agitation. The vertical alignment of the nanotubes and the general morphology are independent of the voltage, and the nanotube arrays are separated from the underlying titanium foil by a dense barrier layer.

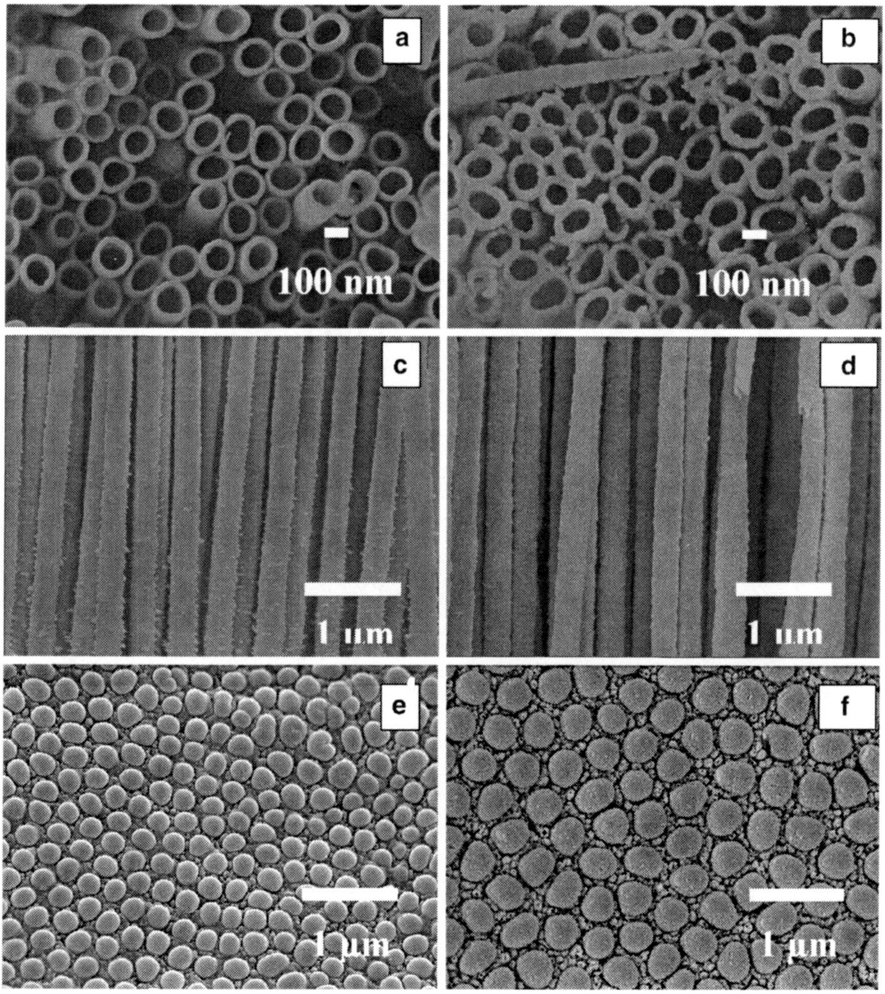

Fig. 1.11 FESEM images of a TiO_2 nanotube array sample grown from a 2.0% HF-DMSO electrolyte for 70 h at: (**a**,**c**,**e**) 40 V and (**b**,**d**,**f**) 60 V. Top surface images are seen in (**a**,**b**), cross sectional images in (**c**,**d**), and tube bottom view images, removed from the underlying barrier layer, are seen in (**e**,**f**) [51, 112]

Anodization parameters include voltage, HF concentration, electrolyte water content, and anodization time. Each plays a key role in determining the nanotube length of the resulting nanotubes. In Fig. 1.12a the length of nanotubes formed on samples with and without a pre-anodization treatment, in effect making a template of the Ti foil surface, is seen to increase with anodization voltage. A maximum length is achieved at 60 V, a dramatically decreased length is achieved at 70 V, and at 80 V the Ti foil sample dissolved into the electrolyte after 24 h. Titanium foils that were pre-anodized in 0.5% HF in water at 20 V for 20 min followed by anodization in 2.0% HF containing DMSO electrolyte showed superior growth rates. The maximum length obtained was 101 μm upon anodizing a pre-anodized sample in 2.0% HF-DMSO at 60 V for 70 h.

To increase nanotube length it is necessary to reduce the chemical dissolution of the oxide at the pore mouth while maintaining active growth at the bottom of the pore.

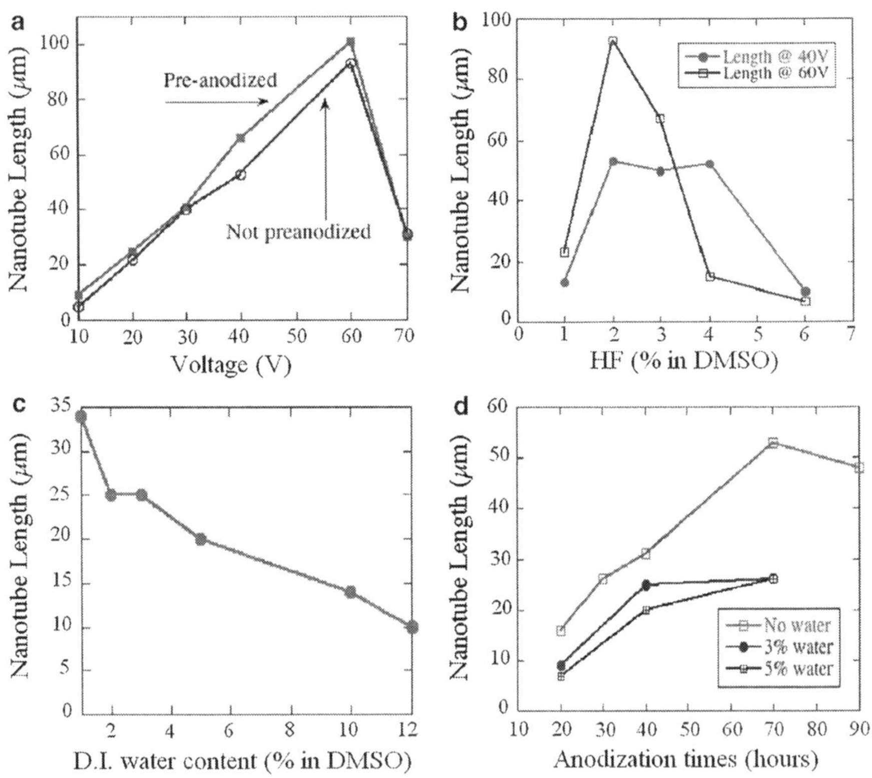

Fig. 1.12 Variation of TiO_2 nanotube array length as a function of: (**a**) applied voltage for a 70 h anodization using 2%HF-DMSO electrolyte with and without a pre-anodization step to template the surface; (**b**) HF concentration in DMSO for 40 and 60 V 70 h anodization; (**c**) variation of nanotube length obtained from a 2.0% HF-DMSO electrolyte containing different deionized water concentrations (40 V, 40 h); and (**d**) duration for a 40 V anodization in 2.0% HF-DMSO electrolytes with 0%, 3% and 5% deionized water content [112]

DMSO, being protophilic, accepts a proton from HF that, in turn, reduces its activity. DMSO modifies the space charge region in the pores and thereby avoids lateral etching. This leads to steady pore growth and decreased chemical etching of the nanotube walls. Figure 1.12b shows the effect of HF concentration on nanotube length. A 2.0% HF DMSO electrolyte yielded the maximum nanotube length at applied voltages of 40 and 60 V. HF concentrations more than 2.0% appear to result in faster etching of the TiO_2. At HF concentrations below 2%, the nanotube growth rate is so small that extended anodization durations are required to achieve the similar tube lengths.

While adding water to the electrolyte ensured field-assisted etching of the Ti foil at the pore bottom, water also assisted chemical dissolution that opened the pore mouth, resulted in thinner nanotube walls, and dissolved the end of the nanotube. Figure 1.12c summarizes the length of the nanotube arrays obtained as a function of water content in 2.0% HF-DMSO electrolytes anodized at 40 V for 40 h. It shows a decreased tube length accompanies increased water content. Addition of water to the electrolyte greatly improved adhesion between the nanotubes and underlying oxide barrier layer, alleviating a problem common to nanotube arrays synthesized in DMSO electrolytes. Figure 1.12d indicates the lengths achieved at 40 V in electrolytes containing 0%, 3% and 5% water as a function of anodization duration. A duration of 70 h yielded tubes of maximum length. Anodization duration also affects the top-surface features of nanotube arrays. Anodization times less than 40 h usually produces tubes clogged with surface debris, that is solution precipitate. Longer anodization times result in more titanium dissolution in the electrolyte, which leads to higher electrolyte conductivity that, in turn, helps prevent precipitation from occurring on the top surface of the nanotubes.

In common with the electrodeposition of ZnO nanostructures using DMSO, as reported by Choi [113, 114], the use of previously used electrolytes affects the length and pore size of the TiO_2 nanotubes achieved by anodization. The most important advantage offered by "used" electrolytes was greatly improved adhesion between the nanotubes and the underlying oxide barrier layer. The conductivity of 2.0% HF-DMSO used for anodization of one Ti sample, 40 V for 70 h, is approximately 100.6 μS/cm in comparison with 8.76 μS/cm for a fresh electrolyte. The electrolyte conductivity increases with extended anodization durations because more titanium ions dissolve in the electrolyte; it was 160.1 μS/cm for a 140 h duration anodization. Compared with fresh electrolytes, previously used electrolytes usually yielded nanotubes of shorter length. For example a fresh 2.0% HF-DMSO electrolyte resulted in 53 μm long nanotubes after a 40 V 24 h anodization and then using the same solution again for a subsequent 40 V 70 h anodization gave nanotubes 50 μm in length.

The higher conductivity of used electrolytes is useful in initiating nanotube growth; therefore, we observed faster initial rates of nanotube formation in used electrolytes. Also, nanotubes obtained from used electrolytes do not suffer from debris, and have open, unclogged pores: there is no need for surface cleaning by sonication. This is probably because a slightly higher chemical etching rate dominates the reaction. Tubes grown from used electrolytes also have thicker walls, approximately 100 nm, that are much rougher, and pore sizes of about 70–90 nm, compared with 120 nm for nanotubes obtained using fresh electrolytes.

1.4.3 Ethylene Glycol Electrolytes

Using ethylene glycol (EG) electrolytes, TiO_2 nanotube arrays were prepared at anodization voltages ranging from 20 to 65 V, with NH_4F concentration ranging from 0.1 to 0.5 wt%, and with H_2O content ranging from 1 to 4% [51, 54]. Table 1.6 summarizes the observations for NH_4F concentrations ranging from 0.25 to 0.5 wt%, and H_2O concentrations from 1 to 3 vol% for a 17 h anodization at 60 V as obtained using a fresh solution and a solution that had been used once under the same conditions. We found that: (i) the length increases with increasing NH_4F concentration up to 0.3 wt%; (ii) for NH_4F concentrations up to 0.3% wt the nanotube length increases with increasing H_2O concentration up to 2 vol%; and (iii) between 0.1 and 0.3 wt% NH_4F and 1–2 vol% H_2O the used solution, in comparison to use of a fresh solution, exhibited an increase in nanotube length ranging from 15 to 70 μm for the same applied potential and duration. While used solutions result in sharply higher growth rates for anodization potentials of 60 V and above, at the lower anodization potentials of 20–40 V the fresh and used solutions produced similar growth rates.

Nanotube lengths obey a power law dependence on applied voltage between 20 and 60 V for the same anodization duration. Table 1.7 shows, for a 17 h anodization, the length produced at 20, 40, 50, 60 and 65 V in a fresh electrolyte mixture of 0.3 wt% NH_4F and 2 vol% H_2O in ethylene glycol were, respectively, 5, 30, 45, 165

Table 1.6 Summary of nanotube length, in μm, obtained by varying the H_2O concentration from 1.0 to 3.0 vol% and NH_4F from 0.1 to 0.50 wt% in ethylene glycol (anhydrous) with respect to fresh and once-used solution [54]

		0.1% wt NH_4F	0.3% wt NH_4F	0.5% wt NH_4F
1% vol	Fresh	54 70	67	47
H_20	Used	70	156	115
2% vol	Fresh	85 70	165	10
H_2O	Used	70	220	45
3% vol	Fresh	–	136	8
H_2O	Used	–	100	66

Table 1.7 Summary of the nanotube inner diameter, outer diameter, and length obtained at different voltages for a 17 h anodization in a fresh, un-used ethylene glycol electrolyte containing 0.3 wt% NH_4F and 2 vol% H_2O [54]

Voltage (V)	Inner diameter (nm)	Outer diameter (nm)	Length (μm)
20	45	65	5
40	70	115	30
50	90	140	45
60	105	155	165
65	135	185	105

and 106 μm. At 80 V potential and 22°C, with a variety of electrolyte combinations the Ti foil sample simply corroded. The FESEM images of Fig. 1.13 illustrate the remarkable self alignment exhibited by the nanotube arrays, in bottom, Fig. 1.13a, top, Fig. 1.13b, and side, Fig. 1.13c and Fig. 1.13d, views of an illustrative sample anodized in 0.3 wt% NH_4F and 2% H_2O in ethylene glycol for 17 h.

The current-time behaviors during fixed-potential anodization of Ti samples in ethylene glycol electrolytes, e.g. fresh 0.3 wt% NH_4F, 2 vol% H_2O, are broadly similar. All exhibit initially stable or near-stable anodization current followed by a rapidly decreasing current until the current reaches a well-defined plateau and then slowly decreases. During the initial high current period gas evolution at the anode is observable. Since gas evolution requires electronic charge transfer, this is indicative that electronic conduction dominates early in the process. After initial formation of an insulating oxide layer the current drops rapidly; in this region electronic conduction decreases blocked by the formed oxide and ionic conduction through the TiO_2 increases. Once the oxide layer is completed over the entire exposed anodic surface electronic conduction through the TiO_2 barrier layer becomes negligible and all conduction is ionic.

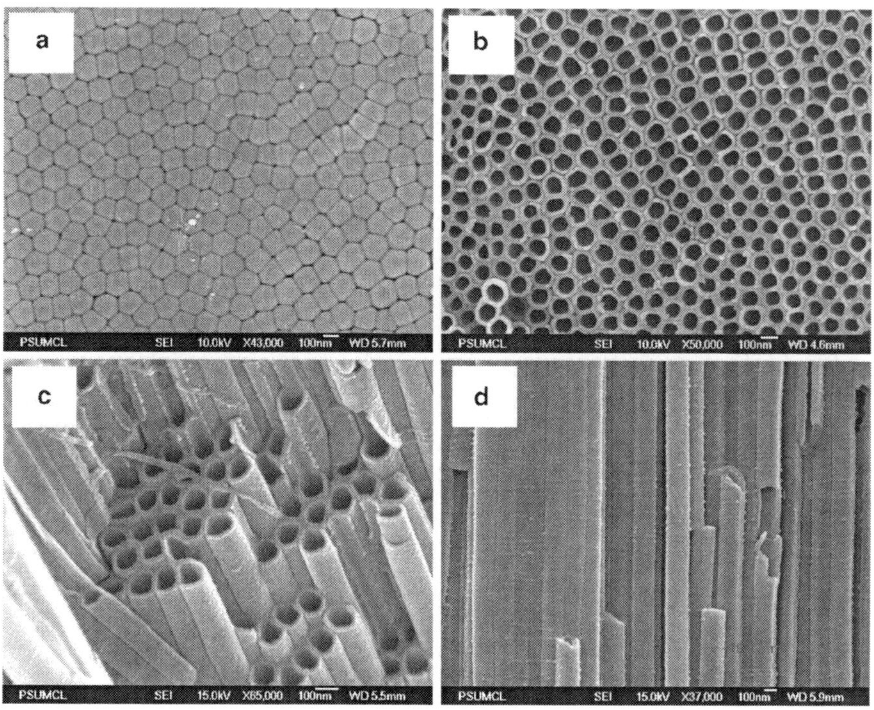

Fig. 1.13 Illustrative FESEM images showing topology of a Ti sample anodized in an electrolyte comprised of 0.3 wt% NH_4F and 2 vol% H_2O in ethylene glycol at (**a–c**) 60 V, and (**d**) 65 V [51, 54]

The reaction rate was examined by anodizing titanium foil samples at 60 V in a fresh (un-used) solution of 0.3% wt NH_4F and 2 vol% H_2O in ethylene glycol for different durations. The samples anodized for 4 h, 17 h, 21 h, 48 h and 96 h exhibited, respectively, lengths of about 58, 160, 188, 289 and 360 μm. The lengths obtained at intervals up to 21 h are proportional to the charge passed during those time intervals and indicate the current efficiency is constant. After accounting for the porosity of the structure and the titanium dioxide dissolved during the formation of the nanotube structure, the effective current efficiency for TiO_2 formation is nearly 100%. This indicates almost no side-reactions occurred during the anodization process; there was negligible bulk chemical dissolution of formed TiO_2 nanotube arrays. The proportionality between the charge passed and the length of formed TiO_2 nanotubes, when adjusted for the porosity factors, also held for anodization potentials lower than 60 V; current efficiency decreased at larger voltages. Figure 1.14 shows cross sectional FESEM images of the samples. The as-anodized samples show surface debris, which is easily removed by ultrasonic cleaning in D.I. water for 15–30 s.

Approximately 1 min of ultrasonic agitation resulted in separation of the nanotube array film from the underlying Ti substrate, resulting in a self-standing membrane of close-packed, vertically oriented TiO_2 nanotubes. Figure 1.15

Fig. 1.14 Cross sectional FESEM images showing the titania nanotube array lengths achieved as a function of anodization duration (60 V, ethylene glycol containing 0.3 wt% NH_4F and 2 vol% H_2O) [54]

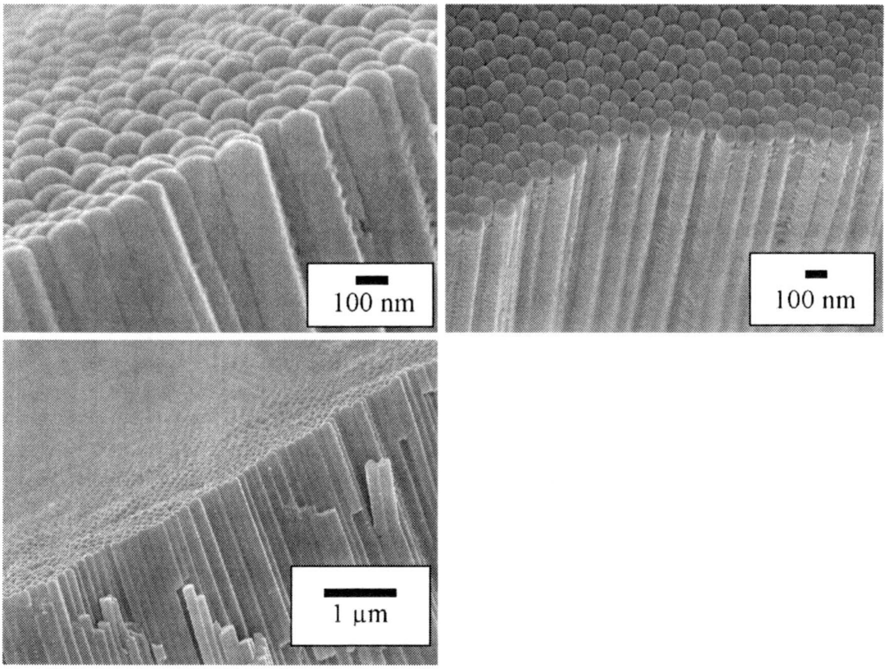

Fig.1.15 FESEM images of different magnifications showing the bottom and partial cross-section of a self-standing nanotube array film, synthesized by 60 V fixed-potential anodization of a Ti sample in an ethylene glycol electrolyte containing 0.3 wt% NH_4F and 2 vol% H_2O in ethylene glycol [54]

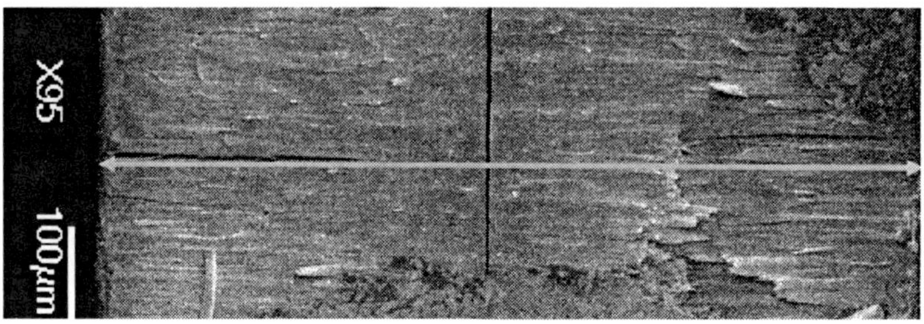

Fig. 1.16 FESEM images of a 720 µm thick membrane synthesized using a fixed 60 V potential for complete, double-sided anodization of a Ti sample after 96 h [54]

shows the back-side FESEM images of such a self-standing membrane array as synthesized in 0.3% wt NH_4F and 2 vol% H_2O ethylene glycol at 60 V; the outer tube diameter is 165 nm. The nanotube arrays, 360 µm long and 165 nm outer diameter, have an aspect ratio of approximately 2,200. During a double-sided anodization process, with both sides of the starting Titanium foil exposed to the anodizing electrolyte, the 250 µm thick Ti foil is becomes fully anodized at about 96 h into a 720 µm thick nanotube film, see Fig. 1.16.

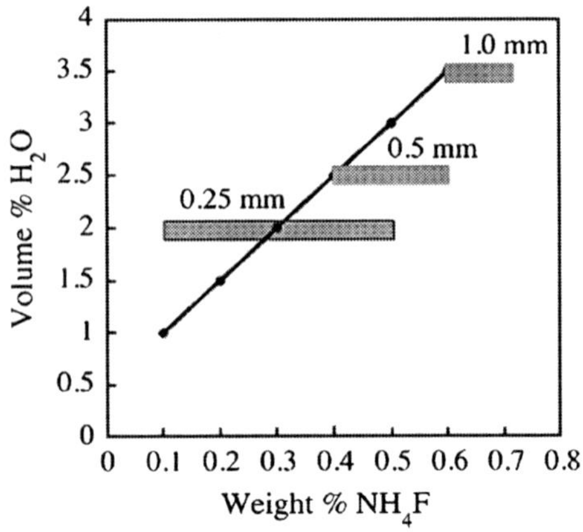

Fig. 1.17 Ratio of wt% NH_4F to vol% H_2O for maximum growth rate in an ethylene glycol electrolyte. The horizontal bars show the range of wt% NH_4F in which complete anodization of 0.25, 0.5, and 1 mm thick Ti foil samples can be achieved for a given concentration of water [115]

A maximum nanotube growth rate was observed at 60 V with ethylene glycol electrolytes. The optimum water concentration for the highest growth rates with different NH_4F concentrations follows the pattern shown in Fig. 1.17. Within a given range of NH_4F and H_2O concentrations the anodic dissolution due to the increased wt% NH_4F is compensated by the increase in H_2O concentration and results in greater growth rates and hence longer nanotubes. Figure 1.17 shows the range of H_2O and NH_4F concentrations for which complete anodization of 0.25 and 0.5 mm Ti foil samples were achieved. From 0.1 to 0.5 wt% NH_4F with 2% water, 0.25 mm foil samples were completely anodized and produced two 320 to 360 μm nanotube arrays separated by a thin barrier layer. The longest tubes were obtained in a solution containing 0.3 wt% NH_4F and 2% H_2O in EG for 96 h. After anodizing 0.5 mm foil in an identical electrolyte for 168 h (7 days), the maximum thickness was about 380 μ m. This suggests the active electrolyte species was completely utilized. Complete anodization of a 0.5 mm foil was achieved in an electrolyte containing 0.4–0.6% NH_4F and 2.5% H_2O in EG (Fig. 1.17); the length of nanotube array on each side was 538 μm and it was achieved by completely anodizing 0.5 mm foil at 60 V for 168 h in 0.4 wt% NH_4F and 2.5% water in EG. As shown in Fig. 1.18, a maximum individual nanotube array length of 1005 μm was obtained upon anodizing 2.0 mm thick Ti foil at 60 V for 216 h (9 days!) in 0.5 wt% NH_4F and 3.0% water in EG. The foil, which was simultaneously anodized on both sides, formed the self-standing nanotube membrane shown in Fig. 1.19; it is over 2 mm thick.

1.4.3.1 Membrane Fabrication

By controlling the various anodization parameters it is possible to vary the tube-to-tube connectivity and hence packing density of the nanotubes within an array. Maximum nanotube packing density is achieved using an ethylene glycol electrolyte; the nanotube coordination number is usually six, i.e. each nanotube is surrounded by six others, with strong bonding between adjacent tubes. This structure

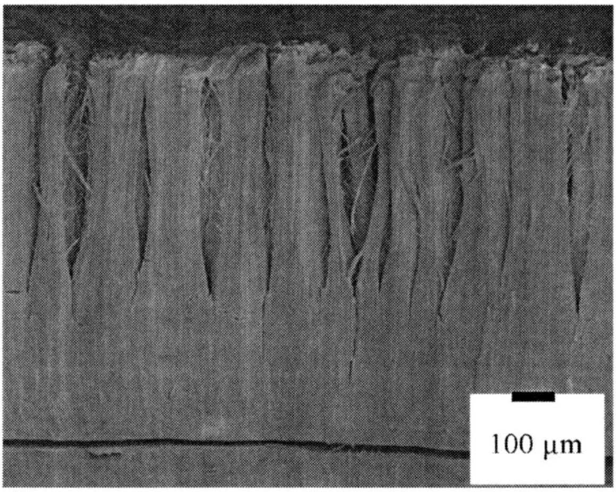

Fig. 1.18 FESEM image of the top half of a completely anodized 2.0 mm thick Ti foil sample. The maximum individual nanotube array length of 1005 μm was obtained by anodizing the Ti foil sample at 60 V for 216 h in 0.5 wt% NH_4F and 3.0% water in ethylene glycol [115]

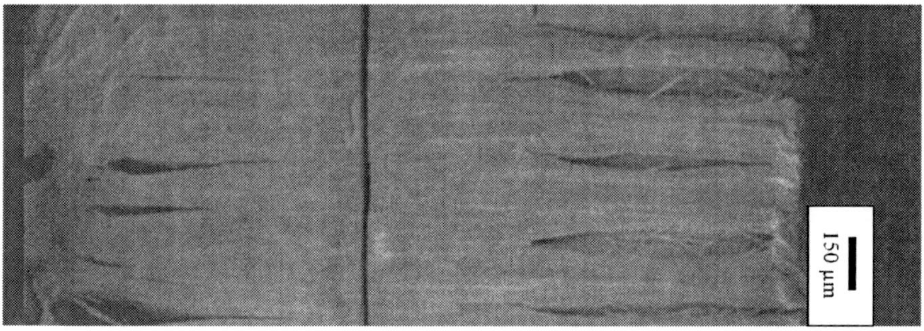

Fig. 1.19 Cross sectional view of self-standing titania membrane over 2 mm thick achieved by anodizing a 2.0 mm thick Ti foil sample at 60 V for 216 h in 0.5 wt% NH_4F and 3.0% water in ethylene glycol [115]

is most useful for achieving a mechanically robust, i.e. non-fragile, membrane of uniform pore size suitable for use in filtration applications [115, 116].

We used an electrolyte composition of 0.3 wt% ammonium fluoride and 2 vol% water in ethylene glycol for membrane fabrication. Anodization was done at room temperature with a platinum foil cathode. A nanotube about 220 μm long with pore size 125 nm and standard deviation of 10 nm, resulted when anodization was performed at 60 V for 72 h. The as-anodized samples were then dipped in ethyl alcohol and subjected to ultrasonic agitation until the nanotube array film separated from the underlying Ti substrate. The compressive stress at the barrier layer-metal interface facilitates detachment from the substrate. Figures 1.20a, b show FESEM images of the membrane top and bottom surface at varying degrees of

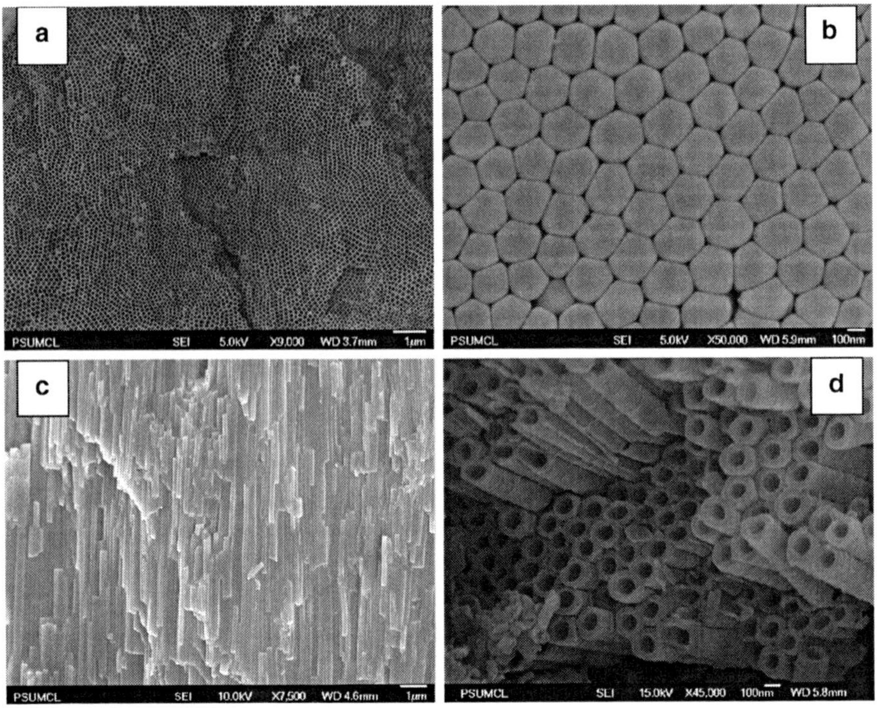

Fig. 1.20 FESEM images of: (**a**) Top side of nanotube array film. (**b**) Back-side or barrier layer side of nanotube array film. (**c,d**) Cross sectional image of mechanically fractured nanotube array film showing its tubular nature [115]

magnification, while Fig. 1.20c, d are cross-sectional images of a mechanically fractured sample.

Figure 1.21a shows the backside, i.e. the barrier layer side of an as-fabricated nanotube array film, see Fig. 1.20, after it is opened with a dilute hydrofluoric acid/ sulfuric acid solution treatment. Figure 1.21b shows a partially opened back-side; the acid rinse is repeated until the pores are completely opened as seen in Fig. 1.21c, after which the membrane is ultrasonically cleaned to remove any etching associated debris.

The planar membranes, perfectly flat while wet, significantly curl after removal from the liquid and dried in air, Fig. 1.22, making them unsuitable for filtering applications. Since the surface tension forces of the solution acting on the membrane were mainly responsible for this behavior, rinsing with low surface tension liquids facilitates their drying flat. Alternatively, critical point drying can be used in which the membrane flatness is preserved, see Fig. 1.23. The figure shows a 200 μm thick membrane after critical point drying. We note the surface of the membrane after drying in this way occasionally shows a nanofiber surface; it can be removed by subjecting the membrane to ultrasonic agitation. Similar to the situation where TiO_2 nanowires are grown by hydrothermal treatment of TiO_2 particles [29],

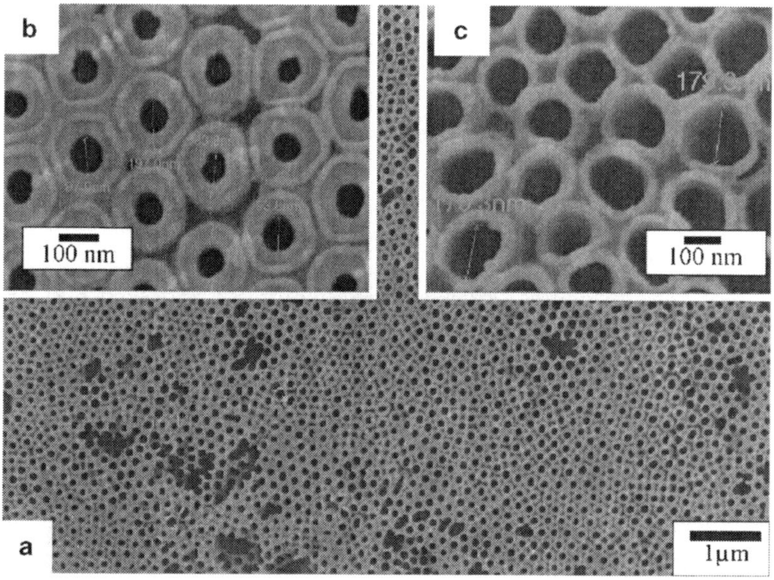

Fig. 1.21 FESEM image shows: (**a**,**b**) bottom of the nanotube array partially opened by chemical etch at different levels of magnification; (**c**) fully opened array bottom [115]

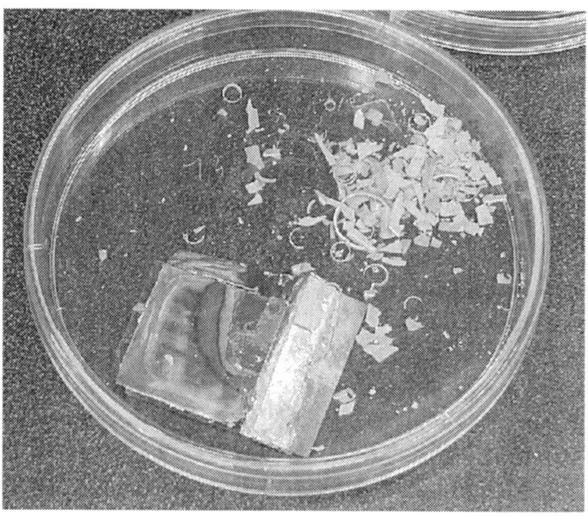

Fig. 1.22 If dried in air the TiO_2 membrane curls and shatters into shards, as shown above after about 24 h of air drying in air; the diameter of Petri dish is 90 mm [115]

we believe that TiO_2 nanoparticles occasionally detach from nanotubes forming wires in the high pressure and temperature conditions used for critical point drying.

In another report [5], 145 μm thick TiO_2 nanotube array films, fabricated by anodization in fluoride-containing ethylene glycol solvent, are separated from the

Fig. 1.23 A stable, mechanically robust nanotube array membrane after critical point drying. The 200 μm thick membrane, 120 pore diameter, is about 2.5 cm × 4.5 cm [115]

Ti substrate by selectively dissolving the Ti by treating in a water-free CH_3OH/Br_2 solution. To open the closed bottoms of the resultant freestanding nanotube layer, the wet nanotube film was exposed (bottom side down) for 30 min to HF vapors. This procedure leads to preferential HF gas condensation at the tube bottoms, causing etching of the backside layer.

1.4.4 Diethylene Glycol Electrolytes

To achieve highly ordered and discrete nanotube arrays, i.e. completely separate and unconnected, without bundling and without suffering from unwanted precipitate on the top surface, Yoriya and co-workers reported the use of diethylene glycol (DEG) for nanotube array synthesis [117]. They investigated tube formation and in particular the dependence of tube separation and crystallization, to be discussed in Chapter 2, in the DEG-based electrolytes. The free self-standing characteristic of the nanotubes offers facile detachment from the underlying oxide barrier layer that is useful for specific applications, such as drug delivery where the tube pores are filled with a drug prior to dispersing them toward a final target [27, 28].

Anodization electrolytes were prepared using 99.7% DEG in combination with either a 48% hydrofluoric acid, a 98% ammonium fluoride, or a tetrabytylammonium fluoride trihydrate (Bu_4NF). Figure 1.24a shows the top view image of nanotube arrays prepared in a DEG+2%HF electrolyte anodized at 40 V for 48 h. The nanotubes are ≈3 μm long with an inner pore diameter of about 140 nm and outer diameter of about 240 nm. The vertically oriented nanotubes are clearly

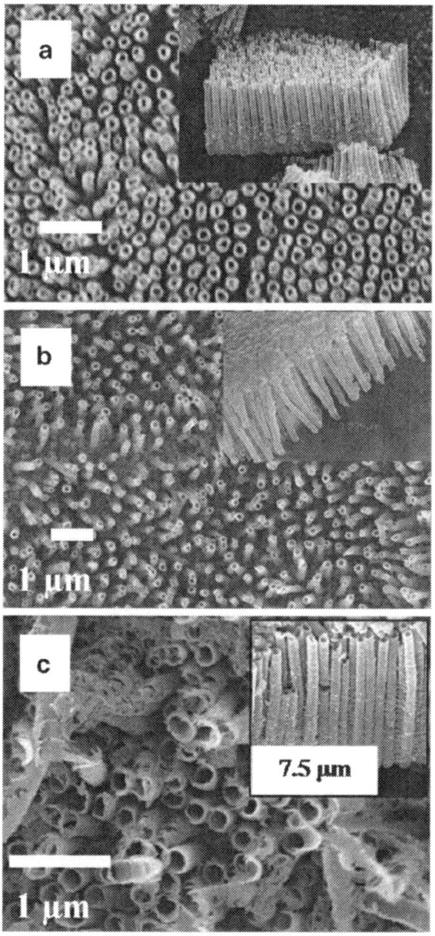

Fig. 1.24 FESEM images of a TiO_2 nanotube array grown from a DEG-2% HF electrolyte after anodizing for 48 h at: (**a**) 40 V, (**b**) 60 V, and (**c**) 100 V. Inset shows the cross-sectional views of nanotubes [117]

separated and the pores fully open. Figure 1.24b shows FE-SEM images of nanotube arrays prepared in a DEG+2%HF electrolyte anodized at 60 V for 48 h. In comparison with the tubes of Fig. 1.24a there is greater tube separation and a slight increase in the pore diameter to about 150 nm with the higher anodization voltage; the tube length remains at about 3 μm. The pore mouths of the as-fabricated tubes are free of precipitate. A further increase of the anodization voltage to 100 V with the DEG+2%HF electrolyte increases the average nanotube pore size to about 190 nm and the length to about 7.5 μm. The tube-to-tube separation is similar for 100 and 60 V samples, as shown in Fig. 1.24c.

Changing the fluoride-bearing species influences the topology of the resulting nanotubes. Figure 1.25 is an illustrative FESEM image of nanotubes obtained from a DEG electrolyte containing 0.3 wt% NH_4F+2% H_2O anodized at 80 V for 48 h. The tubes have a distinct pyramidal shape, a tube length of about 3.5 μm, and an average top pore diameter of about 120 nm. The as-synthesized nanotubes from

Fig. 1.25 FESEM images of a TiO_2 nanotube array samples grown from DEG-0.3 wt% NH_4F+ 2%H_2O electrolyte at 80 V for 48 h [117]

DEG electrolytes containing HF and NH_4F show a noticeable pyramidal shape. We attribute pyramidal shape to change in fluoride ion, F^-, concentration and the thereby changed solution conductivity during anodization. Typically, the electrolyte F^- concentration decreases and electrolyte conductivity increases with anodization duration due to field- and chemical-assisted titanium dissolution. As the electrolyte conductivity increases with anodization duration, a larger fraction of the applied DC potential appears across the oxide layer. Hence, the actual potential applied across the oxide increases with anodization duration. Since the nanotube pore diameter increases with increasing applied potential, the tubes initially show a comparatively small tube diameter that increases with time, thereby producing somewhat pyramidal tubes. Increasing the NH_4F concentration in the electrolyte up to 0.9 wt% yielded only titania flakes. The solubility of such high NH_4F concentrations in DEG is poor, hence water was added to the electrolyte to facilitate NH_4F dissolution. Figure 1.26 shows a nanotube array of about 21 μm obtained in a 7 wt% Bu_4NF (tetrabutylammonium fluoride)+DEG electrolyte, 48 h at 80 V. The pore diameter and the wall thickness are in range of 136–151 nm and 14–19.6 nm, respectively. The general topology of open-pore, smooth, tightly packed nanotubes is similar to that obtained using ethylene glycol electrolytes [51].

The electrolyte containing the largest cation, Bu_4N^+, yielded the longest nanotubes; shorter nanotubes resulted from using the smaller cations NH_4^+ and H^+. The weaker surface charge of the larger cation results in less cation–anion interaction, hence they are more favorable for ionic transport during anodization, which facilitates rapid tube growth. Furthermore, in comparison to H^+ and NH_4^+, chemical etching is lower with the Bu_4N^+ cation due to the comparatively lower ability of Bu_4N^+ to donate a proton for the etching process [53]. The larger Bu_4N^+ cation clearly affects the tube-to-tube spacing and results in a tightly packed tube morphology, see Fig. 1.26. The distinctive morphology of the nanotubes, e.g. shape,

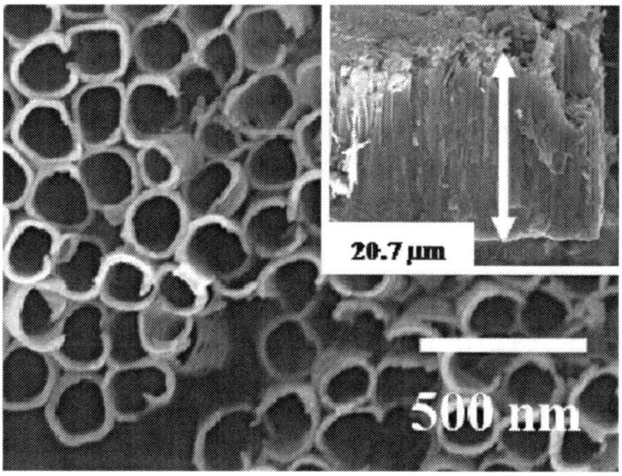

Fig. 1.26 FESEM images of a TiO_2 nanotube array grown from a DEG electrolyte contaiing 0.3 wt% Bu_4NF electrolyte at 80 V for 48 h [117]

surface roughness, tube-to-tube separation and length, are due to several synthesis variables including cation size and their solvent interactions, electrolyte composition, anodization potential, and temperature [118].

1.4.5 Using Glycerol and NH_4F

During anodization of Ti foil in aqueous electrolytes (first generation) and buffered solutions (second generation), regular current oscillations occur that correlate with distance between periodic ridges on the nanotube walls. It was hypothesized that a current transient is produced by a pH burst at the pore tip where the sudden increase in acidity leads to a temporarily increased dissolution rate. It is known that diffusion affects local acidification [45]. A lower diffusion rate, achieved by using a highly viscous, glycerol-based electrolyte, reduces current fluctuations and produces smoother walls. Smooth-wall TiO_2 nanotubes about 6–7 μm long with an average inner diameter of 40–50 nm were produced using a glycerol electrolyte with 0.5 wt% NH_4F [46, 48]. At extended anodization periods tube length increased almost linearly for some 18 h to about 6.1 μm, and then plateaus. Although it was found that the higher the temperature the larger the diameter and longer the tubes, no regular and mechanically stable nanotube structures were produced at bath temperatures higher than 40°C.

A pulse waveform between 20 V with dwell time of about 5 min, and 4 V with a dwell time of about 30 s, successfully produced nanotubes in 0.36M NH_4F in glycerol containing 1% water [119]. In another study, the pH of 75% glycerol and 25% water electrolyte containing 0.5 wt% NH_4F was modified using H_2SO_4 [120]. If the pH was adjusted to 5.6 by adding H_2SO_4 the resulting nanotube length was 950 nm. With addition of 0.1M sodium acetate while maintaining the pH at 5.6, 4.16 μm nanotubes were obtained after anodizing at 20 V for 8 h.

1.4.6 Methanol, Water, and HF

During anodization, hydrogen bubbles are produced on the surface of the cathode. In aqueous solutions these bubbles become relatively large remaining on the electrode surface until they have sufficient buoyancy to break free and rise to the surface. The forming and freeing of large bubbles causes current density fluctuations [121] that are reduced with the addition of methanol to the electrolyte [122]. Since the surface tension of water and pure methanol at 303.15K and atmospheric pressure are, respectively, 71.4×10^{-3} and 21.7×10^{-3}N/m [123] the reduced bubble size appears to be due to a methanol-induced reduction in surface tension.

TiO_2 nanotubes were grown using galvanostatic anodic oxidation in a methanol water electrolyte with 1 wt% HF [121] for 30 min and current densities between 10 and 30 mA/cm^2. The inner diameter of the nanotubes increased with current density from about 30 to about 145 nm. The use of methanol was also investigated at a current density of 10 mA/cm^2. Nanotubes fabricated in 50 vol% methanol were tightly packed but increasing the methanol concentration to more than 80 vol% dramatically reduced the nanotube density; spacing between nanotubes increasing by 300%, as shown in Fig. 1.27. The length of the nanotube arrays was approximately 400 nm. Methanol content affects formation of Ti and F hydrate complexes [43, 44], with a higher methanol content lowering the dissolution rate.

1.5 Nanotube Array Synthesis Using Non-Fluoride Based Electrolytes: The Fourth Generation

In 2005, Nakayama et al. reported the formation of TiO_2 nanotube bundles, i.e. clumps of tubes rather than arrays, by anodization in a perchloric acid solution [78]. In 2007, Richter grew similar nanotube bundles in 0.5M oxalic acid, 0.5M formic acid, and 0.05M sulfuric acid with 0.3–0.6M NH_4Cl added in each case. The nanotubes were found to contain about 20 at.% carbon [79] with less than 5% chlorine, suggesting that the role of chlorine in tube formation was mostly catalytic. It was thought that tubes might be achieved using acids other than oxalic. To investigate this hypothesis, Richter and co-workers anodized Ti foil in electrolytes containing chlorine salts in combination with oxalic, formic, trichloroacetic, gluconic, hydrochloric, and sulfuric acid [80]. Ti foil samples were anodized in a solution with a pH of around 1.5 and a 0.4M chlorine ion content. To compare results relative to the carbon content of the nanotubes, a 0.5M acid concentration for the organic acids was used. Samples were anodized in a standard solution for each of the acids at fixed voltages between 8 and 20 V. For all solutions, anodization at voltages less than about 11 V yielded no nanotubes. For voltages above that threshold, however, TiO_2 nanotubes formed in each of the acids: oxalic, formic, trichloroacetic, gluconic, hydrochloric, and sulfuric. There were no significant morphological differences between the nanotube bundles produced in the different

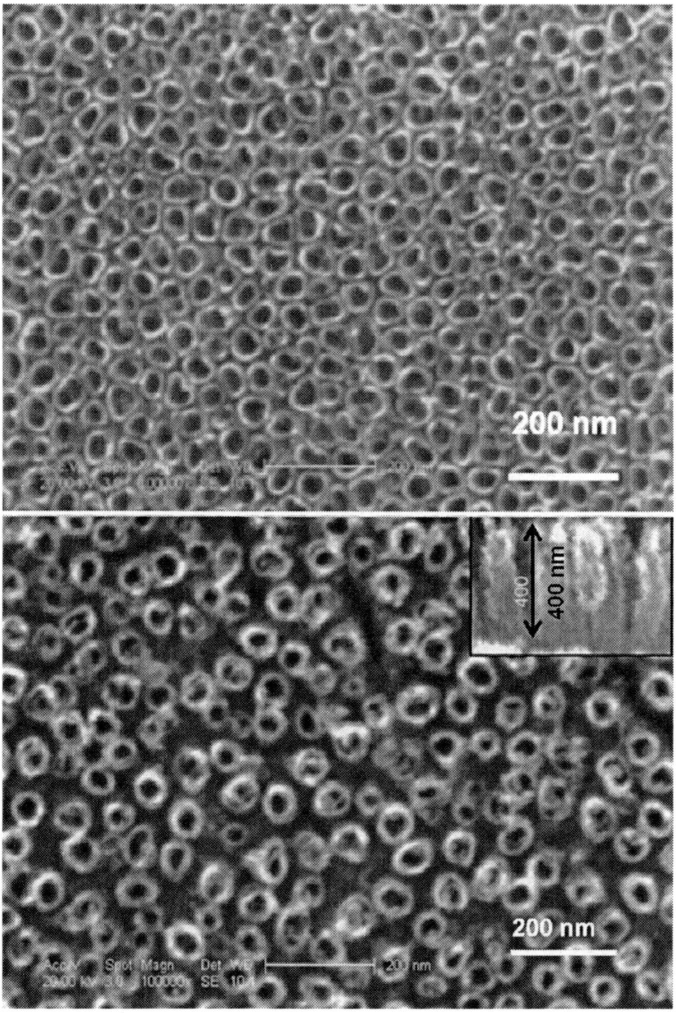

Fig. 1.27 FESEM images of TiO_2 nanotubes anodized at 10 mA/cm^2 for 30 min in: (**a**) methanol and water mixture (50 vol% methanol, 1 mass% HF) (**b**) methanol and water mixture (90 vol% methanol, 1 mass% HF) ([121]; Copyright 2007. Reprinted with permission from Elsevier)

acids; they were typically observed in various "attack areas," i.e., edges or holes that were etched through the surface of the anodized foils. The samples were rapidly etched by the formation of nanotubes in these areas, with bundles of nanotubes continuously released into the electrolyte. The tubes themselves were either arrayed in bundles up to 60 μm long or within irregular agglomerations. There was no obvious dependence of nanotube diameter upon voltage; all tubes diameters were between 15 and 35 nm and with wall thickness of about 5 nm. The length appeared to be more a function of where the rapidly forming tube bundles break off from the substrate, than of anodization time, acidity, or voltage. A significant amount of carbon was found in nanotubes fabricated within each of the organic acids. Tubes grown in trichloroacetic acid contained about 45% carbon but tubes grown in other acids contained about 20% carbon.

Hahn et al. formed bundles of high aspect ratio TiO_2 nanotubes with 10 nm tube walls using anodization of Ti foil under high voltage conditions in perchlorate electrolytes [81]. The bundles of TiO_2 nanotubes consisted of separated tubes with diameters of 20 to 40 nm. TiO_2 nanotubes prepared in 0.1M $HClO_4$ at 30 V for 60 s were 32.9 μm long. Again, growth occurred locally at specific surface locations such as edges or local etching sites (i.e. surface pitting).

1.5.1 Using HCl

Mao and co-workers formed TiO_2 nanotubes [82] using electrolyte solutions between 0.22 and 1.1 wt% (0.05–0.3M) HCl. The anode was titanium foil, the cathode copper. The voltage was increased manually and then cycled twice between a fixed voltage and zero before being left at the fixed voltage, typically 10 V, for periods between 30 min and 2 h. Organized nanotubes with inner diameters of about 10–20 nm were formed in solutions of 0.55 wt% HCl (0.15M); the TiO_2 nanotubes were a few hundred nanometers long. Both lower and higher electrolyte concentrations tended to increase the nanotube diameter but produced less developed and less aligned structures. The diameters of the TiO_2 nanotubes formed in 0.06M, 0.15M and 0.3M HCl were, respectively, about 80, 10 and 30 nm. Allam et al. reported producing TiO_2 nanotubes using a 3.0M HCl electrolyte with applied voltages ranging between 9 and 13 V [83]. The complexity of the reaction of Ti in HCl resulted in a limited window for successful processing [84]. In later work Allam reported efforts to establish a wider processing window by adding H_2O_2 to the electrolyte [85]. They reported measurements using three electrolytic conditions: (1) 10M H_2O_2 aqueous electrolyte with applied voltages of 60 and 80 V, (2) 0.5M HCl+0.1–0.5M H_2O_2 aqueous electrolytes with applied voltage in the range 5–25 V and (3) 0.5M HCl+0.1–0.5M H_2O_2 in ethylene glycol electrolytes containing 2 vol% H_2O with applied voltage between 5 and 20 V [85].

1.5.2 H_2O_2 Aqueous Electrolytes

Hydrogen peroxide, H_2O_2, is a powerful oxidant (E=1.77 V). It increases the oxidation reaction rate and consequently thickens the oxide layer in those cases where the formation potential (U_f) mainly determines the film thickness (L) [86]:

$$L = d_0 + k\ (U_f - U_{ox}) \tag{1.1}$$

d_0 is the initial, or native, oxide thickness, k is the film formation constant dL/dU, U_{ox} is the potential of the oxide, -0.86 V for TiO_2. Hydrogen peroxide introduced into an acidic Ti(IV) solution initially produces an intense yellow color that slowly changes first to green and then to purple. Samuni suggested that the reaction mechanism of Ti with H_2O_2 involves reaction between more than one H_2O_2 molecules with one Ti(IV) ion [87]. Activation energy analysis suggests that two H_2O_2 molecules meet and react at the metal–titania interface, thereby producing a titania layer. Recent theoretical work by Sever and Root postulated that the energy required to bind one H_2O_2 molecule to a Ti(IV) ion is about 0.17 eV [88].

Figure 1.28 shows an FESEM micrograph of a Ti sample anodized in 10M H_2O_2 aqueous solution for 4 h at 80 V. The result is a porous, amorphous titania layer similar to that reported by Wu [89]. Arabatzis and Falaras reported the fabrication of porous TiO_2 using H_2O_2 [90]. Luiz and Costa reported that the main effect of the oxidant occurs on the barrier layer, which they attributed to defect generation [91]. Haber suggested a two-electron model to describe oxide dissolution and reformation at the Ti/TiO_2 surface that may explain the porous-like structure [92]:

$$2TiO_2 + H_2O_2 + 2e^- \rightarrow Ti_2O_3 + H_2O + O_2 \quad (1.2)$$

$$Ti_2O_3 + H_2O_2 \rightarrow 2TiO_2 + H_2O + 2e^- \quad (1.3)$$

Fig. 1.28 Illustrative FESEM image of titania surface achieved by Ti anodization for 4 h using an aqueous electrolyte of 10M H_2O_2 at 80 V [85]

1.5.3 HCl/H_2O_2 Aqueous Electrolytes

Initial investigations into the use of HCl as an electrolyte for Ti anodization showed that the nanotube architectures could not be at achieved at anodization potentials other than 13 V and with HCl concentrations below 3M [83]. In order to establish a wider processing window for achieving nanotube arrays, H_2O_2 was added to the HCl electrolytes. Ti foil samples were anodized in aqueous solutions containing 0.5M HCl with different H_2O_2 concentrations (0.1–0.5M) over the voltage range 5–25 V. For a single anodization step, porous structures or nanotube arrays below 8 V were not obtained; nanoporous structures started to form at 8 V and nanotube arrays at 10 V. However, using a series of anodization treatments of the same sample in new (i.e. fresh, unused) electrolytes, nanotube arrays could even be obtained at 5 V.

Figure 1.29 are FESEM images showing illustrative morphologies of the nanotube arrays achieved by anodization using an aqueous electrolyte of 0.5M HCl+xM H_2O_2 at 10 V for a 1-h anodization; (a) x=0.1M, (b) x=0.4M. The nanotube morphology is starting to form in Fig. 1.29a, while Fig. 1.29b shows a nanotube array architecture having the morphology commonly obtained using fluoride-based electrolytes [93]. We note the "fuzzy" layer atop the tubes is due to precipitation of a chloride layer on the sample surface [80, 81]. The anodization current-time response at 10 V, for 0.5M HCl aqueous electrolytes containing variable H_2O_2 content is essentially identical to that seen when nanotube arrays are achieved via anodization of Ti in aqueous HF solutions [93]. Figure 1.30 is an illustrative FESEM image of nanotube arrays fabricated at 20 V and anodized 1 h in the presence of 0.5M HCl and 0.4M H_2O_2; the nanotube array architecture is clearly evident. H_2O_2 concentrations above 0.4M, or anodization voltages of 30 V or more resulted in sample corrosion. Figure 1.31 summarizes the increase in nanotube length for a 1-h anodization as a function of H_2O_2 concentration and voltage

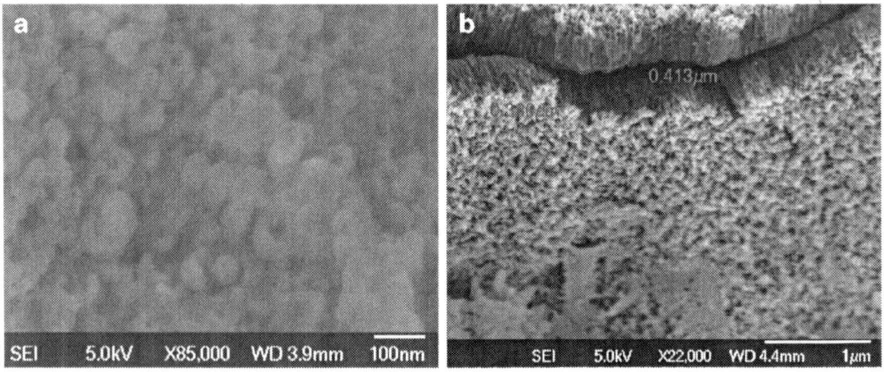

Fig. 1.29 FESEM images of titania nanotube arrays achieved by anodization using an aqueous electrolyte of 0.5M HCl+xM H_2O_2 at 10 V; (**a**) x=0.1M, (**b**) x=0.4M [85]

Fig. 1.30 FESEM image of a TiO_2 nanotube array surface achieved by Ti anodization using aqueous electrolytes of 0.5M HCl+0.4M H_2O_2 at 20 V [85]

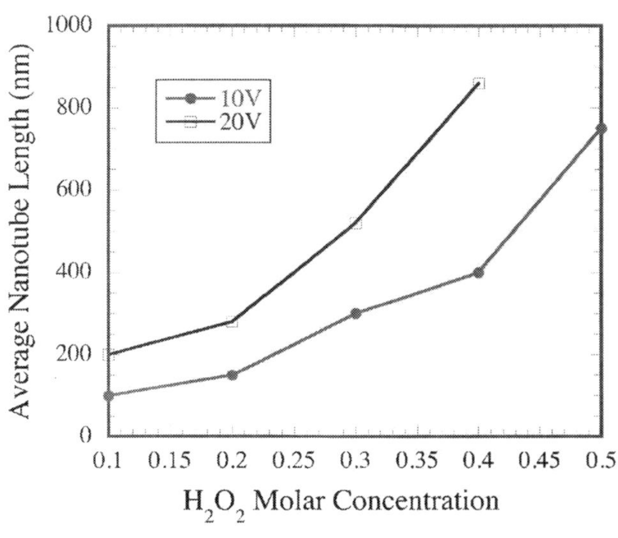

Fig. 1.31 Average TiO_2 nanotube length as a function of H_2O_2 concentration, with anodizing at 10 V for 1 h in aqueous electrolytes containing 0.5M HCl and 0.1–0.5M H_2O_2 [85]

(10 V, 20 V; 0.5 HCl). The dissociation of H_2O_2 to HO_2^- and H^+ ions is a pH dependent acid-base equilibrium:

$$H_2O_2 \leftrightarrow HO_2^- + H^+ \tag{1.4}$$

It has been suggested, but not confirmed, that HO_2^- ions form a peroxide complex with titanium [94] as:

$$Ti + HO_2^- + 3OH^- \leftrightarrow Ti(OH)_2O_2 + H_2O + 4e^- \quad (1.5)$$

The $Ti(OH)_2O_2$ complex hydrolyses to H_2O_2 and $Ti(OH)_4$ and the latter detaches from the metal surface. The $Ti(OH)_2O_2$ complex may also react with HO_2^- ions to form soluble $HTiO_3^-$ ions. Thus, the concentration of HO_2^- is critical during Ti oxidation in H_2O_2-containing electrolytes [94]. Nanotubular architectures are formed using electrolytes containing low H_2O_2 concentrations with chloride ions. This result is in agreement with Richter and co-workers who studied the formation of small diameter (≈30 nm outer diameter) nanotube bundles in a mixture of oxalic acid and chloride-containing electrolytes; they reported that chloride was essential for nanotube formation [80, 95]. However H_2O_2 itself plays an active role and should not be neglected: Hahn and co-workers studied the formation of TiO_2 nanotube arrays in chloride containing electrolytes ($HClO_4$ and $NaClO_4$), in the absence of H_2O_2, and reported that the breakdown of TiO_2 in chloride-containing electrolytes requires potentials of several tens of volts [81]. In contrast, with $HCl+H_2O_2$ electrolytes Allam formed TiO_2 nanotube arrays at 10 V [85] suggesting a possible synergism between the H_2O_2 and chloride ions to enhance the formation process.

1.6 Fabrication of Transparent TiO_2 Nanotubes Arrays

Methods have been developed for fabrication of TiO_2 nanotube arrays from thin films of Ti atop substrates compatible with photolithographic processing. The result

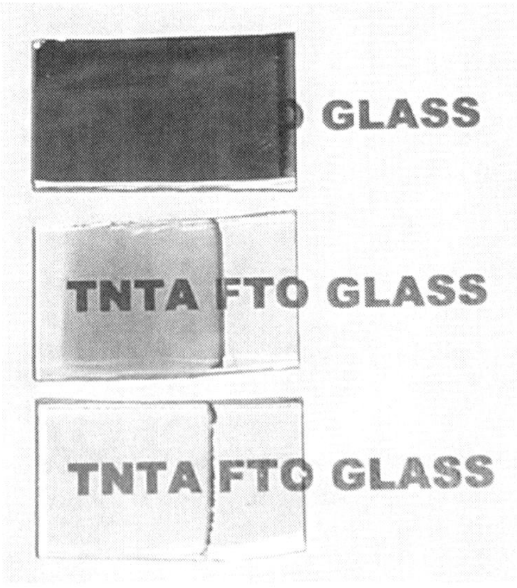

Fig. 1.32 Key stages in fabrication of a transparent nanotube array, TNTA, films: (*top*) sputter deposition of a high quality Ti film; (*middle*) anodization of resulting film, and (*bottom*) heat treatment to oxidize remaining metallic islands [22]

is a transparent nanotube array structure, illustrated in Fig. 1.32 that is promising for a variety of applications, such as anti-reflection coatings, heterojunction solar cells, photoelectrochemical diodes, and hydrogen sensors. Deposition of a high-quality Ti film [124–126] suitable for transformation into a nanotube array is critical to the success of thin-film TiO_2 nanotube array devices [127].

Anodization of single-layer titanium films was not successful because the metal layer at the air-electrolyte interface etched rapidly away and broke electrical contact with the submerged portion of the film undergoing anodization before nanotubes could be formed. It appears the non-uniform electric field distribution at the metal-air-electrolyte interface enhances the chemical etching rate relative to the field-assisted oxidation of the metal. Therefore, a bilayer-film scheme was adopted for thin film deposition and anodization. The first layer, of the desired thickness, was deposited over the entire substrate and then a second layer at least 400 nm thick was deposited atop the first layer over the interface region. Anodization to form nanotube arrays was successfully accomplished by keeping the single layer region completely immersed in the electrolyte and the double-layer region at the air–electrolyte interface. The greater thickness of the film in contact with the electrolyte surface ensured the sustained anodization of the submerged single layer to the desired specifications.

Figure 1.33 shows the typical time-anodization current behavior of a 400 nm Ti thin film deposited by rf sputtering and anodized at 10 V in an HF based electrolyte. At a fixed F^- concentration the dimensions of the tube depend upon the applied voltage; for a fixed anodization potential the dimensions vary with respect to electrolyte F^- concentration. For example, at a concentration of 0.5% HF aqueous

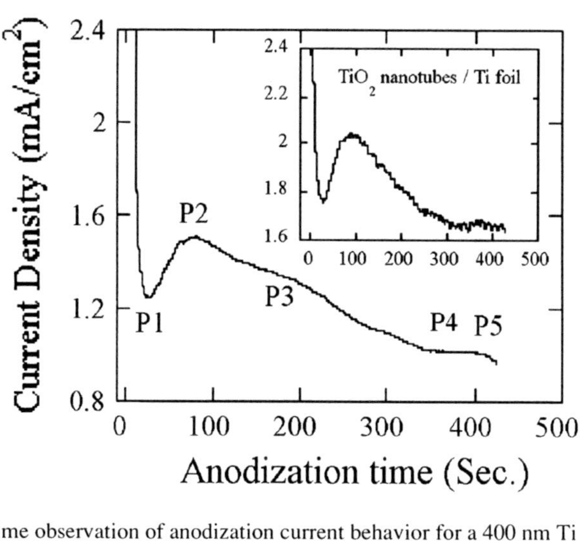

Fig. 1.33 Real time observation of anodization current behavior for a 400 nm Ti thin film at 10 V in the HF based electrolyte of acetic acid with 0.5 vol% HF mixed in ratio of 1:7. Inset shows a typical current density vs. time response observed for a Ti foil with one face protected with polymer coating and anodized at the same potential and electrolyte [127]

electrolyte nanotubes were formed at potentials between 10 and 15 V. The anodization voltage window for successfully achieving the thin film transparent nanotube arrays is 6 to 10 V for 0.25% HF concentration and 10 to 18 V for 1% HF concentration.

Referencing the anodization current of Fig. 1.33, for a 400 nm Ti thin film anodized at 10 V in an HF-based electrolyte, about 25 s after application of the voltage the measured current density reduced from >15 mA/cm^2 to a local-minimum of 1.25 mA/cm^2, point P1 on the plot. The reduced current density resulted from the field-assisted oxidation of the Ti metal surface. The structure of the film at point P1 of Fig. 1.33 is shown in Fig. 1.34a, which shows fine pits or cracks form on the oxide surface that act as pore nucleation sites. These pits and cracks arise from chemical and field-assisted dissolution of the oxide at local points of high energy. The reduced oxide layer thickness at these points decreases the current density; Fig. 1.34b corresponds to point P2 on the plot where the crack/pit density reached saturation and produced a current maximum. Beyond this point the current gradually drops with a corresponding increase in porous structure depth. The porous structure, point P3, is clearly seen in Fig. 1.34c, with pore diameters of about 20 nm. Figure 1.34d, point P4, shows the transition between the porous and nanotube structures. Figure 1.34e, point P5, shows the resulting nanotube-array with pore diameters of 20–30 nm. Nanotube-array length increases to point P4, after about 360 s, as evident from the decrease in the current. Between P4 and P5 the nanotube array length remains essentially constant.

As the anodization proceeds the metal below the oxide barrier layer is progressively consumed. Beyond a certain stage the metal layer becomes thin or discontinuous enough to create highly resistive electrical current pathways. Hence beyond P5 (Fig. 1.33) the current drastically reduces, finally dropping to zero as the metal film becomes completely discontinuous. We make note that only a small dip in current can be seen at P5 as the sample was taken out of the anodization bath at this point. Keeping the sample inside the anodization bath beyond point P5 destroys the nanotube structure due to the its chemical dissolution in HF electrolyte. Real-time observation of the current enabled us to remove samples from the anodization bath as soon as the dip in the current plateau region, just beyond P5, was observed. FESEM images of samples kept beyond point P5 showed the nanotube layer to be severely damaged or completely eliminated. Thus, strict process control is necessary to obtain intact nanotube-arrays without an underlying metal layer. However, if there is a need to keep a continuous metal layer underneath the nanotubes the samples should be removed from the anodization bath at a point between P4 and P5. Close monitoring of the electric current response during potentiostatic anodization helps determine the optimum anodization parameters and serves as a process control tool.

The inset of Fig. 1.33 shows a typical current vs. time plot obtained during anodization (same conditions as thin film) of a 250 μm thick Ti foil with one face protected with a polymer coating. It can be seen that the current-time behavior is no different from that of the thin films. We note that if both sides of the foil are exposed to anodization, the current behavior will be significantly different from that seen in

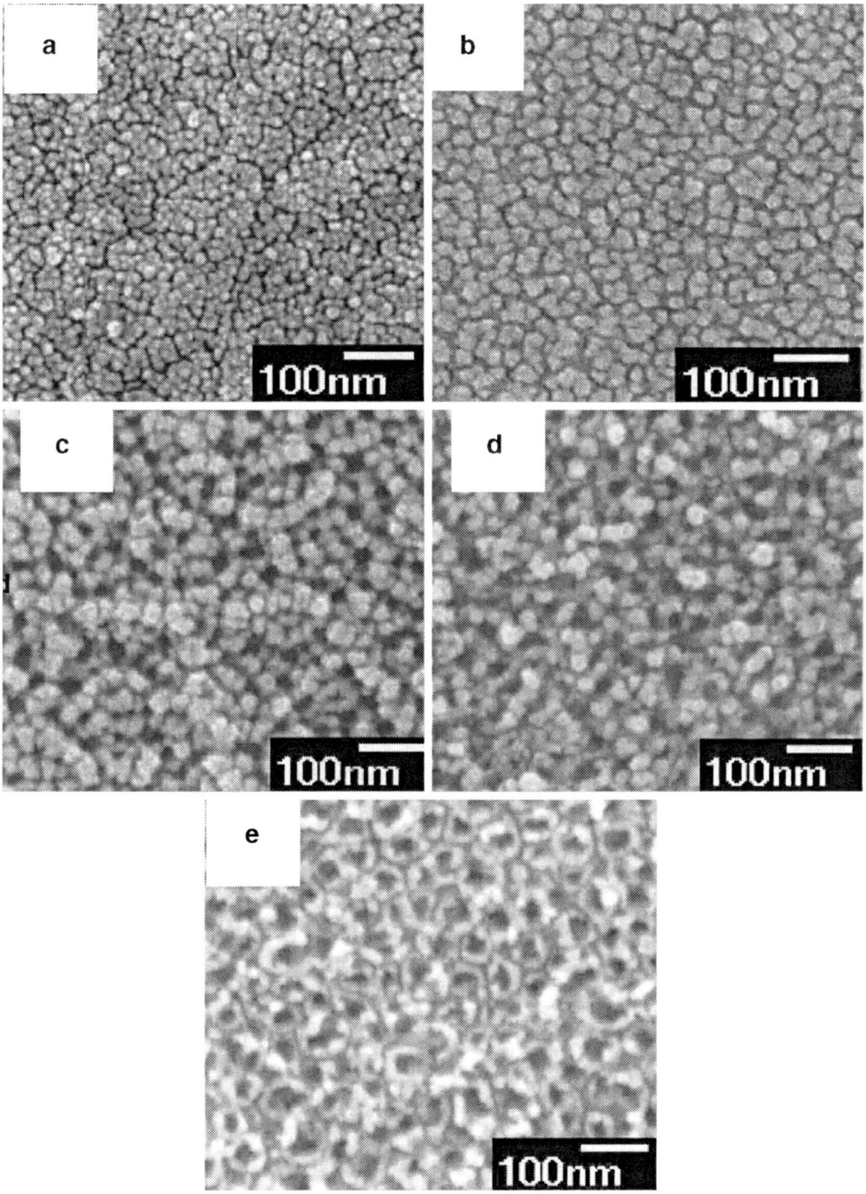

Fig. 1.34 FESEM images of samples taken at points P1, P2, P3, P4, and P5 of Fig. 1.33 [93]

Fig. 1.33. In this case, the changes in the current after the initial dip cannot be discerned due to the anodization process progressing at different rates on both sides of the sample, due to one side directly facing the cathode and the other turned away from it. It was noticed in both thin films and metal foils (Fig. 1.33) that current shows periodical fluctuations of a small magnitude between point P2 and P5;

similar fluctuations were observed by others [45, 128]. Macak and co-workers predicted that these current transients causing inhomogeneities on tube wall are due to pH burst at the pore tip that can suppressed by decreasing the diffusion constant of the electrolyte [47].

In strongly acidic aqueous solutions ($pH < 1$) both the nanotube growth rate and dissolution rate are increased, therefore increasing the anodization time does not appear to increase the nanotube length. For short nanotubes, achieved at 10 V potential, the time required to form the nanotubes is about 6 min. Further increasing the anodization time results in a more uniform nanotubular structure both in shape and size, but with not much change in the nanotube length. Increasing the pH decreases the chemical dissolution rate, and apparently prolongs the time needed to reach equilibrium between the rate of nanotube growth and the dissolution rate; in weak acid electrolytes, therefore the nanotube length appears to be time dependent. In such solutions, even for double-sided anodization of a Ti foil sample, we still observed the same general pattern in current vs. time plot as shown in Fig. 1.33 [129]. However, anodization has to be continued to point P4 for obtaining a complete nanotubular structure. In an aqueous pH 5.0 KF solution, anodization voltage 25 V, the time required to reach point P4 is 17 h and the resulting nanotube length is 6.0 μm [130].

1.7 Mechanistic Model of Nanotube Array Formation by Potentiostatic Anodization

The key processes responsible for anodic formation of nanoporous alumina [131–138] and TiO_2 [49, 50, 139–142] appear to be the same, and are fundamental to the formation of TiO_2 nanotube arrays. The key processes are: (1) oxide growth at the metal surface due to interaction of the metal with O^{2-}or OH^- ions [132]. After the formation of an initial oxide layer, these anions migrate through the oxide layer reaching the metal/oxide interface where they react with the metal. (2) Metal ion (Ti^{4+}) migration from the metal at the metal/oxide interface; Ti^{4+} cations will be ejected from the metal/oxide interface under application of an electric field that move toward the oxide/electrolyte interface. (3) Field-assisted dissolution of the oxide at the oxide/electrolyte interface [132, 133]. Due to the applied electric field the Ti–O bond undergoes polarization and is weakened promoting dissolution of the metal cations. Ti^{4+} cations dissolve into the electrolyte, and the free O^{2-} anions migrate toward the metal/oxide interface, process (1), to interact with the metal [142, 143]. (4) Chemical dissolution of the metal, or oxide, by the acidic electrolyte. Chemical dissolution of TiO_2 in the HF electrolyte plays a key role in the formation of nanotubes rather than simple nanoporous structures.

To help understand the process of nanotube formation, FESEM images of the surface of the samples anodized at 20 V for different durations were taken and analyzed [42]. As the anodization started, the initial oxide layer [143], formed due to interaction of the surface Ti^{4+} ions with oxygen ions (O^{2-}) in the electrolyte, can be seen to uniformly spread across the surface. At the anode oxidation of the metal releases Ti^{4+} ions and electrons, shown as [102]:

$$Ti + H_2O = TiO_2 + 2H^+ + 4e^- \tag{1.6}$$

At the cathode hydrogen evolution occurs:

$$8H^+ + 8e^- \rightarrow 4H_2 \tag{1.7}$$

The overall process of oxide formation is given by:

$$Ti + 2H_2O \rightarrow TiO_2 + 2H_2 \tag{1.8}$$

Fluorine ions can attack the oxide and hydrated layer; or, the ions being mobile in the anodic layer under the applied electric field, react with Ti^{4+} as described by [144]:

$$TiO_2 + 6F^- + 4H^+ \rightarrow TiF_6^{2-} + 2H_2O \tag{1.9}$$

$$Ti(OH)_4 + 6F^- \rightarrow TiF_6^{2-} + 4OH^- \tag{1.10}$$

$$Ti^{4+} + 6F^- \rightarrow TiF_6^{2-} \tag{1.11}$$

In the initial stages of the anodization process field-assisted dissolution dominates chemical dissolution due to the relatively large electric field across the thin oxide layer [141]. Small pits formed due to the localized dissolution of the oxide; represented by the (1.9), act as pore forming centers, after which these pits convert into pores with increasing pore density, uniformly, over the surface. The pore growth occurs due to the inward movement of the oxide layer at the pore bottom (barrier layer) [143, 145]. The Ti^{4+} ions migrating from the metal to the oxide/electrolyte interface dissolve in the HF electrolyte [131, 142, 143]. The rate of oxide growth at the metal/oxide interface and the rate of oxide dissolution at the pore-bottom/electrolyte interface ultimately become equal, thereafter the thickness of the barrier layer remains unchanged although it moves further into the metal increasing the pore depth. FESEM images show the formation of small pits in the inter-pore regions, which eventually lead to pore-separation and tube formation. The thickness of the tubular structure ceases to increase when the chemical dissolution rate of the oxide at the mouth of the tube (nanotube array surface) becomes equal to the rate of inward movement of the metal/oxide boundary at the base of the tube. Higher anodization voltages increase the oxidation and field-assisted dissolution and hence, a greater nanotube layer thickness can be achieved before equilibrating with chemical dissolution.

With anodization onset a thin layer of oxide forms on the titanium surface Fig. 1.35a. Small pits originate in this oxide layer due to the localized dissolution of the oxide Fig. 1.35b making the barrier layer at the bottom of the pits relatively thin which, in turn, increases the electric field intensity across the remaining barrier layer resulting in further pore growth, Fig. 1.35c. The pore entrance is not affected by electric field assisted dissolution and hence remains relatively narrow, while the electric field distribution in the curved bottom surface of the pore causes pore widening, as well as deepening of the pore. The result is a pore with a scallop shape [140, 146]. As the Ti–O bond energy is high (323 kJ/mol), in the case of TiO_2 it is reasonable to assume that only pores having thin walls can be formed due to the relatively low ion mobility and relatively high chemical solubility of the oxide in the electrolyte, hence un-anodized metallic portions can initially exist between the pores. As the pores grow deeper the electric field in these protruded metallic regions

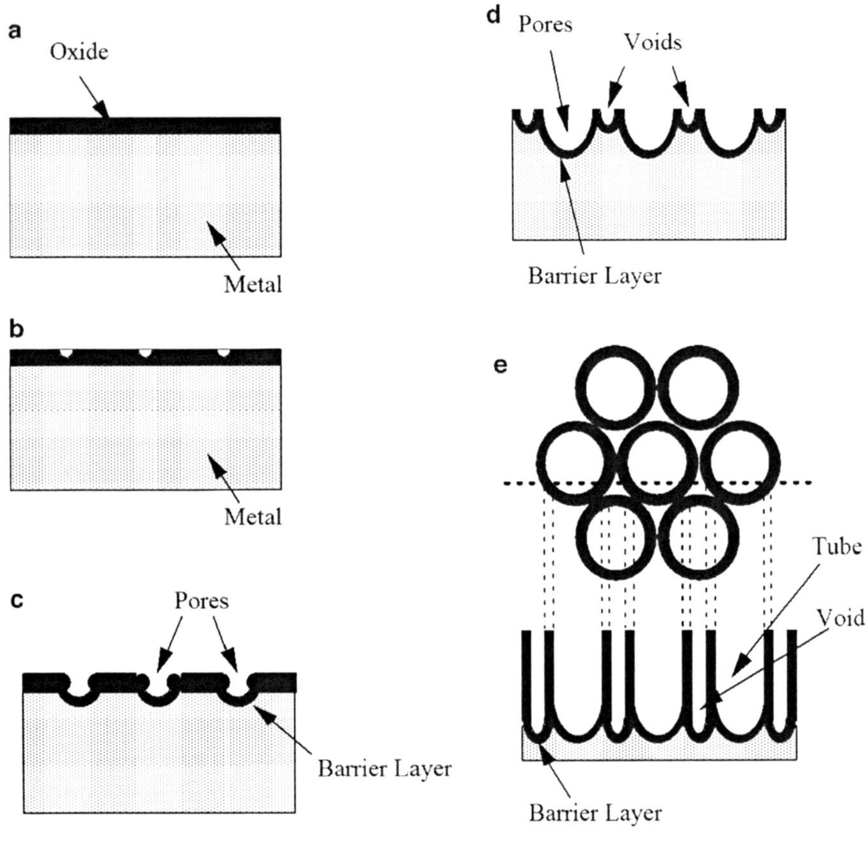

Fig. 1.35 Schematic diagram of nanotube evolution at constant anodization voltage: (**a**) Oxide layer formation, (**b**) pit formation on the oxide layer, (**c**) growth of the pit into scallop shaped pores, (**d**) the metallic region between the pores undergoes oxidation and field assisted dissolution, (**e**) fully developed nanotubes with a corresponding top view [42]

increases, enhancing field assisted oxide growth and oxide dissolution, hence simultaneously with the pores well-defined inter-pore voids start forming Fig. 1.35d. Thereafter, both voids and tubes grow in equilibrium. The nanotube length increases until the electrochemical etch rate equals the chemical dissolution rate of the nanotube top surface. After this point is reached the nanotube length will be independent of the anodization duration, as determined for a given electrolyte concentration and anodization potential.

Chemical dissolution, the key for self-organized formation of the nanotube arrays, reduces the thickness of the oxide layer (barrier layer) keeping the electrochemical etching (field assisted oxidation and dissolution) process active. No nanotubes can be formed if the chemical dissolution is too high or too low. The electrochemical etch rate depends upon anodization potential as well as electrolyte concentrations. If the electrochemical etch proceeds faster than chemical dissolution the thickness of the barrier layer increases, which in turn reduces the electrochemical etching process to the rate determined by chemical dissolution. The chemical dissolution rate is determined by the F^- concentration and solution pH. With increasing F^- and H^+ concentrations chemical dissolution increases. Investigations have shown that for aqueous electrolytes only in a certain F^- concentration range can nanotube arrays be achieved. The anodic potential at which nanotubes are formed is related to the F^- concentration, with higher potentials requiring electrolytes of higher F^- concentration.

In organic electrolytes (like FA, NMF, DMSO and EG), water is usually the source of oxygen in anodizing solutions [147]. While the exact mechanism by which water contributes oxygen to an anodic oxide film is not well-understood, strong evidence has been found for hydroxyl ion injection from the electrolyte into the anodic oxide film during anodization [147, 148]. When more water is present, hydroxyl ions are injected into the body of the oxide layer and affect the structure sufficiently to impede ion transport [149] through the barrier layer, which is necessary for further movement of the metal-oxide interface into the metal. When less water is present, the difficulty in extracting oxygen and/or hydroxyl ions from the solution limits the rate of growth of the overall oxide film. Also, the barrier oxide layer exhibits increased ionic conductivity caused by the non-stoichiometry induced by the reduced hydroxyl ion availability to the oxide [150]. The amount of hydroxyl ion injection is dependent on the solvent structure. It is no coincidence that the solvents where high growth rates of nanotubular TiO_2 films have been obtained, [51–54] namely formamide (~2 μm/min), N-methylformamide(~3 μm/min) and ethylene glycol (~15 μm/min) possess a high degree of structuring. Ethylene glycol has a clearly pronounced spatial net of hydrogen bonds. In DMSO [52], which is not as structured, final nanotube lengths as large as 100 μm are obtained but the growth rates are much slower ($<$1.5 μm per minute). For similar electrolytic compositions, the growth rates are also much smaller in Dimethylformamide (DMF), which is unstructured. When the solvent structure provides an environment in which titanium ions in the surface of the oxide bond to oxygen ions produced at the oxide surface, minimal hydroxyl ion injection occurs and ionic transport through the barrier layer is relatively rapid. With the highly polar FA and NMF electrolytes the dramatically increased nanotube-array length appears to be due to an accelerated rate of nanotube growth from the substrate. The dielectric constant of FA and NMF are, respectively, of 111 and 182.4, much greater than that of water, which has a dielectric constant of 78.39 [151]. For a given potential, higher electrolyte capacitance induces more charges to be formed on the oxide layer improving extraction of the Ti^{4+} ions, while the higher electrolyte polarity allows HF to be easily dissolved facilitating its availability at the TiO_2-electrolyte interface.

In fluoride ion containing formamide electrolytes, the starting anodization current does not drop instantly as observed for aqueous electrolytes, while gas evolution, which is indicative of electronic conduction, is observed at the anode. The anodization current drops steeply thereafter due to the initial formation of an insulating oxide layer, see Figs. 1.14 and 1.15. In this region, electronic conduction decreases due to the blocking action of the formed oxide, and ionic conduction increases. Once the oxide layer is completely formed over the entire exposed surface of the anode electronic conduction becomes negligible, and ionic conduction dominates the mechanistic behavior. Nanotube formation reduces the surface area available for anodization with a correlated decrease in current density, while deepening of the pore occurs.

TiO_2 nanotube growth was studied in fluoride bearing electrolytes containing different cations while maintaining similar, and in some cases identical molar concentrations of the chemical specie. This ensured that each electrolyte had the same concentration of fluoride ions thus allowing us to observe the effect of the cations on nanotube growth. Figure 1.14 shows the real- time potentiostatic anodization behavior of Ti, anodized at 20 V, in 95% formamide+5% water solutions containing 0.27M of NH_4F, Bu_4NF, HF and $BnMe_3NF$. The ionic radius of the cationic species [109] increases in the order

$$H^+ < Na^+ < NH_4^+ < Bu_4N^+ < BnMe_3N^+$$

The conductivities of the various electrolytes increased in the order:

$$H^+ < Na^+ < Bu_4N^+ < NH_4^+ < BnMe_3N^+$$

The Stokes radius of an ion in solution provides information about the solvation of the respective ion and is given by [151]

$$r = \frac{|z|F^2}{6\pi N\eta\lambda_i^0} \tag{1.12}$$

where z is the charge on the ion, F is the Faraday constant and N is Avogadros' number. Equation (1.12) shows the Stokes-law radii r_s of ions to be inversely related to the product of the viscosity η and the limiting equivalent conductance λ_i^0 of the ion. Replacing 6 by 4 in the denominator of (1.12) yields the hydrodynamic radius of the ion. While we have not calculated the limiting equivalent conductance of the electrolytes, a qualitative trend may be established by reviewing the viscosity and conductivity information gathered in Table 1.4. It was also found that the benzyltrimethylammonium fluoride solution is the most conductive among the electrolytes investigated. The solution viscosities increase with the size of the cationic species, with solutions containing Bu_4N^+ and $BnMe_3N^+$ exhibiting much higher viscosities than solutions of the remaining cations. The electrolyte-containing NH_4F has a lower viscosity than the one containing Bu_4NF but is more conductive resulting in a similar value for the viscosity-conductivity product. The NaF electrolyte has a much lower concentration than the other electrolytes and is hence excluded from this comparison. The product of the viscosity and conductivity increases in the following order for the strong electrolytes:

$$H^+ < NH_4^+, Bu_4N^+ < BnMe_3N^+$$

Thus, the resulting stokes radii and hydrodynamic radii may be expected to decrease in the following order:

$$H^+ > NH_4^+, Bu_4N^+ > BnMe_3N^+$$

Thus, we observe that the specie with smallest ionic radii (H^+ and $NH_4{}^+$) have the largest Stokes radii as well as hydrodynamic radii. This suggests that the H^+ and $NH_4{}^+$ cations have larger solvation shells than Bu_4N^+ and $BnMe_3N^+$ ions. This is understandable considering that formamide is a polar protophilic solvent with good hydrogen bond donating ability [151], whereas small cations, such as H^+ and Na^+ are strongly solvated; due to the hydrogen bond donating ability of formamide the F^- ions are strongly solvated as well. From the solvent-berg model, the solvent molecules immediately adjacent to the ion are rigidly bound to it and thus the ion moves as a kinetic unit [151]. Consequently, the conductivity of HF in the formamide–water mixture is limited by the solvent molecule mobility and the anodization currents are relatively small. On the other hand, larger ions such as Bu_4N^+ and $BnMe_3N^+$ are weakly solvated and their higher ionic mobility translates into a higher anodization current. However, at high cation concentrations the viscosity of the solution increases, with a corresponding decrease in solution conductivity. For large cations such as Bu_4N^+ and $BnMe_3N^+$, the Stokes radii are independent of solvent and Walden's rule [152] is obeyed whereby:

$$\lambda^{\infty} \bullet \eta = \text{constant} \tag{1.13}$$

λ^{∞} is the limiting molar conductivity of the cation in the solution and η is the viscosity.

The conductivity of the electrolytes also plays a role in controlling nanotube array growth. Ethylene glycol containing 2% water and 0.35% NH_4F has a conductivity of ≈460 μS/cm, much lower than the conductivity of the formamide based electrolytes (>2000 μS/cm) [54]. The total applied anodization voltage is the sum of the potential difference at the metal-oxide interface, the potential drop across the oxide, the potential difference at the oxide-electrolyte interface, and the potential drop across the electrolyte, which is non-negligible for the ethylene glycol based electrolytes. Therefore, the maximum nanotube length obtained at lower voltages (say 20 V) in a formamide bath could only be achieved at higher voltages in an EG bath. In contrast, with DMSO electrolytes, with relatively small electrolyte conductivities, the increased nanotube-array length appears to be largely due to limited chemical dissolution of TiO_2 by the hydrofluoric acid. For a given rate of pore formation the chemical dissolution of the oxide at the pore mouth by F^- determines the tube length. To control this dissolution reaction, the H^+ ion concentration was reduced by limiting the water content to the level of water contained in the HF containing solution. This water ensured the field-assisted etching of the Ti foil, at the pore bottom, and, additionally, protophilic DMSO accepts a proton from HF reducing its activity. This allowed the DMSO nanotubes to grow deep into the titanium foil without any significant loss from the pore mouth.

Figure 1.36 shows typical anodization current behavior during anodic oxidation of titanium (both sides of the foil sample are being anodized simultaneously) at 40 V in 2.0% HF-DMSO and 4.0% HF-DMSO electrolytes. The anodization current density is less than 2 mA/cm^2, considerably smaller than the anodization currents seen for any other aqueous or polar organic electrolytes that can initially be several tens of mA/cm^2. The slight increase in anodization current in the 4% HF curve (Fig. 1.36) after about 10 h indicates relatively stronger chemical etching compared to that observed in 2.0% HF-DMSO and hence comparatively shorter nanotube lengths. We do not understand the fluctuations seen in the current density,

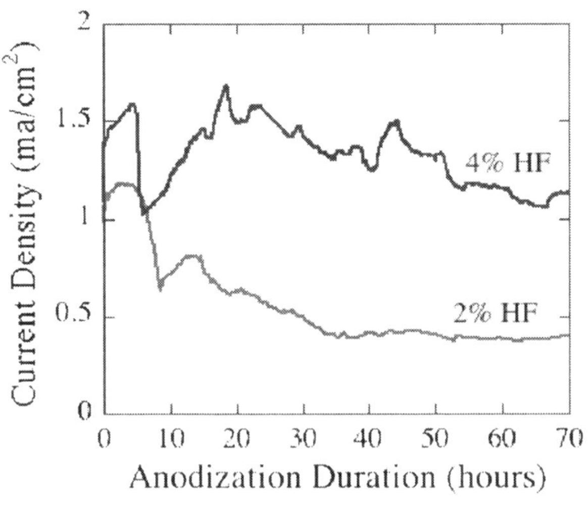

Fig. 1.36 Current density vs. time behavior during anodization of a Ti foil sample at 40 V using a DMSO electrolyte containing 2.0% HF, and 4.0% HF [46]

other than to say they are due to a combination of changes in oxide thickness, electrolyte conductivity, and resulting change in potential drop across the TiO_2-electrolyte interface during anodization. Fresh DMSO electrolytes have low conductivities that increase with anodization duration due to chemical and field assisted dissolution of titanium into the electrolyte. Hence, initially the applied potential drop occurs primarily across the electrolyte between the two electrodes. As the electrolyte conductivity increases with anodization duration the drop in the applied DC potential shifts from the electrolyte to the oxide layer, increasing the rate of oxide formation, resulting in the first "hump" seen in the current density-time curves of Fig. 1.36. A local minimum in the current density is obtained with formation of an oxide layer upon the exposed Ti surface. Then due to non-uniform chemical etching of the oxide layer, i.e. pitting, the anodization current increases until somewhat uniform oxide layer begins to form. Thereafter, the current decreases until an equilibrium is reached between the rate of field assisted oxide growth, field assisted oxide dissolution, and chemical etching. The chemical dissolution is very low in DMSO vs. aqueous electrolytes. Therefore, we expect and observe higher tube lengths from DMSO electrolytes with specific details dependent upon a combination of factors including anodization duration, voltage, HF concentration, and electrolyte condition (fresh or previously used for an anodization).

For nanoporous alumina obtained by anodization, the gradual decrease of the ionic current with time was attributed to diffusion controlled mass-transport of oxygen containing anionic species from the bulk reservoir to the oxide/metal interface [153]. Macak and co-workers [48] claimed that diffusion of the reactants from the pore-tip or of reaction products away from the pore tip was the dominant growth-limiting factor in the anodic formation of TiO_2 nanotubes in viscous electrolytes. A problem with the mass transport explanation is the strong dependence we observe of the nanotube growth rate on anodization voltage. Furthermore, mass transport theory does not account for the higher growth rate in used solutions. Anionic oxygen bearing anionic species consumed during the formation of TiO_2 are depleted in the used electrolyte. Further, the conductivity of the used solution is significantly lower than that of fresh solutions, and the viscosity of used solutions almost the same as fresh indicating reactant diffusion to be similar or lower in the used electrolyte. In a previous report [53],we observed that the higher viscosity solutions of bulkier cations such as tetrabutylammonium exhibited higher nanotube growth rates compared to lower viscosity solutions of smaller cations such as ammonium and sodium. Also, when the concentration of tetrabutylammonium fluoride was increased no new sources of oxygen bearing anionic species were added to the electrolyte; the viscosity of the solution increased and the growth rate increased as well [53]. Further, we observe that a 4 cm^2 piece of Ti foil, 250 μm thick in 100 mL of EG electrolyte was completely converted into TiO_2 after anodization for 4 days. Thus any mass-transport effects, should they be significant, are likely to occur much later in the anodization process, while a potential dependent process dominates earlier stages of the nanotube synthesis process.

For voltage-controlled processes, the growth rate depends on whether the rate-limiting step is solid state transport of ions through the oxide barrier layer or the availability of ionic specie at the oxide-electrolyte interface. Hypotheses assuming that the film growth rate is limited by migration of ionic species through the film under the effect of a high electric field constitute the High Field Model (HFM). When a potential dependent interface reaction is assumed to be rate determining, the Interface Model (IFM) results.

The movement of the Ti–TiO_2 interface into the Ti metal involves ion transport under the influence of an electric field, either by transport of metal cations in the direction of the solution, or oxygen ions (and probably some incorporated anions) in the direction of the metal. For valve metals, under the high field approximation, the ionic current i is related to the anodization potential and the thickness of the barrier layer by an equation of the form [154–156]:

$$i = A \exp\left(\frac{\beta \Delta U}{X_{\text{barrier}}}\right) \tag{1.14}$$

where ΔU is the potential drop across the oxide layer, X_{barrier} is the thickness of the barrier layer and A and β are constants.

Macdonald and co-workers have proposed diagnostic criteria to distinguish between different film growth regimes [157]. For conditions with significant dissolution under a constant polarization, the HFM was shown to give the relation:

$$\sqrt{\frac{-\frac{\mathrm{d}i}{\mathrm{d}t}}{i(i-\delta)}} = k \ln i - k \ln A \tag{1.15}$$

where i is the anodization current, δ is the dissolution current, and A is the high field pre-exponent term from (1.14); k is given by:

$$k = \frac{1}{nF}\sqrt{\frac{V_{\text{ox}} RT}{\alpha^{\text{hf}} sU}} \tag{1.16}$$

where n is the valence, F is Faraday's constant, V_{ox} is the oxide molar volume, R is the ideal gas constant, T is the temperature, U is the applied potential, s is the high field half-jump distance and α^{hf} is the high field symmetry factor [158]. The dissolution current δ is itself a function of the Ti thickness and is hence not constant but varies sufficiently slowly with time that it is considered a constant during the anodization period of 30 h and is taken to be the anodization current at the end of 30 h of anodization. Equation (1.15) results in a linear relationship with a positive slope when the function $f(i, i') = \sqrt{\frac{-\frac{\mathrm{d}i}{\mathrm{d}t}}{i(i-\delta)}}$ is plotted against $\ln i$.

For interfacial kinetics, where a potential dependent interface reaction is assumed to be the rate-determining step, the corresponding equation is:

$$\sqrt{\frac{\frac{-\mathrm{d}i}{\mathrm{d}t}}{i}} = \sqrt{\frac{V_{\text{ox}} \alpha E}{RT}} \tag{1.17}$$

where E is the electric field in the barrier layer and α is a fit parameter for interfacial kinetics. The right hand side of (1.17) is a constant, therefore a horizontal line results when $f(i,i') = \frac{\sqrt{\frac{-di}{dt}}}{i}$ is plotted against $\ln i$. Figure 1.37 shows $f(i,i') = \sqrt{\frac{\frac{-di}{dt}}{i(i-\delta)}}$ vs. $\ln i$ providing an excellent diagnostic for identifying the rate-limiting step. The dashed colored lines represent the values of $f(i, i')$ while the solid black lines are linear fits to the same (R^2=0.99). At constant potential, in the current decay phase, the points in the $f(i, i')$ vs. $\ln i$ are essentially a straight line and indicate the high field model to be applicable.

At higher potentials, the slope is much smaller than at lower potentials. Since the slopes in the $f(i, i')$ vs. ln (i) plot equate to half-jump distances in the high field model [157, 158] and since anionic incorporation can cause the half-jump distance to vary by a factor of two [159], the slope in Fig. 1.37 may be expected to vary by a factor of two. However, in our case, the slope is clearly varying by a factor of three with a consistent trend of lower slopes at higher anodization potentials indicating that interfacial reactions are becoming more significant. However, it is the faster ionic transport at higher potentials [given by the exponential dependence on potential per (1.14)] that allows the interface reaction to become rate-determining. The interface reactions are still not solely rate-determining during the first 30 h of the anodization process, considering the presence of a clear positive slope even at 60 and 65 V. However, as the Ti film becomes increasingly thinner the Ti^{3+} ions will find themselves pulled in opposite directions under the action of the prevailing electric field and diffusive transport of the ions inside the barrier layer may become

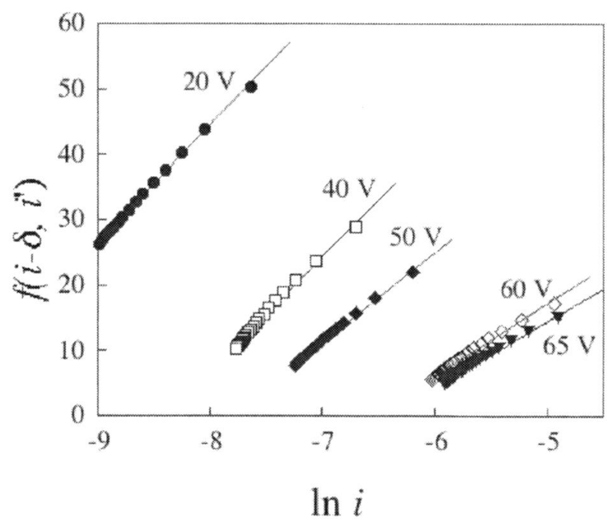

Fig. 1.37 $f(i,i') = \sqrt{\frac{\frac{-di}{dt}}{i(i-\delta)}}$ vs. $\ln i$ for potentiostatic anodization of Ti in ethylene glycol based electrolyte [54]

the rate-limiting step. A change with time of the governing mechanism as the overall reaction rate decreases is likely.

Water is usually the source of oxygen in anodizing solutions [147]. While the exact mechanism by which water contributes oxygen to an anodic oxide film is not well-understood, strong evidence has been found for hydroxyl ion injection [147, 148] from the electrolyte into the anodic oxide film during anodization. When more water is present, hydroxyl ions are injected into the body of the oxide layer and affect the structure sufficiently to impede ion transport [149] though the barrier layer, which is necessary for further movement of the metal-oxide interface into the metal. When less water is present, the difficulty in extracting oxygen and/or hydroxyl ions from the solution limits the rate of growth of the overall oxide film. Also, the barrier oxide layer exhibits increased ionic conductivity caused by the non-stoichiometry induced by the reduced hydroxyl ion availability to the oxide [141]. Thus an optimum water content exists for a maximum nanotube growth rate. A small amount of water is absorbed by the electrolyte from the ambient due to the hygroscopic nature of ethylene glycol. At ammonium fluoride concentrations lower than 0.3%, the reaction rate is low and the water absorbed from the ambient sufficient to meet the demand for oxygen arising from the TiO_2 nanotube formation. NH_4F concentrations greater than 0.3% wt without water addition resulted in corrosion of the foil. For all NH_4F concentrations (0.1–0.5% wt) an increase in nanotube array length as well as a sustainable (non-corroding) reaction rate was obtained by adding D.I water to the NH_4F and EG mixture. For ethylene glycol solutions containing 0.3% NH_4F, the optimum water content for a fast growth rate is 2% as shown in Table 1.6. The amount of hydroxyl ion injection is dependent on the solvent structure. In formamide, chain structure and ring-dimer structure are combined with hydrogen bonding to form three dimensional networks [150]. In NMF, linear but short chain structures predominate while ethylene glycol has a clearly pronounced spatial net of hydrogen bonds [151, 160]. We suggest that it is no coincidence that the organic solvents where high growth rates of nanotubular TiO_2 films have been obtained [51, 52], namely formamide (~2 μm/h), N-methyl-formamide (~3 μm/h) and ethylene glycol (~15 μm/h) possess a high degree of structuring [161]. For similar electrolytic compositions, the growth rates are much smaller in Dimethylformamide (DMF), which is unstructured. When the solvent structure provides an environment in which titanium ions in the surface of the oxide bond to oxygen-bearing ions at the oxide surface, minimal hydroxyl ion injection occurs and ionic transport through the barrier layer is relatively rapid [159].

The conductivity of the electrolytes also plays a role in controlling nanotube array growth. Ethylene glycol containing 2% water and 0.35% NH_4F was found to have a conductivity of 460 $\mu S\ cm^{-1}$, which is much lower than the conductivity of the formamide-based electrolytes (>2000 $\mu S\ cm^{-1}$). The total applied anodization voltage is the sum of the potential difference at the metal-oxide interface, the potential drop across the oxide, the potential difference at the oxide-electrolyte interface, and the potential drop across the electrolyte, which is non-negligible for the ethylene glycol-based electrolytes. Since the anodization current decreases monotonically with time, the potential drop across the solution, given by IR_s

where I is the anodization current and R_s is the solution resistance, also drops with time. Assuming the interface potential remains the same, this results in potential increasing across the oxide layer with time and serves to counteract the decrease in electric field in the oxide layer due to thickening of the barrier layer, thus extending the time-window where rapid nanotube growth occurs.

References

1. Adachi M, Murata Y, Harada M, Yoshikawa S (2000) Formation of titania nanotubes with high photo-catalytic activity. Chem Lett 29:942–943
2. Chu SZ, Inoue S, Wada K, Li D, Haneda H, Awatsu S (2003) Highly porous (TiO_2-SiO_2-TeO_2)/Al_2O_3/TiO_2 composite nanostructures on glass with enhanced photocatalysis fabricated by anodization and sol-gel process. J Phys Chem B 107:6586–6589
3. Lai Y, Sun L, Chen Y, Zhuang H, Lin C, Chin JW (2006) Effects of the structure of TiO_2 nanotube array on Ti substrate on its photocatalytic activity. J Electrochem Soc 153:D123–D127
4. Macak JM, Zlamal M, Krysa J, Schmuki P (2007) Self-organized TiO_2 nanotube layers as highly efficient photocatalysts. Small 3:300–304
5. Albu SP, Ghicov A, Macak JM, Hahn R, Schmuki P (2007) Self-organized, free-standing TiO_2 nanotube membrane for flow-through photocatalytic applications. Nano Lett 7:1286–1289
6. Varghese OK, Gong D, Paulose M, Ong KG, Dickey EC, Grimes CA (2003) Extreme changes in the electrical resistance of titania nanotubes with hydrogen exposure. Adv Mater 15:624–627
7. Mor GK, Carvalho MA, Varghese OK, Pishko MV, Grimes CA (2004) A room-temperature TiO_2-nanotube hydrogen sensor able to self-clean photoactively from environmental contamination. J Mater Res 19:628–634
8. Varghese OK, Mor GK, Grimes CA, Paulose M, Mukherjee N (2004) A titania nanotube-array room-temperature sensor for selective detection of hydrogen at low concentrations. J Nanosci Nanotechnol 4:733–737
9. Paulose M, Varghese OK, Mor GK, Grimes CA, Ong KG (2006) Unprecedented ultra-high hydrogen gas sensitivity in undoped titania nanotubes. Nanotechnol 17:398–402
10. Varghese OK, Yang X, Kendig J, Paulose M, Zeng K, Palmer C, Ong KG, Grimes CA (2006) A transcutaneous hydrogen sensor: from design to application. Sensor Lett 4:120–128
11. de Taconni NR, Chenthamarakshan CR, Yogeeswaran G, Watcharenwong A, de Zoysa RS, Basit NA, Rajeshwar K (2006) Nanoporous TiO_2 and WO_3 films by anodization of titanium and tungsten substrates: influence of process variables on morphology and photoelectrochemical response. J Phys Chem B 110:25347–25355
12. Varghese OK, Paulose M, Shankar K, Mor GK, Grimes CA (2005) Water-photolysis properties of micron-length highly-ordered titania nanotube-arrays. J Nanosci Nanotechnol 5:1158–1165
13. Xie Y, Zhou LM, Huang H (2006) Enhanced photoelectrochemical current response of titania nanotube array. Mater Lett 60:3558–3560
14. Mohapatra SK, Misra M, Mahajan VK, Raja KS (2007) Design of a highly efficient photoelectrolytic cell for hydrogen generation by water splitting: application of $TiO_{2-x}C_x$ nanotubes as a photoanode and Pt/TiO_2 nanotubes as a cathode. J Phys Chem C 111:8677–8685
15. Sun WT, Yu Y, Pan HY, Gao XF, Chen Q, Peng LM (2008) CdS quantum dots sensitized TiO_2 nanotube-array photoelectrodes. J Am Chem Soc 130:1124–1125

16. Bae S, Shim E, Yoon J, Joo H (2008) Enzymatic hydrogen production by light-sensitized anodized tubular TiO_2 photoanode. Sol Energy Mater Sol Cells 92:402–409
17. Mor GK, Shankar K, Paulose M, Varghese OK, Grimes CA (2007) High efficiency double heterojunction polymer photovoltaic cells using highly ordered TiO_2 nanotube arrays. Appl Phys Lett 91:152111 (3pp)
18. Shankar K, Mor GK, Prakasam HE, Varghese OK, Grimes CA (2007) Self-assembled hybrid polymer-TiO_2 nanotube array heterojunction solar cells. Langmuir 23:12445–12449
19. Yu BY, Tsai A, Tsai SP, Wong KT, Yang Y, Chu CW, Shyue JJ (2008) Efficient inverted solar cells using TiO_2 nanotube arrays. Nanotechnol 19:255202 (5pp)
20. Uchida S, Chiba R, Tomiha M, Masaki N, Shirai M (2002) Application of titania nanotubes to a dye-sensitized solar cell. Electrochem 70:418–420
21. Adachi M, Murata Y, Okada I, Yoshikawa Y (2003) Formation of titania nanotubes and applications for dye-sensitized solar cells. J Electrochem Soc 150:G488–G493
22. Mor GK, Shankar K, Paulose M, Varghese OK, Grimes CA (2006) Use of highly-ordered TiO_2 nanotube arrays in dye-sensitized solar cells. Nano Lett 6:215–218
23. Paulose M, Shankar K, Varghese OK, Mor GK, Hardin B, Grimes CA (2006) Backside illuminated dye-sensitized solar cells based on titania nanotube array electrodes. Nanotechnol 17:1446–1448
24. Zhu K, Neale NR, Miedaner A, Frank AJ (2006) Enhanced charge-collection efficiencies and light scattering in dye-sensitized solar cells using oriented TiO_2 nanotubes arrays. Nano Lett 7:69–74
25. Shankar K, Mor GK, Prakasam HE, Yoriya S, Paulose M, Varghese OK, Grimes CA (2007) Highly-ordered TiO_2 nanotube arrays up to 220 μm in length: use in water photoelectrolysis and dye-sensitized solar cells. Nanotechnol 18:065707 (11pp)
26. Shankar K, Bandara J, Paulose M, Wietasch H, Varghese OK, Mor GK, LaTempa TJ, Thelakkat M, Grimes CA (2008) Highly efficient solar cells using TiO_2 nanotube arrays sensitized with a donor-antenna dye. Nano Lett 8:1654–1659
27. Popat KC, Eltgroth M, LaTempa TJ, Grimes CA, Desai TA (2007) Titania nanotubes: a novel platform for drug-eluting coatings for medical implants. Small 3:1878–1881
28. Popat KC, Eltgroth M, LaTempa TJ, Grimes CA, Desai TA (2007) Decreased Staphylococcus epidermis adhesion and increased osteoblast functionality on antibiotic-loaded titania nanotubes. Biomater 28:4880–4888
29. Roy SC, Paulose M, Grimes CA (2007) The effect of TiO_2 nanotubes in the enhancement of blood clotting for the control of hemorrhage. Biomater 28:4667–4672
30. Park J, Bauer S, von der Mark K, Schmuki P (2007) Nanosize and vitality: TiO_2 nanotube diameter directs cell fate. Nano Lett 7:1686–1691
31. Hoyer P (1996) Formation of a titanium dioxide nanotube array. Langmuir 12:1411–1413
32. Lakshmi BB, Dorhout PK, Martin CR (1997) Sol-gel template synthesis of semiconductor nanostructures. Chem Mater 9:857–862
33. Imai H, Takei Y, Shimizu K, Matsuda M, Hirashima H (1999) Direct preparation of anatase TiO_2 nanotubes in porous alumina membranes. J Mater Chem 9:2971–2972
34. Michailowski A, AlMawlwai D, Cheng GS, Moskovits M (2001) Highly regular anatase nanotubule arrays fabricated in porous anodic templates. Chem Phys Lett 349:1–5
35. Jung JH, Kobayashi H, van Bommel KJC, Shinkai S, Shimizu T (2002) Creation of novel helical ribbon and double-layered nanotube TiO_2 structures using an organogel template. Chem Mater 14:1445–1447
36. Kobayashi S, Hamasaki N, Suzuki M, Kimura M, Shirai H, Hanabusa K (2002) Preparation of helical transition-metal oxide tubes using organogelators as structure-directing agents. J Am Chem Soc 124:6550–6551
37. Tian ZRR, Voigt JA, Liu J, McKenzie B, Xu HF (2003) Large oriented arrays and continuous films of TiO_2-based nanotubes. J Am Chem Soc 125:12384–12385
38. Kasuga T, Hiramatsu M, Hoson A, Sekino T, Niihara K (1998) Formation of titanium oxide nanotube. Langmuir 14:3160–3163

39. Chen Q, Zhou WZ, Du GH, Peng LH (2002) Trititanate nanotubes made via a single alkali treatment. Adv Mater 14:1208–1211
40. Yao BD, Chan YF, Zhang XY, Zhang WF, Yang ZY, Wang N (2003) Formation mechanism of TiO_2 nanotubes. Appl Phys Lett 82:281–283
41. Gong D, Grimes CA, Varghese OK, Hu W, Singh RS, Chen Z, Dickey EC (2001) Titanium oxide nanotube arrays prepared by anodic oxidation. J Mater Res 16:3331–3334
42. Mor GK, Varghese OK, Paulose M, Mukherjee N, Grimes CA (2003) Fabrication of tapered, conical-shaped titania nanotubes. J Mater Res 18:2588–2593
43. Cai Q, Paulose M, Varghese OK, Grimes CA (2005) The effect of electrolyte composition on the fabrication of self-organized titanium oxide nanotube arrays by anodic oxidation. J Mater Res 20:230–236
44. Ruan C, Paulose M, Varghese OK, Grimes CA (2006) Enhanced photoelectrochemical-response in highly ordered TiO_2 nanotube-arrays anodized in boric acid containing electrolyte. Sol Energy Mater Sol Cells 90:1283–1295
45. Macak JM, Tsuchiya H, Schmuki P (2005) High-aspect-ratio TiO_2 nanotubes by anodization of titanium. Angew Chem Int Ed 44:2100–2102
46. Macak JM, Tsuchiya H, Taveira L, Aldabergerova S, Schmuki P (2005) Smooth anodic TiO_2 nanotubes. Angew Chem Int Ed 44:7463–7465
47. Balaur E, Macak JM, Tsuchiya H, Schmuki P (2005) Wetting behaviour of layers of TiO_2 nanotubes with different diameters. J Mater Chem 15:4488–4491
48. Macak JM, Schmuki P (2006) Anodic growth of self-organized anodic TiO_2 nanotubes in viscous electrolytes. Electrochim Acta 52:1258–1264
49. Zwilling V, Aucouturier M, Darque-Ceretti E (1999) Anodic oxidation of titanium and TA6V alloy in chromic media. An electrochemical approach. Electrochim Acta 45:921–929
50. Zwilling V, Darque-Ceretti E, Boutry-Forveille A, David D, Perrin MY, Aucouturier M (1999) Structure and physicochemistry of anodic oxide films on titanium and TA6V alloy. Surf Interface Anal 27:629–637
51. Paulose M, Shankar K, Yoriya S, Prakasam HE, Varghese OK, Mor GK, Latempa TJ, Fitzgerald A, Grimes CA (2006) Anodic growth of highly ordered TiO_2 nanotube arrays to 134 μm in length. J Phys Chem B 110:16179–16184
52. Yoriya S, Prakasam HE, Varghese OK, Shankar K, Paulose M, Mor GK, Latempa TJ, Grimes CA (2006) Initial studies on the hydrogen gas sensing properties of highly-ordered high aspect ratio TiO_2 nanotube-arrays 20 μm to 222 μm in length. Sens Lett 4:334–339
53. Shankar K, Mor GK, Fitzgerald A, Grimes CA (2007) Cation effect on the electrochemical formation of very high aspect ratio TiO_2 nanotube arrays in formamide–water mixtures. J Phys Chem C 111:21–26
54. Prakasam HE, Shankar K, Paulose M, Grimes CA (2007) A new benchmark for TiO_2 nanotube array growth by anodization. J Phys Chem C 111:7235–7241
55. Mor GK, Shankar K, Paulose M, Varghese OK, Grimes CA (2005) Enhanced photocleavage of water using titania nanotube arrays. Nano Lett 5:191–195
56. Patermarakis G, Moussoutzanis K (1995) Mathematical models for the anodization conditions and structural features of porous anodic Al_2O_3 films on aluminum. J Electrochem Soc 142:737–743
57. Woo-Jin L, Alhoshan M, Smyrl WH (2006) Titanium dioxide nanotube arrays fabricated by anodizing processes electrochemical properties. J Electrochem Soc 153:B499–B505
58. Perathoner S, Passalacqua R, Centi G, Su DS, Weinberg G (2007) Photoactive titania nanostructured thin films: synthesis and characteristics of ordered helical nanocoil array. Catal Today 122:3–13
59. Bai J, Zhou B, Li L, Liu Y, Zheng Q, Shao J, Zhu X, Cai W, Liao J, Zou L (2008) The formation mechanism of titania nanotube arrays in hydrofluoric acid electrolyte. J Mater Sci 43:1880–1884
60. Quan X, Yang S, Ruan X, Zhao H (2005) Preparation of titania nanotubes and their environmental applications as electrode. Environ Sci Technol 39:3770–3775

61. Zhao J, Wang X, Chen R, Li L (2005) Fabrication of titanium oxide nanotube arrays by anodic oxidation. Solid State Commun 34:705–710
62. Thompson GE, Furneaux RC, Wood GC, Richardson JA, Goode JS (1978) Nucleation and growth of porous anodic films on aluminum. Nature 272:433–435
63. Mor GK, Varghese OK, Paulose M, Grimes CA (2003) A self-cleaning, room-temperature titania-nanotube hydrogen gas sensor. Sens Lett 1:42–46
64. Vitiello RP, Macak JM, Ghicov A, Tsuchiya H, Dick LFP, Schmuki P (2006) N-Doping of anodic TiO_2 nanotubes using heat treatment in ammonia. Electrochem Commun 8:544–548
65. Prida VM, Hernandez-Velez M, Pirota KR, Menendez A, Vazquez M (2005) Synthesis and magnetic properties of Ni nanocylinders in self-aligned and randomly disordered grown titania nanotubes. Nanotechnol 16:2696–2702
66. Macak JM, Tsuchiya H, Berger S, Bauer S, Fujimoto S, Schmuki P (2006) On wafer TiO_2 nanotube-layer formation by anodization of Ti-films on Si. Chem Phys Lett 428:421–425
67. Nainville I, Lemarchand A, Badiali JP (1996) Growth and morphology of thick films formed on a metallic surface. Electrochim Acta 41:1855–1862
68. Tsuchiya H, Macak JM, Taveira L, Balaur E, Ghicov A, Sirotna K, Schmuki P (2005) Self-organized TiO_2 nanotubes prepared in ammonium fluoride containing acetic acid electrolytes. Electrochem Commun 7:576–580
69. Tsuchiya H, Macak JM, Ghicov A, Taveira L, Schmuki P (2005) Self-organized porous TiO_2 and ZrO_2 produced by anodization. Corros Sci 47:3324–3335
70. Ghicov A, Tsuchiya H, Macak JM, Schmuki P (2005) Titanium oxide nanotubes prepared in phosphate electrolytes. Electrochem Commun 7:505–509
71. Tian T, Xiao XF, Liu RF, She HD, Hu XF (2007) Study on titania nanotube arrays prepared by titanium anodization in NH_4F/H_2SO_4 solution. J Mater Sci 42:5539–5543
72. Bauer S, Kleber S, Schmuki P (2006) TiO_2 nanotubes: tailoring the geometry in H_3PO_4/HF electrolytes. Electrochem Commun 8:1321–1325
73. Premchand YD, Djenizian T, Vacandio F, Knauth P (2006) Fabrication of self-organized TiO_2 nanotubes from columnar titanium thin films sputtered on semiconductor surfaces. Electrochem Commun 8:1840–1844
74. Yu XF, Li YX, Ge WY, Yang QB, Zhu NF, Zadeh KK (2006) Formation of nanoporous titanium oxide films on silicon substrates using an anodization process. Nanotechnol 17:808–814
75. Bockris JO (1947) Electrolytic polarization: 1. Overpotential of hydrogen on some less common metals at high current densities. Influence of current density and time. Trans Faraday Soc 43:417–429
76. Nozik AJ (1978) Photoelectrochemistry: applications to solar energy conversion. Ann Rev Phys Chem 29:189–222
77. Allam NK, Grimes CA (2008) Effect of cathode material on the morphology and photoelectrochemical properties of vertically oriented TiO_2 nanotube arrays. Sol Energy Mater Sol Cells 92:1468–1475
78. Nakayama K, Kubo T, Tsubokura A, Nishikitani Y, Masuda H (2006) Anodic formation of high-aspect-ratio titania nanotubes. ECS Meeting Abstracts 502:819
79. Richter C, Wu Z, Panaitescu E, Willey R, Menon L (2007) Ultra-high aspect ratio titania nanotubes. Adv Mater 19:946–948
80. Richter C, Panaitescu E, Willey R, Menon L (2007) Titania nanotubes prepared by anodization in fluorine-free acids. J Mater Res 22:1624–1631
81. Hahn R, Macak JM, Schmuki P (2007) Rapid anodic growth of TiO_2 and WO_3 nanotubes in fluoride free electrolytes. Electrochem Commun 9:947–952
82. Chen X, Schriver M, Suen T, Mao SS (2007) Fabrication of 10 nm diameter TiO_2 nanotube arrays by titanium anodization. Thin Solid Films 515:8511–8514
83. Allam NK, Grimes CA (2007) Formation of vertically oriented TiO_2 nanotube arrays using a fluoride free HCl aqueous electrolyte. J Phys Chem C 111:13028–13032

84. Frayret JP, Pointeau R, Caprani A (1981) Anodic behavior of titanium in acidic chloride containing media (HCl-NaCl) – influence of the constituents of the medium.1. Study of the stationary current calculation of the overall reaction orders. Electrochim Acta 26:1783–1788
85. Allam NK, Shankar K, Grimes CA (2008) Photoelectrochemical and water photoelectrolysis properties of ordered TiO_2 nanotubes fabricated by Ti anodization in fluoride-free HCl electrolytes. J Mat Chem 18:2341–2348
86. Schultze JW, Lohrengel MM, Ross D (1983) Nucleation and growth of anodic oxide films. Electrochim Acta 28:973–984
87. Samuni A (1972) Precursors of metal-complexed hydroperoxyl radical. J Phys Chem 76:2207–2213
88. Sever RR, Root TW (2003) DFT study of solvent coordination effects on titanium-based epoxidation catalysts. part two: reactivity of titanium hydroperoxo complexes in ethylene epoxidation. J Phys Chem B 107:4090–4099
89. Wu JM (2004) Low-temperature preparation of titania nanorods through direct oxidation of titanium with hydrogen peroxide. J Crystal Growth 269:347–355
90. Arabatzis IM, Falaras P (2003) Synthesis of porous nanocrystalline TiO_2 foam. Nano Lett 3:249–251
91. Raja KS, Misra M, Paramguru K (2005) Formation of self-ordered nanotubular structure of anodic oxide layer on titanium. Electrochim Acta 51:154–165
92. Haber F (1900) On autoxidation and its connection to the theory of the ions and the galvanic cells. Z Electrochem 7:441–448
93. Mor GK, Varghese OK, Paulose M, Ong KG, Grimes CA (2006) Fabrication of hydrogen sensors with transparent titanium oxide nanotube-array thin films as sensing elements. Thin Solid Films 496:42–48
94. Ramo J, Saarinen K, Sillanpaa M (2002) Uniform corrosion of titanium in alkaline hydrogen peroxide conditions: influence of transition metals and inhibitors calcium and silicate. Mater Corros 53:898–901
95. Panaitescu E, Richter C, Menon L (2008) A study of titania nanotube synthesis in chloride-ion-containing media. J Electrochem Soc 155:E7–E13
96. Cai Q, Yang L, Yu Y (2006) Investigations on the self-organized growth of TiO_2 nanotube arrays by anodic oxidization. Thin Solid Films 515:1802–1806
97. Taveira LV, Macák JM, Tsuchiya H, Dick LFP, Schmuki P (2005) Initiation and growth of self-organized TiO_2 nanotubes anodically formed in NH4F/(NH4) 2SO4 electrolytes. J Electrochem Soc 152:B405–B410
98. Lu N, Quan X, Li JY, Chen S, Yu HT, Chen GH (2007) Fabrication of boron-doped TiO_2 nanotube array electrode and investigation of its photoelectrochemical capability. J Phys Chem C 111:11836–11842
99. Taveira LV, Macák JM, Sirotna K, Dick LFP, Schmuki P (2006) Voltage oscillations and morphology during the galvanostatic formation of self-organized TiO_2 nanotubes. J Electrochem Soc 153:B137–B143
100. Macak JM, Sirotna K, Schmuki P (2005) Self-organized porous titanium oxide prepared in Na_2SO_4/NaF electrolytes. Electrochim Acta 50:3679–3684
101. Hahn R, Ghicov A, Salonen J, Lehto VP, Schmuki P (2007) Carbon doping of self-organized TiO_2 nanotube layers by thermal acetylene treatment. Nanotechnol 18:105604 (4pp)
102. Jaroenworaluck A, Regonini D, Bowen CR, Stevens R, Macro DA (2007) Micro and nanostructure of TiO_2 anodised films prepared in a fluorine-containing electrolyte. J Mater Sci 42:6729–6734
103. Yang DJ, Kim HG, Cho SJ, Choi WY (2008) Thickness-conversion ratio from titanium to TiO_2 nanotube fabricated by anodization method. Mater Lett 62:775–779
104. Yang DJ, Kim HG, Cho SJ, Choi WY (2008) Vertically oriented titania nanotubes prepared by anodic oxidation on Si substrates. IEEE Trans Nanotechnol 7:131–134

105. Macak JM, Taveira LV, Tsuchiya H, Sirotna MJ, Schmuki P (2006) Influence of different fluoride containing electrolytes on the formation of self-organized titania nanotubes by Ti anodization. J Electroceram 16:29–34
106. Christophersen M, Carstensen J, Voigt K, Foll H (2003) Organic and aqueous electrolytes used for etching macro- and mesoporous silicon. Phys Status Solidi A 197:34–38
107. Liu Z, Zhang X, Nishimoto S, Jin M, Tryk DA, Murakami T, Fujishima A (2008) Highly ordered TiO_2 nanotube arrays with controllable length for photoelectrocatalytic degradation of phenol. J Phys Chem C 112:253–259
108. Furneaux RC, Rigby WR, Davidson AP (1989) The formation of controlled-porosity membranes from anodically oxidized aluminum. Nature 337:147–149
109. Varela H, Torresi RM, Buttry DA (2000) Mixed cation and anion transport during redox cycling of a self-doped polyaniline derivative in nonaqueous media. J Electrochem Soc 147:4217–4223
110. Pletnev MA, Shirobokov IB, Ovechkina OE, Reshetnikov SM (1995) Effect of tetraalkylammonium salts on the cathodic evolution of hydrogen in concentrated acidic bromide solutions. Prot Met 31:317–320
111. Oguzie EE, Okolue BN, Ebenso EE, Onuoha GN, Onuchukwu AI (2004) Evaluation of the inhibitory effect of methylene blue dye on the corrosion of aluminium in hydrochloric acid. Mater Chem Phys 87:394–401
112. Yoriya S, Paulose M, Varghese OK, Mor GK, Grimes CA (2007) Fabrication of vertically oriented TiO_2 nanotube arrays using dimethyl sulfoxide electrolytes. J Phys Chem C 111:13770–13776
113. López CM, Choi KS (2005) Enhancement of electrochemical and photoelectrochemical properties of fibrous Zn and ZnO electrodes. Chem Commun 26:3328–3330
114. Antato C, López CM, Choi KS (2007) Synthesis and characterization of polycrystalline Sn and SnO_2 films with wire morphologies. Electrochem Commun 9:1519–1524
115. Paulose M, Prakasam HE, Varghese OK, Peng L, Popat KC, Mor GK, Desai TA, Grimes CA (2007) TiO_2 nanotube arrays of 1000 μm length by anodization of titanium foil: phenol red diffusion. J Phys Chem C 111:14992–14997
116. Wang J, Lin Z (2008) Freestanding TiO_2 nanotube arrays with ultrahigh aspect ratio via electrochemical anodization. Chem Mater 20:1257–1261
117. Yoriya S, Mor GK, Sharma S, Grimes CA (2008) Synthesis of ordered arrays of discrete, partially crystalline titania nanotubes by Ti anodization using diethylene glycol electrolytes. J Mater Chem 18:3332–3336
118. Grimes CA (2007) Synthesis and application of highly ordered arrays of TiO_2 nanotubes. J Mater Chem 17:1451–1457
119. Chanmanee W, Watcharenwong A, Chenthamarakshan RC, Kajitvichyanukul P, de Tacconi NR, Rajeshwar K (2007) Titania nanotubes from pulse anodization of titanium foils. Electrochem Commun 9:2145–2149
120. Yin Y, Jin Z, Hou F, Wang X (2007) Synthesis and morphology of TiO_2 nanotube arrays by anodic oxidation using modified glycerol-based electrolytes. J Am Ceram Soc 90:2384–2389
121. Kaneco S, Chen Y, Westerhoff P, Crittenden JC (2007) Fabrication of uniform size titanium oxide nanotubes: impact of current density and solution conditions. Scri Mater 56:373–376
122. Fraunholcz N (2004) Separation of waste plastics by froth flotation – a review Part 1. Miner Eng 17:261–268
123. Jasper JJ (1972) The surface tension of pure liquid compounds. J Phys Chem Ref Data 1:841–1010
124. Feng X, Macak JM, Schmuki P (2007) Robust self-organization of oxide nanotubes over a wide pH range. Chem Mater 19:1534–1536
125. Watcharenwong A, Chanmanee W, de Tacconi NR, Chenthamarakshan CR, Kajitvichyanukul P, Rajeshwar K (2007) Self-organized TiO_2 nanotube arrays by anodization of Ti substrate: effect of anodization time, voltage and medium composition on oxide morphology and photoelectrochemical response. J Mater Res 22:3186–3195

126. Yu-Xin Y, Zheng-Guo J, Feng H (2007) Fabrication and properties of TiO_2 nanotube arrays using glycerol–DMSO–H_2O electrolyte. Acta Phys Chim Sin 23:1797–1802
127. Mor GK, Varghese OK, Paulose M, Grimes CA (2005) Transparent highly ordered TiO_2 nanotube arrays via anodization of titanium thin films. Adv Funct Mater 15:1291–1296
128. Beranek R, Hildebrand H, Schmuki P (2003) Self-organized porous titanium oxide prepared in H_2SO_4/HF electrolytes. Electrochem Solid State Lett 6:B12–B14
129. Paulose M, Mor GK, Varghese OK, Shankar K, Grimes CA (2006) Visible light photoelectrochemical and water-photoelectrolysis properties of titania nanotube arrays. J Photochem Photobiol A 178:8–15
130. Patermarakis G, Karayannis HS (1995) The mechanism of growth of porous anodic Al_2O_3 films on aluminum at high film thicknesses. Electrochim Acta 40:2647–2656
131. Leenheer AJ, Miedaner A, Curtis CJ, van Hest MFAM, Ginley DS (2007) Fabrication of nanoporous titania on glass and transparent conducting oxide substrates by anodization of titanium films. J Mater Res 22:681–687
132. Parkhutik VP, Shershulsky VI (1992) Theoretical modeling of porous oxide-growth on aluminum. J Phys D 25:1258–1263
133. Macdonald DD (1993) On the formation of voids in anodic oxide-films on aluminum. J Electrochem Soc 140:L27–L30
134. Diggle JW, Downie TC, Goulding CW (1970) Dissolution of porous oxide films on aluminium. Electrochim Acta 15:1079–1084
135. Wood GC, O'Sullivan JP (1970) Anodizing of aluminium in sulphate solutions. Electrochim Acta 15:1865–1870
136. Patermarakis G, Lenas P, Papayiannis G (1991) Kinetics of growth of porous anodic Al_2O_3 films on Al metal. Electrochim Acta 36:709–725
137. Li AP, Muller F, Birner A, Nielsch K, Gosele U (1998) Hexagonal pore arrays with a 50–420 nm interpore distance formed by self-organization in anodic alumina. J Appl Phys 84:6023–6026
138. Jessensky O, Muller F, Gosele U (1998) Self-organized formation of hexagonal pore arrays in anodic alumina. Appl Phys Lett 72:1173–1175
139. Delplancke JL, Winand R (1988) Galvanostatic anodization of titanium.2. Reactions efficiencies and electrochemical-behavior model. Electrochim Acta 33:1551–1559
140. Sul YT, Johansson CB, Jeong Y, Albrektsson T (2001) The electrochemical oxide growth behaviour on titanium in acid and alkaline electrolytes. Med Eng Phys 23:329–346
141. Hwang BJ, Hwang JR (1993) Kinetic-model of anodic-oxidation of titanium in sulfuric acid. J Appl Electrochem 23:1056–1062
142. Siejka J, Ortega C (1977) O-18 study of field-assisted pore formation in compact anodic oxide-films on aluminum. J Electrochem Soc: Solid State Sci Technol 124:883–891
143. Thompson GE (1997) Porous anodic alumina: fabrication, characterization and applications. Thin Solid Films 297:192–201
144. Lohrengel MM (1993) Thin anodic oxide layers on aluminum and other valve metals – high-field regime. Mater Sci Eng R Rep 11:243–294
145. Pakes A, Thompson GE, Skeldon P, Morgan PC (2003) Development of porous anodic films on 2014–T4 aluminium alloy in tetraborate electrolyte. Corros Sci 45:1275–1287
146. Chen SG, Paulose M, Ruan C, Mor GK, Varghese OK, Kouzoudis D, Grimes CA (2006) Electrochemically synthesized CdS nanoparticle-modified TiO_2 nanotube-array photoelectrodes: preparation, characterization, and application to photoelectrochemical cells. J Photochem Photobiol 177:177–184
147. Melody B, Kinard T, Lessner P (1998) The non-thickness-limited growth of anodic oxide films on valve metals. Electrochem Solid State Lett 1:126–129
148. Krembs GM (1963) Residual tritiated water in anodized tantalum films. J Electrochem Soc 110:938–940
149. Li YM, Young L (2001) Non-thickness-limited growth of anodic oxide films on tantalum. J Electrochem Soc 148:B337–B342

150. Lu Q, Hashimoto T, Skeldon P, Thompson GE, Habazaki H, Shimizu K (2005) Nanoporous anodic niobium oxide formed in phosphate/glycerol electrolyte. Electrochem Solid State Lett 8:B17–B20
151. Izutsu K (2002) Electrochemistry in Nonaqueous Solutions. Wiley-VCH, Germany.
152. Rice MJ, Kraus CA (1953) The viscosity and the conductance-viscosity product of electrolyte solutions in bromine. Proc Natl Acad Sci 39:1118–1124
153. Lee W, Ji R, Gosele U, Nielsch K (2006) Fast fabrication of long-range ordered porous alumina membranes by hard anodization. Nature Mater 5:741–747
154. Guntherschulze A, Betz H (1952) Electrolytkondensatoren. Hebert Cram, Stuttgart
155. Cabrera N, Mott NF (1948) Theory of the oxidation of metals. Rep Prog Phys 12:163–184
156. Verwey EJW (1935) Electrolytic conduction of a solid insulator at high fields – The formation of the anodic oxide film on aluminium. Phys 2:1059–1063
157. Zhang L, Macdonald DD, Sikora E, Sikora J (1998) On the kinetics of growth of anodic oxide firms. J Electrochem Soc 145:898–905
158. Olsson COA, Verge MG, Landolt D (2004) EQCM study of anodic film growth on valve metals. J Electrochem Soc 151:B652–B660
159. Cattarin S, Musiani M, Tribollet B (2002) Nb electrodissolution in acid fluoride medium – Steady-state and impedance investigations. J Electrochem Soc 149:B457–B464
160. Kalugin ON, Lebed AV, Vyunnik IN (1998) Properties of 1–1 electrolytes solutions in ethylene glycol at temperatures from 5 to 175 degrees C – Part 2 Limiting ion conductances and ion-molecule interactions. J Chem Soc Faraday Trans 94:2103–2107
161. Shirobokov IB, Pletnev MA, Povolyako TA, Reshetnikov SM (1995) On the role of the structurization of solvent in the inhibition of cathodic evolution of hydrogen. Prot Met 31:516–519

Chapter 2
Material Properties of TiO_2 Nanotube Arrays: Structural, Elemental, Mechanical, Optical, and Electrical

2.1 Introduction

In this chapter, we examine the structural, elemental, crystallization, optical, electrical, and mechanical properties of the anodization-synthesized titania nanotube arrays.

2.2 Structural and Elemental Characterization

It is known that the as-fabricated nanotube arrays have an amorphous crystallographic structure. Upon annealing at elevated temperatures in an oxygen ambient, the nanotube walls transform into anatase phase, and a layer of metal underneath the nanotubes converts into rutile [1–9]; the observed crystalline phases are polycrystalline. We make note of a publication where the authors mistook the diffraction pattern of a selected small area, determined using transmission electron microscopy (TEM), as representing a single-crystal nanotube [10]. Titania properties depend on the crystallinity and isomorph type, and hence the utility of their application also varies. For example, anatase phase is preferred in charge-separating devices such as dye-sensitized solar cells (DSCs) and in photocatalysis, while rutile is used predominantly in gas sensors and as dielectric layers. Of the titania polymorphs, rutile has minimum free energy, and hence given the necessary activation energy, other polymorphs including anatase transform into rutile through a first-order phase transformation. The temperature at which metastable anatase converts to rutile depends upon several factors including the presence of impurities, feature size, texture, and strain. Hence with sintering, porosity and/or surface area reduction occur due to nucleation-growth type of phase transformations [11–13].

Glancing angle X-ray diffraction (GAXRD) patterns of a 20 V HF aqueous electrolyte sample annealed at different temperatures in dry oxygen are shown in Fig. 2.1 [1]. The anatase phase begins to appear at 280°C. As the sample annealed at

C.A. Grimes and G.K. Mor, *TiO_2 Nanotube Arrays*,
DOI 10.1007/978-1-4419-0068-5_2, © Springer Science+Business Media, LLC 2009

250°C was amorphous, only reflections from the titanium support can be seen, but it is clear that the sample begins to crystallize in anatase phase at a temperature between 250 and 280°C. At a temperature near 430°C, the rutile phase appears in the X-ray diffraction pattern. Beyond this temperature, the rutile (110) peak grows, whereas the anatase (101) peak from the nanotube arrays, only a few hundred nanometers long, diminishes. Complete transformation to rutile occurs in the temperature range 620–680°C. It can also be seen from Fig. 2.1 that the reflection from the titanium support is reduced at temperatures between 430 and 580°C, as the rutile barrier layer develops, fully vanishing at around 680°C.

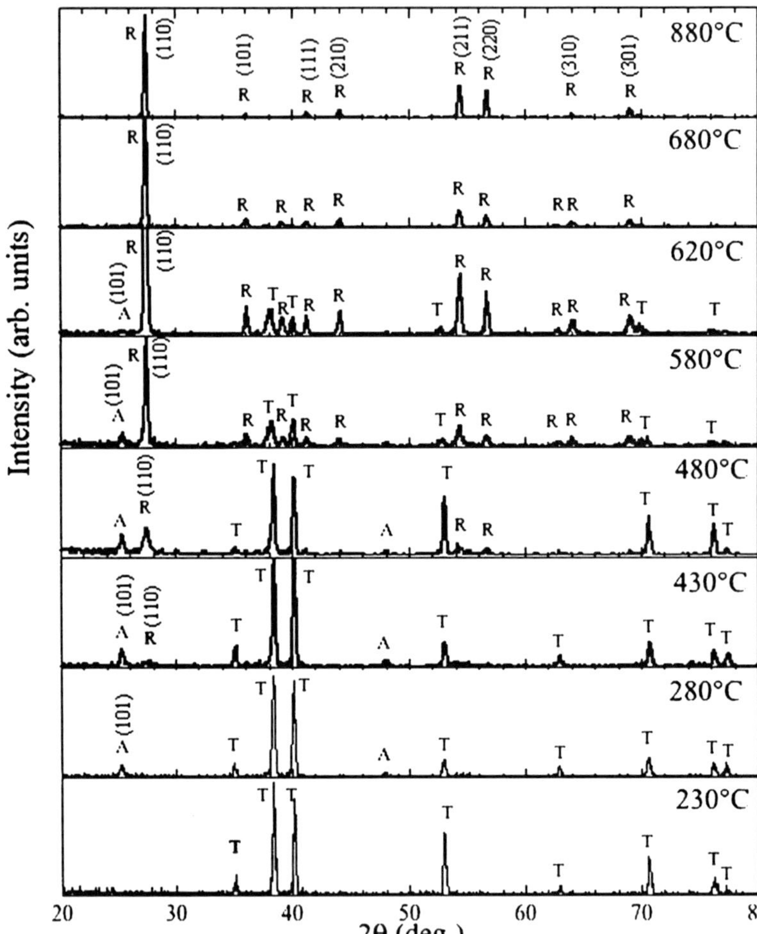

Fig. 2.1 GAXRD patterns of the nanotube samples annealed at temperatures ranging from 230 to 880°C in dry oxygen ambient for 3 h. A, R, and T represent anatase, rutile, and titanium, respectively [1]

The anatase grain size initially increases with temperature, decreasing between ~480 and 580°C, then again increasing beyond 580°C. The rutile grain size progressively increases with temperature after its nucleation. For a 3 h anneal, at 430°C a rutile fraction of 31% compared to anatase was formed, increasing to 75% at 480°C, and 92% at 580°C. For nanotube arrays atop Ti foil, 20 V HF aqueous electrolyte, the nanotube array structure was found to be stable up to around 580°C (for a slightly shorter 10-V sample, it is about 500°C). No discernible change in the pore diameter or wall thickness was observed even after annealing for 3 h at this temperature. Depending upon the sample, it was observed that at temperatures between 550 and 580°C, small protrusions, due to underlying oxide growth, come out through the porous structure. A 20-V sample was annealed in oxygen ambient to 820°C at a heating and cooling rate of 10°C/min; the tubular structure was disturbed, but porosity remained with the walls of the nanotubes coalesced to form a worm-like appearance. High-resolution TEM (HRTEM) images of the walls indicated a crystallite length of between ~35 and ~12 nm width; all wall crystallites were found to be anatase. On comparing energy dispersive X-ray spectra (EDS) of the as-synthesized nanotubes, and nanotubes fired at 580°C in an oxygen ambient, the relative intensity of oxygen peak with respect to titanium $K\alpha$ peak increased on annealing in presence of oxygen, indicating improvement in the sample stoichiometry [1].

We find that thin-film samples with a continuous metal layer underneath the nanotubes behave in a way similar to that of the foil samples, with both rutile and anatase phases coexisting at 480°C. Thin-film samples with a discontinuous metal film underneath the tubes, e.g., taken out of the anodization bath at P5 of Fig. 1.33, generally show rutile but may not be as dependent upon the amount of remaining metal. These results indicate that rutile phase grows at the interface between the barrier layer and titanium metal, where the metal is thermally oxidized. The physical constraints imposed by the size of the nanotube walls make it difficult for the anatase crystals situated there to undergo phase transformation to rutile.

X-ray patterns of samples obtained in H_3BO_3–HNO_3–HF and HNO_3–HF baths, annealed at 550°C for 6 h with a heating and cooling rate of 1°C/min in oxygen ambient, are similar to the one observed for short nanotubes obtained via HF aqueous bath [14]. After annealing, the phase-structure of the architecture can be viewed as an anatase nanotube array atop a rutile barrier layer. In comparison, a TiO_2 film made by 550°C thermal annealing is primarily rutile phase with traces of anatase phase. The normalized reference intensity ratio (RIR) method was used to estimate the weight fraction of anatase, rutile, and titanium in the resulting samples [15]. The calculated RIR result of the H_3BO_3–HNO_3–HF prepared sample is anatase 33.6%, rutile 58.7%, and titanium 7.7%. The calculated RIR result of the HNO_3–HF-prepared sample is anatase 1.7%, rutile 66.5%, and titanium 31.7%. Considering a similar X-ray sampling depth for both samples, the higher weight percentage of titanium in the HNO_3–HF-anodized sample indicates a thinner barrier layer and shorter nanotube-array length.

X-ray photoelectron spectroscopy (XPS) was used to characterize the stoichiometry of as-prepared titania nanotube arrays, anodized at 20 V in an aqueous 0.5% HF electrolyte. Figure 2.2 shows the XPS spectra of Ti 2p and O 1s on the nanotube

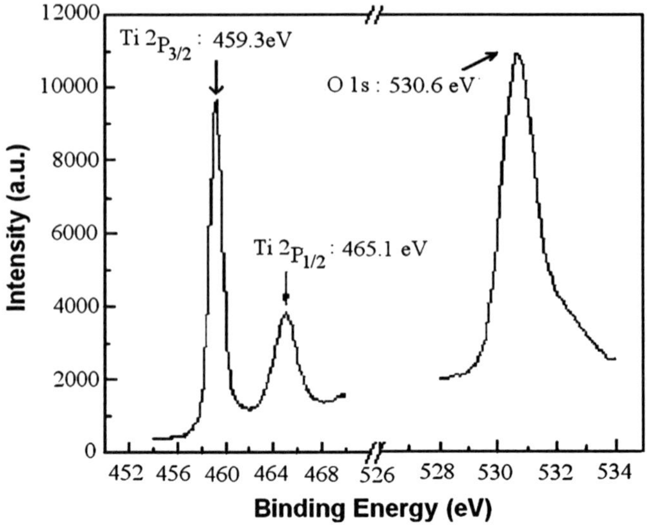

Fig. 2.2 High-resolution XPS spectra of the Ti 2p and O 1s region of as-synthesized nanotube array anodized at 20 V ([16]; Reprinted by permission of ECS –The Electrochemical Society)

array surface. Ti $2p_{3/2}$ and $2p_{1/2}$ peaks were observed at 459.3 and 465.1 eV, respectively [16], indicating Ti^{4+} [17]. The O 1s peak has a slightly asymmetric tail on the high-binding-energy side, which might be attributed to hydroxyl groups at the outermost surface as well as the adsorbed water resulting from moisture adsorption in air [16, 18].

A XPS scan for the HNO_3–HF sample, after 550°C annealing, indicated the elements Ti (23.0%), O (64.3%), N (1.0%), F (0.5%), and C (8.0%). For the H_3BO_3–HNO_3–HF electrolyte sample, 550°C annealed, XPS indicated Ti (27.4%), O (65.3%), N (0.3%), F (0.8%), and C (4.7%). Both carbon and some of the oxygen can be viewed as surface contaminates, while the small amounts of N and F originate from the anodization electrolytes. XPS analysis of the H_3BO_3–HNO_3–HF sample before annealing specifies Ti (26.9%), O (60.2%), N (1.7%), F (6.6%), and C (4.5%). Chemical state analysis indicates that the sample is comprised of Ti^{4+} bonded with oxygen (TiO_2), contaminated with N, F, and C compounds. The O1s spectra of the samples (in boric acid bath) showed a single peak at 530.8 eV. However, in the HNO_3–HF sample there is an indication of a second peak at 532 eV, revealing the presence of two forms of oxygen [14]. The Ti2$p_{3/2}$ peak has a binding energy of 459.0 eV for both samples, indicating that Ti present in the samples is in the form of TiO_2. For these measurements, the sampling depth of the X-rays is 8 nm, and therefore the Ti substrate cannot be detected. The position of $2p_{3/2}$ peak is consistent with the formation of a crystalline TiO_2 [19, 20].

GAXRD patterns of 6-μm nanotube array samples fabricated using KF (or NaF, the two acids result in equivalent architectures) based electrolytes, with sodium hydroxide used to adjust the pH, and annealed at temperatures up to 700°C are

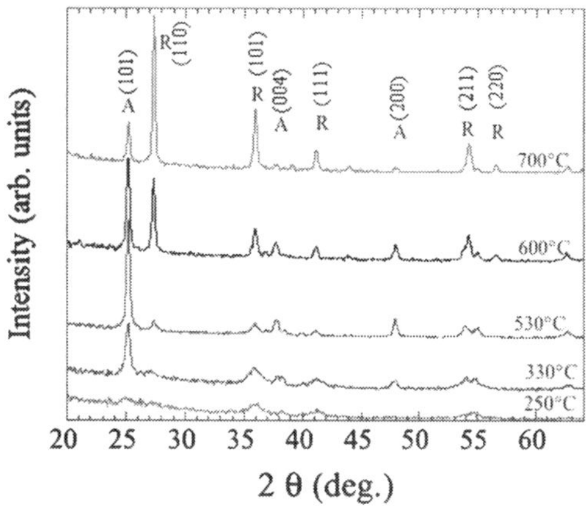

Fig. 2.3 Glancing angle X-ray diffraction patterns of a 6-μm-long nanotube array sample as a function of annealing temperature (oxygen ambient) [21]

shown in Fig. 2.3 [2, 5, 21]. Crystallization occurs in the anatase phase at a temperature near 280°C, the same as for nanotube samples prepared using an HF electrolyte [1]. Electrolyte concentration and/or pH appear to have no influence on the crystallization temperature of the nanotubes. The rutile phase appears at 530°C, with the signal increasing with annealing temperature. No sign of nanotube disruption was observed in the field emission scanning electron microscopy (FESEM) images of the samples annealed at 580°C. This is in contrast with shorter nanotubes, HF grown at pH < 1, which started disintegrating at this temperature [1]. Yang and coworkers made similar observations during the crystallization of 4.2-μm-long nanotubes (pore size of 80 nm) in the temperature range 300–800°C [9], with the length and average diameter of the nanotubes showing no discernible changes after calcination up to 500°C. However, the nanotube length decreased to 3 μm after calcination at 550°C. After calcination at higher temperatures, 600 and 700°C, the length of the nanotubes decreased to 2.8 and 1.5 μm, respectively, and at the same time a thick oxide layer grew between the titanium substrate and the nanotube arrays. At 700°C, small protrusions emerged through the nanotubes, resulting in their cracking. The nanotubular structure of the samples annealed 800°Ccompletely collapsed.

A TEM image of a single nanotube grown from a sample prepared at pH 5.0 and annealed at 600°C, see Fig. 2.4, indicates the presence of anatase, but no rutile, phase in the walls [21, 22]. The TEM data in conjunction with the GAXRD patterns confirm the crystallization model proposed in our earlier work on short nanotubes prepared using HF electrolyte [1, 3]. According to this model, anatase crystals are formed at the nanotube–Ti substrate interface as a result of elevated temperature metal oxidation. The rutile crystallites originate in the oxide layer (formed by the oxidation of titanium metal) through phase transformation of anatase crystallites existing in the region. The constraints imposed by the nanotube walls prevent anatase crystallites in the nanotube walls from transitioning to the rutile phase [2, 21, 23]. Therefore, nanotubes, annealed at temperatures between 530 and 580°C, can be considered as anatase crystallites stacked in cylindrical shape on a rutile foundation.

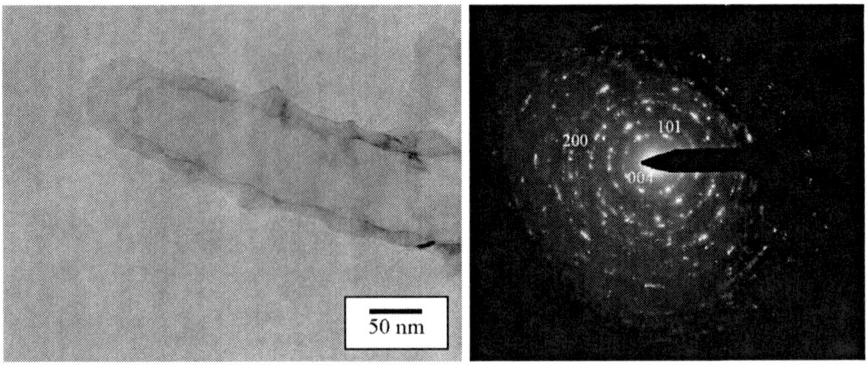

Fig. 2.4 TEM image of a single nanotube from a sample prepared at pH 5.0 and annealed at 600°C, with corresponding diffraction pattern [21]

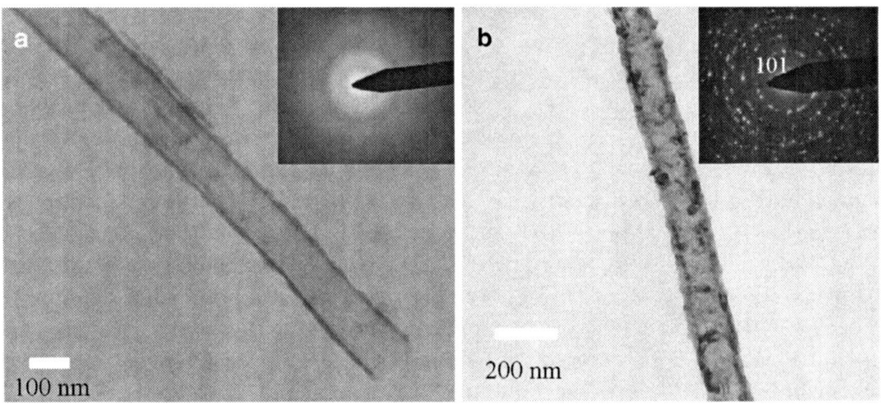

Fig. 2.5 High-resolution TEM images of DMSO 40-V nanotube sample: (**a**) as-prepared, and (**b**) after 550°C anneal in oxygen for 6 h. The corresponding selected area diffraction patterns are given in the *insets* [24]

Figure 2.5 shows HRTEM images of a portion of a nanotube~45 μm long grown at 40 V in DMSO containing 2% HF [24]. Figure 2.5a is an image of an as-prepared nanotube, and Fig. 2.5b that of a nanotube annealed at 550°C in oxygen. The corresponding diffraction patterns from the nanotubes are given in the inset. Clearly, the as-prepared nanotubes are amorphous. Consistent with the earlier studies regarding the nanotube array crystallization [1], the sample annealed at 550°C shows only the anatase phase in the nanotube's walls. GAXRD analysis confirms the TEM studies that the as-anodized samples are amorphous. Figure 2.6 is GAXRD of a 40-V ethylene glycol, 14-μm-long 4-h 580°C annealed nanotube array sample exhibiting anatase peaks [24, 25]. While the barrier layer underlying the nanotube array remains rutile [1], no rutile can be seen in the spectra since the tubes are of such great length. Identical GAXRD spectra are obtained for any of the nanotube array samples fabricated using the polar organic electrolytes described in Chap. 1.

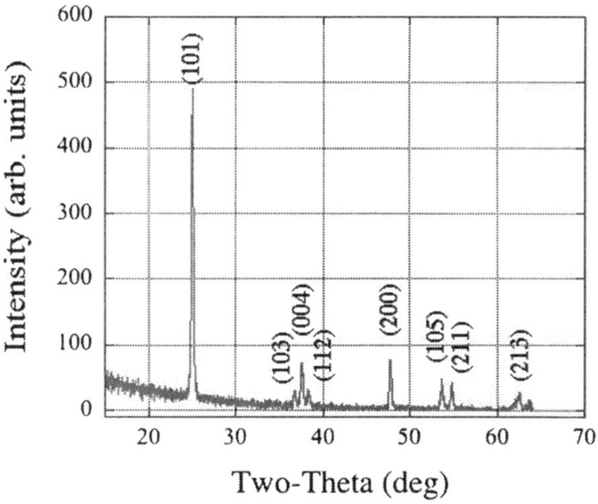

Fig. 2.6 GAXRD pattern of a nanotube array sample anodized at 40 V in an ethylene glycol electrolyte containing 0.25% wt NH_4F, annealed in oxygen at 580°C for 4 h [24]

2.2.1 *Anodic Formation of Crystalline Metal Oxide Nanotubes*

Highly crystalline structures offer unique advantages over amorphous architectures by providing a direct and rapid pathway for charge transport, thus decreasing the carrier path length, which in turn reduces recombination losses [27, 28]. The need for high-temperature crystallization limits the use of nanotube arrays with temperature-sensitive materials, such as polymers. One approach to achieving crystalline nanotube arrays is to start with a crystalline layer, and then anodize the film to achieve crystalline nanotubes [26]. Titanium foil samples were first treated with an oxidizing agent-containing electrolyte, e.g., H_2O_2 at 80°C or $(NH_4)_2S_2O_8$ at 120°C, then anodized at constant voltage in a fluoride-containing electrolyte [26]. A schematic of this two-step process is shown in Fig. 2.7. The two steps are not symmetrical; treating as-synthesized amorphous TiO_2 nanotube arrays with the oxidizing agent-containing electrolyte destroys the nanotubes.

After the initial treatment of the Ti foil sample in a peroxide electrolyte at 80°C, the resulting films were then anodized in aqueous solutions containing NH_4F [26]. Figure 2.8 shows the GAXRD results for an as-anodized sample. The inset of Fig. 2.8 shows an FESEM top-view image of the nanotube array architecture. Figure 2.9a shows a TEM micrograph for a similarly fabricated nanotube array sample, and Fig. 2.9b the corresponding diffraction pattern which indicates a best fit to anatase. Figure 2.9c gives the intensity from a rotationally averaged pattern with expected Bragg lines from anatase [26].

The 80°C peroxide pretreatment of the Ti foil produces a crystalline oxide layer~1.2 μm thick. Subsequent anodization in a fluoride-containing electrolyte

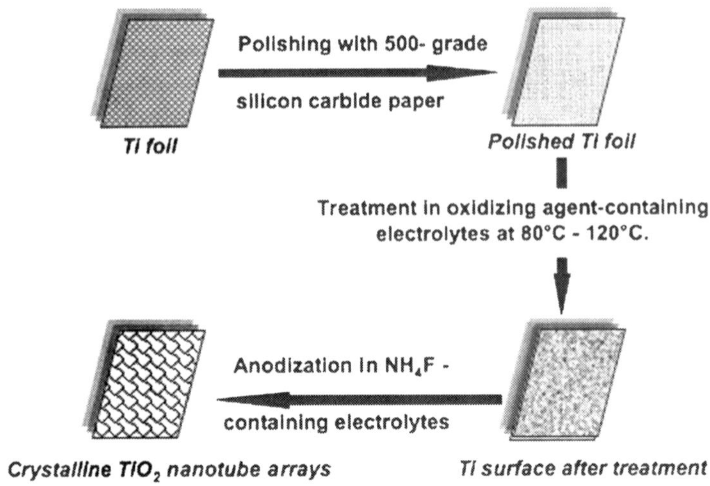

Fig. 2.7 Schematic presentation of the two-step fabrication process used to directly synthesize crystalline TiO_2 nanotube arrays [26]

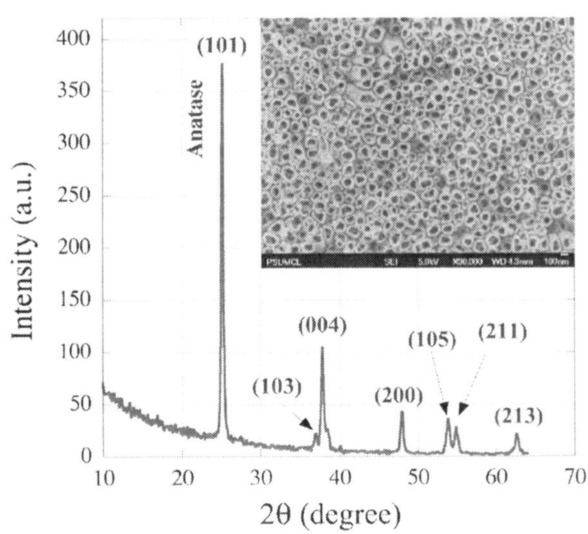

Fig. 2.8 GAXRD pattern of well-developed nanotube arrays formed by anodizing a peroxide-treated Ti foil sample in an aqueous electrolyte containing 0.25 M NH_4F+0.1 M H_3PO_3+0.05 M H_2O_2;the *inset* shows a FESEM top-view image of this same sample [26]

initially produces a crystalline nanotube array structure by structuring of the crystalline oxide layer. As the anodization continues with time, due to field-assisted oxidation and dissolution, as well as chemical dissolution, the Ti–oxide interface gradually moves deeper into the Ti metal and the outer surface dissolves. Therefore, with increasing anodization duration (longer than 3 h) the nanotube growth

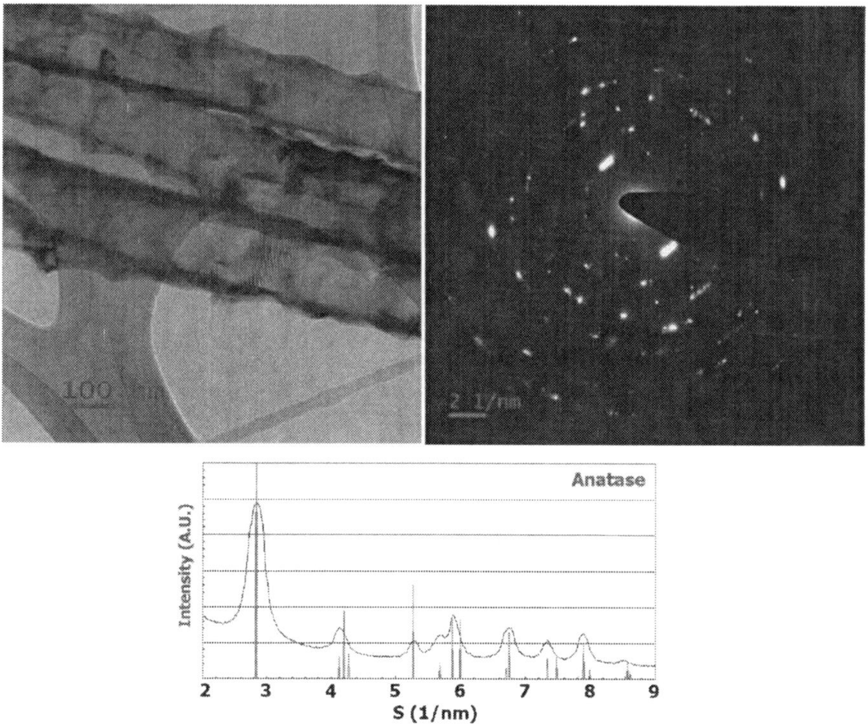

Fig. 2.9 (**a**) TEM image of nanotubes formed by described technique; (**b**) corresponding selected area diffraction pattern; and (**c**) intensity from rotationally averaged pattern compared with expected Bragg lines for anatase [26]

proceeds through the crystalline oxide layer initially present, and is then replaced with an amorphous nanotube array structure; hence the GAXRD-observed crystallinity of the nanotube arrays decreases for extended anodization durations. This anodization time restriction limits the crystalline tube length to about 800 nm. Strategies to overcome this limitation include the use of stronger oxidizing agents to produce a thicker crystalline oxide layer, which is then converted into nanotubular architecture by subsequent anodization. To this end, we immersed Ti foil samples in ammonium persulfate, $(NH_4)_2S_2O_8$, which is a stronger oxidizing agent, at 120°C for 3 h. The 120°C ammonium persulfate pretreatment of the Ti foil produces a crystalline oxide layer approximately 1.6μm thick; temperatures lower than 120°C resulted in thinner films, and higher temperatures resulted in an insulator-like film which we were unable to anodize. Subsequent anodization of the resulting samples in aqueous solutions containing NH_4F enables fabrication of nanotube arrays 1.4 μm in length. Figure 2.10 shows the GAXRD results for an as-anodized persulfate sample, indicating that the nanotube arrays are largely anatase, and the inset of Fig. 2.10 shows an FESEM top-view image of the nanotube array architecture achieved with the persulfate sample [26].

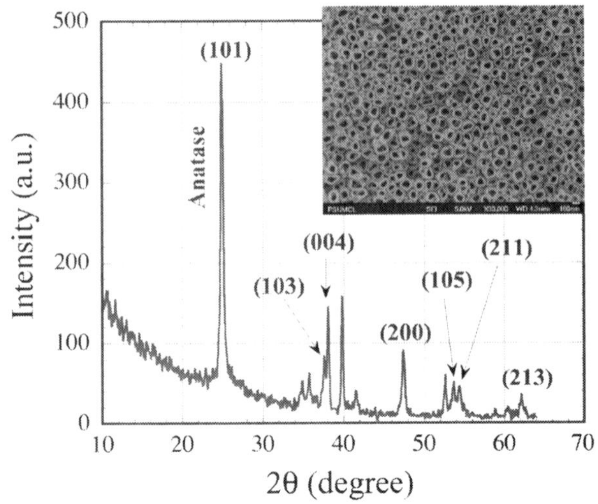

Fig. 2.10 GAXRD pattern of well-developed nanotube array sample formed by anodizing a persulfate-treated Ti foil sample in an aqueous electrolyte containing 0.25 M NH_4F+0.1 M H_3PO_3+0.05 M H_2O_2; *inset* shows FESEM top-view image of similar sample [26]

2.2.2 *Improved Crystallization via Solvothermal Treatment*

Thin-wall nanotube array architectures offer high surface area, facile diffusion of the photogenerated holes to the electrolyte, and directed charge transfer of the photogenerated electrons to the back contact for hydrogen evolution [29]. A high degree of crystallization appears a prerequisite for useful photocatalytic materials, with crystallinity playing a key role in charge separation and transport [30, 31]. In general, higher crystallinity implies fewer amorphous regions and grain boundaries, and hence a lower density of trap states. The anodically grown titania nanotube arrays are amorphous and hence exhibit negligible photocatalytic activity. High-temperature crystallization anneals, in an oxygen ambient to maintain stoichiometry, result in the growth of a dense rutile layer from the metal film underlying the nanotube arrays, introducing a significant series resistance that hinders transport of the photogenerated electrons to the back contact [4]. Solvothermal methods are well known for crystal size enhancement [32, 33]. Reactions take place in a closed system under autogeneous pressure, which can be varied by choice of solvent, reaction temperature, and volume filling of the reaction vessel [34, 35]. When solvothermal treatments were performed on water-immersed titania nanotube array samples, we observed disorientation of the vertically aligned nanotubes and detachment of the tubes from the underlying substrate. These problems were eliminated when ethanol [36, 37]was used instead of water, with the samples exposed to the resulting ethanol vapor rather than keeping them in physical contact with the liquid medium.

The ethanol vapor treatment was carried out in a 23-ml Teflon-lined stainless steel chamber. Optimal enhanced crystallization conditions were found with a

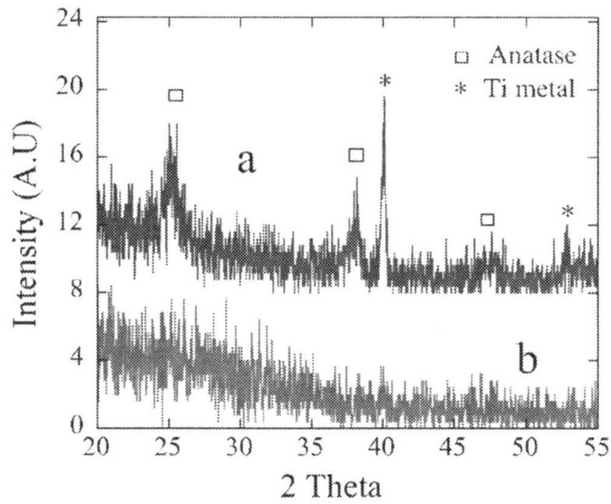

Fig. 2.11 GAXRD of (**a**) ethanol vapor-treated titania nanotube array sample; (**b**) as-anodized amorphous titania nanotube array sample

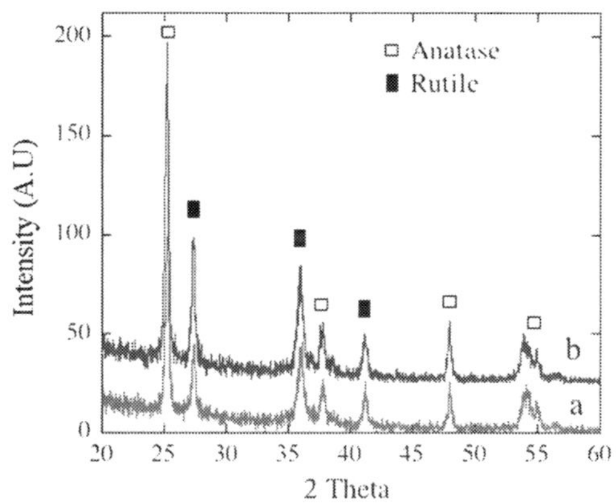

Fig. 2.12 GAXRD patterns of (**a**) 580°C annealed titania nanotube array sample without solvothermal treatment; (**b**) sample thermally annealed and then subjected to ethanol vapor treatment

reaction temperature of 140°C and 25% filling volume, corresponding to a reaction chamber pressure of approximately 50 psi. Hence for the treatment of all samples, 6 ml of ethanol was placed within the Teflon beaker, with the nanotube array sample mounted on the top of a grooved Teflon stub to avoid direct contact with the liquid medium. The sealed chamber was then kept at 140°C for 18 h.

Figure 2.11 shows the GAXRD patterns of an as-anodized titania nanotube array sample and a similar sample after being subjected to the ethanol vapor treatment. It is evident that the 140°C vapor treatment results in some minimal degree of crystallization. Figure 2.12 shows the GAXRD patterns of a thermally annealed sample and a vapor-treated, thermally annealed sample (samples of the same dimensions and identical instrument conditions were used for recording the spectra). The patterns from both samples show identical peak positions and the presence of both anatase and rutile phases. All major reflections from the vapor-treated, thermally annealed sample are stronger than those of the untreated sample, indicating a significant increase in the degree of crystallization with the ethanol vapor treatment.

2.2.3 *Partially Crystalline Anatase Phase Nanotubes by Anodization*

As-synthesized nanotube arrays are commonly amorphous. Hence Fig. 2.13, which is the GAXRD pattern of an as-anodized sample prepared from DEG+0.5 wt% NH_4F+2% H_2O electrolyte anodized at 80 V for 48 h at room temperature, is of considerable interest since the anatase (101) peak indicates partial crystallization of the tubes [38]. The anatase (101) intensity significantly increases with an increase of NH_4F concentration to 0.9 wt%, with a distinct anatase (004) peak, see Fig. 2.14. However, at this higher NH_4F concentration, the electrolyte is not able to maintain the nanotube structure, and only flakes are obtained.

To achieve the nanotube structure in crystalline form, optimization of a combination of anodization parameters is required. Crystallization is strictly determined by migration of Ti^{4+} outward from the substrate and O^{2-}/OH^{-} inward to form crystalline oxide in a ratio that favors crystal growth [39]. Additionally, because crystallization of anodic titania can be correlated with ionic transport during the film growth [40, 41], it appears that the chemical properties of the solvent

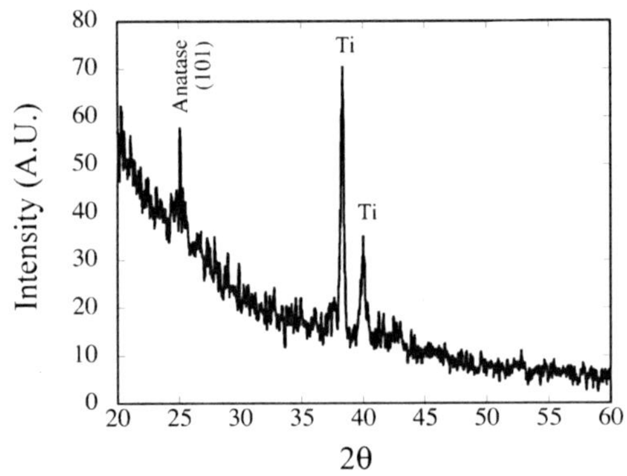

Fig. 2.13 GAXRD pattern of an as-anodized sample obtained from DEG –0.5 wt% NH_4F–2%H_2O electrolyte at room temperature [38]

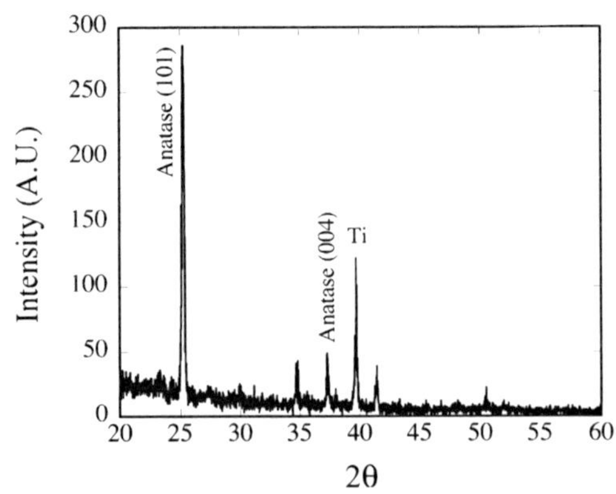

Fig. 2.14 GAXRD pattern of an as-anodized sample obtained from DEG –0.9 wt% NH_4F–2%H_2O electrolyte at room temperature [38]

diethylene glycol, such as inter- and intra-molecular hydrogen bonding [42, 43], and its physical properties, such as viscosity, play an as yet undetermined role.

2.3 Characterization of Doped Titania Nanotubes

Titania nanotube arrays appear a useful catalyst, under bandgap illumination, for water photoelectrolysis [44–48]. Semiconducting TiO_2 (3.2 eV bandgap) absorbs UV light, which comprises only~4% of the solar energy spectrum. Means to shift the absorption properties of TiO_2 have historically focused on sensitization of the TiO_2 with low bandgap semiconductors having suitable energy level positions with respect to that of TiO_2, or by doping with foreign elements such as C, N, F, B, P, and S [49–53].

2.3.1 Carbon Incorporation Within the Nanotubes

Park et al. prepared carbon-doped titania ($TiO_{2-x}C_x$) nanotubular arrays by reduction of carbon monoxide [45]. Raja et al. reported $TiO_{2-x}C_x$ nanotubular arrays prepared by acetylene reduction [46]. Hahn et al. treated the nanotubes at 500°C under a mixed flux of N_2 and acetylene to incorporate carbon [49]. Mohapatra and coworkers reported carbon-doped titania nanotube arrays prepared by the sonoelectrochemical anodization method using an ethylene glycol–ammonium fluoride solution and annealing the as-anodized sample under reducing atmosphere (H_2), which converts the amorphous nanotube arrays to photoactive anatase phase as well as helps in doping of the carbon (from the reduction of ethylene glycol) to give the $TiO_{2-x}C_x$-type photoanode [50, 54]. Shankar and coworkers subjected nanotube arrays to a flame annealing process [55]; an XPS scan of these samples revealed the presence of carbon in all samples; a summary of the carbon content and carbon state information is provided in Table 2.1 [55]. Fluorine was present in all samples, at

Table 2.1 Carbon content and chemical state information (from XPS) of flame annealed TiO_2 nanotube arrays [56]

Sample	Depth (nm)	Total C (atom%)	C–C	C–O	COO
Short NT	0	0.7	–	–	–
	100	0.3	–	–	–
Flame-annealed short NT	0	3.3	2.5	0.4	0.5
	100	2.8	1.9	0.6	0.3
Flame-annealed long NT	0	5.6	3.9	0.8	0.9
	100	5.2	3.8	0.9	0.6
Long NT	0	3.5	2.3	0.5	0.7
	100	3.0	1.9	0.6	0.5
Flame-annealed Ti foil	0	4.0	2.7	0.6	0.7
	100	3.8	2.5	0.7	0.5

a surface concentration of nearly 2 atomic%, decreasing to about 0.2 atom% in the interior [56]. Based on the analysis of the C1s peak, incorporated carbon was present in C–C (285.3 eV), CO (286.5 eV), COO (289.0 eV), and C–N bonds. The Ti–C signal at 281.9 eV was not observed. The carbon content of the short nanotubes, which is initially quite small, becomes appreciable upon flame annealing [55]. A significant amount of carbon (~3%) is present in the long nanotube sample even prior to flame annealing, which is attributed to the presence of a large number of organic ions such as citrate and tetrabutylammonium in the anodizing bath. In long nanotubes, flame annealing introduces additional carbon into a structure where carbon pre-exists in appreciable quantities. Hence, flame-annealed long nanotubes have the highest carbon content (>5%) of the samples studied. While these various routes, and those mentioned below, are able to dope the titania nanotube arrays, the resulting photocurrents offer no significant improvement beyond TiO_2.

2.3.2 Nitrogen Incorporation Within the Nanotubes

Substitutional N-doping has been suggested as a route for lowering the bandgap of titania [56–59]. Nitrogen ions substitute oxygen atoms in the TiO_2 lattice and thus the corresponding N(2p) states are located above the valence band edge. Mixing of N(2p) states with O(2p) states results in a reduction of the bandgap of the N-doped TiO_2. Common approaches to achieving N-doped TiO_2 nanotubes are N_2/H_2 gas plasma treatment, annealing in ammonia gas, N-ion bombardment, N-ion doping during the fabrication of nanotubes via anodization, etc. [52, 60–63].

Efforts have been made to nitrogen dope nanotube arrays in situ during the growth and fabrication of nanotubes [63]. Titanium foils were potentiostatically anodized at 25 V in an electrolyte of pH 3.5 containing 0.4M ammonium nitrate NH_4NO_3 and 0.07M HF acid; with reference to Fig. 2.15, Sample A was removed after 17 s of anodization, while Sample B was anodized for 240 s. Sample C was

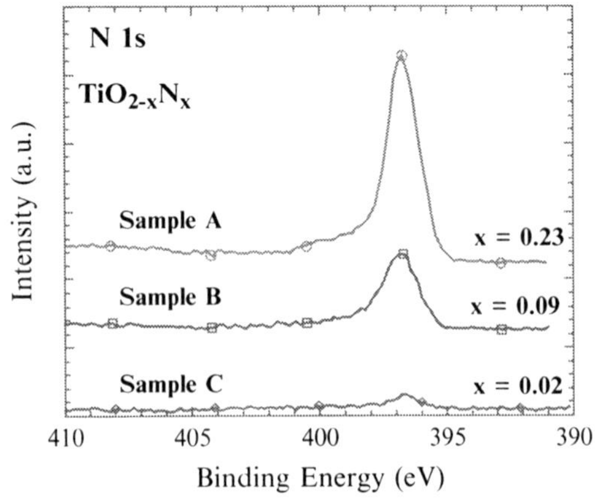

Fig. 2.15 N 1s XPS spectra for samples A–C with respective nitrogen doping levels [63]

anodized for 6 h at 20 V in an electrolyte of pH 3.5 containing 2.5M $NH_4NO_3$and 0.07M HF. Such anodization chemistry restricts the electrolytic ions to nitrogen and fluorine bearing species, allowing control of the possible elements that can be incorporated into the anodic titania films. The potential and pH regimes chosen were such as to facilitate nanotube array formation. The maximum current at the onset of the anodization was limited by the compliance of the power supply used to perform the anodization. In the first 25 s after application of the voltage, the measured current density reduced from 4,120 mA/cm^2 to a local minimum between 15 and 25 mA/cm^2, with the field-assisted oxidation of the Ti metal surface reducing the current. In the potential range under consideration, this behavior is typical for the anodization of Ti in fluoride ion-containing acidic electrolytes; however, the magnitude of the anodization currents is much greater. The larger anodization currents are attributed to the stronger oxidizing and etching action of the nitrate ion-containing electrolyte. High-resolution N 1s XPS spectra of Samples A (x=0.23), B (x=0.09), and C (x=0.02) (see [63])are shown in Fig. 2.15. XPS data confirm that all the incorporated nitrogen is substitutional on the oxygen site. The nitrogen peak at 396.8 eV was observed and assigned to atomic β–N, indicating a chemically bound N^- state [64, 65]. Fluorine was present in the amorphous as-anodized samples, with the final concentration of incorporated F^- sensitive to the annealing conditions. Annealing processes (in air) lasting longer than 6 h at temperatures above 600°C resulted in fluorine atoms being completely resubstituted by oxygen. The depth profile of a 250-nm-thick film with a surface nitrogen concentration x=0.05 (the sample was anodized at 20 V for 120 s in a pH 4.5 electrolyte containing 0.4 M ammonium nitrate and 0.07 M hydrofluoric acid then annealed per the other samples) indicated that the nitrogen doping is inhomogeneous with the maximum nitrogen being incorporated close to the surface, and then linearly decreasing with increasing depth inside the film. The precise reactions involving the decomposition of ammonium ions and nitrate ions at the anodic surface to form N-doped titania are unclear. However, the anodization of aluminum in nitric acid has been studied previously and is known to be relatively complex [66]. A study of the interaction of aluminum with nitrate ions in thin oxide films formed in nitrate ion-containing electrolytes indicated that the adsorption of nitrate ions on the oxidized surface of aluminum was followed by their reduction inside the oxide film [67]. Parhutik and coworkers [68] reported the incorporation of electrolyte anions in the anodic film formed by anodization of Al in HNO_3 solution. Furthermore, it was reported that the anion concentration in the growing oxide reaches a maximum value at the moment when intensive pore growth starts and the oxide is thin, i.e., when the anodizing time is very brief [66, 68]. Similar behavior was observed for Ti anodization, with maximum nitrogen incorporation occurring in a film anodized for a mere 17 s (Sample A). Thus, a trade-off exists between the morphology and the level of nitrogen doping. Shorter anodization periods result in higher concentrations of incorporated nitrogen, whereas longer anodization periods are required for evolution of the nanotube-array architecture.

Nitrogen doping by ion implantation of self-organized anodic titania nanotubes combined with a subsequent anneal to remove induced defects has been investigated [61, 62, 69]. Ochsner et al. [69] formed nanotube array films in a H_2SO_4/HF electrolyte, which were then annealed to an anatase structure, implanted with nitrogen using ion bombardment at 60 keV and a nominal dose of 1×10^{16} ions/cm^2, and then finally re-annealed to anatase. Images do not show any detectable morphological difference with the different steps. XRD results presented for the films before doping, after nitrogen implantation, and after re-annealing showed that ion implantation leads to amorphization of the structure and that re-annealing at 450°C for 3 h recovers the structure to anatase [61]. XPS spectra of the implanted TiO_2 tubes exhibited N 1s peaks in the implanted and annealed nanotube layer (annealed at 450°C for 3 h) after 20 nm by Ar^+ sputtering [61]. Two peaks were distinguished: one at 400 ± 0.2 eV and another one at 396 ± 0.2 eV. The peak detected at 400 eV is assigned to a well-screened γ–N state (essentially adsorbed N), and the one detected at 396 eV is assigned to an atomic β–N state such as TiN [70].Therefore, one can deduce from XPS results that nitrogen is successfully implanted in the structure in a chemically bonded state.

2.3.3 Boron-Doped Nanotubes

Boron doping was reported by use of electrodeposition, where the anodized nanotube sample was subjected to a current density of 10 μA/cm^2 for 27 min in a 0.1 M H_3BO_3 electrolyte [53, 71]. XPS spectrum showed B 1s peak at 191.4 eV for the boron-doped sample. According to the standard B 1s binding energies in TiB_2 (187.5 eV, Ti–B) and B_2O_3 (193.0 eV, B–O) [72], the B 1s binding energy of 191.4 eV might be assigned to the mixed state Ti–O–B of TiB_2 and B_2O_3, which is consistent with the published value of the B 1s signal for boron-doped TiO_2 in XPS analysis [71–73].

2.3.4 Organic Bath

XPS was used to determine the elemental composition of the nanotube array samples fabricated in an organic anodization bath, and results are summarized in Table 2.2 [74]. The nanotube array samples are predominately titanium and oxygen, with traces of fluorine and sulfur due to solvent incorporation in the anodic films. It was

Table 2.2 XPS results of titanium etched in fluorinated 1:1 DSMO and ethanol at 20 V for 48 h, one of the samples annealed at 550°C for 6 h [74]

Atomic con. (%)	Ti	O	F	N	C	S
Nanotubes	23.9	53.4	13.7	0.9	6.8	1.1
Annealed nanotubes	26.6	64.3	1.6	0.5	5.8	1.2

believed that surface contamination is the likely source for the nitrogen and carbon found in the samples. Chemical state analysis for titanium indicates that the sample is Ti^{4+} bonded with oxygen (TiO_2). Compared with samples fabricated in aqueous electrolytes, the atomic concentration of fluorine (13%) is considerably increased using the organic electrolyte [74]. However, the fluorine concentration is dramatically reduced in the annealed samples, falling to 1.6%; GAXRD gives no indication of TiO_xF_y or TiO_xS_y in the samples. Hence, while the results of XPS and XRD indicate a considerable amount of solvent is trapped in the amorphous anodic films, the trapped elements such as F, C, and S do not enter the rutile or anatase lattice.

2.3.5 *CdS-Coated Nanotubes*

In the GAXRD pattern of CdS-coated TiO_2 nanotube array films (HF fabricated, a few hundred nanometers in length) annealed at 350°C for 1 h, Chen and coworkers [75] observed a prominent TiO_2 Bragg peak along with weak Bragg reflections at 2θ values of 26.55, 30.75, 44.04, 52.16, and 54.67°, corresponding to the (111), (200), (220), (311), and (222) Bragg reflections of cubic CdS, respectively. The general scan spectrum of XPS of CdS–TiO_2 electrodes showed sharp peaks for Ti, O, Cd, S, and also C. The Cd 3d core level XPS spectrum has two peaks at 405.3 eV (3 $d_{5/2}$) and 411.9 eV (3 $d_{3/2}$), which are in good agreement with published values [76]. The S 2p core level spectrum indicated that there are two chemically distinct species in the spectrum. The peak at 161.9 eV is for sulfide, the structure occurs because of a split between $2p_{3/2}$ and $2p_{1/2}$; the split is near 1.18 eV and the area ratio is 2:1, in excellent agreement with published values of the S 2p signal for CdS [14]. Measured atomic concentrations of the as-prepared samples suggest that when the sulfate/O ratio is 1, the sulfide/Cd ratio is 0.86; this means that the CdS nanoparticles obtained are slightly Cd rich, which is expected for CdS under normal synthesis conditions [77].

2.4 Optical Properties of Titania Nanotubes Arrays

2.4.1 *Finite Difference Time Domain Simulation of Light Propagation in Nanotube Arrays*

Titania nanotube arrays can be grown over a wide range of pore diameters, wall thicknesses, and lengths, and with each topology showing different light absorption and photocatalytic properties leading to different values of photoconversion efficiency [78, 79]. Knowledge of the light absorption properties of the various nanotube-array geometries prior to sample fabrication would be desirable. Therefore, the electromagnetic computational technique known as *finite difference time*

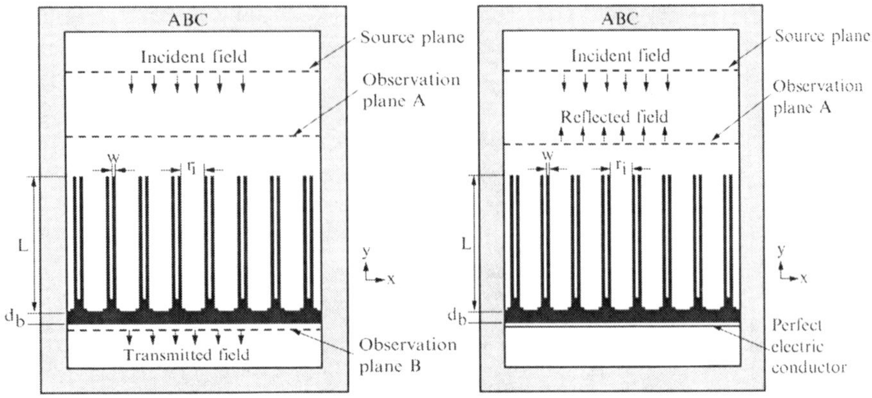

Fig. 2.16 Geometry of two-dimensional FDTD models used for determining the propagation of a transverse electromagnetic (TEM) wave through: (**a**)aself-standing titania nanotube array film (Type-I), and (**b**) titania nanotube array film on titanium substrates (Type-II) [81]

domain (FDTD) was used to simulate the light-absorbing properties of the nanotube arrays as a function of feature size and wavelength [80, 81]. The simulations were performed for titania nanotube array films with no metal layer underneath the nanotubes (transparent, Type-I) and also for the nanotubes grown on titanium foil (opaque, Type-II). Note that in the former case, Type-I, the glass substrates were not included in the simulations and hence the nanotube film can be considered self-standing. Figure 2.16 shows, respectively, the two models used to represent self-standing titania nanotube array films (transparent, Type-I) and nanotube array films on titanium foil (opaque, Type-II). The distance between two adjacent tubes was taken as 10 nm. The FDTD space was terminated with an absorbing boundary condition (ABC) made of a uniaxial perfect matching layer [82] to eliminate the reflection of any fields from these boundaries. The Type-II (Fig. 2.16b) model contains a perfect electrical conducting layer at the bottom of the nanotube array to represent the titanium layer. Therefore, in the case of Type-I, transmittance and in the case of Type-II, reflectance are used to determine the absorbance of light by the nanotube array. In all simulations reported to date, the distance between two adjacent tubes was taken as 10 nm. The validity of the FDTD simulations were established by comparison of the calculated and experimentally measured transmittance of a Type-I film of different porosities, see Fig. 2.17.

The measured absorbance spectrum of a titania nanotube array (200 nm length, 22 nm pore, 13 nm wall thickness, and 100 nm barrier layer thickness) was compared with the simulated results [83]. Both curves were found similar, except that the absorption edge of measured spectra is shifted slightly to the higher wavelength region compared to the simulated spectra. This is due to the fact that the barrier layer has rutile crystallites and the nanotube walls consist of anatase crystallites. The bandgap of the rutile is lower (3.0 eV) compared to the anatase (3.2 eV). The rutile phase at the barrier layer leads to the shifting of the absorption edge to higher wavelength, a property not taken into account by the FDTD simulations.

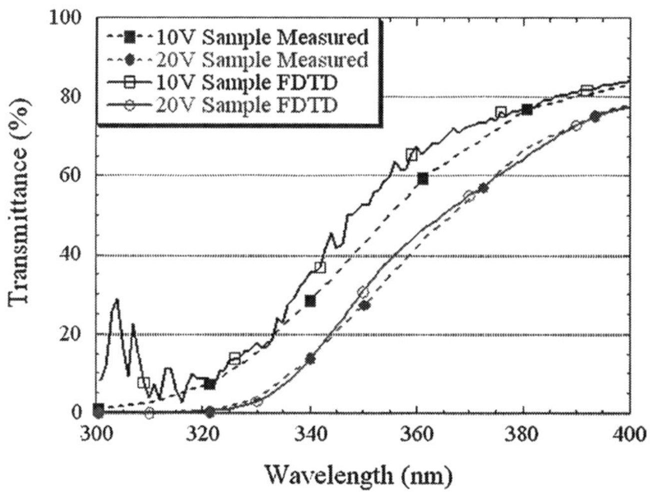

Fig. 2.17 Measured and FDTD calculated transmittance of a Type-I film made using different anodization voltages. The 10-V sample is 200 nm long, and the 20-V sample is 360 nm long [81]

As an informative demonstration, Fig. 2.18 shows the propagation of a transverse electromagnetic (TEM) wave through a titania nanotube array on Ti foil (Type-II). A derivative Gaussian wave (center frequency $= 8 \times 10^{14}$ Hz, bandwidth $= 2 \times 10^{14}$ Hz) is used as the excitation source. The tube length, pore diameter, wall thickness, and barrier layer thickness are, respectively, 1,000, 100, 20, and 100 nm. Figure 2.18a shows the wave originating from the source and moving towards the nanotubes. When the wavefront hits the top surface of the nanotube array, Fig. 2.18b, most of the incident energy is transmitted into the nanotubes, with a negligible portion reflecting back. The reflected wave can be seen in Fig. 2.18b, c as a faint horizontal line on the top of the nanotubes. The wave dissipates as it travels through the nanotube array to reach the barrier layer, Fig. 2.18c. Figure 2.18d–f shows the wave reflecting back from the conducting Ti layer at the bottom of the nanotube array. Note that the reflected wave contains multiple wavefronts, as the derivative Gaussian pulse contains radiation over a wide frequency range; the individual frequencies travel at different velocities through the barrier layer and nanotube array due to the frequency-dependent variation in the titania permittivity.

With respect to the applied properties of Type-II samples, it should be noted that to induce crystallinity the nanotube array samples are annealed at elevated temperatures in an oxygen environment. The diffusion of oxygen into Ti foil is consistent with the Fick's second law, and hence a gradient in the oxide composition exists from the top of the barrier layer to the Ti metal [83]. Consequently, there is a gradient in the complex permittivity spectrum of the oxide layer underneath the nanotubes and hence light is bent before it is reflected back from the metal. This gradient was considered during the simulation process by linearly increasing the permittivity values of the barrier layer; so the permittivity at the bottom of barrier layer is 10 times larger than at the top. Hence when light is reflected back from the

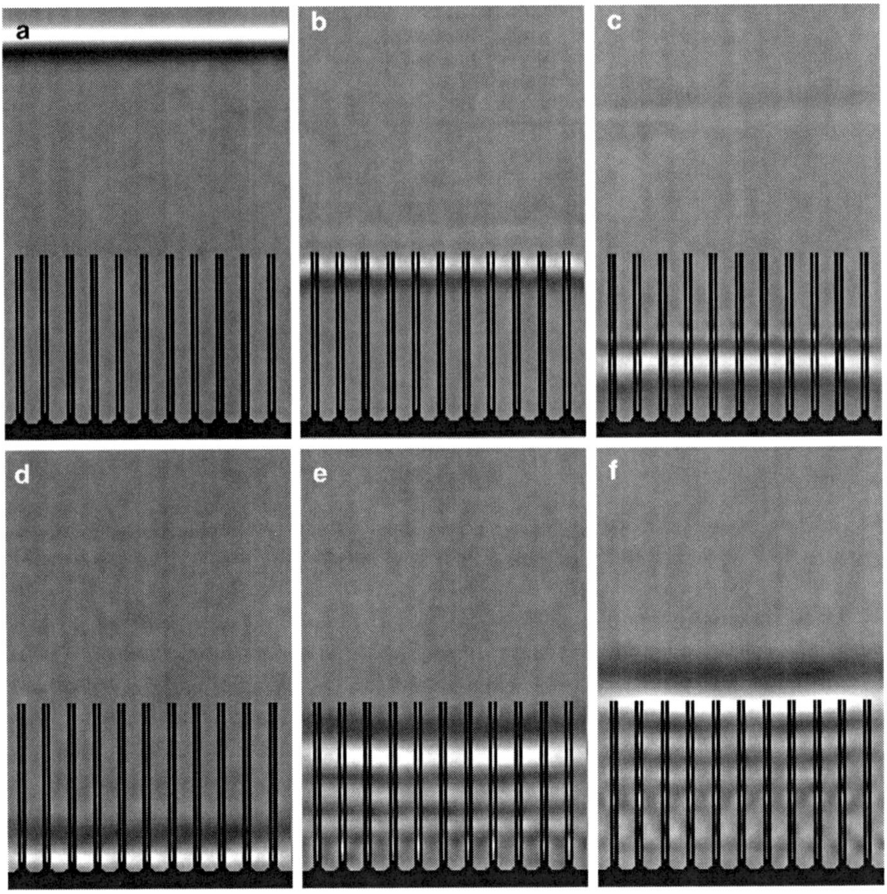

Fig. 2.18 Propagation of light through Ti-foil backed titania nanotube array at (**a**) 4.67, (**b**) 9.34, (**c**) 11.68, (**d**) 14.01, (**e**) 16.35, and (**f**) 17.51 fs (femtoseconds). The variation in field strength is represented by the different shades of gray [81]

metal, it is more readily absorbed, and therefore the intensity of the rejected light is very low, on a unit length basis, compared to that of the transmitted wave in Type-I samples. As a result, a clear difference in the absorbance can be seen between the Type-II samples (Fig. 2.19) and the Type-I samples (Fig. 2.20) [81]. The increased light absorption in Type-II samples makes them more suitable for water photolysis experiments, while the Type-I samples are better suited for application in solar cells.

The transmittance of light through self-standing TiO_2 nanotube array films (Type-I) is calculated as a function of tube length while keeping wall thickness, pore diameter, and barrier layer thickness constant. Figure 2.21 plots the transmittance of the film as a function of excitation wavelength and tube length for nanotubes of length 200 nm, pore size 22 nm, and barrier layer thickness 100 nm. The transmittance reaches a value over 95% at wavelengths greater than 380 nm.

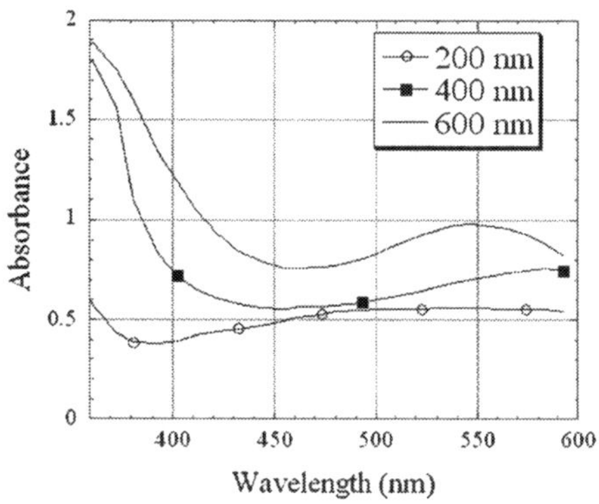

Fig. 2.19 Absorbance of titania nanotube films as a function of tube length for arrays grown on Ti foil (Type-II); the inner diameter of the tube is 20 nm, wall thickness 10 nm, and barrier layer thickness 100 nm

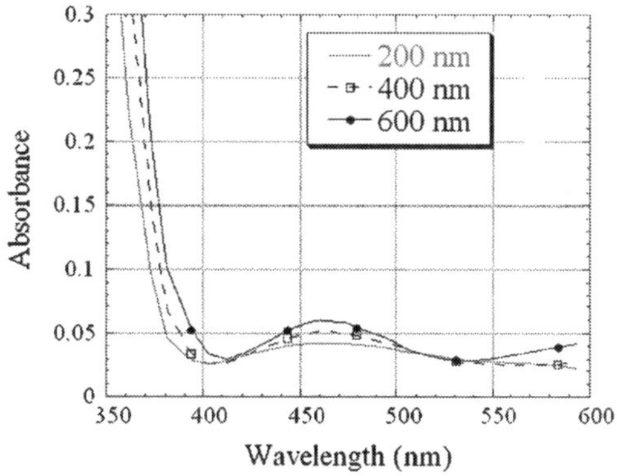

Fig. 2.20 Absorbance of titania nanotube films as a function of tube length for arrays grown on Ti foil (Type-I); the inner diameter of the tube is 20 nm, wall thickness 10 nm, and barrier thickness 100 nm [81]

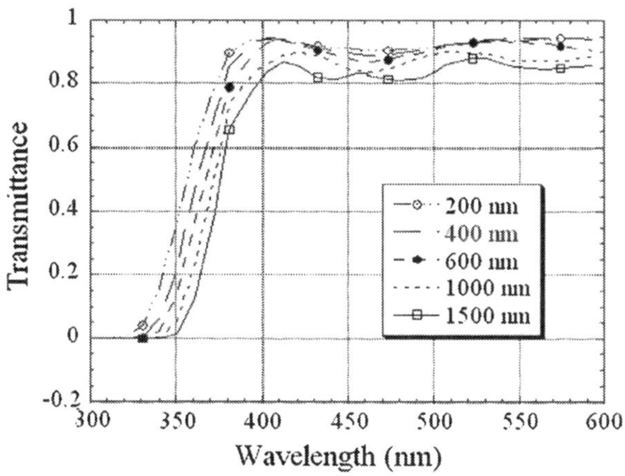

Fig. 2.21 Transmittance of TiO_2 nanotube array films as a function of tube length (Type-I; tube's inner diameter is 20 nm, wall thickness 10 nm, and barrier thickness 100 nm) [81]

The spacing between the interference patterns, created by the interaction of the transmitted wave and the wave reflected back from the top of the nanotubes, reduces with increasing nanotube length [84–86]. In the region below about 330 nm, the absorption is so high that the nanotube length has little influence. Here the light is completely absorbed by the nanotubes within a path length of a few tens of nanometers. Above this wavelength region, the transmitted fields depend on the nanotube length. It was found that for a given nanotube length, wall thickness, and barrier layer thickness, the transmittance increases slightly with increasing pore size. With the increase in porosity, the air column volume increases and the solid material volume decreases, yielding reduced effective refractive indices.

2.4.2 Measured Optical Properties

Figure 2.22 shows the transmittance spectrum of a transparent titania nanotube array film on a glass substrate [3]. The optical behavior of the TiO_2 nanotube arrays are quite similar to those reported for mesostructured titanium dioxide [87–89]. The difference in the envelope magnitude encompassing the interference fringe maxima and minima is relatively small compared to that observed in titania films deposited by rf sputtering, e-beam, and sol-gel methods [90–92].

The absorbance (or optical density) of the films were estimated from the transmittance T using the relation: $A = -\log(T)$. Here, we assumed that all the incident light is either transmitted or absorbed, with reflection or scattering being negligible. The Napierian absorption coefficient of the sample was calculated using

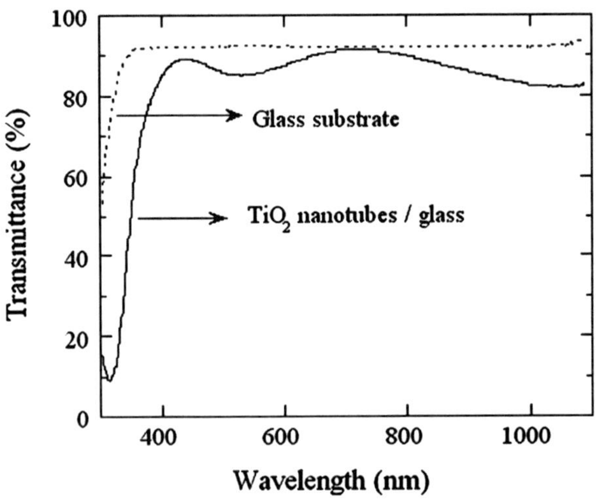

Fig. 2.22 Transmittance spectra of glass (Corning 2947) substrate, and 450°C annealed nanotubular titania film atop the same glass (Corning 2947) substrate [3]

Lamberts law, where d is the thickness of film, which can be determined using the relation

$$d = \frac{\lambda_1 \lambda_2}{2[\lambda_2 n(\lambda_1) - \lambda_1 n(\lambda_2)]} \tag{2.1}$$

where λ_1 and λ_2 are the wavelengths corresponding to the two adjacent maxima or minima, and $n(\lambda_1)$ and $n(\lambda_2)$ are the refractive indices at λ_1 and λ_2, respectively. The refractive indices of the titania nanotube film were calculated using the transmittance spectrum in the range 380–1,100 nm by employing Manifacier's envelope method [93]:

$$n(\lambda) = \sqrt{S + \sqrt{S^2 - n_0^2(\lambda) n_S^2(\lambda)}} \tag{2.2}$$

$$S = \frac{1}{2}\left[n_0^2(\lambda) + n_S^2(\lambda)\right] + 2 n_0 n_S \frac{T_{max}(\lambda) - T_{min}(\lambda)}{T_{max}(\lambda) \times T_{min}(\lambda)} \tag{2.3}$$

where n_0 and n_s are the refractive indices of air and film, respectively, T_{max} is the maximum envelope, and T_{min} is the minimum envelope. From the transmittance spectrum, the refractive index of glass is calculated as a function of wavelength using the relation

$$n_s(\lambda) = \frac{1}{T_s(\lambda)} + \sqrt{\frac{1}{T_s^2(\lambda)} - 1} \tag{2.4}$$

where T_s is the transmittance of glass.

Figure 2.23 shows the refractive index of the thin-film titania nanotubes and, for comparison, the glass substrate, calculated using (2.3) and (2.4) [3]. The optical behavior of the TiO_2 nanotube arrays is quite similar to that reported for mesostructured titanium dioxide. The average refractive index of the nanotube array (450°C annealed) was found to be 1.66 in the visible range, 380–800 nm. The thickness, as calculated by inserting the values of refractive indices and the wavelength corresponding to two consecutive maxima or minima (Fig. 2.23) in (2.2), was found to be 340 nm. This agrees with the value of 300 nm for the total thickness of the nanotube array including the barrier layer, determined from FESEM images.

The porosity of the nanotube array architecture was determined from the relation [94]

$$Porosity(\%) = \left[1 - \frac{n^2 - 1}{n_d^2 - 1}\right] \times 100 \tag{2.5}$$

where n (=1.66) and n_d (= 2.5) are the refractive indices of the nanotube structure (annealed at 450°C) and nonporous anatase films, respectively. The porosity of the

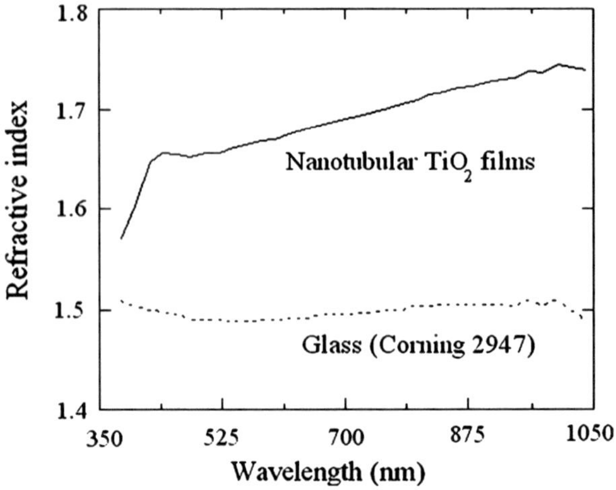

Fig. 2.23 Refractive index variation of 450°Cannealed nanotubular titania film and, for comparison, a glass (Corning 2947) substrate, in the range 380–1,050 nm. The TiO_2 film has an average refractive index in the visible range of 1.66

nanotube structure was calculated as 66.5%, which is close to the calculated value of 67% for nanotube arrays grown on titanium foil using a 10-V anodization potential [3, 95]. The low refractive index is due to the high porosity of the nanotube architectures, with nanotube diameters much less than the wavelength of light in the visible range, which reduces the light reflection from the surface of the array. The absorption coefficient α and the bandgap E_g are related through the equation [96]

$$(\alpha h v)^s = hv - E_g \tag{2.6}$$

where v is the frequency, h is Plank's constant, and s=0.5 for indirect bandgap material. The Tauc plot, $\sqrt{\alpha hv}\ vs. hv$, obtained after substituting the value of α in this equation is shown in Fig. 2.24 [3]. The optical bandgap, obtained by dropping a line from the maximum slope of the curve to the x-axis, is 3.34 eV. XRD results showed only anatase phase in the transparent titania nanotube array film. The reported bandgap value of anatase phase in bulk is 3.2 eV [97]. A slight blue shift in the value might be due to a quantization effect in the nanotubular film where the wall thickness is about 12 nm. A band tail to 2.4 eV is observed. The degree of lattice distortion is likely to be relatively higher for nanotube array films, thus causing aggregation of vacancies acting as trap states along the seams of nanotube walls leading to a lower band-to-band transition energy.

As seen from the structural studies, nanotube array films that retain an underlying metal layer have anatase-phase nanotube walls and a rutile-phase barrier layer. The absorbance spectra of these opaque films were compared with those of nanotube array films grown on metal foils (both annealed at 480°C) in the wavelength range 320–800 nm, as shown in Fig. 2.25 [3]. Although there is no significant difference between the behavior of thin film and foil samples, the shift in the

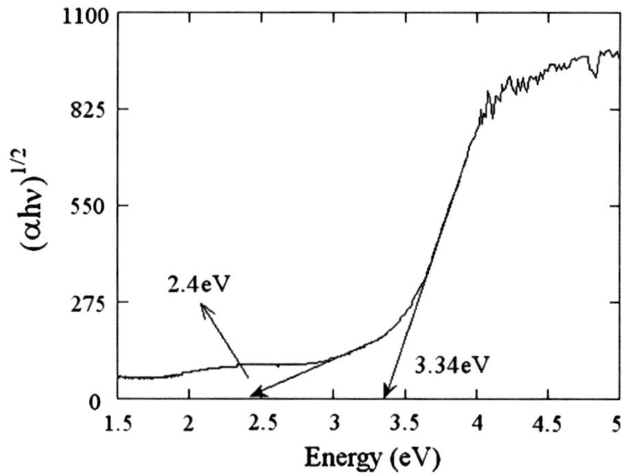

Fig. 2.24 $\sqrt{\alpha\hbar\omega}$ *vs.* $\hbar\omega$ plot of a 450°C annealed nanotube array film. An indirect bandgap of 3.34 eV and band tailing up to 2.4 eV is observed in the film [3]

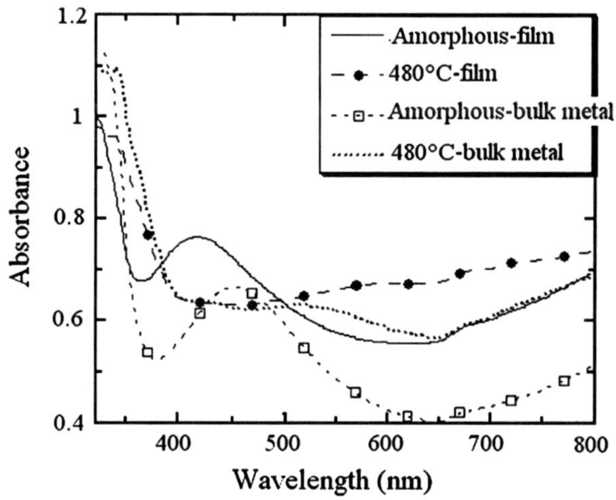

Fig. 2.25 Diffuse reflectance spectra of thin film (with a significant underlying Ti layer) and bulk metal samples, as-anodized (amorphous) and 480°C annealed. Both film and foil samples were prepared in the same electrolyte and annealed under identical conditions [3]

absorption edge towards higher wavelengths on annealing the samples is evident from this figure. The presence of rutile phase, which has a bandgap of ~3.0 eV, makes the absorption edge close to 400 nm [97].

The UV–vis spectra of the electrochemically doped nitrogen and fluorine-doped TiO_2 thin films revealed that the presence of fluorine did not result in a discernable change in optical absorption, whereas N-doping exhibited slightly higher optical absorption in the wavelength range from 400 to 510 nm [63]. The optical absorption

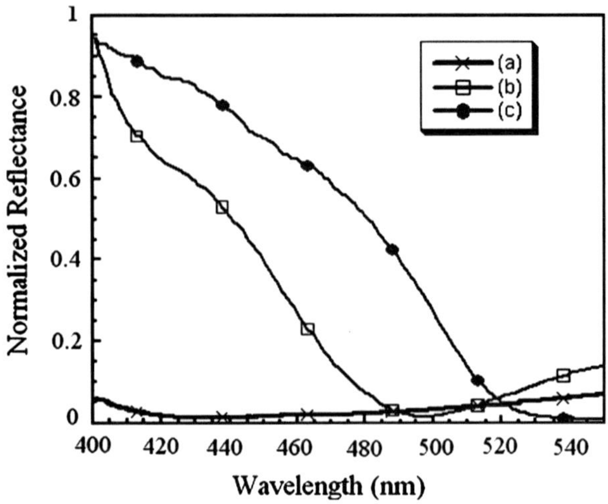

Fig. 2.26 Normalized visible reflectance spectra of CdS–TiO_2 nanotube array electrodes. (**a**): TiO_2 nanotube array electrode; (**b**) 20 min –0.5 V electrodeposited CdS-modified TiO_2 nanotube array electrode, as fabricated; (**c**) electrode of (**b**) after annealing at 350°C for 60 min in N_2 [75]

is a function of both film thickness and nitrogen concentration. The film with the highest nitrogen concentration, Sample A with $x=0.23$, is also the thinnest film owing to the fact that it was anodized for only 17 s [63]. The improvement in optical absorption is manifested most clearly for films of similar thickness. All N-doped films exhibited a shift in the primary absorption threshold, with the magnitude of this shift increasing with the concentration of incorporated nitrogen.

Normalized visible reflectance spectra of a plain TiO_2 nanotube array electrode as well as CdS-modified TiO_2 nanotube array electrodes are shown in Fig. 2.26 [75]. The reflectance onset was determined by linear extrapolation from the inflection point of the curve toward the baseline. The CdS coating on nanotube array has red-shifted the absorption edge into the visible region, with the absorption tail extending to 500 nm; the bandgap calculated from this reflectance edge is about 2.53 eV [75]. After annealing (N_2, 350°C, 1 h) its absorption behavior has further red-shifted, with the reflectance tail extending to 515 nm, with a calculated bandgap edge of 2.41 eV; the absorption edge corresponds to a nanoparticle size of approximately 10 to 20 nm [98, 99]. With annealing, the CdS particles aggregate and cause the spectrum to red-shift, a behavior previously attributed to the formation of valence-band tail states [100].

2.4.3 Ellipsometric Measurements

Ellipsometric methods have been used to study the growth process of TiO_2 nanotube array films [101]. Changes in ellipsometric parameters (Δ,Ψ) were measured

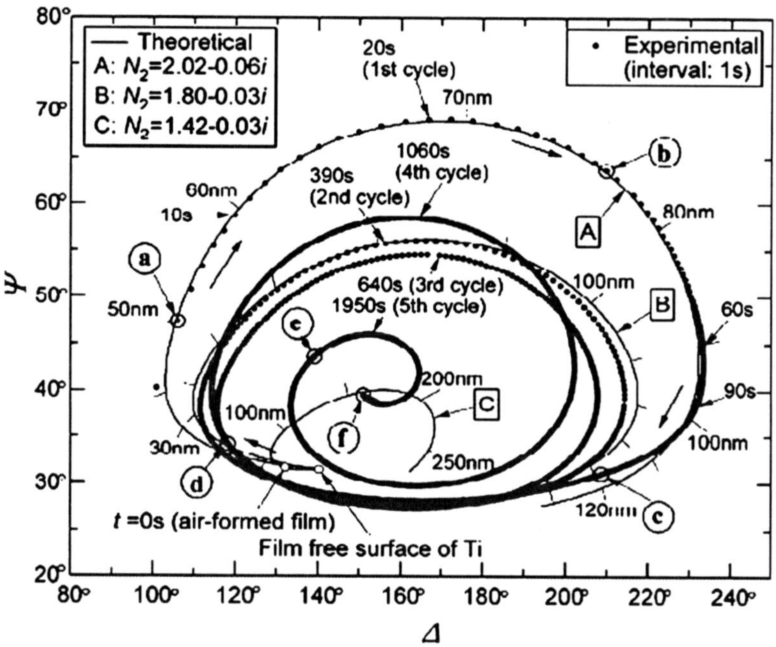

Fig. 2.27 Experimental Δ–Ψ curve for growth of TiO_2 film on Ti at 20 V in 0.5 wt%NH_4F, and theoretical Δ–Ψ curves ([101]; Reprinted by permission of ECS –The Electrochemical Society)

as a function of time during anodization of Ti at 20 V in 0.5 wt%NH_4F. Figure 2.27 shows the experimental Δ,Ψ plot in 0.5 wt%NH_4F and three theoretical Δ–Ψ curves *A*, *B*, and *C* calculated for the growth of single-layer films [101]. The experimental Δ–Ψ plot starts at an initial point determined by the air-formed film and traces out a circular curve that passes repeatedly near the initial point. The experimental Δ–Ψ point for the initial anodization period of 60 s moves at a high speed through the first cycle and fits the theoretical Δ–Ψ curve *A* calculated for a film with N_2=2.02–0.06i. This indicates that a single-layer film grows rapidly in this period. Beyond 60 s, the experimental Δ–Ψ plot deviates from the theoretical curve *A*, moving inside theoretical curve *A*. The experimental Δ–Ψ plot of the next three cycles does not fit a theoretical curve for growth of a single-layer film having a fixed optical constant, but is close to the theoretical curve *B* calculated for a film with N_2= 1.80–0.03i, which is small compared to the N_2 value used for the calculation of the theoretical curve *A* (N_2=2.02–0.06i). The experimental Δ–Ψ plot of the fifth cycle deviates from that of the antecedent four cycles, spiraling inward. The experimental plot in this cycle cannot fit a theoretical curve for the growth of a single-layer film but may be simulated by a model in which the refractive index of the film decreases as low as that used for the calculation of the theoretical curve *C* (N_2=1.42–0.03i).

An optical model was suggested to compare the theoretical Δ–Ψ curves for growth of anodic TiO_2 nanotube layers with the experimental Δ–Ψ curves [101].

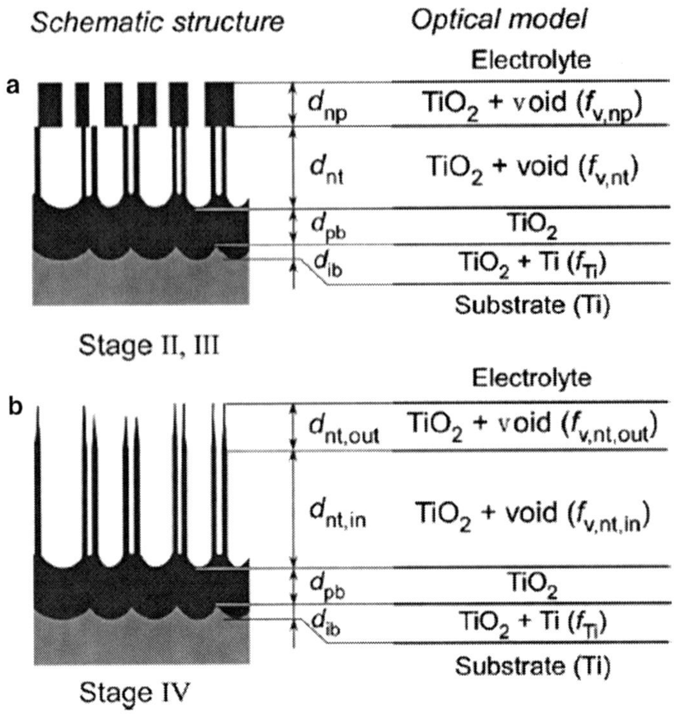

Fig. 2.28 Schematic structure and optical models of TiO_2 nanotube layers formed on Ti in NH_4F solutions ([101]; Reprinted by permission of ECS –The Electrochemical Society)

Figure 2.28a shows a four-layer optical model employed to calculate the theoretical curves for the films growing in Stage II. Nanoporous and nanotube layers are modeled as a mixture of TiO_2 and voids, which are filled with electrolytes. The optical constants of these two layers can be calculated as a function of the void fraction f_v by using the Bruggeman effective medium approximation (EMA) [102]. Because a barrier layer at the bottom of nanotubes is scalloped, this layer can be divided into upper and lower parts, which are called the *pure barrier layer* and *interface barrier layer*, respectively [103]. The pure barrier layer is regarded as pure TiO_2 and the interface barrier layer as a mixture of TiO_2 and substrate material (Ti). The optical constant of the interface barrier layer can be calculated using the EMA model for a given Ti fraction f_{Ti}. In the Bruggeman EMA, the effective dielectric functionε of a layer consisting of two materials A and B is given by solving the following equations [104]:

$$\left[f_A \frac{(\varepsilon_A - \varepsilon)}{(\varepsilon_A + 2\varepsilon)}\right] + \left[f_B \frac{(\varepsilon_B - \varepsilon)}{(\varepsilon_B + 2\varepsilon)}\right] = 0 \tag{2.7}$$

$$f_A + f_B = 1 \tag{2.8}$$

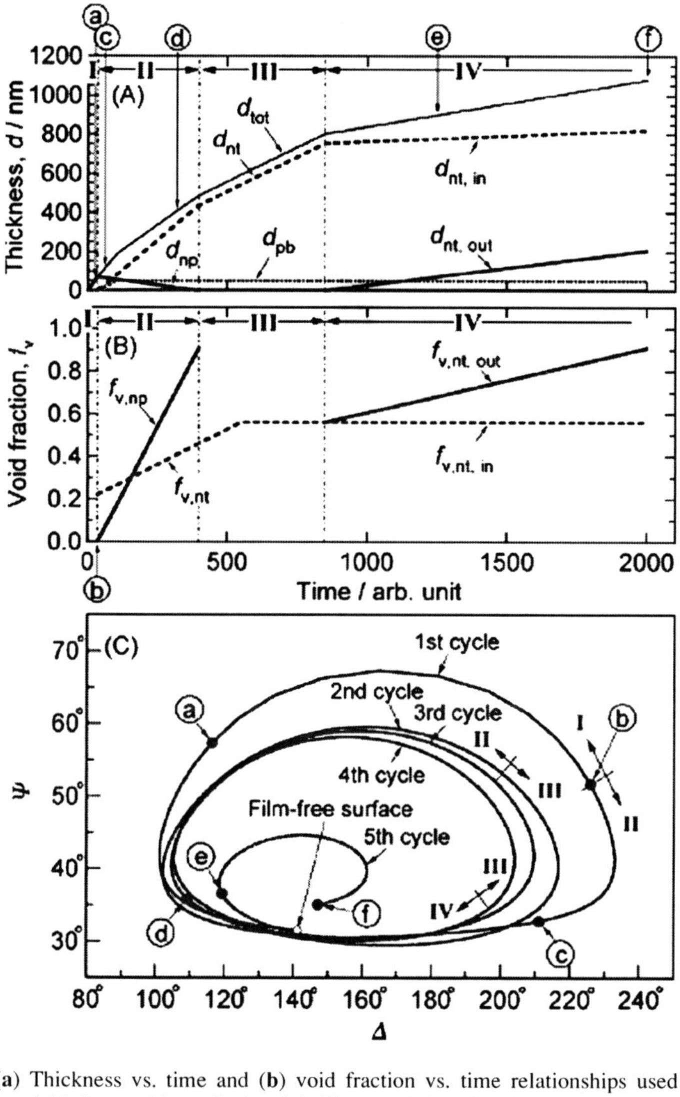

Fig. 2.29 (**a**) Thickness vs. time and (**b**) void fraction vs. time relationships used for model calculation, and (**c**) the resulting calculated Δ–Ψ curve ([101]; Reprinted by permission of ECS – The Electrochemical Society)

where ε_A and ε_B are the dielectric functions of constituents A and B, respectively. f_A and f_B represent the volume fractions of these constituents. The dielectric function can be related to the optical constant N by the following simple equation: $\varepsilon = N^2$. The films in the last growth Stage IV can also be expressed by the four-layer optical model as shown in Fig. 2.28b [101]. In this case, the nanotube layer is separated into two layers with different void fractions $f_{v,nt,out}$ and $f_{v,nt,in}$.

Figure 2.29 exhibits a set of thickness vs. time relations (top), and void fraction vs. time relations (middle) used in a model calculation, and the resulting calculated

Δ–Ψ trajectory (bottom) [101]. The following assumptions were used for the calculation: (i) the initial TiO_2 barrier layer grows to 70 nm; (ii) the optical constant of TiO_2 material is 2.20–0.00i; (iii) the void fraction of the nanoporous layer $f_{v,np}$ increases linearly from 0 to 0.91 with decreasing thickness of this layer d_{np} from 70 to 0 nm; (iv) the void fraction of the nanotube layer $f_{v,nt}$ increases linearly from 0.22 to 0.56 with increasing thickness of this layer d_{nt} from 0 to 520 nm; (v) the dissolution of the nanotube mouths becomes significant when a certain period of time passes after the surface nanoporous layer disappears and the void fraction of the resulting outer nanotube layer $f_{v,nt,out}$ increases linearly from 0.56 to 0.91 with time; (vi) the thicknesses of the pure and interface barrier layers increase with time in the initial period of oxidation and then remain constant at 50 and 5 nm, respectively; and (vii) the Ti fraction of the interface layer f_{Ti} is kept constant at 0.25. The calculated Δ–Ψ trajectory (Fig. 2.29c) is similar to the experimental Δ–Ψ plot shown in Fig. 2.27 [101].

2.4.4 Raman Spectra Measurements

The different phases of titanium dioxide, i.e., anatase, rutile, and brookite, have characteristic Raman peaks, and hence Raman spectra can be used to demonstrate the effects of calcination temperature on the phase of nanotube array on the Ti substrate. Figure 2.30 shows the Raman spectra of titania nanotube arrays after

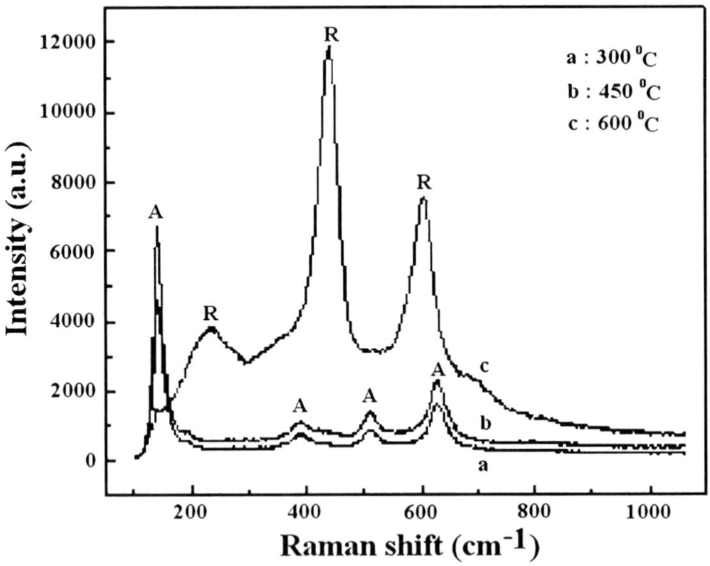

Fig. 2.30 Raman spectra of the TN array calcined at various temperatures, where the peaks representing anatase and rutile are labeled A and R, respectively ([16]; Reprinted by permission of ECS – The Electrochemical Society)

calcination at temperatures ranging from 300 to 600°C [16]. Peaks representing the anatase and the rutile form of TiO_2 are labeled A and R, respectively. Amorphous TiO_2 has no Raman peaks. When the sample was calcined at 300°C, the typical peaks 144, 197, 400, 515, and 640 cm^{-1} corresponding to anatase phase appeared [105]. When the calcination temperature was increased from 300 to 450°C, the peak intensity of anatase increased, which is ascribed to the gradual growth of crystallites and increase in the degree of crystallization. While the temperature was increased to above 450°C, the anatase phase began to transform to the more stable phase of rutile. The Raman spectra showed that rutile (143, 235, 447, and 612 cm^{-1}) became the main phase at 600°C [105]. On observing the Raman spectra of the three samples prepared by different voltages and calcined at 450°C, it was found that the intensity ratio of the anatase peak to the rutile peak increased with the increase in anodization voltage, indicating that the content of anatase phase prepared at 20 V is higher than that of the other two samples [16].

2.5 Electrical Property Measurements

2.5.1 *Photocurrent Transient Measurements*

The electronic properties of TiO_2 nanotube layers were characterized using photocurrent transient measurements [106]. Generated photocurrent transients, the shape of which depends on the crystal state, were recorded when a light pulse at a wavelength of 350 nm was applied for 10 s [106]. For the annealed nanotube layer, an initial rapid increase of the current is recorded, followed by a gradual decrease of the photocurrent with time. The same transient behavior can be observed on compact layers. This behavior was described in terms of a classical onset of recombination. The simplest kinetic approach is based on an exponential law [106]

$$I_{ph}(t) = \left\{ I_{ph}(t=0) - I_{ph}(t=\infty) \right\} \exp(-kt) + I_{ph}(t=\infty) \tag{2.9}$$

where k denotes a rate constant. By fitting (2.9) to the experimental data points, the rate constants yield $k=0.315$, 0.393 and 0.14 s^{-1} for the as-anodized compact layer, the annealed compact layer, and the annealed nanotube layers, respectively. The rate constant for the annealed compact layer is comparable to that for the as-formed compact layer, contradicting what one might anticipate. For the as-anodized nanotube layer, the photocurrent continuously increases with time during illumination; it appears trap-filling results in a photocurrent increase. A number of localized states are introduced in the TiO_2 bandgap after anodization, which act as recombination or trap sites. The as-anodized nanotube layer could contain deeper traps due to fluoride-induced etching of the formed oxide. When trap-filling behavior is dominant, during illumination over time the traps are filled with photogenerated charge carriers, which can enhance charge carrier lifetimes. Therefore, the photocurrent gradually increases as the number of filled traps increases. The process may be described by the following kinetic equation [106]:

$$I_{ph}(t) = \{1 - \exp(-kt)\} \cdot \{I_{ph}(t = \infty) - I_{ph}(t = 0)\} + I_{ph}(t = 0) \qquad (2.10)$$

where k denotes a rate constant. Upon fitting (2.10) to the observed data points, the rate constants yield $k=0.49\ s^{-1}$ for the as-anodized nanotube layer.

2.5.2 Capacitance Measurements

The semiconducting properties of TiO_2 nanotube array/electrolyte interfaces were analyzed by means of impedance spectroscopy near the flat band potential. Three different samples were used; compact layers: 1 M H_3PO_4 (10 V, 10 min, oxide thickness~25 nm); short e-tubes: 1 M H_3PO_4+0.3% HF (10 V, 2 h, thickness ~500 nm, pore size 75 nm); and long e-tubes: 1 M NaH_2PO_4+0.5% HF (20 V, 2 h, thickness ~1 μm, pore size 150 nm) [107, 108]. Thermal annealing of the oxide layers was carried out at 450°C in air during 3 h using a heating and cooling rate of 20°C/s. Acetate and borate buffer solutions of pH 2.73 – 9 were used for the capacitance analysis. Figure 2.31 shows that the cathodic polarization curves performed on compact and porous oxides are characterized by a current plateau followed by an exponential increase at $U<-0.5$ V [107]. This behavior change may be ascribed to the disappearance of the Schottky barrier control with the onset

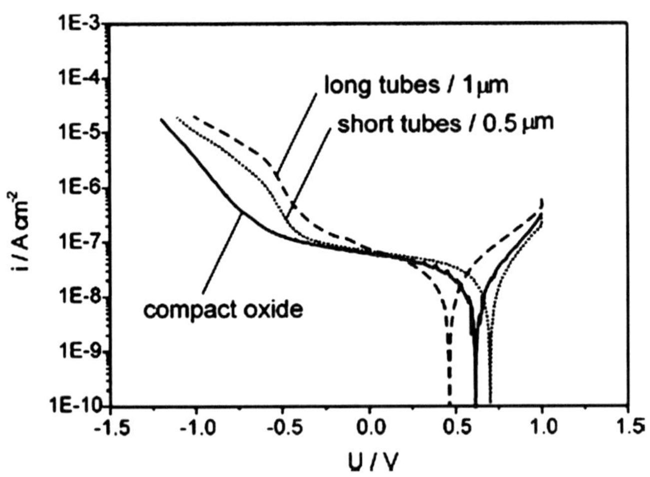

Fig. 2.31 Polarization curves obtained in acetate buffer solution of pH 6 on compact and porous oxides, with a voltage sweep rate of ν=5 mVs^{-1}([107]; Copyright 2007. Reprinted with permission from Elsevier)

of a kinetic-controlled electron transfer. However, the presence of an anodic current at potentials between 0.5 and 1 V indicates the start of oxygen evolution promoted by hole injection from the valence band. A marked difference between the current densities of the compact and porous oxides can be observed at a potential $U < -0.5$ V, which is mainly related to the larger active area of the nanotubes. However, practically no differences can be discerned at $U > -0.5$ V, indicating a common controlling step in the charge-transfer mechanism. It is suggested that these features are probably a consequence of a two-film structure, a compact base oxide and a nanotube array overlayer [107].

2.5.2.1 Mott–Schottky Plots: Analysis of Interfacial Properties

The Mott–Schottky equation is a common tool used for characterization of the semiconductor/electrolyte interface [109]. The flat band potential U_{fb} and the donor concentration N_D in a specific electrochemical system can be determined by plotting ${C_{\mathrm{sc}}}^{-2}$ vs. potential U according to

$$C_{\mathrm{sc}}^{-2} = \left(\frac{2}{\varepsilon\varepsilon_0 e N_{\mathrm{D}}}\right)\left(U - U_{\mathrm{fb}} - \frac{kT}{e}\right) \tag{2.11}$$

where C_{sc} represents the differential capacitance of the space charge layer. The presence of a high concentration of multiple donor levels in the bandgap promotes indirect tunneling of electrons through the semiconductor layer, resulting in a variation of the slope of (2.11) with frequency. This frequency dependence is determined by the distribution of relaxation times for electron emission, which in turn depends on the position of the states relative to the conduction band [110]. At sufficiently low frequencies, the response of a large fraction of the frequency dependent donors is expected, avoiding interferences arising from the double layer response at higher frequencies. Figure 2.32 shows the Mott–Schottky plots obtained at $f = 3$ Hz, where two main features can be seen [107, 108]. First, all curves show a marked change of slope, which is associated with band state ionization. Then, the porous oxides show a marked shift of the linear extrapolation to $C^{-2} = 0$ towards more positive values as compared with the compact oxide. According to (2.11), values of $U_{\mathrm{fb}} = -0.725$ V and $U_{\mathrm{fb}} = -0.33$ V can be calculated for the compact oxide layer and the 1-μm long nanotube layer, respectively [107]. Donor concentrations $N_{\mathrm{D}} = 4.19 \times 10^{19}\,\mathrm{cm}^{-3}$ and $N_{\mathrm{D}} = 1.63 \times 10^{20}\,\mathrm{cm}^{-3}$(referenced to the geometrical area) for the compact and the 1-μm long tubes oxide layers, respectively, were estimated from the slope near the flat band potential. Typical values reported for the donor concentration are within 10^{19}–$10^{20}\,\mathrm{cm}^{-3}$ as dependent on sample preparation conditions [111–113]. In spite of very different surface areas, similar slopes are found in both the compact and nanotube array layers, reflecting the concept of a porous oxide layer as two oxide films of different capacitance, the lower of which dominates the total response.

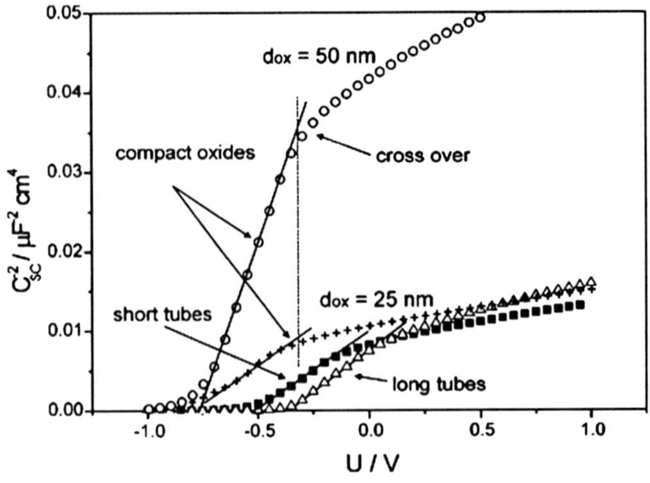

Fig. 2.32 Mott–Schottky plots obtained on compact and porous oxides in acetate buffer solution of pH 6. f=3 Hz; $U_{s,i}$ =1 V ([107]; Copyright 2007. With permission from Elsevier)

The width of a depletion layer in a semiconductor is given by

$$L_D = \left(\frac{2\varepsilon\varepsilon_0}{eN_D}\right)^{1/2} \left(U - U_{fb} - \frac{kT}{e}\right)^{1/2} \tag{2.12}$$

where ε is the dielectric constant of TiO_2, taken as 42. This expression predicts a space charge layer thickness of 7 nm at the crossover potential for the 50-nm-thick compact layer formed at 20 V [107]. An alternative interpretation of the impedance responses seen for the anodic semiconductor oxides was introduced by application of amorphous semiconductor theory [114, 115]. According to this model, the capacitance will be given by

$$C_{sc}^{-1} = \frac{1}{\sqrt{\varepsilon\varepsilon_0 e N_D}} \ln \frac{\psi_S}{\psi_C} \tag{2.13}$$

where $\psi_S$is the band bending and ψ_C corresponds to the maximum polarization for which states still respond. Figure 2.33 shows the inverse of capacitance corresponding to compact and porous oxides [107]; expression (2.13) (solid lines) accurately simulates experimental results after adopting the following values: for compact oxide N_D=3.1 × 10^{20}cm^{-3}eV^{-1}, U_{fb}=−0.85 V and ψ_C=0.15 V; for short tubes N_D=6.5 × 10^{20}cm^{-3}eV^{-1}, U_{fb}=−0.72 V and ψ_C=0.30 V; for long tubes N_D=6.5 × 10^{20}cm^{-3}eV^{-1}, U_{fb}=−0.50 V and ψ_C=0.27 V [107]. It was noted that only the bottom of the porous film responds to the perturbation, and that the pore walls present a highly conducting medium with rather a large contribution due to surface effects (potential drop at the oxide/solution interface). The amorphous-semiconductor theory works well for annealed TiO_2 layers which points to a large

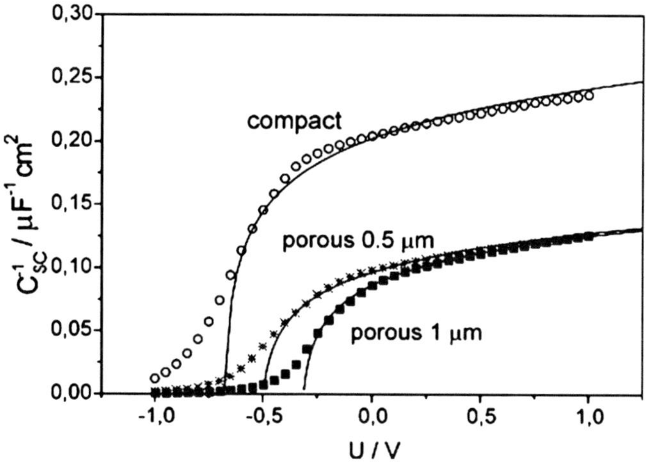

Fig. 2.33 Potential dependence of the inverse of capacitance for annealed compact and porous oxides obtained in acetate buffer solution at $f=3$ Hz. (—) Fitting curves after (5.5) for the amorphous semiconductor theory ([107]; Copyright 2007. Reprinted with permission from Elsevier)

number of bandgap states, which may be related to mobile defects at $U>U_{fb}$. Boschloo and coworkers [116, 117] reported the presence of electron traps approximately 0.5 eV below the conduction band as a means to explain the appearance of a photocurrent maximum on zinc-porphyrin-sensitized TiO_2 electrodes. Since the electrode has a large internal area consisting of numerous interconnected nanocrystallites, a high electron trap density is expected at the oxide/electrolyte interface and grain boundaries. Although the chemical origin of these traps is not known, it was suggested that they arise from an incompletely coordinated Ti surface [118].

Interfacial properties of the oxide were further analyzed by the potential dependence of the surface capacitance of compact and porous layers obtained under different solutions and illumination conditions [107]. Without illumination, the Mott–Schottky plots were shifted to more positive potentials with decreasing pH. The pH-dependence of the flat band potential is commonly related to the potential drop at the double layer as a consequence of acid–base reactions

$$\mathrm{TiO_2}_{\,\mathrm{surface}} + \mathrm{H}^+ \leftrightarrow \mathrm{TiO(OH)}_{\,\mathrm{surface}} \quad (2.14)$$

$$\mathrm{TiO(OH)}_{\,\mathrm{surface}} + \mathrm{H}^+ \leftrightarrow \mathrm{TiO(OH_2)}_{\mathrm{surface}} \quad (2.15)$$

In the absence of surface states and adsorbed ions, the following pH dependence of the flat band potential is expected:

$$U_{\mathrm{fb}} = U_{\mathrm{fb,iep}} - 2.3kT\,(pH - pH_{\mathrm{iep}}) \quad (2.16)$$

where $U_{\mathrm{fb,iep}}$ and $\mathrm{pH_{iep}}$ are the flat band potential and the pH at the isoelectric point, respectively, which was found to be 5.8 for anatase [116]. As can be seen in Fig. 2.34,

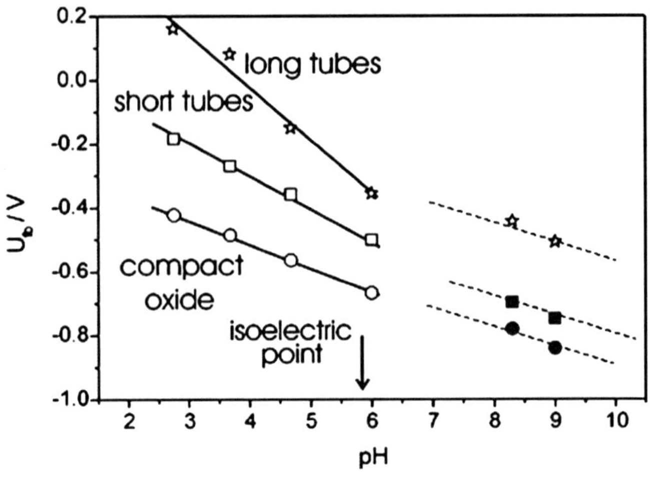

Fig. 2.34 pH dependence of the flat band potential for the different oxides obtained by extrapolation of the low band-bending parts to $C^{-2}=0$ ([107]; Copyright 2007. Reprinted with permission from Elsevier)

this point defines two zones with large deviations of (2.16) at $pH < pH_{iep}$. Slope values of 105 and 166 mV/pH for tube layers 0.5 and 1 μm in length, respectively, are observed, indicating a strong influence of the surface chemistry [107]. Surface charging of TiO_2 surfaces in aqueous electrolytes involves not only chemistry of the terminal oxo and hydroxo groups, (2.14) and (2.15), but also the uptake of protons at the band edge trap sites near the flat band potential [119, 120]. Thus, it is feasible that the value of the flat band potential will be mainly determined by equilibration of the conduction band with the trap-based redox couple, which can be written as

$$Ti^{4+}O_2 + H^+ + e^- \rightarrow Ti^{3+}O(OH) \tag{2.17}$$

The displacement towards more positive pH values for the porous layers was interpreted as the natural consequence of a larger density of surface charge and the related Fermi level shift caused on the pore bottoms. It was suggested that the different slopes observed around the isoelectric point were related to pH-dependent surface charging [107], which is related to hydrogen loading in the subsurface region with the formation of incompletely coordinated Ti^{3+} surfaces [121], and interstitial proton and hydroxyl bonds (Ti–O–H) whose formation is favored at $pH < pH_{iep}$.

2.5.2.2 Surface State Model

The pH dependence of the Mott–Schottky plots showed different slopes near the isoelectric point, which may be related to the nature of the surface states, see Fig. 2.34 [107]. It was suggested that surface state formation is related to incomplete coordination of surface Ti atoms [114], which seems to be favored at $pH < pH_{iep}$. The change of solution pH also affects the donor concentration as can be inferred from the pH dependence of the Mott–Schottky plot slopes. This effect is more pronounced in regions of high band bending, and even more pronounced for oxide layers comprised of short nanotubes [107]. This effect cannot be associated with the reductive effects of hydrogen intercalation, which is observed at a potential more negative than the flat band condition with a consequent marked increase of the donor concentration [116, 117];hence it appears the effects of pH are largely related to modifications of the surface chemistry [107].

The impedance behavior of the porous structures were interpreted by use of a two-oxide-layer model, with an underlying layer whose behavior can be considered similar to the compact oxide, and where the wall of the oxide tubes presents a higher donor concentration with a consequent higher electronic conductivity than the underlying layer, see Fig. 2.35 [108]. According to this model, with the system under electronic equilibrium, the introduction of the pore walls with a higher donor concentration will shift the energy levels of the compact layer toward higher energies. This, in turn, is reflected by a positive shift of the observed flat band potential [108]. The total capacitance of the oxide will be given by [108]

$$\frac{1}{C_{\text{ox}}} = \frac{1}{C_{\text{sc(bottom)}}} + \frac{1}{C_{\text{sc(walls)}}} \tag{2.18}$$

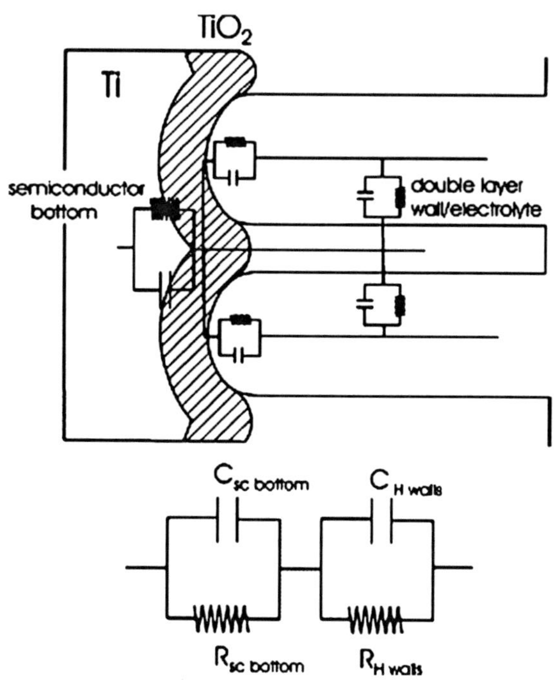

Fig. 2.35 Equivalent circuit representing the impedance response of the porous TiO_2 layers ([108]; Copyright Springer-Verlag 2006. With kind permission from Springer Science+Business Media)

Taking into account that C_{sc} is proportional to (N_D) in the depletion region, and that $N_{walls} \gg N_{bottom}$, the observed capacitance response will be dominated by the underlying oxide [108]. Due to the high conductivity of the polycrystalline walls, they can be regarded as an electronic conductor connecting the base semiconductor with the double layer (in solution) [108]. Therefore, the observed response corresponds largely to that of the bottom oxide layer, while the influence of the nanotube arrays is manifest in enhanced surface chemistry effects. The annealing process decreases the defect concentration, and as a consequence the flat band potential shifts toward more positive potentials [108].

The presence of a porous structure on the pH dependence of the flat band potential can also be related to the double-layer semiconductor structure [108]. Here, it is expected that the shift of the Fermi level of the compact oxide will be controlled by the surface chemistry of the pore walls. Thus the noticeable potential shift at pH<6 in the porous oxide (cf. Fig. 2.34) can be explained in terms of the acid–base reactions of the –OH groups on the surface of the pore walls and the fact that the potential drop seen by the underlying oxide is given by [108]

$$\Delta\phi = \frac{\Delta q A_{pores}}{C A_{bottom}} \tag{2.19}$$

since part of the surface charge generated in the pore walls must be compensated in the bottom oxide. The presence of these groups would lead to the formation of surface states, the charging of which results in Fermi level pinning.

2.5.2.3 Photoeffects

Surface states lead to trap site formation. Therefore, depending on the relative rates at which the surface states charge and discharge, an additional capacitive element will be built at the interface giving rise to the appearance of a capacitance peak [107]. Electrochemical and spectroscopic experiments have indicated the presence of two intraband surface states in nanotube array films acting as electron traps [116, 118, 121]. These are, respectively, related to adsorbed peroxy or peroxide radicals (located at midgap) and to incomplete coordinated surface Ti atoms in the solid lattice (found to be about 0.8 eV below the conduction band edge). The enhancement of photocurrent after hydrogen loading was observed by Weber et al. [122] on sputtered TiO_2 films,which was attributed to the filling of electronic traps by reduction of H^+ at the electrode/electrolyte interface. Reductive processes involving hydrogen intercalation were observed at potentials more negative than the flat band condition, with a consequent marked increase of the donor concentration [107, 116, 123, 124].

2.6 Mechanical Properties

The mechanical properties of a material are critical to its successful application [125–127]. This is particularly true in bone-implant materials, where long-term *in vivo* structural stability is crucial. Here, bone resorption occurs due to the mismatch in mechanical properties, such as elastic modulus, between bone and the implant material, causing implant loosening and eventual failure [128]. Considering the drug elution properties of the nanotube array surfaces, discussed in Chap. 6, it is important to characterize the mechanical behavior of these materials for potential use as implants. Furthermore, it is important to establish the relationships between microstructure and mechanical behavior in an effort to elucidate deformation mechanisms.

Mechanical characterization of thin coatings is challenging because of the small length scales involved. In this regard, nanoindentation is an appropriate technique for these materials because of its low load (1 μN) and small displacement (1 nm) resolution [129]. Nanoindentation is an effective technique for probing mechanical properties of thin (<10 μm) coated systems including soft coatings on hard substrates [130, 131], hard coatings on soft substrates [132–134], and multilayered systems [132, 135, 136]. This technique has also been used to investigate the mechanical response of TiO_2 with nanophase microstructure [137, 138] and of thin TiO_2 coatings on Ti substrates [133, 134]. Crawford and coworkers carried out a systematic investigation of the microstructure and mechanical behavior of nanotube array films [125], focusing on the relationship between microstructure characterization and deformation behavior by nanoindentation of TiO_2 nanotube array films on Ti substrate. Ti foil was anodized for 2 h in NaF-containing electrolyte (0.1 mol/l F^-), at a pH of 4.5, and constant potential of 20 V. A uniform structure of thin-walled TiO_2 nanotubes was obtained with tube diameters ranging between 35 and 70 nm, tube wall thicknesses between 10 and 18 nm, and tube lengths between 230 and 670 nm [125].

Figure 2.36a shows representative load vs. displacement curves for the samples with varying anodization time [125]. Figure 2.36b shows the extracted apparent Young's modulus and hardness, respectively, obtained from the continuous stiffness measurement (CSM) system [125]. From inspection of Fig. 2.36b, total nanoindentation depth is much larger than the thickness of the coating. In order to gain insight into the underlying deformation mechanisms, SEM of the indentations was conducted. It was noted that coating delamination occurred on unloading only in thin coatings (230–250 nm) and did not occur in thicker coatings (600–650 nm). An additional characteristic of indentations in thicker coatings was wear markings on the indentation surface. These markings suggest some degree of frictional sliding between the indenter and the TiO_2 surface. From FESEM micrographs and the indentation data, two main deformation processes were suggested, densification of the porous nanotube structure and wear of the dense surface.

The deformation mechanisms were correlated to the measured modulus as a function of depth [125], in which three distinct regions were identified [125].

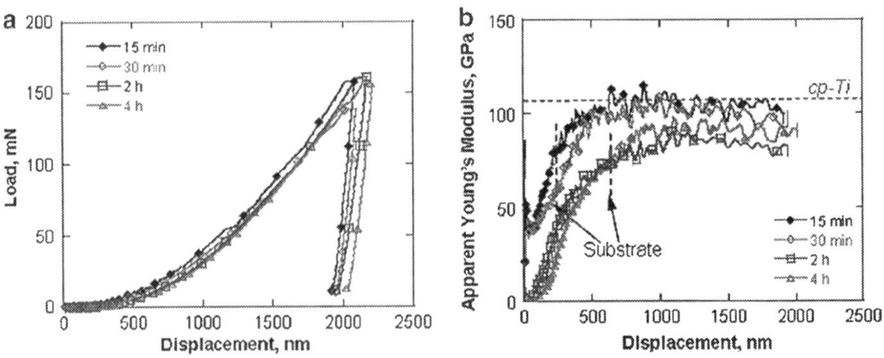

Fig. 2.36 Characteristic (**a**) load and (**b**) apparent Young's modulus vs. nanoindentation displacement (depth) for Ti samples anodized in NaF-containing electrolytes (0.1 mol/l F^-), at pH 4.5, and constant potential of 20 V for various times ([125]; Copyright 2007. Reprinted with permission from Elsevier)

Region I was characterized by a linear increase in modulus with increasing time. In this region, the increasing modulus is primarily due to increased densification of the TiO_2 nanotubes. A parabolic increase in the indentation modulus was observed in Region II. In this region, the indentation depth is comparable to the thickness of the TiO_2 coating. Thus, increases in modulus were considered as the result of an increasing contribution from the substrate. In addition, there are minor contributions of densification of the outer regions of the indentation and wear between the indenter surface and the dense TiO_2 surface. As the TiO_2 coating under the indenter becomes increasingly dense, the modulus increases. Finally, Region III described the region where the coating has become nearly fully dense. This region was characterized by a plateau in indentation modulus and termed as the composite modulus, as it is a combination of moduli of the dense coating and the Ti substrate. Assuming the volume fraction of pores in the coating of about 72% (estimated from image analyses) and a modulus for dense amorphous TiO_2 of 130–150 GPa (for the tubes themselves) [139], theYoung's modulus for the TiO_2 nanotubes was calculated to be approximately 36–43 GPa [125].

References

1. Varghese OK, Gong D, Paulose M, Ong KG, Grimes CA, Dickey EC (2003) Crystallization and high-temperature structural stability of titanium oxide nanotube arrays. J Mater Res 18:156–165
2. Cai Q, Paulose M, Varghese OK, Grimes CA (2005) The effect of electrolyte composition on the fabrication of self-organized titanium oxide nanotube arrays by anodic oxidation. J Mater Res 20:230–236
3. Mor GK, Varghese OK, Paulose M, Grimes CA (2005) Transparent highly ordered TiO_2 nanotube arrays via anodization of titanium thin films. Adv Funct Mater 15:1291–1296

4. Mor GK, Varghese OK, Paulose M, Shankar K, Grimes CA (2006) A review on highly ordered, vertically oriented TiO_2 nanotube arrays: fabrication, material properties, and solar energy applications. Sol Energy Mater Sol Cells 90:2011–2075
5. Zhao J, Wang X, Sun T, Li L (2005) In situ templated synthesis of anatase single-crystal nanotube arrays. Nanotechnol 16:2450–2454
6. Ghicov A, Tsuchiya H, Macak JM, Schmuki P (2006) Annealing effects on the photoresponse of TiO_2 nanotubes. Phys Stat Sol 203:R28–R30
7. Hahn R, Ghicov A, Salonen J, Lehto VP, Schmuki P (2007) Carbon doping of self-organized TiO_2 nanotube layers by thermal acetylene treatment. Nanotechnol 18:105604 (4pp)
8. Macak JM, Ghicov A, Hahn R, Tsuchiya H, Schmuki P (2006) Photoelectrochemical properties of N-doped self-organized titania nanotube layers with different thicknesses. J Mater Res 21:2824–2828
9. Yang Y, Wang X, Li L (2008) Crystallization and phase transition of titanium oxide nanotube arrays. J Am Ceram Soc 91:632–635
10. Zhao J, Wang X, Sun T, Li L (2007) Crystal phase transition and properties of titanium oxide nanotube arrays prepared by anodization. J Alloys Compd 434–435:792–795
11. Whittemore OJ, Sipe JJ (1974) Pore growth during initial stages of sintering ceramics. Powder Technol 9:159–164
12. Kumar KNP, Keizer K, Burggraaf AJ (1993) Textural evolution and phase-transformation in titania membranes 1. Unsupported membranes. J Mater Chem 3:1141–1149
13. Kumar KNP, Keizer K, Burggraaf AJ, Okubo T, Nagamoto H (1993) Textural evolution and phase-transformation in titania membranes 2. Supported membranes. J Mater Chem 3:1151–1159
14. Macak JM, Tsuchiya H, Schmuki P (2005) High-aspect-ratio TiO_2 nanotubes by anodization of titanium. Angew Chem Int Ed 44:2100–2102
15. Cullity BD, Stock SR (2001) Elements of X-ray diffraction. Prentice-Hall, Englewood Cliffs, NJ
16. Lai Y, Sun L, Chen Y, Zhuang H, Lin C, Chin JW (2006) Effects of the structure of TiO_2 nanotube array on Ti substrate on its photocatalytic activity. J Electrochem Soc 153:D123–D127
17. Chastain J (1992) Handbook of X-ray photoelectron spectroscopy. Perkin-Elmer, Eden Prairie, MN
18. McCafferty E, Wightman JP (1998) Determination of the concentration of surface hydroxyl groups on metal oxide films by a quantitative XPS method. Surf Interface Anal 26:549–564
19. Saha NC, Tomkins HC (1992) Titanium nitride oxidation chemistry – an X-ray photoelectron-spectroscopy study. J Appl Phys 72:3072–3079
20. Marino CEB, Nascente PAP, Biaggio SR, Rocha-Filho RC, Bocchi N (2004) XPS characterization of anodic titanium oxide films grown in phosphate buffer solutions. Thin Solid Films 468:109–112
21. Varghese OK, Paulose M, Shankar K, Mor GK, Grimes CA (2005) Water-photolysis properties of micron-length highly-ordered titania nanotube-arrays. J Nanosci Nanotechnol 5:1158–1165
22. Paulose M, Mor GK, Varghese OK, Shankar K, Grimes CA (2006) Visible light photoelectrochemical and water-photoelectrolysis properties of titania nanotube arrays. J Photochem Photobiol A 178:8–15
23. Ohya Y, Saiki H, Tanaka T, Takahashi Y (1996) Microstructure of TiO_2 and ZnO films fabricated by the sol-gel method. J Am Ceram Soc 79:825–830
24. Shankar K, Mor GK, Prakasam HE, Yoriya S, Paulose M, Varghese OK, Grimes CA (2007) Highly-ordered TiO_2 nanotube-arrays up to 220 μm in length: use in water photoelectrolysis and dye-sensitized solar cells. Nanotechnol18:065707 (11pp)
25. Yoriya S, Prakasam HE, Varghese OK, Shankar K, Paulose M, Mor GK, Latempa TA, Grimes CA (2006) Initial studies on the hydrogen gas sensing properties of highly-ordered high aspect ratio TiO_2 nanotube-arrays 20 μm to 222 μm in length. Sens Lett 4:334–339

26. Allam NK, Shankar K, Grimes CA (2008) General method for the anodic formation of crystalline metal oxide nanotube arrays without the use of thermal annealing. Adv Mater 20:3942–3946
27. Nishikiori H, Qian W, El-Sayed MA, Tanaka N, Fujii T (2007) Change in titania structure from amorphousness to crystalline increasing photoinduced electron-transfer rate in dye-titania system. J Phys Chem C 111:9008–9011
28. Kondo JN, Domen K (2008) Crystallization of mesoporous metal oxides. Chem Mat 20: 835–847
29. Grimes CA (2007) Synthesis and application of highly ordered arrays of TiO_2 nanotubes. J Mat Chem 17:1451–1457
30. Lee J, Orilall MC, Warren SC, Kamperman M, Disalvo FJ, Wiesner U (2008) Direct access to thermally stable and highly crystalline mesoporous transition-metal oxides with uniform pores. Nature Mater 7:222–228
31. Ohtani B, Ogawa Y, Nishimoto S (1997) Photocatalytic activity of amorphous-anatase mixture of titanium (IV) oxide particles suspended in aqueous solutions. J Phys Chem B 101:3746–3752
32. Demazeau G (2008) Solvothermal reactions: an original route for the synthesis of novel materials. J Mater Sci 43:2104–2114
33. Cushing BL, Kolesnichenko VL, O'Connor CJ (2004) Recent advances in the liquid-phase syntheses of inorganic nanoparticles. Chem Rev 104:3893–3946
34. Roy R (1994) Accelerating the kinetics of low-temperature inorganic syntheses. J Solid State Chem 111:11–17
35. Walton RI (2002) Subcritical solvothermal synthesis of condensed inorganic materials. Chem Soc Rev 31:230–238
36. Mahltig B, Gutmann E, Meyer DC, Reibold M, Dresler B, Günther K, Faßler D, Böttcher H (2007) Solvothermal preparation of metallized titania sols for photocatalytic and antimicrobial coatings. J Mater Chem 17:2367–2374
37. Chen D, Jiao XF, Ritchie RO (2000) Effects of grain-boundary structure on the strength, toughness, and cyclic-fatigue properties of a monolithic silicon carbide. J Am Ceram Soc 83:2079–2081
38. Yoriya S, Mor GK, Sharma S, Grimes CA (2008) Synthesis of ordered arrays of discrete, partially crystalline titania nanotubes by Ti anodization using diethylene glycol electrolytes. J Mater Chem 18:3332–3336
39. Habazaki H, Shimizu K, Nagata S, Skeldon P, Thompson GE, Wood GC (2002) Ionic transport in amorphous anodic titania stabilised by incorporation of silicon species. Corros Sci 44:1047–1055
40. Habazaki H, Shimizu K, Nagata S, Skeldon P, Thompson GE (2007) Fast migration of fluoride ions in growing anodic titanium oxide. Electrochem Commun 9:1222–1227
41. Yu T, Bu H, Chen J (1986) The effect of units derived from diethylene glycol on crystallization kinetics of poly(ethylene terephthalate). Macromol Chem 187:2697–2709
42. Kinart CM, Cwiklinska A, Maj M, Kinart WJ (2007) Thermodynamic and physicochemical properties of binary mixtures of sulfolane with ethylene glycol, diethylene glycol, triethylene glycol, and tetraethylene glycol systems at 303.15 K. Fluid Phase Equilib 262:244–250
43. Cocchi M, Marchetti A, Pigani L, Sanna G, Tassi L, Ulrici A, Vaccari G, Zanardi C (2000) Density and volumetric properties of ethane-1, 2-diol plus di-ethylene glycol mixtures at different temperatures. Fluid Phase Equilib 172:93–104
44. Mor GK, Shankar K, Paulose M, Varghese OK, Grimes CA (2005) Enhanced photocleavage of water using titania nanotube arrays. Nano Lett 5:191–195
45. Park JH, Kim S, Bard AJ (2006) Novel carbon-doped TiO_2 nanotube arrays with high aspect ratios for efficient solar water splitting. Nano Lett 6:24–28
46. Raja KS, Misra M, Mahajan VK, Gandhi T, Pillai P, Mohapatra SK (2006) Photo-electrochemical hydrogen generation using band-gap modified nanotubular titanium oxide in solar light. J Power Sources 161:1450–1457

47. Paulose M, Shankar K, Yoriya S, Prakasam HE, Varghese OK, Mor GK, Latempa TJ, Fitzgerald A, Grimes CA (2006) Anodic growth of highly ordered TiO_2 nanotube arrays to 134 μm in length. J Phys Chem B 110:16179–16184
48. Zhu K, Neale NR, Miedaner A, Frank AJ (2006) Enhanced charge-collection efficiencies and light scattering in dye-sensitized solar cells using oriented TiO_2 nanotubes arrays. Nano Lett 7:69–74
49. Hahn R, Ghicov A, Salonen J, Lehto V, Schmuki P (2007) Carbon doping of self-organized TiO_2 nanotube layers by thermal acetylene treatment. Nanotechnol 18:105604 (4pp)
50. Mohapatra SK, Misra M, Mahajan VK, Raja KS (2007) Design of a highly efficient photoelectrolytic cell for hydrogen generation by water splitting: application of $TiO_{2-x}C_x$ nanotubes as a photoanode and Pt/TiO_2 nanotubes as a cathode. J Phys Chem C 111: 8677–8685
51. Tang XH, Li DY (2008) Sulfur doped highly ordered TiO_2 nanotubular arrays with visible light response. J Phys Chem C 112:5405–5409
52. Vitiello RP, Macak JM, Ghicov A, Tsuchiya H, Dick LFP, Schmuki P (2006) N-Doping of anodic TiO_2 nanotubes using heat treatment in ammonia. Electrochem Commun 8:544–548
53. Lu N, Quan X, Li JY, Chen S, Yu HT, Chen GH (2007) Fabrication of boron-doped TiO_2 nanotube array electrode and investigation of its photoelectrochemical capability. J Phys Chem C 111:11836–11842
54. Mohapatra SK, Misra M, Mahajan VK, Raja KS (2007) A novel method for the synthesis of titania nanotubes using sonoelectrochemical method and its application for photoelectrochemical splitting of water. J Catal 246:362–369
55. Shankar K, Paulose M, Mor GK, Varghese OK, Grimes CA (2005) A study on the spectral photoresponse and photoelectrochemical properties of flame-annealed titania nanotube arrays. J Phys D 38:3543–3549
56. Choi Y, Umebayashi T, Yamamoto S, Tanaka S (2003) Fabrication of TiO_2 photocatalysts by oxidative annealing of TiC. J Mater Sci Lett 22:1209–1211
57. Asahi R, Morikawa T, Ohwaki T, Aoki K, Taga Y (2001) Visible-light photocatalysis in nitrogen-doped titanium oxides. Science 293:269–271
58. Lindgren T, Mwabora JM, Avendano E, Jonsson J, Hoel A, Granvist CG, Lindquist SE (2003) Photoelectrochemical and optical properties of nitrogen doped titanium dioxide films prepared by reactive DC magnetron sputtering. J Phys Chem B 107:5709–5716
59. Kosowska B, Mozia S, Morawski A, Grznil B, Janus M, Kalucki K (2005) The preparation of TiO_2-nitrogen doped by calcination of TiO_2 center dot xH(2) O under ammonia atmosphere for visible light photocatalysis. Sol Energy Mater Sol Cells 88:269–280
60. Wang Y, Feng C, Jin Z, Zhang J, Yang J, Zhang S (2006) A novel N-doped TiO_2 with high visible light photocatalytic activity. J Mol Catal A 260:1–3
61. Ghicov A, Macak JM, Tsuchiya H, Kunze J, Haeublein V, Frey L, Schmuki P (2006) Ion implantation and annealing for an efficient N-doping of TiO_2 nanotubes. Nano Lett 6:1080–1082
62. Ghicov A, Macak JM, Tsuchiya H, Kunze J, Haeublein V, Kleber S, Schmuki P (2006) TiO_2 nanotube layers: dose effects during nitrogen doping by ion implantation. Chem Phys Lett 419:426–429
63. Shankar K, Tep KC, Mor GK, Grimes CA (2006) An electrochemical strategy to incorporate nitrogen in nanostructured TiO_2 thin films: modification of bandgap and photoelectrochemical properties. J Phys D 39:2361–2366
64. Chourasia AR, Chopra DR (1992) X-ray photoelectron spectra of TiN. Surf Sci Spectra 1:233–237
65. Chourasia AR, Chopra DR (1995) X-ray photoelectron study of Tin/SiO_2 and Tin/Si interfaces. Thin Solid Films 266:298–301
66. Yakovleva NM, Anicai L, Yakovlev AN, Dima L, Khanina EY, Buda M, Chupakhina EA (2002) Structural study of anodic films formed on aluminum in nitric acid electrolyte. Thin Solid Films 416:16–23

67. Augustynski J, Berthou H, Painot J (1976) XPS study of interactions between aluminum metal and nitrate ions. Chem Phys Lett 44:221–224
68. Parkhutik VP, Makushok YE, Kudryavtsev VI, Sokol VA, Khodan AN (1987) X-ray photoelectron study of the formation of anodic oxide-films on aluminum in nitric-acid. Sov Electrochem 23:1439–1444
69. Öchsner R, Kluge A, Ryssel H (1989) A versatile ion implanter for material modification. Nucl Instrum Methods B 37–38:504–507
70. Saha NC, Tompkins HG (1992) Titanium nitride oxidation chemistry – an X-ray photoelectron-spectroscopy study. J Appl Phys 72:3072–3079
71. Li J, Lu N, Quan X, Chen S, Zhao H (2008) Facile method for fabricating boron-doped TiO_2 nanotube array with enhanced photoelectrocatalytic properties. Ind Eng Chem Res 47:3804–3808
72. Chen D, Yang D, Wang Q, Jiang ZY (2006) Effects of boron doping on photocatalytic activity and microstructure of titanium dioxide nanoparticles. Ind Eng Chem Res 45:4110–4116
73. Zhao W, Ma WH, Chen CC, Zhao JC, Shuai ZG (2004) Efficient degradation of toxic organic pollutants with $Ni_2O_3/TiO_{2-x}B_x$under visible irradiation. J Am Chem Soc 126:4782–4783
74. Ruan C, Paulose M, Varghese OK, Mor GK, Grimes CA (2005) Fabrication of highly ordered TiO_2 nanotube arrays using an organic electrolyte. J Phys Chem B 109:15754–15759
75. Chen SG, Paulose M, Ruan C, Mor GK, Varghese OK, Kouzoudis D, Grimes CA (2006) Electrochemically synthesized CdS nanoparticle-modified TiO_2 nanotube-array photoelectrodes: preparation, characterization, and application to photoelectrochemical cells. J Photochem Photobiol 177:177–184
76. Kundu M, Khosravi AA, Kulkarni SK (1997) Synthesis and study of organically capped ultra small clusters of cadmium sulphide. J Mater Sci 32:245–258
77. Savelli M, Bougnot J (1979) Topics in applied physics, vol 31. Solar Energy Conversion, Springer, Berlin
78. Mor GK, Varghese OK, Paulose M, Shankar K, Grimes CA (2005) Effect of anodization bath chemistry on photochemical water splitting using titania nanotubes. Mater Res Soc Symp Proc 836:L1.9.1
79. de Taconni NR, Chenthamarakshan CR, Yogeeswaran G, Watcharenwong A, de Zoysa RS, Basit NA, Rajeshwar K (2006) Nanoporous TiO_2 and WO_3 films by anodization of titanium and tungsten substrates: Influence of process variables on morphology and photoelectrochemical response. J Phys Chem B 110:25347–25355
80. Taflove A (1995) Computational electrodynamics: the finite-difference time-domain method. Artech House Inc, Boston
81. Ong KG, Varghese OK, Mor GK, Grimes CA (2005) Numerical simulation of light propagation through highly-ordered titania nanotube arrays: Dimension optimization for improved photoabsorption. J Nanosci Nanotechnol 5:1–7
82. Roden JA, Gedney SD (1997) Efficient implementation of the uniaxial-based PML media in three-dimensional nonorthogonal coordinates with the use of the FDTD technique. Microwave Opt Technol Lett 14:71–75
83. Wittberg TN, Wolf JD, Keil RG, Wang PS (1983) Low-temperature oxygen diffusion in alpha titanium characterized by auger sputter profiling. J Vac Sci Technol A 1:475–478
84. Negishi N, Takeuchi K, Ibusuki T (1998) Surface structure of the TiO_2 thin film photocatalyst. J Mater Sci 33:5789–5794
85. Amanullah FM, Al-Mobarak MS, Al-Dhafiri AM, Al-Shibani KM (1999) Development of spray technique for the preparation of thin films and characterization of tin oxide transparent conductors. Mater Chem Phys 59:247–253
86. Yang P, Liou KN, Mishchenko MI, Gao BC (2000) Efficient finite-difference time-domain scheme for light scattering by dielectric particles: application to aerosols. Appl Opt 39:3727–3737
87. Mor GK, Shankar K, Varghese OK, Grimes CA (2004) Photoelectrochemical properties of titania nanotubes. J Mater Res 19:2989–2996

88. Zheng S, Gao L, Zhang QH, Sun J (2001) Synthesis, characterization, and photoactivity of nanosized palladium clusters deposited on titania-modified mesoporous MCM-41. J Solid State Chem 162:138–141
89. Vogel R, Meredith P, Kartini I, Harvey M, Riches JD, Bishop A, Heckenberg N, Trau M, Rubinsztein-Dunlop H (2003) Mesostructured dye-doped titanium dioxide for micro-optoelectronic applications. Chem Phys Chem 4:595–603
90. Liu FM, Wang TM (2002) Surface and optical properties of films grown by radio frequency nanocrystalline anatase titania reactive magnetron sputtering. Appl Surf Sci 195:284–290
91. Oh SH, Kim DJ, Hahn SH, Kim EJ (2003) Comparison of optical and photocatalytic properties of TiO_2 thin films prepared by electron-beam evaporation and sol-gel dip-coating. Mater Lett 57:4151–4155
92. Asanuma T, Matsutani T, Liu C, Mihara T, Kiuchi M (2004) Structural and optical properties of titanium dioxide films deposited by reactive magnetron sputtering in pure oxygen plasma. J Appl Phys 95:6011–6016
93. Manifacier JC, Gasiot J, Fillard JP (1976) Simple method for determination of optical-constants n, k and thickness of a weakly absorbing thin-film. J Phys E 9:1002–1004
94. Vogel R, Meredith P, Kartini I, Harvey M, Riches JD, Bishop A, Heckenberg N, Trau M, Dunlop HR (2003) Mesostructured dye-doped titanium dioxide for micro-optoelectronic applications. Chem Phys Chem 4:595–603
95. Mor GK, Carvalho MA, Varghese OK, Pishko MV, Grimes CA (2004) A room-temperature TiO_2-nanotube hydrogen sensor able to self-clean photoactively from environmental contamination. J Mater Res 19:628–634
96. Yoldas BE, Partlow DP (1985) Formation of broad-band antireflective coatings on used silica for high-power laser applications. Thin Solid Films 129:1–14
97. Tauc J (1970) Absorption edge and internal electric fields in amorphous semiconductors. Mater Res Bull 5:721–729
98. Sant PA, Kamat PV (2002) Interparticle electron transfer between size-quantized CdS and TiO_2 semiconductor nanoclusters. Phys Chem Chem Phys 4:198–203
99. Henglein A (1989) Small-particle research – physicochemical properties of extremely small colloidal metal and semiconductor particles. Chem Rev 89:1861–1873
100. Kokai J, Rakhshani AE (2004) Photocurrent spectroscopy of solution-grown CdS films annealed in $CdCl_2$ vapour. J Phys D 37:1970–1975
101. Joo S, Muto I, Hara N (2008) In situ ellipsometric analysis of growth processes of anodic TiO_2 nanotube films. J Electrochem Soc 155:C154–C161
102. Bruggeman DAG (1935) Calculation of various physics constants in heterogenous substances. I. Dielectricity constants and conductivity of mixed bodies from isotropic substances. Ann Phys 24:636–664
103. Stein N, Rommelfangen M, Hody V, Johann L, Lecuire JM (2002) In situ spectroscopic ellipsometric study of porous alumina film dissolution. Electrochim Acta 47:1811–1817
104. Wang J, Lin Z (2008) Freestanding TiO_2 nanotube arrays with ultrahigh aspect ratio via electrochemical anodization. Chem Mater 20:1257–1261
105. Bersani D, Antonioli G, Lottici PP, Lopez T (1998) Raman study of nanosized titania prepared by sol-gel route. J Non-Cryst Solids 234:175–181
106. Tsuchiya H, Macak JM, Ghicov A, Räder AS, Taveira L, Schmuki P (2007) Characterization of electronic properties of TiO_2 nanotube films. Corros Sci 49:203–210
107. Muñoz AG (2007) Semiconducting properties of self-organized TiO_2 nanotubes. Electrochim Acta 52:4167–4176
108. Muñoz AG, Chen Q, Schmuki P (2007) Interfacial properties of self-organized TiO_2 nanotubes studied by impedance spectroscopy. J Solid State Electrochem 11:1077–1084
109. Sato N (1998) Electrochemistry at metal and semiconductor electrodes. Elsevier, Amsterdam, The Netherlands
110. Dafonseca C, Ferreira MG, Belo MD (1994) Modeling of the impedance behavior of an amorphous-semiconductor schottky-barrier in high depletion conditions –application to the study of the titanium anodic oxide electrolyte junction. Electrochim Acta 39:2197–2205

111. Oliva FY, Avalle LB, Santos E, Cámara OR (2002) Photoelectrochemical characterization of nanocrystalline TiO_2 films on titanium substrates. J Photochem Photobiol A 146:175–188
112. Dolata M, Kedzierzawski P, Augustynski J (1996) Comparative impedance spectroscopy study of rutile and anatase TiO_2 film electrodes. Electrochem Acta 41:1287–1293
113. Simons W, Pauwels L, Hubin A (2002) Impedance spectroscopy to characterise an anodised titanium substrate in contact with silver complexing agents: elements for an optimal parameter estimation. Electrochim Acta 47:2169–2175
114. Di Quarto F, La Mantia F, Santamaría M (2005) Physicochemical characterization of passive films on niobium by admittance and electrochemical impedance spectroscopy studies. Electrochim Acta 50:5090–5102
115. Cohen JD, Lang DV (1982) Calculation of the dynamic-response of schottky barriers with a continuous distribution of gap states. Phys Rev B 25:5321–5350
116. Boschloo GK, Goossens A, Schoonman J (1997) Photoelectrochemical study of thin anatase TiO_2 films prepared by metallorganic chemical vapor deposition. J Electrochem Soc 144:1311–1317
117. Boschloo GK, Goossens A (1996) Electron trapping in porphyrin-sensitized porous nanocrystalline TiO_2 electrodes. J Phys Chem 100:19489–19494
118. Redmond G, Fitzmaurice D, Grätzel M (1993) Effect of surface chelation on the energy of an intraband surface-state of a nanocrystalline TiO_2 film. J Phys Chem 97:6951–6954
119. Lyon LA, Hupp JT (1999) Energetics of the nanocrystalline titanium dioxide aqueous solution interface: approximate conduction band edge variations between H–0=−10 and H−=+26. J Phys Chem 103:4623–4628
120. Kavan L, Kratochvilova K, Grätzel M (1995) Study of nanocrystalline TiO_2 (anatase) electrode in the accumulation regime. J Electroanal Chem 394:93–102
121. Gutierrez C, Salvador P (1982) Bandgap at the normal-TiO_2 electrolyte interface. J Electroanal Chem 138:457–463
122. Weber MF, Schumacher LC, Dignam MJ (1982) Effect of hydrogen on the dielectric and photo-electrochemical properties of sputtered TiO_2 films. J Electrochem Soc 129:2022–2028
123. Ghicov A, Tsuchiya H, Hahn R, Macak JM, Muñoz AG, Schmuki P (2006) TiO_2 nanotubes: H^+ insertion and strong electrochromic effects. Electrochem Commun 8:528–532
124. Fabregat-Santiago F, Barea EM, Bisquert J, Mor GK, Shankar K, Grimes CA (2008) High carrier density and capacitance in TiO_2 nanotube arrays induced by electrochemical doping. J Am Chem Soc 130:11312–11316
125. Crawford GA, Chawla N, Das K, Bose S, Bandyopadhyay A (2007) Microstructure and deformation behavior of biocompatible TiO_2 nanotubes on titanium substrate. Acta Biomater 3:359–367
126. Katz JL (1996) Application of materials in medicine and dentistry: orthopedic applications. In: Ratner BD, Hoffman AS, Schoen FJ, Lemons JE (eds) Biomaterials Science: An Introduction to Materials in Medicine. San Diego, Academic Press, pp 335–346
127. Park JB, Lakes RS (1992) Biomaterials: an introduction. Plenum Press, New York, pp 79–115
128. Huiskes R, Weinans H, Vanrietbergen B (1992) The relationship between stress shielding and bone-resorption around total hip stems and the effects of flexible materials. Clin Orthop Relat Res 274:124–134
129. Oliver WC, Pharr GM (1992) An improved technique for determining hardness and elastic-modulus using load and displacement sensing indentation experiments. J Mater Res 7:1564–1567
130. Tsui TY, Pharr GM (1999) Substrate effects on nanoindentation mechanical property measurement of soft films on hard substrates. J Mater Res 14:292–2301
131. Tsui TY, Vlassak J, Nix WD (1999) Indentation plastic displacement field: Part I. The case of soft films on hard substrates. J Mater Res 14:2196–2203
132. Tsui TY, Vlassak J, Nix WD (1999) Indentation plastic displacement field: Part II. The case of hard films on soft substrates. J Mater Res 14:2204–2209

133. Bahr DF, Woodcock CL, Pang M, Weaver KD, Moody NR (2003) Indentation induced film fracture in hard film–soft substrate systems. Int J Fract 119:339–349
134. Pang M, Bahr DF (2001) Thin-film fracture during nanoindentation of a titanium oxide film–titanium system. J Mater Res 16:2634–2643
135. Deng X, Cleveland C, Karcher T, Koopman M, Chawla N, Chawla KK (2005) Nanoindentation behavior of nanolayered metal–ceramic composites. J Mater Eng Perform 14:417–423
136. Deng X, Chawla N, Chawla KK, Koopman M, Chu JP (2005) Mechanical behavior of multilayered nanoscale metal–ceramic composites. Adv Eng Mater 7:1099–1108
137. Mayo MJ, Siegel RW, Narayanasamy A, Nix WD (1990) Mechanical properties of nanophase TiO_2 as determined by nanoindentation. J Mater Res 5:1073–1081
138. Han Y, Hong SH, Xu KW (2002) Porous nanocrystalline titania films by plasma electrolytic oxidation. Surf Coat Tech 154:314–318
139. Fischer-Cripps AC (2004) Nanoindentation. Springer, New York

Chapter 3
TiO_2 Nanotube Arrays: Application to Hydrogen Sensing

3.1 Introduction

In this chapter we consider application of TiO_2 nanotube arrays to hydrogen gas sensing. Hydrogen, a combustible, widely used industrial gas has great potential for use as a carbon-free chemical fuel. The use of hydrogen, or where hydrogen is an undesired contaminant, requires a monitor suitable for detection of meaningful concentrations. Furthermore, quantification of ppm – ppb hydrogen gas concentrations has medical relevance as an indicator of lactose intolerance [1–3], fructose malabsorption [4–8], microbial activity [9], bacterial growth [10–12], fibromyalgia [13], diabetic gastroparesis [14–16], and neonatal necrotizing enterocolitis (NEC) [17–21]. The pathogenesis of neonatal NEC results in the production of hydrogen gas, which accumulates as bubbles in the sub-mucosal area of the bowel wall [18]. Hydrogen is absorbed into the blood stream and excreted transcutaneously, as well as via the lungs into the exhaled breathe [19–21]. For monitoring of NEC in preterm infants, it appears a clinically useful hydrogen sensor must be capable of detecting transcutaneous hydrogen at levels of approximately 25 ppm to 1 ppm, while the sensitivity of the infants' skin requires the use of unheated sensors.

In 2003, Varghese and co-workers [22, 23] reported high-temperature hydrogen sensing using TiO_2 nanotube arrays a few hundred nm in length, anodized in HF aqueous electrolytes [24–26]. In this early work it was noted that due to chemisorption of molecular hydrogen, the electrical resistance of the TiO_2 nanotubes was highly sensitive to hydrogen as well as fully reversible; for example, for short tubes of 76 nm pore size, there is a variation in measured resistance of 10^3 upon exposure to 500 ppm hydrogen at 290°C. It was found that the hydrogen sensitivity of the nanotubes was highly dependent upon nanotube diameter, with 22 nm pore size nanotubes being about 200 times more sensitive than 76 nm pore size nanotubes, although the smaller pore size nanotubes have a larger surface area by only 30%. The nanotube morphology and the wall thickness are found to invoke unique responses, as the wall thickness is comparable to the Debye length. Thus, the

C.A. Grimes and G.K. Mor, *TiO_2 Nanotube Arrays*,
DOI 10.1007/978-1-4419-0068-5_3, © Springer Science+Business Media, LLC 2009

space charge layers are strongly modified by the dimensional features and the charge transfer associated with the chemisorptions at such length scales.

The demand for highly sensitive, selective and stable hydrogen sensors has increased in recent years. Various types of hydrogen sensor technologies [27] such as Schottky junction [31–38], fiber optic [32–34], catalytic [35–37], electrochemical [38–41], field effect transistor (FET) [42–44], oxide semiconductor [45–47] and combinations of these, are being developed, or have been developed. Oxide semiconductor gas sensors are relatively simple and, hence, involve lower costs. However, they are commonly operated at elevated temperatures, typically several hundred degrees Celsius, to enhance gas sensitivities and enable fast and reversible reactions. Elevated operating temperatures are not favorable for many applications, such as those involving flammable environments, biomedical applications, and those requiring low-power operation. Only a few efforts employing metal oxide semiconductors for room temperature hydrogen sensing can be found in the literature [48–55].

A critical concern of any sensor platform is the potential for unwanted contamination, or poisoning, which introduces spurious measurements and generally ends its useful lifetime. A sensor used in a noncontrolled environment faces potential contamination from, for example, volatile organic vapors. An important advance in sensor technology would be a sensor able to self-clean, thereby extending its useful lifetime and minimizing the potential for spurious measurements. To that end, the effect of contamination on the TiO_2 nanotube array hydrogen sensors is overcome by ultraviolet (UV) illumination of the nanotube array surface [56]. TiO_2 is a semiconductor characterized by a filled valence band and an empty conduction band. When a photon with energy hν matches or exceeds the bandgap energy an electron – hole pair are produced. The valence band holes are powerful oxidants, while conduction band electrons are good reducers. Most organic photodegradation reactions utilize the oxidizing power of electron holes, either directly or indirectly, generally producing CO_2 and H_2O without production of potentially toxic side products [57]. Factors improving photocatalytic efficiency are (i) the presence of electron-hole pairs on the surface, (ii) longer electron-hole recombination lifetimes, (iii) geometries providing higher surface-to-volume ratios, and (iv) a predominance of the anatase crystalline phase. Items (i)–(iii) are directly related to the material architecture [58], while crystallinity (iv) is controlled through subsequent annealing of the initially amorphous nanotubes. Room temperature TiO_2 nanotube array hydrogen sensors were used to demonstrate the self-cleaning capabilities [53, 59]. Their photocatalytic properties are such that the hydrogen-sensing capabilities of the sensors are largely recovered by UV light exposure after being completely extinguished by the rather extreme means of sensor contamination: Coating of the sensor in motor oil.

Metal oxide semiconductors do not have the ability to selectively respond to a particular gas from a mixture and hence, chemical modification, such as doping or surface functionalization, must be used [60–64]. With the advent of nanotechnology, various nanostructures of functional metal oxides have been developed for environmental sensing, and it has become evident that nanoscale architectural features have the ability to yield superior and often unexpected gas-dependent electrical behavior [65–74]. TiO_2 nanotube arrays of up to a few microns in length, pore diameters from 22 to 110 nm, and wall thicknesses from 9 to 34 nm show an unprecedented change in electrical resistance in response to low levels of hydrogen at room temperature [75, 76]. Results demonstrate the ability of nanodimensional structures to possess extraordinary properties that micro/macroscale structures have never exhibited, which can open new avenues in nanotechnology [77].

The ability to measure low hydrogen concentrations from an unheated, room temperature sensor enables a variety of new applications, of particular interest to the authors measurement of transcutaneous gas concentrations. Transcutaneous gas monitoring has been used as a means for diagnosing disease, as well as monitoring treatment effectiveness. For example, transcutaneous carbon dioxide is routinely monitored for bronchopulmonary dysplasia, apnea, upper airway obstruction, and respiratory problems associated with hypercapnia [78]. TiO_2 nanotube arrays were used for the measurement of transcutaneous hydrogen concentrations [79] as indicators of lactose intolerance [80–82]. Currently, lactose intolerance is diagnosed with the Lactose Tolerance Test [83, 84], where a patient's blood samples are taken over a 2-h period to measure the blood glucose levels after consumption of lactose to indicate how well the body is able to digest lactose. Alternatively, the Hydrogen Breath Test [80] can be performed to measure the amount of hydrogen present in the patient's breath since undigested lactose in the colon is fermented by bacteria and produces various gases including hydrogen, that are absorbed from the intestines, carried through the bloodstream to the lungs, and exhaled. As indicated by the enabling work of Gerlach [85], a portion of hydrogen in the bloodstream is released through the skin, thus allowing hydrogen detection with a noninvasive, transcutaneous hydrogen sensor.

3.2 High Temperature Sensors using TiO_2 Nanotube Arrays

TiO_2 nanotube arrays were grown by anodization of Ti foil using an aqueous electrolyte of 0.5% hydrofluoric acid [22, 24]. Nanotube samples were prepared using 20, 14 and 10 V anodization voltages having, respectively, pore sizes of 76, 53 and 22 nm; wall thicknesses of 17, 15 and 13 nm; and tube lengths 400, 260 and 200 nm. Prior to electrical measurements, all samples were annealed at 500°C for 6 h in oxygen ambient. The samples under test were placed within a quartz tube furnace. Two platinum contact pads were deposited on the nanotube array. Before measurements, the test chamber was evacuated and then nitrogen gas was passed while the sample under test was heated to the desired temperature. The test gases examined, oxygen, carbon dioxide, ammonia, carbon monoxide and hydrogen, were mixed in appropriate ratios with nitrogen to create the desired test gas ambient. No appreciable variation in the electrical resistance of the nanotubes was observed due to either carbon monoxide or ammonia. In the presence of oxygen, the conductance of the nanotubes decreased by a factor of approximately

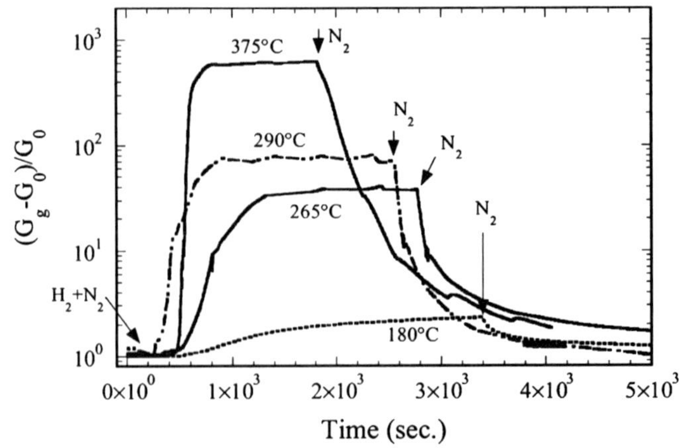

Fig. 3.1 Normalized change in electrical conductance of 76 nm pore size TiO_2 nanotube array to 1,000 ppm hydrogen at different temperatures. G_g is the conductance in the presence of the test gas and G_0 the base resistance associated with a nitrogen atmosphere [22]

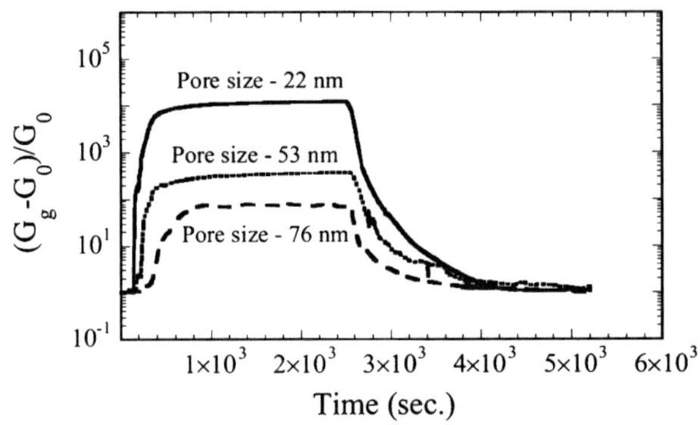

Fig. 3.2 Normalized change in electrical conductance of TiO_2 nanotube arrays of 76, 53, and 22 nm pore size to 1,000 ppm hydrogen at 290°C [22]

two; original values were not regained even after several tens of hours in a nitrogen environment.

Fig. 3.1 shows the normalized change in electrical conductance of a 76 nm pore size nanotube array as a function of temperature, as it is switched from a nitrogen ambient to a nitrogen plus 1,000 ppm hydrogen ambient and then back to nitrogen. With reference to Fig. 3.1, G_g is the conductance in the presence of the test gas and G_0 the base conductance associated with a nitrogen atmosphere. The magnitude of the conductance variation increases with temperature, starting from a marginal variation at 180°C to a variation of 3 orders of magnitude at approximately 400°C. The variation in electrical conductance of 76, 53, and 22 nm diameter TiO_2 nanotube arrays with exposure to 1,000 ppm hydrogen at 290°C is shown in Fig. 3.2. It is clear that the conductance variation is more prominent at smaller

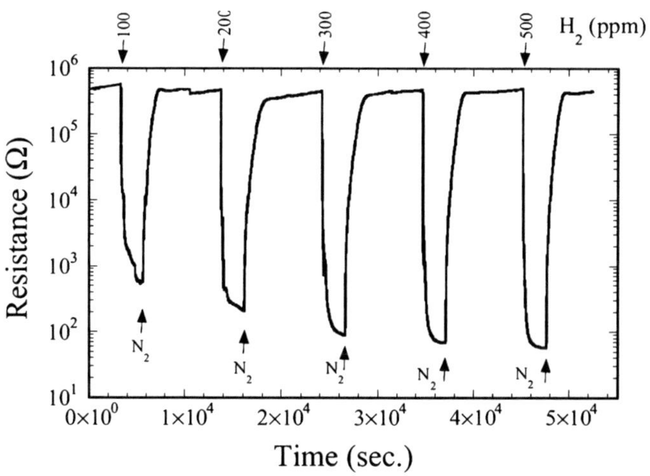

Fig. 3.3 Electrical resistance of 22 nm pore size TiO_2 nanotube array when exposed to different hydrogen concentrations at 290°C. The nanotube response is completely reversible without hysteresis or drift [22]

nanotube diameters. The 22 nm diameter nanotube sample shows over a four order of magnitude change in conductance, a factor of 10 greater than the maximum hydrogen sensitivity limit previously reported of 1,400 [47]. Assuming the density of the nanotube is 90% that of single crystal TiO_2, the calculated surface area of the 76 nm diameter nanotube array is approximately 19 m^2/g, and that of the 22 nm array 38 m^2/g. While reduction of nanotube diameter from 76 to 22 nm increases the surface area by a factor of two the hydrogen sensitivity is increased by approximately a factor of 125. Fig. 3.3 shows the variation in resistance of a 22 nm diameter nanotube sample at 290°C when the hydrogen concentration is cycled from 0 to 500 ppm in discrete steps of 100 ppm. A three order of magnitude variation in the resistance is seen with exposure of only 100 ppm hydrogen. It is clear from Fig. 3.3 that the nanotubes respond without hysteresis.

The results indicate that the mechanism behind this unusual enhancement of conductance in the TiO_2 nanotubes with hydrogen exposure is neither the reduction of the TiO_2 lattice [86, 87] nor removal of chemisorbed oxygen from the nanotube surface by the hydrogen [88]. The platinum electrodes suggest the availability of hydrogen atoms (H^*), ions (H^+) and various other forms of hydrogen such as H^- ions and H_3 on the surface of the nanotubes due to the well known spill-over mechanism of hydrogen by platinum [89, 90]. Although the diffusion of the "spilt" hydrogen atoms into the TiO_2 interstitial positions and their subsequent ionization would enhance the conductivity [91], the fast response and lack of measurement hysteresis suggest that this mechanism does not contribute significantly to the overall changes in nanotube resistivity. It appears the dominant mechanism behind the observed electrical behavior of the nanotubes is chemisorption of the spilled-over hydrogen atoms [24, 89] on the nanotube surface, which

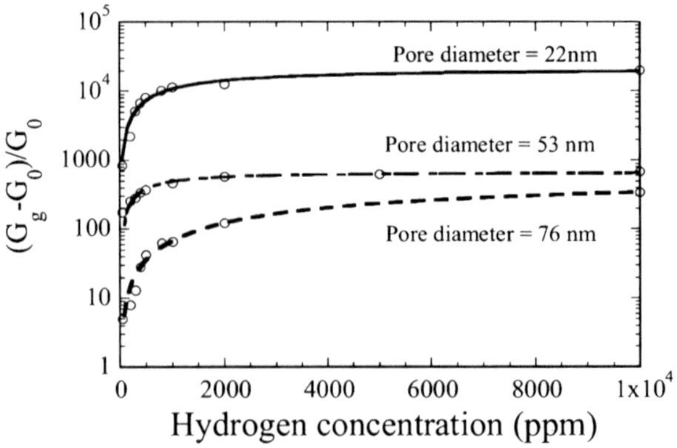

Fig. 3.4 Conductance isotherms of 76, 53, and 22 nm diameter samples at 290°C. The open circles represent experimental data and lines the isotherm fit given by (3.1) [22]

make a partial charge transfer to the TiO_2, thereby creating an electron accumulation layer on the nanotube surface that enhances the electrical conductance.

Fig. 3.4 shows the conductance isotherms of the 76, 53 and 22 nm pore size samples measured at 290°C. The open circles represent measured values and the lines curve fits from a hydrogen "spill-over" model [89]. As shown, the electrical conductance of the nanotubes is sensitive to hydrogen concentrations ranging from 10 ppm, the lower limit measured, to 1%. The curve fit of the data in Fig. 3.4 relates the maximum (saturated) conductance variation to hydrogen concentration using:

$$\frac{G_{\mathrm{gs}} - G_{\mathrm{o}}}{G_{\mathrm{o}}} = \frac{kbP_{\mathrm{p}}}{1 + bP_{\mathrm{p}}} \tag{3.1}$$

G_{gs} denotes the saturation conductance value for a given concentration, P_{p} denotes partial pressure, b is a fitting parameter, and k a real-number scaling factor. Equation (3.2.1) closely resembles the Langmuir isotherm modified to include only chemisorption associated with charge transfer [89]. Assuming that the catalytic activity of platinum and the potential barrier for the spilt-over hydrogen to diffuse onto the TiO_2 remain constant with varying hydrogen concentration fitting parameter b can be written as:

$$b = \frac{\alpha P_{\mathrm{p}}}{\sqrt{2\pi mkT}} \frac{e^{(E^0_{\mathrm{ads}} - eV_0)/kT}}{\vartheta_{\mathrm{eff}}\, \sigma_{\mathrm{sites}}} \frac{1 + 2e^{(E_{\mathrm{F}} - E^{\mathrm{el}}_{\mathrm{chem}} - eV_0)/kT}}{e^{-(E_{\mathrm{G}} + E^{\mathrm{el}}_{\mathrm{chem}})/kT}} \tag{3.2}$$

where α is the probability that a gas molecule that has arrived at an adsorption center on the surface will be adsorbed, m the mass of the adsorbed species, E^0_{ads} is the adsorption energy of neutral species, V_0 is the surface potential created by the adsorbed hydrogen, ϑ_{eff} is the rate constant of desorption, σ_{sites} is the surface density of adsorption sites, E_{F} is the Fermi energy of TiO_2, $E^{\mathrm{el}}_{\mathrm{chem}}$ is the energy of chemisorbed charged species, E_G is the band gap of TiO_2, T is the temperature in Kelvin, and k is Boltzmann's constant.

The modified Langmuir isotherm of (3.2.1) was developed for hydrogen desorption in atomic form, hence the resemblance with this isotherm suggests the same phenomenon in the TiO_2 nanotubes. Hydrogen dissociation takes place at the platinum electrodes and at crystal defects on the nanotube surface [91, 92]. From the electrodes the dissociated species travel upon the nanotube surface a distance on the order of mm. During chemisorptions, hydrogen acts as a surface state having an energy level near the TiO_2 conduction band; electron transfer from hydrogen to the TiO_2 creates a space charge layer of potential V_0 at the surface reducing the electrical resistance of the nanotube. The band bending and creation of the surface space charge layer continues until the Fermi level of TiO_2 equilibrates with the energy level of the surface H^+ adsorbate layer. Hydrogen desorption, in atomic form, takes place by transferring an electron back to the adsorbate with a corresponding increase in electrical resistance.

Calculations based upon FESEM images indicate a factor of two increase in surface area with reduction in nanotube pore size from 76 to 22 nm, which is not enough to explain the factor of $\approx$125 change in hydrogen sensitivity. It was believed that the nanotube wall thickness, and the number of contact points between the nanotubes are additional factors that play a role in determining the nanotube sensitivity. For example with reduction of the wall thickness to less than the Debye length, the space charge layer can extend throughout the wall resulting in a significant reduction in resistance consistent with observations in nano-crystalline TiO_2 [93]. The nanotubes are in contact with each other providing a high resistance path for electron travel. On the creation of the space charge layer due to hydrogen adsorption, the neck points become highly conductve relative to the rest of the nanotube (see Fig. 3.7). For a constant conductivity with nanotube diameter, the greater the number of contact points the greater will be the resistance change upon exposure to hydrogen. Therefore, the smaller diameter tubes, with thinner walls and a greater number of contact points will exhibit higher sensitivities than their larger diameter counterparts.

3.3 Self-Cleaning Room-Temperature Hydrogen Sensors

TiO_2 nanotube-array sensors ($\approx$250 nm in length, 22 nm pore size, 13 nm wall thickness) were prepared by anodizing a titanium sheet at 10 V in a 1:7 acetic acid and 0.5% hydrogen fluoride electrolyte solution [53, 59]. The addition of acetic acid to the predominately HF solution results in more mechanically robust samples without changing the shape or size of the resulting nanotubes. The as-prepared nanotubes were then annealed at 500°C for 6 h in oxygen. The sensor measurement geometry is shown in Fig. 3.5. The sensors have one electrical contact on their surface, a rf-sputtered platinum pad about 2.0 mm in diameter, and an electrical contact to the underlying titanium foil. A 3.0 mm diameter aluminum disk was

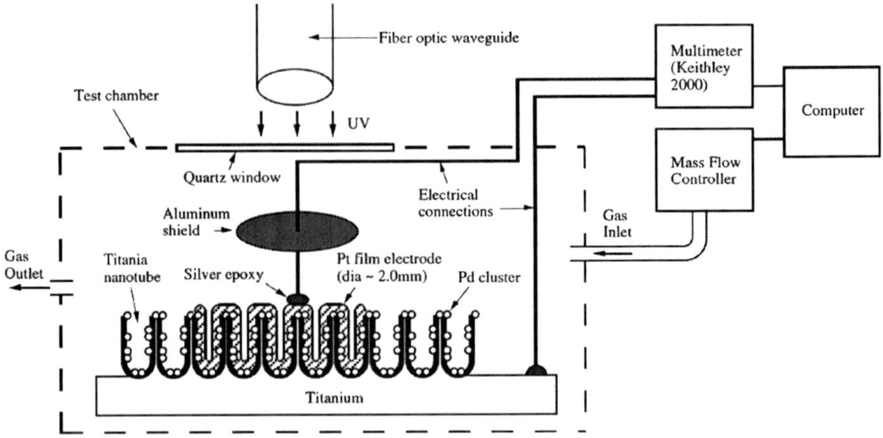

Fig. 3.5 Schematic diagram of the set up used for investigating the self-cleaning capability of TiO_2 based room temperature hydrogen gas sensor [53]

placed approximately 3.0 mm directly above the platinum contact, shading it from the UV illumination to prevent degradation of the electrical contact. All measurements were done at room temperature, i.e., 24°C.

The self-cleaning sensor experiments were performed in a 60 cm^3 Plexiglas test chamber, with an opening for introducing the contaminant, in this case motor oil, onto the sensor surface, and a quartz window for passing the UV illumination onto the sensor surface. Different grades of motor oil, 20W-50, 20W-40, 5W-30 and 10W-40 were used as sensor contaminants representing a rather extreme challenge to the sensor. The contaminated sensors were uniformly illuminated by 365 nm UV light, directed from the source (Hoya Schott, Model-HLS 210U, 150W mercury xenon lamp) to the sensor surface through the use of a fiber optic waveguide to eliminate the possibility of heating the sensor surface. The intensity of the 365 nm UV light measured at the sensor surface was 270 mW/cm^2. The electrical resistance of a sensor was measured using a computer-controlled multimeter. A mass flow controller was used to regulate the H_2+N_2 gas mixture and compressed air flow through the test chamber. Stearic acid and cigarette smoke were also examined as sensor contaminates, with the sensor able to self-clean from these challenges in a fashion similar to that seen for the motor oil contamination. A drop of motor oil upon the sensor was able to extinguish the hydrogen-sensing capabilities of the sensor on a repeatable basis.

Fig. 3.6 shows the real-time electrical resistance of an illustrative hydrogen sensor operated at room temperature in response to different atmospheres, contamination, and UV light exposures. After flushing the test chamber with compressed air at a flow rate of 1,000 sccm, the sensor was then exposed to a hydrogen-nitrogen mixture of 1,000 ppm hydrogen. After reaching the saturation resistance, the gas was switched back to air with the sensor returning to its original state, see Fig. 3.6(a). Prior to sensor contamination, Fig. 3.6(a), a 175,000% change in sensor

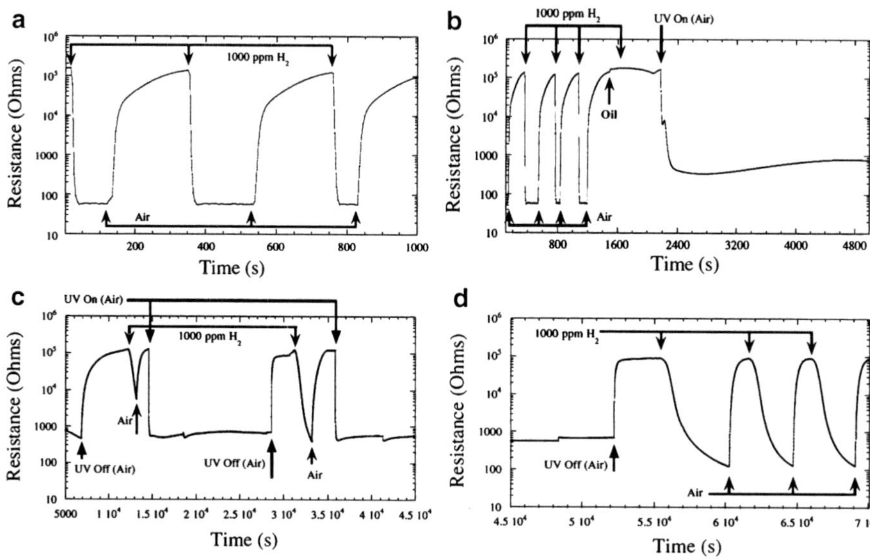

Fig. 3.6 A plot of real time variation of resistance change before, during and after cleaning the contaminant, motor oil 10W30, with UV exposure. The plot, broken into four parts for clarity, shows: **(a)** The original sensor behavior from time 10 to 1,000 s. **(b)** Behavior of sensor over time 100 to 6,000 s during which the sensor is contaminated with oil losing its hydrogen sensing capabilities, and is initially exposed to UV light. **(c)** The behavior of sensor from time 5,000 to 45,000 s. At time 7,000 s the UV is turned off, with the sensor regaining its nominal starting resistance of approximately 100 kΩ, at which point it is exposed to 1,000 ppm hydrogen and shows relative change in resistance of approximately 50. The sensor is then again exposed to UV, from roughly time 15,000 to 29,000 s. After this second UV exposure the sensor is again exposed to 1,000 ppm hydrogen, showing an approximate factor of 500 change in electrical resistance. The sensor is once again exposed to UV, from time 36,000 s. **(d)** Sensor behavior from time 45,000 to 70,000 s continues with UV exposure of the sensor to time 52,000 s, after which the sensor is repeatedly cycled between air and 1,000 ppm hydrogen showing a relative change in impedance of approximately 1,000. Compared to the hydrogen sensitivity of a non-contaminated sensor, the relative response of the "recovered" sensor is with-in a factor of two [59]

resistance was observed upon exposure to 1,000 ppm hydrogen. While in air, see Fig. 3.6(b), the sensor was then contaminated with an approximately 0.06 mm layer of 10W30 (Penzoil) motor oil after which the sensor demonstrated virtually no change in electrical resistance with hydrogen exposure over a period of about 10 min, see Fig. 3.6(b) at time ≈1,600 s. The described TiO_2 nanotube hydrogen sensors are able to accurately measure hydrogen concentrations in electrical transformer oil, an application useful for prediction of transformer failure [94], and that the solubility of hydrogen in oil is approximately 7% by volume. However, in actual practice, it is very challenging to achieve hydrogen levels in oil above a few tens of ppm. Hence, while hydrogen can diffuse into oil where it would then – at least conceptually, be measured, with the oil-contamination of the sensor held in a hydrogen environment little change in the sensor resistance was seen from 1,600 to 2,200 s. The oil-contaminated region on the sensor, Fig. 3.6(b), was uniformly illuminated with UV light in the presence of air. The recovery of the hydrogen-sensing capability of the sensor after UV illumination of 1 h duration can be seen in Fig. 3.6(b-c), 8 h duration in Fig. 3.6(b), and 10 h duration in Fig. 3.6(c-d). As seen from Fig. 3.6(d), at the end of the UV-driven sensor-cleaning period the measured relative change in electrical resistance with exposure to 1,000 ppm hydrogen was 100,000%. The clean and recovered sensor have similar resistance values with exposure to air; the recovered sensor has a 1,000 ppm hydrogen resistance value of approximately 100Ω compared to the 60Ω value of the sensor prior to contamination.

The electrical resistance of the sensor was noted to drop rapidly with UV illumination due to photogeneration of charge carriers; yet, after the UV light source is turned off, a relatively long time is required for resistance to be regained. The light source has a shutter that closes when the light source is turned off, and charge carrier lifetimes do not run into the minutes, so another mechanism underlies the relatively slow recovery in electrical resistance when the UV illumination ends. It was suggested that adsorbed oxygen plays a major role in manipulating the conductivity. On UV illumination, the chemisorbed oxygen must be desorbed increasing the conductivity. Hence, the conductivity increase with UV exposure is partly due to the photogenerated current, and partly to the electrons donated by the desorbed oxygen. On removing the UV illumination, oxygen will be re-adsorbed and hence, the electrons will be extracted from the sensor. However, the process of oxygen re-adsorption is slow, in the order of several minutes; hence, it takes a relatively long time to regain the original sensor resistance after the UV illumination is removed.

Multiple experiments were performed using 20W40, 10W40, and 5W30 grade motor oils as sensor contaminates. The approximate thickness of the oil layer required to completely extinguish the sensor response to hydrogen was found to be approximately 0.03, 0.06 and 0.09 mm respectively corresponding to oil grades 20W40, 10W40 and 5W30. The stepwise recovery of the sensors contaminated with different oil grades is shown by plotting relative resistance change (R_{air}/R_H) vs. UV exposure duration in Fig. 3.7; the data points represent the average of three different sensors contaminated with the same grade oil. Here, R_{air} is the resistance of the clean sensor in air, generally equivalent to that of the contaminated sensor in air, and R_H is the steady state resistance of the sensor exposed to 1,000 ppm hydrogen. The value R_{air}/R_H of the clean sensor is displayed at time equal – 1 h, that of the contaminated sensor is shown at time equal zero. The plot clearly demonstrates the successive recovery of the R_{air}/R_H value towards that of the clean sensor with incremental UV exposure. It was also found that sensor poisoning by the lower weight oil, 5W30, requires a longer exposure duration to achieve recovery, possibly due to the relatively thicker film required to contaminate the sensor. However, the gas-sensing properties of sensors poisoned by commercially available light-weight spray oil, WD-40 (San Diego, CA), could not be recovered with UV exposure. It is possible the WD-40 spray-oil contained products, such as phosphorous or sulfur, that formed oxidation products, which, in turn, deactivated the photo-oxidation capabilities of TiO_2 [95].

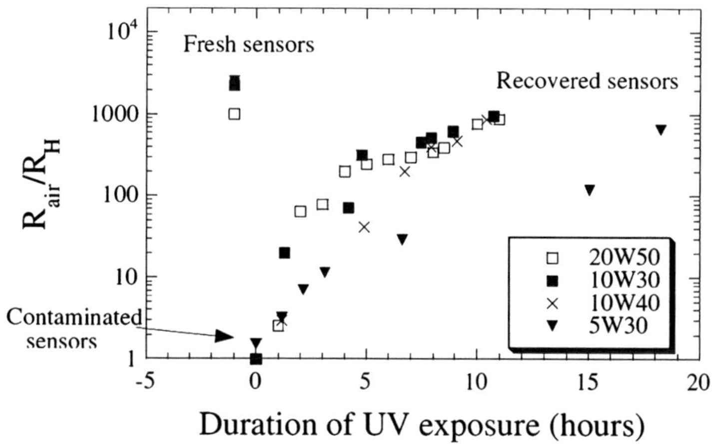

Fig. 3.7 A semi-log plot showing the stepwise improvement in sensor response with UV illumination of sensors contaminated with 20W40, 10W30, 10W40, and 5W30 motor oils. Here the relative change in resistance of a sensor is measured with respect to 1,000 ppm hydrogen as a function of duration of UV exposure. The fresh sensor is shown at arbitrary time while the dirty sensor is at 0 h. The data points represent the average response of five different sensors exposed to each contaminate (grade of motor oil) [59]

It is well established that the presence of oxygen and water plays a crucial role in the photocatalytic cleaning of TiO_2 [96], hence the sensor was exposed to air during UV exposure to facilitate removal of the oil from the sensor surface. In fact, it should be noted that the UV exposure with the sensor kept in a nitrogen atmosphere did not result in sensor recovery. If the relevant redox potential of the contaminant does not lie within the bandgap of TiO_2, the organic contaminants cannot be oxidized by photogenerated electron-hole pairs on the surface of TiO_2. However, since the potential of water and oxygen exists within the bandgap of TiO_2, the photo-generated holes in the valence band can oxidize water to produce a highly reactive hydroxyl radical and the photo-generated electrons in the conduction band can reduce oxygen to form highly reactive super-oxide ions, see (3.3–3.5), which then assist in oxidizing the organic species.

$$TiO_2 + hv \rightarrow h^+ + e^- \quad (3.3)$$

$$H_2O + h^+ \rightarrow {}_{\bullet}OH + H^+ \quad (3.4)$$

$$O_2 + e^- \rightarrow O_{2\bullet}^- \quad (3.5)$$

The overall quantum efficiency for steady-state photolysis depends on the interfacial charge transfer determined in the competition between charge carrier recombination and interfacial charge transfer. An increase in either the charge carrier recombination lifetime or the interfacial electron transfer rate is expected to result in higher quantum efficiencies of the photodegradation reactions. Earlier work [97] has shown that Pd clusters on TiO_2 facilitate separation of photogenerated charge carriers thereby increasing their recombination life time, and increases the rate of electron transfer to oxygen [97–99]. Wang and co-workers [100] reported an increase in photocatalytic activity with a decrease in TiO_2 particle size from 21 to 11 nm; however, the photocatalytic activity decreased when the size was further reduced to 6 nm. The results indicate that, depending upon the nanoarchitecture, there exists an optimal particle in nanocrystalline TiO_2 for maximum photocatalytic activity. Zhang and co-workers [101] explained these observations by correlating an increase in the rate of interfacial charge carrier transfer with decrease in particle size; on reaching a certain limiting particle size, surface recombination exceeds the interfacial charge transfer process and therefore, the photocatalytic activity is reduced.

3.4 Room-Temperature Hydrogen Sensors of Enhanced Sensitivity

3.4.1 TiO_2 Nanotube Arrays on Ti Foil

Nanotube arrays were fabricated by anodizing a titanium foil in an electrolyte solution consisting of 13 volume% acetic acid and 0.5 volume% hydrofluoric acid (~50% concentration) in water [24, 54]. Sample crystallization was achieved by annealing amorphous as-anodized nanotubes at 500°C for 6 h, with a heating and cooling rate of 1°C/min, in oxygen ambient. For hydrogen-sensing measurements the annealed samples were coated with a palladium layer of thickness 10 nm deposited by thermal evaporation. Platinum electrodes with 40 nm thickness and about 2.0 mm in diameter were then sputter-coated onto the sample. Copper leads were attached to the platinum electrodes by silver epoxy. A schematic diagram of the measurement setup is given in Fig. 3.8. There is a difference in the measurement of this sensor from the self-cleaning sensors. Here, hydrogen sensing occurs through observing the change in resistance of a vertical stack of nanotubes, connected through barrier layer and also through adjacent tubes somewhere along the length, lying between two platinum pads whereas in the self-cleaning sensors, the change in resistance upon hydrogen exposure represents the change in resistance of nanotubes along their length and of the barrier layer lying between platinum pad and titanium (i.e., underneath the barrier layer). Experiments were conducted by alternately passing air and a mixture of N_2 and H_2 at the desired concentration levels. Sensing measurements were performed in a sealed Plexiglas chamber of volume ~110 cm^3. The total gas flow rate was kept constant at 500 sccm for all experiments.

The typical behavior of these sensors on exposure to 20–1,000 ppm H_2 concentrations is shown in Fig. 3.9; measurements were done by keeping the ambient switching intervals the same for all concentrations. As the response time at low

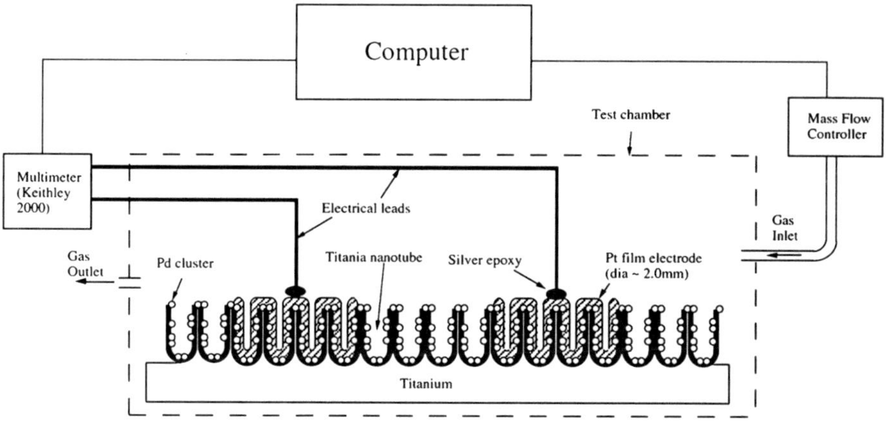

Fig. 3.8 Schematic diagram of experimental set-up used for gas sensing experiments [54]

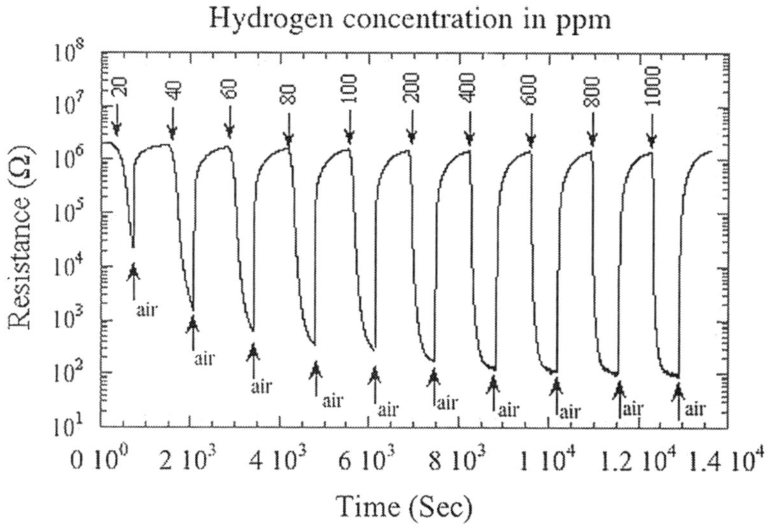

Fig. 3.9 Typical sensor response at 25°C to hydrogen concentrations in the 20–1,000 ppm range [54]

hydrogen concentrations was slightly higher than the one at high concentrations, the sensor resistance in the low concentration range did not reach the saturation value before switching back to air from hydrogen ambient. As seen in this plot, the initial resistance is in the order of $10^6\Omega$, and falls to below $10^2\Omega$ when exposed to 1,000 ppm H_2. Typical 90% response times are $\approx$15 s. The sensor recovers completely after H_2 flow is terminated in each case; no detectable drift was observed. The sensitivity increases quite rapidly with H_2 concentration in the $\leq$200 ppm regime; this behavior is plotted in Fig. 3.10. Thereafter, the change in

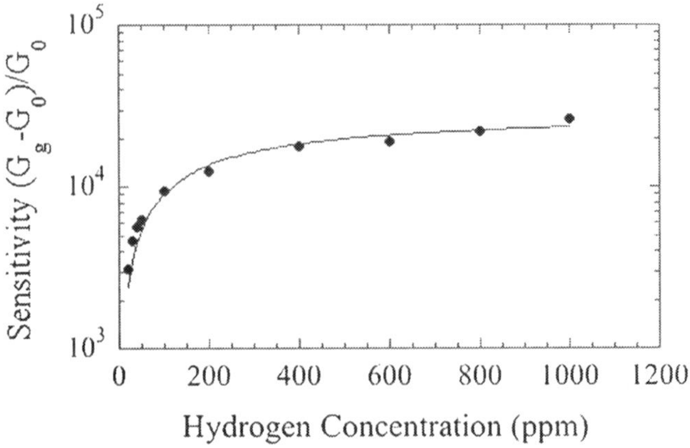

Fig. 3.10 Sensitivity of the sensor at 25°C to different hydrogen concentrations [54]

sensitivity with a further increase in concentration is relatively less. Thus the dynamic range can be said to extend to approximately 1,000 ppm. Fig. 3.10 reveals that the sensitivity S, $S=(G_g-G_0)/G_0$ where G_0 and G_g are the conductance in air and N_2+H_2 mixture respectively, approaches 10^4 at 100 ppm H_2. The points for this plot were collected by noting the saturation resistance of the sensor at each hydrogen gas concentration. The measured hydrogen sensitivities are superior to any reported in the literature for resistive type metal oxide semiconductors at room temperature. Fonash et al. [48], used a Pd/TiO_x/Si heterostructure for H_2 sensing and obtained a sensitivity of 10^4 for 400 ppm hydrogen, with the TiO_x layer deposited on silicon by low-pressure chemical vapor deposition. Basu and Dutta [102] observed a three-order of magnitude variation in resistance for 2,000 ppm of hydrogen in Pd/ZnO/Zn heterojunctions. Recently, Nakagawa et al. [51] used nanocrystalline tungsten trioxide with Pt catalyst to achieve a sensitivity of 10^6 for 1% hydrogen; however, 85% humidity and elevated temperatures between 100°C and 150°C were required for the sensor to completely reverse within a reasonable time interval.

To confirm that measurement-drift is indeed minimal, as suggested by Fig. 3.9, the sensor was subjected to repeated cycling between air and a 1,000 ppm N_2–H_2 mixture. The results are shown in Fig. 3.11. As seen, there is no significant drift in resistance (both initial and final) with repeated exposure to a high concentration of hydrogen. Furthermore the sensor did not show any measurable resistance variation in the presence of 2,000 ppm carbon monoxide and 10% carbon dioxide atmospheres.

One motivation for development of an ambient temperature hydrogen gas sensor is application in the biomedical field per transcutaneous measurements. Consequently, the performance of the sensors was tested at 36°C, roughly body temperature. As shown in Fig. 3.12, a slight increase in sensitivity was observed at 36°C in comparison to 25°C results. The sensitivity to 20 ppm H_2 at 36°C (Fig. 3.12) is

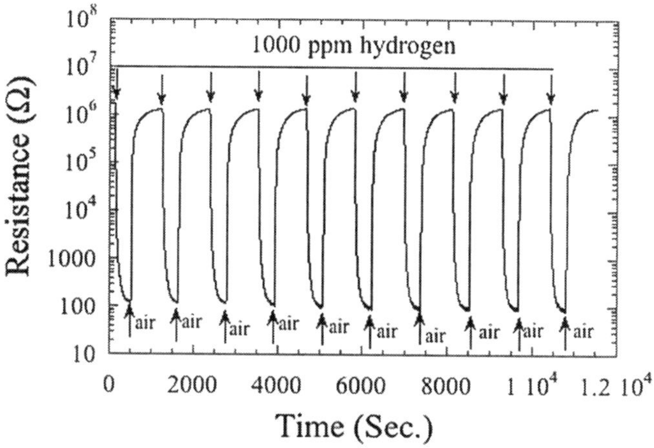

Fig. 3.11 Sensor response at room temperture upon repeated cycling to 1,000 ppm hydrogen. No drift is observed in both the initial and final resistance values [54]

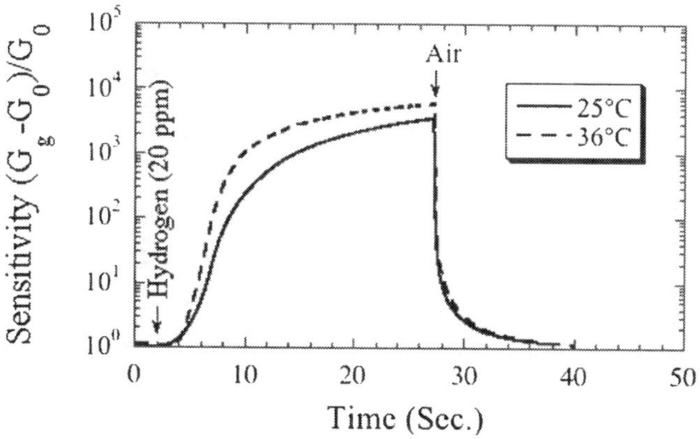

Fig. 3.12 Response of the sensor to 20 ppm hydrogen at room temperature (25°C) and ≈body temperature (36°C). The response time and sensitivity of the sensor are enhanced at 36°C, while the recovery time is approximately the same [54]

more than three orders in magnitude, confirming the utility of these sensors for biomedical applications where hydrogen concentrations are generally present in the low ppm range.

The catalytic properties of palladium for hydrogen dissociation as well as its hydrogen adsorption-desorption characteristics are well known [103]. Palladium and platinum form a Schottky barrier with TiO_2 as the work function of Pd (5.12 eV) and Pt (5.65 eV) [104] are higher than that of TiO_2 ($\approx$4.6 eV) [105]. For a sensor in air, oxygen is chemisorbed on the surface of these metals increasing their work function, in turn further increasing the barrier height [106]. Thus, the

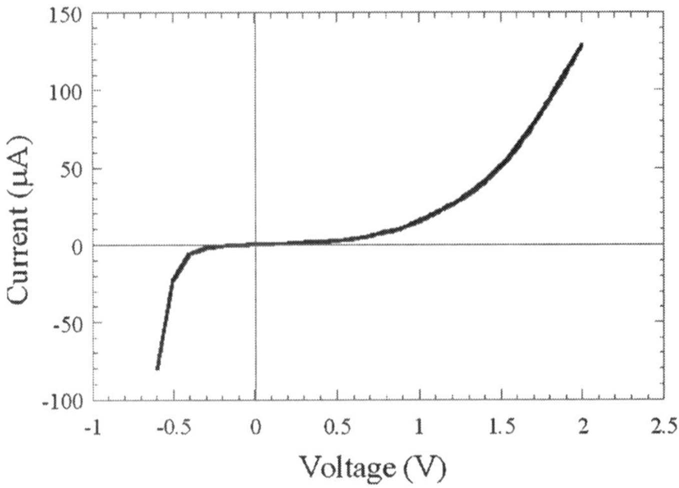

Fig. 3.13 I–V plot of the sensor in air. The supporting Ti foil and Pt contact were used as the electrodes. The diode-like behavior confirms the presence of Schottky barriers at metal-semiconductor interfaces [54]

nanotube samples exhibit a higher resistance in air. The room temperature current–voltage (I–V) characteristics of the nanotubes in air are shown in Fig. 3.13, obtained using a scanning probe microscope. For this measurement, the titanium substrate was taken as one electrode as titanium (work function ~4.33 eV) forms an ohmic contact with TiO_2 and the Pt electrode deposited on the nanotubes was taken as the other. A diode type I–V behavior can be seen in Fig. 3.13, which shows the presence of Schottky barrier at the electrode nanotube contact region. It is well known that Pd dissociates hydrogen into atomic form [106, 107]. The activated hydrogen removes adsorbed oxygen, and is adsorbed onto the Pd and Pt surfaces [107] of the sensor. Hydrogen adsorption reduces the Pd and Pt work functions. Therefore, the height of the potential barrier at the metal–TiO_2 interface is lowered, which in turn reduces the electrical resistance of the sensor. In this manner, Pd and Pt enhance the H_2 sensitivity of the nanotubes.

Close inspection of Fig. 3.11 reveals that the sensor resistance decreases to a few tens of ohms on exposure to 1,000 ppm hydrogen. This result implies that the intrinsic resistance of TiO_2 is changing in the presence of hydrogen, and that the change in barrier height at the metal-semiconductor interface is not the only factor that controls the resistance response of the sensor. Thus it can be concluded that the dissociated hydrogen gets adsorbed directly onto the nanotube walls and inter-tubular contact regions exposed to the ambient. The adsorbed hydrogen creates an electron space charge layer on the walls and the tube-to-tube contact regions, reducing the resistance of the tubes significantly [45, 89]. The effect is quite pronounced in the present case since the small thickness (~13 nm) of the nanotube walls may result in the overlap of neighboring space charge regions; thus the entire volume of the inter-tubular region may experience low resistivity.

3.4.2 *Transparent Hydrogen Sensors*

Transparent TiO_2 nanotube arrays films, 22 nm pore size, were fabricated on glass substrates by anodizing 400 nm thick rf sputter deposited Ti films in a HF and acetic acid electrolyte, and then annealing them at ≈450°C in oxygen for 6 h [55]. A 10 nm discontinuous palladium layer was evaporated atop the nanotube-array surface. The total thickness of the nanotube-array oxide structure is 300 nm, with 200–220 nm long nanotube arrays over a thermally oxidized layer of Ti of 80–100 nm. The typical room temperature behavior of this device on exposure to 1,000 ppm H_2 is shown in Fig. 3.14. In this resistance versus time plot, the initial resistance is of the order of $10^9\,\Omega$ in nitrogen, and drops to about $10^5\,\Omega$ when exposed to 1,000 ppm of hydrogen. The response time, that required to reach 10% of the initial resistance value upon gas exposure, is about 30 s. The sensor completely recovers after H_2 flow is terminated. It is interesting to note that the resistance change upon hydrogen exposure (10^4) is not as large as that observed in sensors having a metal (Ti) layer underneath the nanotubes (10^7, see [108]), presumably due to the thinner rutile barrier layer underlying the nanotube array film.

The palladium overlayer helps in lowering the nascent operating temperature of the sensor from 290°C to room temperature. Palladium is known to dissociate hydrogen molecules into atomic form, which immediately diffuses into the nanotube walls. The absorbed hydrogen creates an electron space charge layer on the walls and the tube-to-tube contact regions, reducing the resistance of the tubes [89]. The thin-film, transparent hydrogen sensors are mechanically robust, not at all susceptible to shock, and have successfully been operated between 25°C and 250°C. These sensors show negligible change in resistance to the reducing gases including carbon monoxide, methane and ammonia [108]. They can also measure hydrogen concentrations accurately in transformer oil, an application useful for prediction of transformer failure.

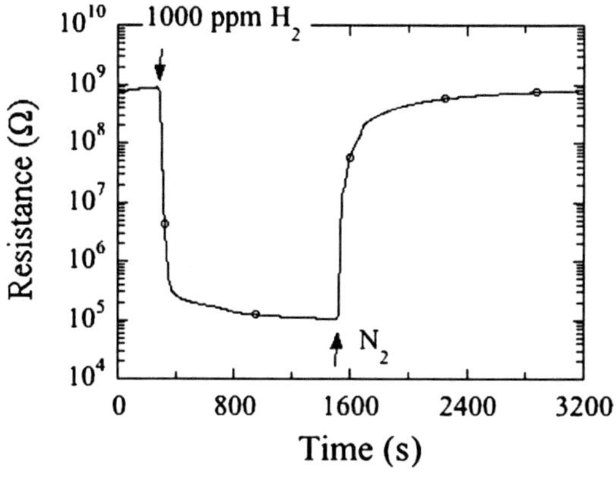

Fig. 3.14 Room temperature response of transparent TiO_2 nanotube array hydrogen sensor, on glass, to 1,000 ppm hydrogen concentration in H_2+N_2 gas mixture, and recovery in N_2 [55]

3.5 Extreme Hydrogen Gas Sensitivities at Room Temperature

TiO_2 nanotube-arrays were prepared by anodization of titanium foil in an electrolyte containing 1.0 M sodium hydrogen sulfate monohydrate, 0.1 M potassium flouride and 0.2 M sodium citrate tribasic dihydrate, with electrolyte pH adjusted by addition of sodium hydroxide [77]. Nanotube lengths varied from 380 nm to 6 μm and pore diameters from 30 to 110 nm as dependent on the electrolyte pH (1.11 to 5) and the anodization potential (10–25 V); the anodization was conducted for 17 h for all samples. The samples were crystallized by annealing at temperatures ranging from 370°C to 630°C in oxygen for 6 h. For the electrical measurements platinum circular electrodes (500 μm diameter, 100 nm thick, 1 mm edge-to-edge spacing) were sputter coated onto the nanotube-array samples; increasing the electrode separation increases the baseline resistance, but does not affect sensitivity. After fixing the samples to TO-8 steel headers (Electrovac GesmbH, Klosterneuberg, Austria), gold wire (1 mil) was bonded on the platinum pads with a Mech-El-907 ultrasonic wire bonder (Mech-El Industries, Woburn, MA).

A 10 V anodization in pH 1.1 results in a nanotube array of 350 nm length and 30 nm pore diameter. A 25 V anodization in pH 5.0 results in a nanotube array of 6.5 μm length with a 110 nm pore diameter. The greatest hydrogen sensitivities are achieved with the 30 nm pore diameter samples fabricated at 10 V, consequently the majority of our discussion is focused on these samples. Table 3.1 summarizes the maximum room temperature hydrogen sensitivity, log(S), of the 30 nm pore diameter nanotube arrays in response to 1,000 ppm hydrogen, as a function of length which, in turn, is determined by electrolyte pH. It was found that in crystallizing the initially amorphous nanotube array samples maximum hydrogen sensitivity was achieved with a 480°C annealing. It can be seen from Table 3.1 that the highest resistance variation, 8.7 orders of magnitude, was obtained for nanotube samples prepared using electrolyte pH 4.0 (length ~1 μm, pore diameter ~30 nm and wall thickness ~13 nm). To the knowledge of the authors this is the largest change in electrical resistance of any material, to any gas, at any temperature [77]. Assuming a 10Ω measurement resolution, linear extrapolation of the sensitivity indicates the ability of the nanotube-array to detect hydrogen at 20 ppt (parts per trillion). The lower limit of our test chamber is about 20 ppm, at which 7.2 orders of magnitude variation in electrical resistance to 20 ppm hydrogen was seen.

Table 3.1 Summary of the nanotube dimensions under different voltage and pH conditions and their maximum resistance variation to 1,000 ppm H_2 at room temperature [77]

pH	10 V (30 nm pore size)		
	Length (nm)	Log(S) no Pd	Log(S) with Pd
1.1	350	7.2	7.3
3.0	730	8.0	8.7
4.0	990	8.7	8.3
4.5	1,400	8.0	8.3
5.0	2,000	–	–

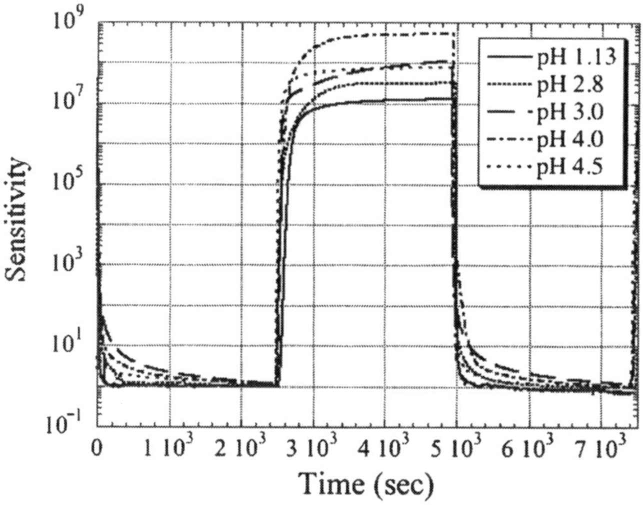

Fig. 3.15 Variation in hydrogen sensitivity *S* of 30 nm pore diameter nanotube-array samples as a function of electrolyte pH [77]

Fig. 3.15 shows the variation in hydrogen sensitivity of the 10 V samples as a function of electrolyte pH; note the pH 5.0 sample was not tested, as nanotube array formation occurs on only $\approx$25% of the film. The typical response of a pH 4.0 sample on switching the ambient atmosphere between air and 1,000 ppm hydrogen in nitrogen is shown in Fig. 3.16. On exposure to hydrogen a rapid reduction in resistance is observed; there is no indication of measurement hysteresis. Sensitivity *S* is not greatly influenced by nanotube length; for the 10 V samples a sensitivity shift of 7 orders to 8.7 orders is found with an increase in length from 380 nm to 1 μm. It was found that 6 μm long samples (25 V, pH 5.0) showed less sensitivity, with significantly longer response/recovery times due to the time required for hydrogen to diffuse inside the longer pores. The sensitivity of the nanotube-arrays is in stark contrast to the hydrogen sensitivity of a crystalline TiO_2 film made by thermal oxidation of a Ti film which demonstrates, to 1,000 ppm hydrogen at room temperature, a resistance variation of $\approx$400%. As palladium is known to help hydrogen detection in metal oxide materials, some nanotube array samples were coated with a 10 nm Pd layer and their response to hydrogen tested. The Pd layer degraded the sensitivity of the highly sensitive samples due to a drop in the base resistance; samples of relatively lower performance typically saw a modest sensitivity increase (Table 3.1).

It behoves to consider how hydrogen interacts with the TiO_2 nanotube architecture, a material which is essentially all surface and no bulk, to achieve such remarkable changes in electrical resistance. At room temperature there is little reduction of the surface or bulk oxide [109, 110], while the fast response and recovery without hysteresis (Fig. 3.16) rules out a significant contribution from diffusion of hydrogen into the TiO_2 lattice [111]. Oxygen in air may be

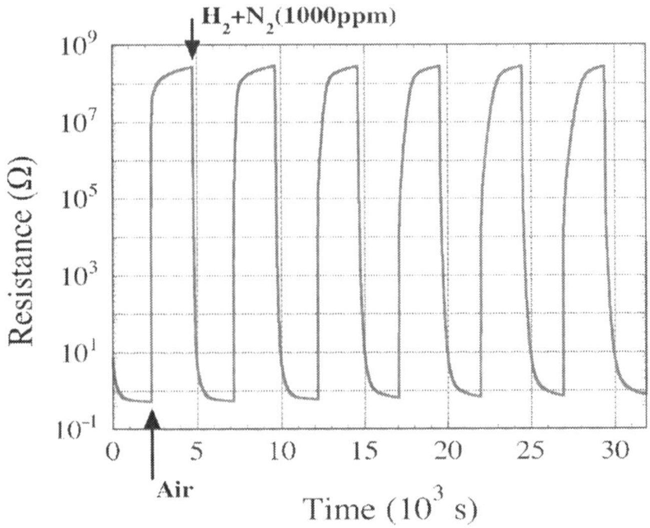

Fig. 3.16 Room temperature resistance variation of a nanotube-array sample prepared in a pH 4.0 electrolyte using 10 V (30 nm pore diameter), annealed at 480°C, alternately exposed to air and 1,000 ppm hydrogen in nitrogen [77]

chemisorbed in the form of O_2^- on the nanotube surface by trapping electrons from its conduction band leading to an enhanced base resistance [112, 113]. Hydrogen can remove this chemisorbed oxygen thus reducing the resistance [114]. To understand the role of chemisorbed oxygen, initially air was passed through the test chamber, then nitrogen, then a hydrogen-nitrogen mixture, and finally nitrogen. It was observed that nitrogen reduced the 1,000 ppm hydrogen sensitivity to approximately six orders of magnitude while only slightly reducing the base resistance, and greatly extended the needed recovery time. The high sensitivity in the absence of oxygen indicates that direct chemisorption of hydrogen on the nanotubes is the dominant mechanism leading to the tremendous reduction in nanotube electrical resistance, while the presence of oxygen facilitates removal of the chemisorbed hydrogen. Platinum is known to be a catalyst that activates hydrogen by adsorbing and dissociating hydrogen molecules which are then spilled over to the semiconducting material for chemisorption [89, 115]. Replacing the Pt electrodes with gold reduced the 1,000 ppm sensitivity of the pH 4.0 10 V nanotube sample to $\approx$5 orders of magnitude with a considerably slower response time. While it is evident that the platinum electrodes facilitate the high sensitivity by providing spilt-over hydrogen for chemisorption, it is not the only cause of observed nanotube response.

Hydrogen activation appears to occur on the walls of the undoped nanotubes at highly active surface states provided by nanoscale surface defects. The dissociated hydrogen species form OH groups with the surface oxygen accompanied by electron transfer to the TiO_2 conduction band and formation of an electron rich region within the nanotube walls. To ensure that the nanotube samples were undoped, samples annealed at 480°C were studied using XPS. In addition to the

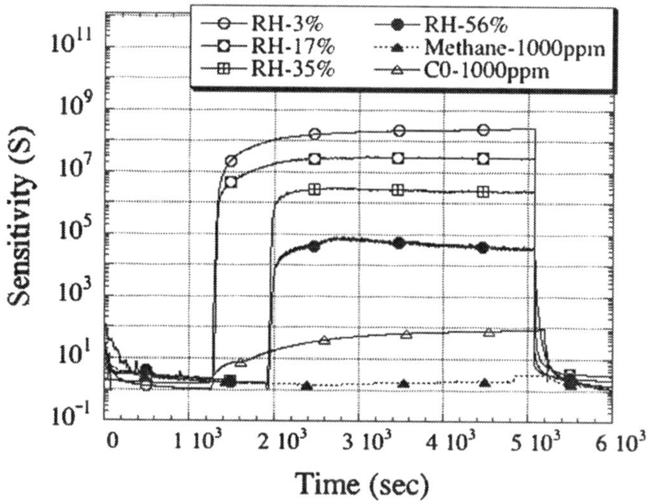

Fig. 3.17 Hydrogen sensitivity (1,000 ppm) of a 10 V nanotube-array sample prepared in a pH 4.0 electrolyte as a function of relative humidity. The sensitivity of the sample to carbon monoxide and methane in dry ambient is also shown [77]

Ti and O peaks, the XPS spectra showed a weak carbon (C 1s) peak. Depth-profiling showed carbon was present inside the sample at low concentrations. The carbon peak is believed to come from adventitious carbon or hydroxyl and carboxyl groups. This study, however, does not show any effect of C on the hydrogen sensitivity. Fig. 3.17 shows the response of 1 μm length and 30 nm pore diameter nanotubes to 1,000 ppm hydrogen in humid ambients. In humid conditions physisorption of water molecules takes place on a layer of initially chemisorbed OH radicals at the nanotube surface which reduces the nanotube base resistance [116]. The adsorbed water molecules block the active sites where hydrogen chemisorption occurs, thus reducing sensitivity. Fig. 3.17 also shows the response of the nanotube sample to 1,000 ppm carbon monoxide and 1,000 ppm methane; although both gases are strongly reducing, the sensitivity to these gases is negligible.

It was believed that the crystallized nanoscale walls and inter-tubular connecting points play critical roles in determining the remarkable hydrogen sensitivities of the TiO_2 nanotube arrays. The adsorption of oxygen in air takes place on either side of the nanotube walls creating an electron depletion region. The width of the space charge layer L is given by $L=L_D\,[2eV/kT]^{1/2}$, where $L_D=[\varepsilon_0\varepsilon kT/2e^2N_D]^{1/2}$ is the Debye length, eV is the barrier height, kT is the thermal energy and N_D the ionized donor density. In metal oxides the space charge layer extends to a few tens of nanometers [117]. If the nanotube wall half-thickness $t/2$ is significantly greater than the width of the spacecharge region, as shown in Fig. 3.18(a), oxygen removal by hydrogen and subsequent hydrogen chemisorption will have little effect on device resistance and hence, high sensitivity cannot be expected. In contrast, when $t/2$ is comparable to or less than the space charge region the shift in the electrical resistance on exposure to hydrogen can be very high, Fig. 3.18(b,c), with

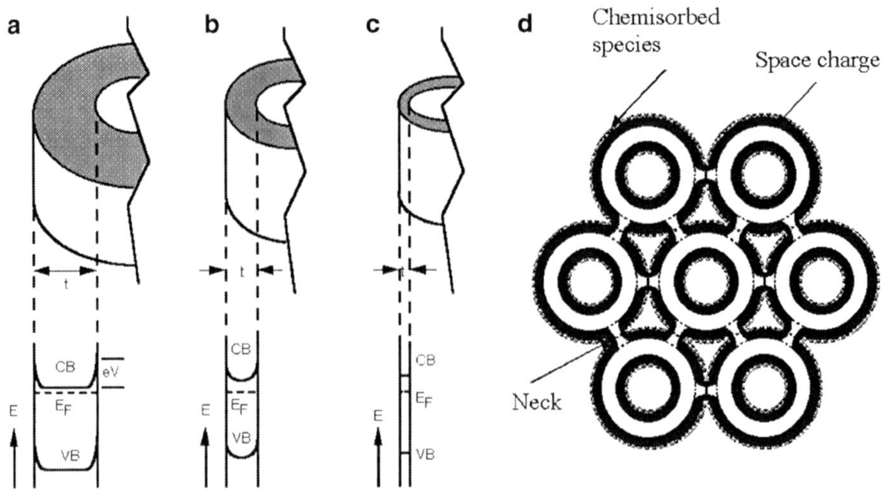

Fig. 3.18 The influence of nanotube wall thickness on band bending due to oxygen chemisorption: (**a**) When nanotube wall half-thickness ($t/2$) is much greater than the space charge layer; (**b**) when comparable; and (**c**) when $t/2$ is less than the width of the space charge region; (**d**) Schematic illustration of nanotubes, top view, and the tube-to-tube connecting points corresponding to case shown in (**a**); [77]

a flat-band condition existing when the wall thickness is less than the width of the space charge region. The nanotube sample showing the highest sensitivity had a wall thickness of ≈ 13 nm, corresponding to the geometry of Fig. 3.18(b). The inter-wall connecting points also appear to play a significant role in enabling the ultra-high hydrogen sensitivity. The oxygen adsorption and its removal by hydrogen atoms as well as chemisorption of hydrogen at these constricted points, Fig. 3.18(d), regulates the current passing from nanotube to nanotube.

3.6 Transcutaneous Hydrogen Monitoring using TiO_2 Nanotube Arrays

It can be seen from Fig. 3.16 that the highest hydrogen sensitivity, 8.7 orders of magnitude in response to 1,000 ppm hydrogen at room temperature, was obtained for nanotube samples prepared using electrolyte pH 4.0 (length $\sim 1\,\mu$m, pore diameter ~ 30 nm and wall thickness ~ 13 nm). These samples were crystallized at 480°C for 6 h in oxygen ambient. These sensors were used for transcutaneous hydrogen measurements. Since metal oxides, to some extent, respond to a variety of environmental parameters, i.e., demonstrate unwanted cross-sensitivities, it is necessary to use other sensors, and cross correlate the responses of these different sensors to obtain an absolute measurement of the target gas from within a complex environment. Therefore, we incorporated a humidity-temperature sensor along with the hydrogen sensor. The packaged sensor array is displayed in Fig. 3.19.

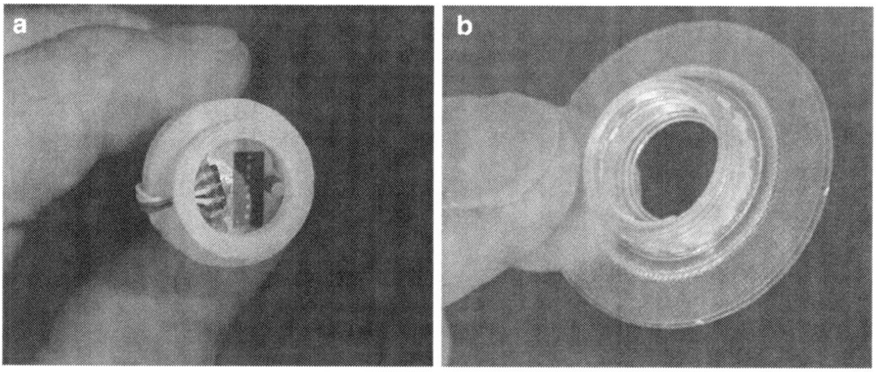

Fig. 3.19 (**a**) Packaging of the sensor, comprised of a hydrogen sensor and a humidity-temperature sensor. The connector twists onto the adhesive-backed ring. (**b**) Vent holes (not visible) on the side prevent the accumulation of water vapor and gases in the housing [79]

The hydrogen sensor and humidity-temperature sensor (SHT15 from Sensorian Inc) are safely secured in a white plastic housing, which is secured to the adhesive-backed ring that is attached to the skin of the patient undergoing the test.

A further complicating factor was the baseline resistance of the nanotube array sensors. The baseline resistance of metal oxides can be well over 100 GΩ, an impedance so large that it is virtually impossible to make interface electronics. A post-anodization acid rinse could be used to controllably reduce the thickness of the oxide barrier layer between the nanotubular region and the underlying metal substrate, thus lowering the baseline resistance of the sensor. A robust electronic interface suitable for handling sensor resistances from 50 GΩ to the metallic range was developed. Fig. 3.20 shows a block diagram of the sensor interface electronics. For sensor measurement, a constant current I is applied to the sensor and the resulting dc voltage V is measured. The sensor resistance R is then calculated by $R = V/I$. In order to track the sensor resistance over a wide range (from 0 to 50 GΩ), the microcontroller is programmed to precisely and automatically tune the dc current value as well as the amplifier gain for optimal measurement accuracy. The measured data, including resistance of the hydrogen sensor and the humidity and temperature measurements (a digital humidity and temperature sensor that directly sends digitized measurements to the microcontroller), are sent to a PC for storage and processing. Fig. 3.21 shows the transcutaneous hydrogen measurement system.

3.6.1 Cross Interference and Calibration

To determine which parameters affect the resistance of the hydrogen sensor, applied for measurement of transcutaneous hydrogen, the sensor was placed inside the test chamber and the test environment (humidity, a variety of interfering gases,

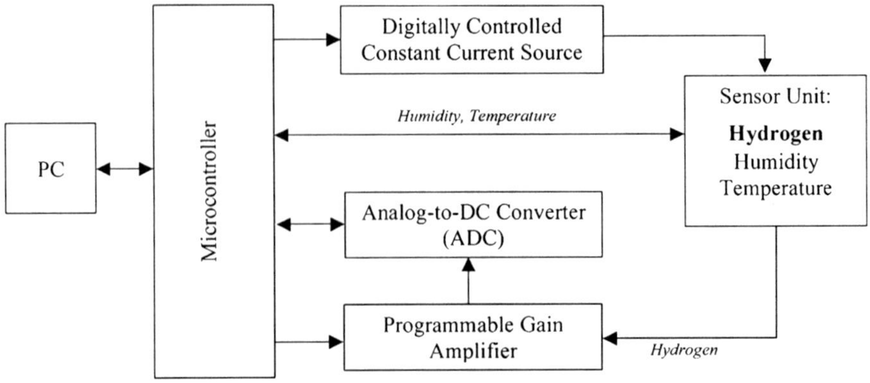

Fig. 3.20 Block diagram of the sensor interface electronics. The system generates a constant current to the hydrogen sensor, and measures the resulting voltage to determine the resistance. The response of a humidity-temperature sensor is also measured [79]

and temperature), was varied to observe the change in the sensor resistance. The parameters tested were temperature, humidity, oxygen, carbon dioxide, carbon monoxide, and methane.

Temperature: As shown in Fig. 3.22, temperature has a large impact on the resistance of the hydrogen sensor. Nevertheless, in practice, the operating temperature of the transcutaneous sensor is near or at body temperature. As indicated in the Fig. 3.22, the change in the sensor sensitivity is only $\approx$0.08 order of magnitude (from 3,800 to 4,600) from 25° to 37°C. As a result, although the temperature of the hydrogen sensor was also measured during the experiments, no calibration was performed.

Humidity: Fig. 3.17 shows the effect humidity has on sensor performance at 1,000 ppm hydrogen. Between relative humidity of 3% and 56%, the sensitivity of the hydrogen sensor varies from 2.5×10^8 to 7×10^4, a change of more than 3 orders of magnitude. Therefore, to ensure measurement integrity, the humidity level inside the sensor package was also measured and used for calibration.

Oxygen: The addition of oxygen into the test environment reduces the sensitivity of the sensor as indicated in Fig. 3.23. However, results show that the difference in the sensor sensitivity is only about 1 order of magnitude between zero (a pure nitrogen environment) and 20% oxygen (atmospheric condition). As a result, variations in oxygen concentration do not represent a significant cross-sensitivity to the measurement since the hydrogen sensor was usually exposed to a near-atmospheric condition, with vent holes in the transcutaneous sensor housing allowing for the free flow-through of gases.

Carbon Dioxide: The resistance of the hydrogen sensor did not change when it was switched from a pure nitrogen environment and nitrogen with 10% CO_2. This is expected because carbon dioxide is neither a strong reducing gas nor a strong oxidizing gas.

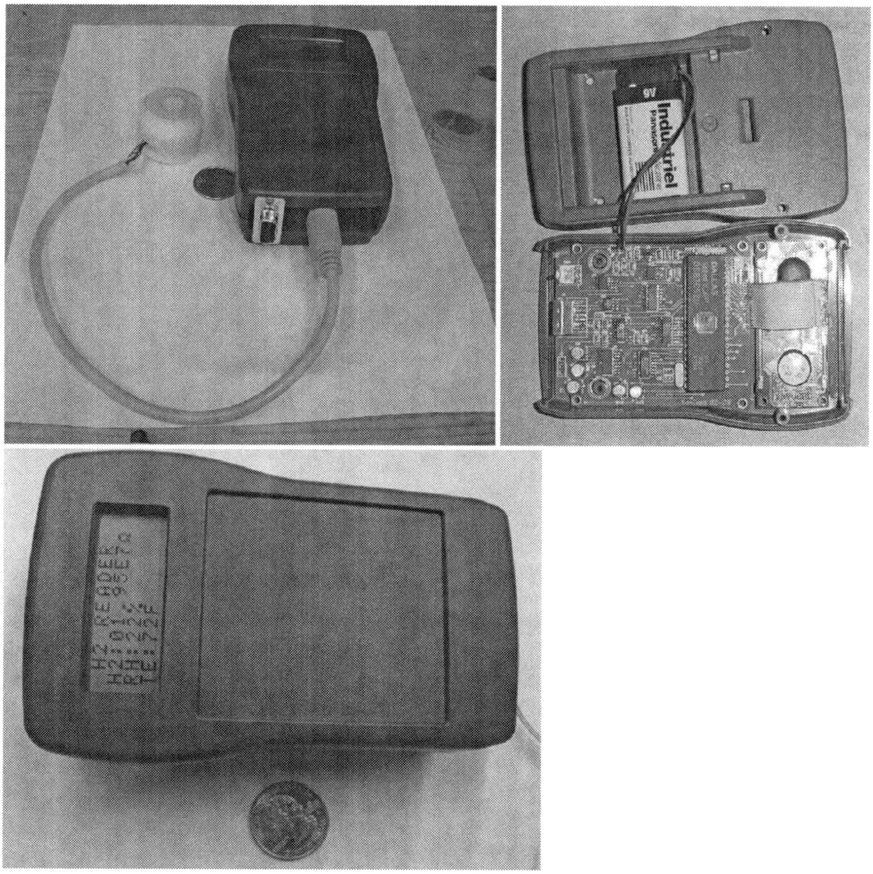

Fig. 3.21 Photos showing packaged sensor with cable connecting sensor to electronics, as well as internal view of sensor electronics [79]

Carbon Monoxide: Fig. 3.24 plots the change in the sensitivity when the sensor is exposed to 1,000 ppm carbon monoxide. Results show the sensitivity of the sensor varies less than two orders of magnitude when exposed to 1,000 ppm of carbon monoxide. This is more than four times smaller than hydrogen, which can result in a change of 8.7 orders of magnitude. The response of the sensor towards carbon monoxide does not cause any concern for transcutaneous measurements, since it is highly unlikely for carbon monoxide to be released by a patient.

Methane: Fig. 3.24 plots the sensitivity of the sensor when it was exposed to 1,000 ppm of methane. Results show that 1,000 ppm of methane causes a change in electrical resistance of about a half order of magnitude, much smaller than the effect of hydrogen.

Evaluating the performance of the hydrogen sensor against various gases and operating conditions, it was concluded that for transcutaneous hydrogen measurements the only potentially interfering parameter of significance is humidity.

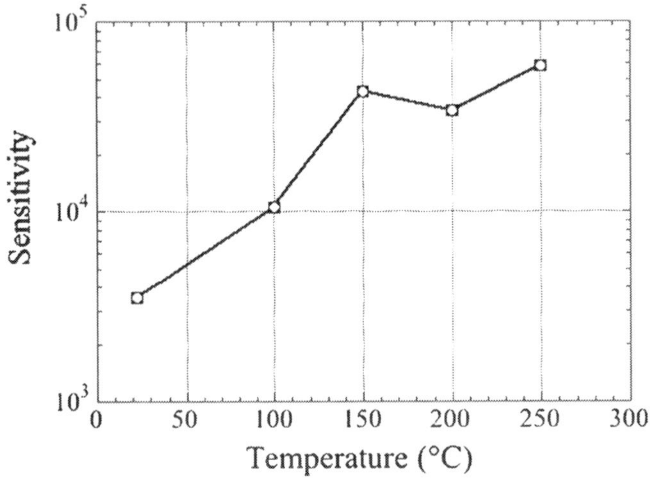

Fig. 3.22 The sensitivity of a 200 nm long nanotube-array hydrogen sensor increases with temperature [79]

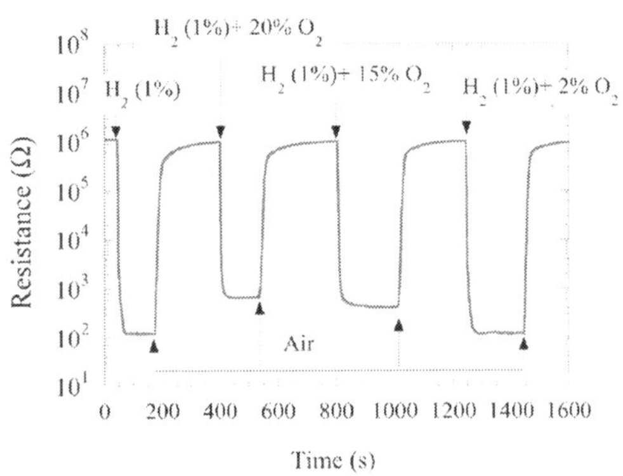

Fig. 3.23 (*left*): The sensitivity of the hydrogen sensor reduces with increasing oxygen concentration

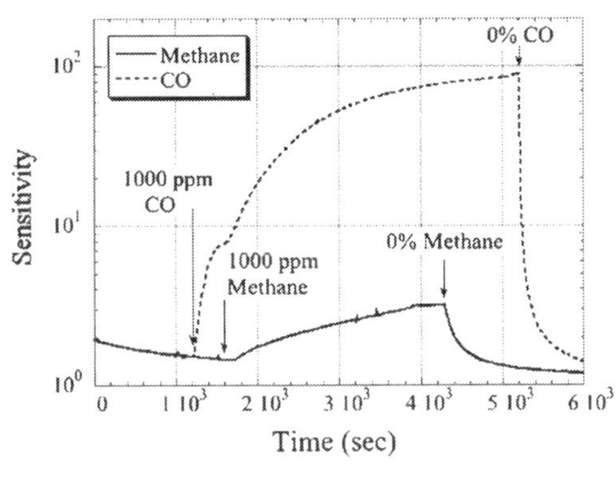

Fig. 3.24 (*right*): Introduction of 1,000 ppm carbon monoxide causes the sensor sensitivity to increase for about two orders of magnitude, while methane only changes the sensor sensitivity by half order of magnitude [79]

Hence, a calibration lookup table was generated to compensate for the effect of humidity. Before use, the resistance of the same sensor was measured as a function of humidity and hydrogen concentration, and the results were stored in a database. When taking a measurement, the operating software first obtained the humidity and hydrogen sensor measurements (resistance) from the microcontroller. It then determined the closest humidity and resistance values by going through the lookup table, and found the corresponding hydrogen concentration [79].

3.6.2 Transcutaneous Hydrogen and Lactose Intolerance

The transcutaneous hydrogen sensor was initially evaluated in the Pediatric Gastroenterology Clinic of the Penn State Children's Hospital. The transcutaneous hydrogen sensor was placed on the forearm of children and adult volunteers who were referred for breath hydrogen testing for the evaluation and diagnosis of possible lactose intolerance. In these patients, the sensor has documented the transcutaneous excretion of hydrogen. Moreover, in those patients with documented lactose intolerance, there is a good correlation between the transcutaneous sensor readings and the exhaled breath hydrogen levels as measured by the commercial gas chromatography instrument (Quintron Instrument Company, Milwaukee, Wisconsin). Negative results in some patients helped to reassure them that the novel hydrogen sensor is not responding to some volatile molecule that might be excreted through the skin. Fig. 3.25 shows illustrative measurement of transcutaneous hydrogen levels using the described transcutaneous sensor, note the sensor measurement is continuous in operation, and the hydrogen levels (ppm) in exhaled breath as measured using a gas chromatograph, from a lactose-intolerant adult volunteer who drank a small sample of milk. Fig. 3.26 shows the decrease in the sensor measurement, which indicates an increase in transcutaneous hydrogen, coincides with an increase in exhaled breath hydrogen concentration.

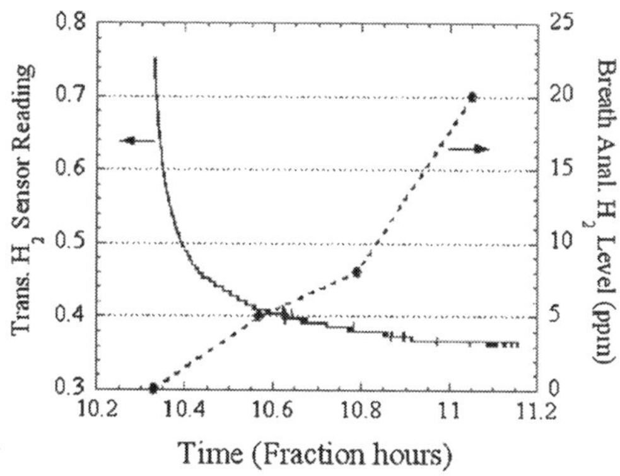

Fig. 3.25 (*left*): Measurement of transcutaneous hydrogen concentrations using the described sensor (*y*-axis on LHS), and exhaled breath hydrogen levels (*y*-axis on RHS) measured by a gas chromatograph from a lactose-intolerant adult volunteer

Fig. 3.26 (*right*): Correlation between the response of the transcutaneous hydrogen sensor and exhaled hydrogen concentration [79]

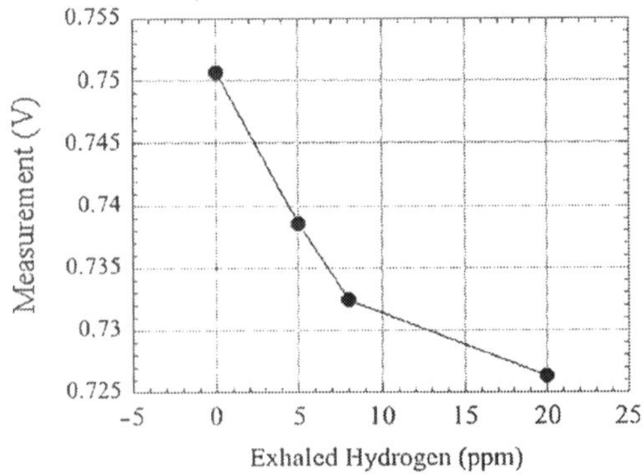

A transcutaneous hydrogen sensor of such great sensitivity offers the potential for monitoring the onset of other diseases such as NEC, a devastating disease of uncertain etiology and pathogenesis that causes high levels of mortality and morbidity in about 10% of preterm infants in neonatal intensive care units throughout the world [118]. Elevated levels of exhaled breath hydrogen gas have been shown to precede the onset of the clinical and radiographic signs of NEC [19, 119]. However, from a practical standpoint, it is difficult to measure gas levels in exhaled breath of preterm infants because of their small tidal volumes and rapid respiratory rates. In neonatal intensive care units, the use of endotracheal tubes, nasal CPAP systems and nasal cannula flow systems are all barriers to the successful collection of undiluted end expiratory gas samples for gas analysis. Therefore, transcutaneous measurement of hydrogen gas, using the sensors described within this chapter, in preterm infants has the potential to facilitate the early identification of NEC, and hence its treatment.

References

1. Rizkalla SW, Luo J, Kabir M, Chevalier A, Pacher N, Slama G (2000) Chronic consumption of fresh but not heated yogurt improves breath-hydrogen status and short-chain fatty acid profiles: A controlled study in healthy men with or without lactose maldigestion. Am J Clin Nutr 72:1474–1479
2. Chong SKF, Ramadan AB, Livesey E, Wood G (2002) The use of a portable breath hydrogen analyser in screening for lactose intolerance in paediatric patients with chronic abdominal pain or chronic diarrhoea. Gastroenterology 122:A363–A363
3. Kanabar D, Randhawa M, Clayton P (2001) Improvement of symptoms in infant colic following reduction of lactose load with lactase. J Hum Nutr Diet 14:359–363
4. Li DY, Barnes Y, Cuffari C (2002) Who should request a breath hydrogen test? A five-year feasibility, sensitivity of clinical suspicion and cost-effectiveness analysis. Gastroenterology 122:A574–A574
5. Duro D, Rising R, Cedillo M, Lifshitz E (2002) Association between infantile colic and carbohydrate malabsorption from fruit juices in infancy. Pediatr Res 109:797–805

6. Moukarzel AA, Lesicka H, Ament ME (2002) Irritable bowel syndrome and nonspecific diarrhea in infancy and childhood – Relationship with juice carbohydrate malabsorption. Clin Pediatr 41:145–150
7. Lebenthal-Bendor Y, Theuer R, Lebenthal A, Tabi I, Lebenthal E (2001) Malabsorption of modified food starch (acetylated distarch phosphate) in normal infants and in 8–24-month-old toddlers with non-specific diarrhea, as influenced by sorbitol and fructose. Acta Paediatr 90:1368–1372
8. Ledochowski M, Widner B, Murr C, Sperner-Unterweger B, Fuchs D (2001) Fructose malabsorption is associated with decreased plasma tryptophan. Scand J Gastroenterol 4:367–371
9. Backus RC, Puryear LM, Crouse BA, Biourge VC, Rogers QR (2002) Breath hydrogen concentrations of cats given commercial canned and extruded diets indicate gastrointestinal microbial activity vary with diet type. J Nutr 6:1763S–1766S
10. Riordan SM, McIver CJ, Duncombe VM, Thomas MC, Bolin TD (2000) Evaluation of the rice breath hydrogen test for small intestinal bacterial overgrowth. Am J Gastroenterol 95:2858–2864
11. Bauer TM, Schwacha H, Steinbruckner B, Brinkmann FE, Ditzen AK, Kist M, Blum HE (2000) Diagnosis of small intestinal bacterial overgrowth in patients with cirrhosis of the liver: Poor performance of the glucose breath hydrogen test. J Hepatol 33:382–386
12. Funayama Y, Sasaki I, Naito H, Fukushima K, Shibata C, Masuko T, Takahashi K, Ogawa H, Sato S, Ueno T, Noguchi M, Hiwatashi N, Matsuno S (1999) Monitoring and antibacterial treatment for postoperative bacterial overgrowth in Crohn's disease. Dis Colon Rectum 42:1072–1077
13. Pimentel M, Chow EJ, Lin HC (2000) Comparison of peak breath hydrogen production in patients with irritable bowel syndrome, chronic fatigue syndrome and fibromyalgia. Gastroenterology 118:A413–A413
14. Burge MR, Tuttle MS, Violett JL, Stephenson CL, Schade DS (2000) Breath hydrogen testing identifies patients with diabetic gastroparesis. Diabetes Care 23:860–861
15. Chiloiro M, Darconza G, Piccioli E, De Carne M, Clemente C, Riezzo G (2001) Gastric emptying and orocecal transit time in pregnancy. J Gastroenterol 36:538–543
16. Stordal K, Nygaard EA, Bentsen B (2001) Organic abnormalities in recurrent abdominal pain in children. Acta Paediatr 90:638–642
17. Bisquera JA, Cooper TR, Berseth CL (2002) Impact of necrotizing enterocolitis on length of stay and hospital charges in very low birth weight infants. Pediatrics 109:423–428
18. Engel RR, Virnig NL (1973) Origin of mural gas in necrotizing enterocolitis. Pediatr Res 7:292–292
19. Godoy G, Truss C, Philips J, Young M, Coffman K, Cassady G (1986) Breath hydrogen excretion in infants with necrotizing enterocolitis. Pediatr Res 20:A348–A348
20. Garstin WIH, Boston VE (1987) Sequential assay of expired breath hydrogen as a means of predicting necrotizing enterocolitis in susceptible infants. Pediatr Res 22:208–210
21. Cheu HW, Brown DR, Rowe MI (1989) Breath hydrogen excretion as a screening test for the early diagnosis of necrotizing enterocolitis. Am J Dis Children 143:156–159
22. Varghese OK, Gong D, Paulose M, Ong KG, Dickey EC, Grimes CA (2003) Extreme changes in the electrical resistance of titania nanotubes with hydrogen exposure. Adv Mater 15:624–627
23. Varghese OK, Gong D, Paulose M, Ong KG, Grimes CA (2003) Hydrogen sensing using titania nanotubes. Sens Actuators B 93:338–344
24. Gong D, Grimes CA, Varghese CA, Hu W, Singh RS, Chen Z, Dickey EC (2001) Titanium oxide nanotube arrays prepared by anodic oxidation. J Mater Res 16:3331–3334
25. Varghese OK, Paulose M, Gong D, Grimes CA, Dickey EC (2003) Crystallization and high-temperature structural stability of titanium oxide nanotube arrays. J Mater Res 18:156–165
26. Mor GK, Varghese OK, Paulose M, Mukherjee N, Grimes CA (2003) Fabrication of tapered, conical-shaped titania nanotubes. J Mater Res 18:2588–2593

27. Christofides C, Mandelis A (1990) Solid-state sensors for trace hydrogen gas-detection. J Appl Phys 68:R1–R30
28. Ruths PF, Ashok S, Fonash SJ (1981) A study of Pd-Si MIS schottky-barrier diode hydrogen detector. IEEE Trans Electron Devices 28:1003–1009
29. Schalwig J, Muller G, Karrer U, Eickhoff M, Ambacher O, Stutzmann M, Gorgens L, Dollinger G (2002) Hydrogen response mechanism of Pt-GaN Schottky diodes. Appl Phys Lett 80:1222–1224
30. Roy S, Jacob C, Lang C, Basu S (2003) Studies on Ru/3C-SiC Schottky junctions for high temperature hydrogen sensors. J Electrochem Soc 150:H135–H139
31. Cheng S-Y (2003) A hydrogen sensitive Pd/GaAs Shottky diode sensor. Mater Chem Phys 78:525–528
32. Butler MA (1991) Fiber optic sensor for hydrogen concentrations near the explosive limit. J Electrochem Soc 138:L46–L47
33. Sekimoto S, Nakagawa H, Okazaki S, Fukuda K, Asakura S, Shingemori T, Takahashi S (2000) A fiber-optic evanescent-wave hydrogen gas sensor using palladium-supported tungsten oxide. Sens Actuators B 66:142–145
34. Sutapun B, Tabib-Azar M, Kazemi A (1999) Pd-coated elastooptic fiber optic Bragg grating sensors for multiplexed hydrogen sensing. Sens Actuators B 60:27–34
35. Matsumiya M, Shin W, Izu N, Murayama N (2003) Nano-structured thin-film Pt catalyst for thermoelectric hydrogen gas sensor. Sens Actuators B 93:309–315
36. Katti VR, Debnath AK, Gadkari SC, Gupta SK, Sahni VC (2002) Passivated thick film catalytic type H-2 sensor operating at low temperature. Sens Actuators B 84: 219–225
37. Luo RX, Chen LH, Chen AF, Liu CC (1991) A novel catalytic sensor for monitoring the concentration of mixed combustible gases. Sci China A 34:1500–1507
38. Maffei N, Kuriakose AK (1999) A hydrogen sensor based on a hydrogen ion conducting solid electrolyte. Sens Actuators B 56:243–246
39. Katahira K, Matsumoto H, Iwahara H, Koide K, Iwamoto T (2001) A solid electrolyte hydrogen sensor with an electrochemically-supplied hydrogen standard. Sens Actuators B 73:130–134
40. Lu G, Miura N, Yamazoe N (1996) High temperature hydrogen sensor based on stabilized zirconia and a metal oxide electrode. Sens Actuators B 35:130–135
41. Miura N, Harada T, Shimizu Y, Yamazoe N (1990) Cordless solid-state hydrogen sensor using proton-conductor thick film. Sens Actuators B 1:125–129
42. Lundstroem I, Shivaraman S, Svensson C, Lundkvist L (1975) Hydrogen-sensitive mos field-effect transistor. Appl Phys Lett 26:55–57
43. Miura N, Harada T, Yoshida N, Shimizu Y, Yamazoe N (1995) Sensing characteristics of ISFET-based hydrogen sensor using proton conductive thick film. Sens Actuators B 25:499–503
44. Fomenko S, Gumenjuk S, Podlepetsky B, Chuvashov V, Safronkin G (1992) The influence of technological factors on the hydrogen sensitivity of mosfet sensors. Sens Actuators B 10:7–10
45. Mor GK, Varghese OK, Paulose M, Shankar K, Grimes CA (2006) A review on highly ordered, vertically oriented TiO_2 nanotube arrays: Fabrication, material properties, and solar energy applications. Sol Energy Mater Sol Cells 90:2011–2075
46. Hyodo T, Nishida N, Shimizu Y, Egashira M (2002) Preparation and gas sensing properties of thermally stable mesoporous SnO_2. Sens Actuators B 83:209–215
47. Chaudhary VA, Mulla IS, Vijaymohanan K (1999) Selective hydrogen sensing properties of surface functionalized tin oxide. Sens Actuators B 55:154–160
48. Fonash SJ, Li Z, Oleary MJ (1985) An extremely sensitive heterostructure for parts per million detection of hydrogen in oxygen. J Appl Phys 58:4415–4419
49. Reddy CVG, Manorama SV (2000) Room temperature hydrogen sensor based on SnO_2: La_2O_3. J Electrochem Soc 147:390–393

50. Basu S, Dutta A (1994) Modified heterojunction based on zinc-oxide thin-film for hydrogen gas-sensor application. Sens Actuators B 22:83–87
51. Nakagawa H, Yamamoto N, Okazaki S, Chinzei T, Asakura S (2003) A room-temperature operated hydrogen leak sensor. Sens Actuators B 93:468–474
52. Yamamoto N, Tonomura S, Matsuoka T, Tsubomura H (1980) Study on a palladium-titanium oxide schottky diode as a detector for gaseous components. Surf Sci 92:400–406
53. Mor GK, Varghese OK, Paulose M, Grimes CA (2003) A Self-cleaning Room Temperature Titania-Nanotube Hydrogen Gas Sensor. Sens Lett 1:42–46
54. Varghese OK, Mor GK, Grimes CA, Paulose M, Mukherjee N (2004) A titania nanotube array room temperature sensor for selective detection of hydrogen at low concentrations. J Nanosci Nanotechnol 4:733–737
55. Mor GK, Varghese OK, Paulose M, Grimes CA (2006) Fabrication of hydrogen detectors with transparent titanium oxide nanotube array thin films as sensing elements. Thin Solid Films 496:42–48
56. Huiberts JN, Griessen R, Rector JH, Wijngaarden RJ, Dekker JP, de Groot DG, Koeman NJ (1996) Yttrium and lanthanum hydride films with switchable optical properties. Nature 380:231–234
57. Pick MA, Davenport JW, Strongin M, Dienes GJ (1979) Enhancement of hydrogen uptake rates for Nb and Ta by thin surface overlayers. Phys Rev Lett 43:286–289
58. Beydoun D, Amal R, Low G, McEvoy S (1999) Role of nanoparticles in photocatalysis. J Nanopart Res 1:439–458
59. Mor GK, Carvalho MA, Varghese OK, Pishko MV, Grimes CA (2004) A room-temperature TiO_2-nanotube hydrogen sensor able to self-clean photoactively from environmental contamination. J Mater Res 19:628–634
60. Moseley PT (1992) Materials selection for semiconductor gas sensors. Sens Actuators B 6:149–156
61. Madou MJ, Morrison SR (1989) Chemical sensing with solid state devices. Academic, New York
62. Wang CC, Akbar SA, Madou MJ (1998) Ceramic based resistive sensors. J Electroceram 2:273–282
63. Harrison PG, Willett MJ (1988) The mechanism of operation of tin(iv) oxide carbon-monoxide sensors. Nature 332:337–339
64. Shimizu Y, Kuwano N, Hyodo T, Egashira M (2002) High H_2 sensing performance of anodically oxidized TiO_2 film contacted with Pd. Sens Actuators B 83:195–201
65. Varghese OK, Grimes CA (2004) Metal oxide nanostructures as gas sensors. Encyclopedia of Nanoscience and Nanotechnology. (Ed) Nalwa HS, American Scientific Publishers, Valencia, CA. USA 5:505–521
66. Kolmakov A, Klenov DO, Lilach Y, Stemmer S, Moskovits M (2005) Enhanced gas sensing by individual SnO_2 nanowires and nanobelts functionalized with Pd catalyst particles. Nano Lett 5:667–673
67. Arnold MS, Avouris P, Pan ZW, Wang ZL (2003) Field-effect transistors based on single semiconducting oxide nanobelts. J Phys Chem B 107:659–663
68. Wu NL, Wang SY, Rusakova IA (1999) Inhibition of crystallite growth in the sol-gel synthesis of nanocrystalline metal oxides. Science 285:1375–1377
69. Zhang DH, Liu ZQ, Li C, Tang T, Liu XL, Han S, Lei B, Zhou CW (2004) Detection of NO_2 down to ppb levels using individual and multiple In_2O_3 nanowire devices. Nano Lett 4:1919–1924
70. Comini E, Faglia G, Sberveglieri G, Pan ZW, Wang ZL (2002) Stable and highly sensitive gas sensors based on semiconducting oxide nanobelts. Appl Phys Lett 81:1869–1871
71. Liu JF, Wang X, Peng Q, Li YD (2005) Vanadium pentoxide nanobelts: Highly selective and stable ethanol sensor materials. Adv Mater 17:764–767
72. Wan Q, Li QH, Chen YJ, Wang TH, He XL, Li JP, Lin CL (2004) Fabrication and ethanol sensing characteristics of ZnO nanowire gas sensors. Appl Phys Lett 84:3654–3656

73. Wang SH, Chou TC, Liu CC (2003) Nano-crystalline tungsten oxide NO_2 sensor. Sens Actuators B 94:343–351
74. Wang ZL (2004) Functional oxide nanobelts: Materials, properties and potential applications in nanosystems and biotechnology. Ann Rev Phys Chem 55:159–196
75. Mor GK, Shankar K, Paulose M, Vaeghese OK, Grimes CA (2005) Enhanced photocleavage of water using titania nanotube arrays. Nano Lett 5:191–195
76. Varghese OK, Paulose M, Shankar K, Mor GK, Grimes CA (2005) Water-photolysis properties of micron-length highly-ordered titania nanotube-arrays. J Nanosci Nanotechnol 5:1158–1165
77. Paulose M, Varghese OK, Mor GK, Grimes CA, Ong KG (2006) Unprecedented ultra-high hydrogen gas sensitivity in undoped titania nanotubes. Nanotechnol 17:398–402
78. Capovilla J, VanCouwenberghe C, Miller WA (2000) Noninvasive blood gas monitoring. Crit Care Nurs Q 23:79–86
79. Varghese OK, Yang X, Kendig J, Paulose M, Zeng K, Palmer C, Ong KG, Grimes CA (2006) A transcutaneous hydrogen sensor: From design to application. Sens Lett 4:120–128
80. Hammerschmidt DE (2004) Breath hydrogen and lactose intolerance. J Lab Clin Med 144:279–279
81. Pimentel M, Kong Y, Park S (2003) Breath testing to evaluate lactose intolerance in irritable bowel syndrome correlates with lactulose testing and may not reflect true lactose malabsorption. Am J Gastroenterol 98:2700–2704
82. Adam AC, Rubio-Texeira M, Polaina J (2004) Lactose: The milk sugar from a biotechnological perspective. Crit Rev Food Sci Nutr 44:553–557
83. Goldberg DM (1987) The enzymology of intestinal disease. Clin Biochem 20:63–72
84. Cavalli-Sforza LT, Menozzi P, Strata A (1983) A model and program for study of a tolerance curve: Application to lactose absorption tests. Int J Bio-Med Comput 14:31–41
85. Gerlach P, das Hautathmen U (1851) Arch Anat Physiol 431–479
86. Shannon RD (1964) Phase transformation studies in TiO_2 supporting different defect mechanisms in vacuum-reduced+hydrogen-reduced rutile. J Appl Phys 35:3414–3418
87. Kim KH, Oh EJ, Choi JS (1984) Electrical-conductivity of hydrogen-reduced titanium-dioxide (rutile). J Phys Chem Solids 45:1265–1269
88. Walton RM, Dwyer DJ, Schwank JW, Gland JL (1998) Gas sensing based on surface oxidation reduction of platinum-titania thin films II. The role of chemisorbed oxygen in film sensitization. Appl Surf Sci 125:199–207
89. Roland U, Braunschweig T, Roessner F (1997) On the nature of spilt-over hydrogen. J Mol Catal A 127:61–84
90. Knotek ML (1980) Characterization of hydrogen species on TiO_2 by electron-stimulated desorption. Surf Sci 91:L17–L22
91. Raupp GB, Dumesic JA (1985) Adsorption of CO, CO_2, H_2, and H_2O on titania surfaces with different oxidation-states. J Phys Chem 89:5240–5246
92. Gopel W, Rocker G, Feierabend R (1983) Intrinsic defects of TiO_2 (110) – interaction with chemisorbed O_2, H_2, CO, and CO_2. Phys Rev B 28:3427–3438
93. Knauth P, Tuller HL (1999) Electrical and defect thermodynamic properties of nanocrystalline titanium dioxide. J Appl Phys 85:897–902
94. Bodzenta J, Burak B, Gacek Z, Jakubik WP, Kochowski S, Urbanczyk M (2002) Thin palladium film as a sensor of hydrogen gas dissolved in transformer oil. Sens Acuators B 87:82–87
95. Abdullah M, Low GKC, Matthews RW (1990) Effects of common inorganic anions on rates of photocatalytic oxidation of organic-carbon over illuminated titanium-dioxide. J Phys Chem 94:6820–6825
96. Zeman P, Takabayashi S (2002) Self-cleaning and antifogging effects of TiO_2 films prepared by radio frequency magnetron sputtering. J Vac Sci Tech A 20:388–393
97. Zheng S, Gao L, Zhang QH, Sun J (2001) Synthesis, characterization, and photoactivity of nanosized palladium clusters deposited on titania-modified mesoporous MCM-41. J Solid State Chem 162:138–141

98. Wang CM, Heller A, Wang GH, CM HA, Gerischer H (1992) Palladium catalysis of O_2 reduction by electrons accumulated on TiO_2 particles during photoassisted oxidation of organic-compounds. J Am Chem Soc 114:5230–5234
99. Papp J, Shen HS, Kershaw R, Dwight K, Wold A (1993) Titanium(iv) oxide photocatalysts with Palladium. Chem Mater 5:284–288
100. Wang CC, Zhang ZB, Ying JY (1997) Photocatalytic decomposition of halogenated organics over nanocrystalline titania. Nanostruct Mater 9:583–586
101. Zhang ZB, Wang CC, Zakaria R, Ying JY (1998) Role of particle size in nanocrystalline TiO_2-based photocatalysts. J Phys Chem B 102:10871–10878
102. Basu S, Dutta A (1997) Room-temperature hydrogen sensors based on ZnO. Mater Chem Phys 47:93–96
103. Zuttel A, Nutzenadel C, Schmid G, Emmenegger C, Sudan P, Schlapbach L (2000) Thermodynamic aspects of the interaction of hydrogen with Pd clusters. Appl Surf Sci 162:571–575
104. Michaelson HB (1977) Work function of elements and its periodicity. J Appl Phys 48:4729–4733
105. Henrich VE, Cox PA (1994) The surface science of metal oxides. Cambridge University Press, New York
106. Huang WX, Zhai RS, Bao XH (2000) Investigation of oxygen adsorption on Pd (100) with defects. Appl Surf Sci 158:287–291
107. Kobayashi H, Kishimoto K, Nakato Y (1994) Reactions of hydrogen at the interface of palladium titanium-dioxide schottky diodes as hydrogen sensors, studied by work function and electrical characteristic measurements. Surf Sci 306:393–405
108. Varghese OK, Mor GK, Paulose M, Grimes CA (2005) A titania nanotube-array room-temperature sensor for selective detection of low hydrogen concentrations. Mater Res Soc Symp Proc 835:A3.1.1/K4.1.1
109. Eder D, Kramer R (2003) Stoichiometry of "titanium suboxide" – Part 2 Electric properties. Phys Chem Chem Phys 5:1314–1319
110. Walton RM, Dwyer DJ, Schwank JW, Gland JL (1998) Gas sensing based on surface oxidation reduction of platinum-titania thin films I. Sensing film activation and characterization. Appl Surf Sci 125:187–198
111. Birkefeld LD, Azad AM, Akbar SA (1992) Carbon-monoxide and hydrogen detection by anatase modification of titanium-dioxide. J Am Chem Soc 75:2964–2968
112. Madou MJ, Morrison SR (1989) Chemical sensing with solid state devices. Academic, New York
113. Wang CC, Akbar SA, Madou MJ (1998) Ceramic based resistive sensors. J Electroceram 2:273–282
114. Bates JB, Wang JC, Perkins RA (1979) Mechanisms for hydrogen diffusion in TiO_2. Phys Rev B 19:4130–4139
115. Roland U, Salzer R, Braunschweig T, Roessner F, Winkler H (1995) Investigations on hydrogen spillover.1. electrical-conductivity studies on titanium-dioxide. J Chem Soc-Faraday Trans 91:1091–1095
116. Morimoto T, Nagao M, Tokuda F (1969) Relation between amounts of chemisorbed and physisorbed water on metal oxides. J Phys Chem 73:243–248
117. Ogawa H, Niashikawa M, Abe A (1982) Hall measurement studies and an electrical-conduction model of tin oxide ultrafine particle films. J Appl Phys 53:4448–4455
118. Bisquera JA, Cooper TR, Berseth CL (2002) Impact of necrotizing enterocolitis on length of stay and hospital charges in very low birth weight infants. Pediatrics 109:423–428
119. Engel RR, Virnig NL (1973) Origin of mural gas in necrotizing enterocolitis. Pediatr Res 7:292–294

Chapter 4
TiO_2 Nanotube Arrays: Application to Photoelectrochemical Water Splitting

4.1 Introduction

The key to a hydrogen economy is a means to generate hydrogen efficiently and inexpensively on a renewable basis, that is, without the use of fossils fuel. While there are different means to generate hydrogen, photoelectrolysis is often seen as a technique by which we might put the hydrogen in the hydrogen economy. A photoelectrochemical (PEC) system combines the harvesting of solar energy with electrolysis of water. When a semiconductor of proper characteristics is immersed in an aqueous electrolyte and irradiated with sunlight, sufficient energy is generated to split water into hydrogen and oxygen. However, there are three major challenges for the solar production of hydrogen: (1) *Stability*. Metal oxides are the most photochemically stable semiconductors in aqueous solution, but their band gaps are either too large (~3 eV) to absorb a significant fraction of incident solar energy, or their semiconductor characteristics (e.g., charge transport) are poor. (2) *Band-gap*. Considering the water splitting energy of 1.23 eV and overpotential losses, the semiconductor(s) should have a bandgap greater than 1.7 eV. However, semiconductors with such relatively low bandgaps have been found to lack stability during water splitting. (3) *Energy levels*. Even though a semiconductor electrode may generate sufficient energy to drive an electrochemical reaction, the band edge positions may prevent it from doing water splitting. For spontaneous water splitting, water oxidation and reduction potentials must lie between the valence and conduction band edges. Materials known to do this, for example, $SrTiO_3$, have such large bandgaps they absorb only in the UV region.

In this chapter, we describe the use of TiO_2 nanotube arrays as photoanodes for water photoelectrolysis. The nanotube array architecture possesses high surface to volume ratios, which enables a large internal surface area to be in intimate contact with the electrolyte, and excellent charge transfer properties [1–3]. However, TiO_2 is responsive to UV light, which accounts for only a small fraction ($\approx 4\%$) of the sun's energy. Visible light comprises approximately 45% hence any shift in the optical response of TiO_2 from the UV to the visible spectral range while

C.A. Grimes and G.K. Mor, *TiO_2 Nanotube Arrays*,
DOI 10.1007/978-1-4419-0068-5_4, © Springer Science+Business Media, LLC 2009

maintaining the intrinsic properties of TiO_2, for example the excellent charge transfer properties and photocorrosion stability, will have a positive impact on the photocatalytic and photoelectrochemical utility of the material. Modifying the band structure of TiO_2 to make it more responsive to visible wavelengths has been an enduring activity and, to date, an unsuccessful effort. Compositional doping of TiO_2 has been the approach generally taken for bandgap engineering. When employing dopants to change the optical response of a material, it is desirable to maintain the integrity of the host material crystal structure while changing its electronic structure. The crystal structure of the material is directly related to the ratio of cation and anion size in the crystal lattice. It appears to be relatively easier to replace Ti^{4+} in TiO_2 with any cation than to substitute O^{2-} with any other anion due to the difference in the charge states and ionic radii. While several authors have reported that transition metal ion doping decreases the photothreshold energy of TiO_2, there is also an increase in thermal instability and a decrease in charge carrier lifetimes [4, 5] which limits overall photoconversion efficiencies. In this chapter, we review efforts in nanotube array bandgap engineering to absorb visible light with, of course, application to water splitting. Anionic doping of TiO_2 consists of the substitution of a nonmetal atom, such as nitrogen [6–8], carbon [9, 10], sulfur, and fluorine [11], for oxygen. A large part of the existing literature agrees that anionic doping produces enhancement in the visible light absorption properties of TiO_2 photocatalysts. Yet as far as the photoelectrochemical response of anionically doped TiO_2 is concerned, the improvement of the visible photocurrent response is at best modest in such reports [12–15]. We also consider coupling of the TiO_2 nanotubes with low band gap materials, such as CdS, CdTe, and CdSe, as another approach to enhance visible light absorption. We then discuss a strategy of combining TiO_2 with semiconducting metal oxides with smaller band-gaps that are processing compatible with TiO_2 as a means to improve the visible response. Presently, we have focused our attention on α-Fe_2O_3 and copper oxides (CuO and Cu_2O) forming ternary oxide (Cu-Ti-O and Fe-Ti-O) nanotube architectures. In these architectures, TiO_2 confers both structural support and chemical stability to the narrow band-gap oxide material. Even when the two oxide phases are segregated, the ternary oxide nanotube architecture shows good photoelectrochemical response in the visibility, likely due the lower recombination resulting from the fast capture of photogenerated charge carriers by close-lying solution ions.

4.2 Photoelectrolysis Cell

Photoelectrolysis is the general term used to describe semiconductor-based water splitting by the use of a photoelectrochemical cell (PEC) [1], as depicted in Fig. 4.1. One approach is to couple a photovoltaic system and an electrolyzer as a single system. Photovoltaic cells can be combined in series to generate the potential needed to split water and then connected to H_2 and O_2 producing electrodes. With semiconductor–liquid junctions, the water splitting potential is generated directly at the

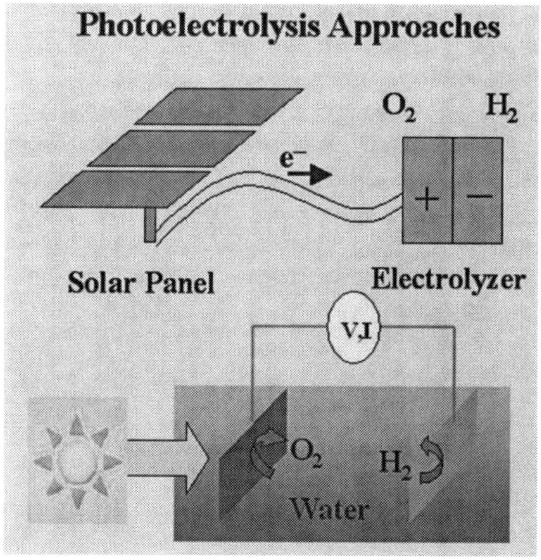

Fig. 4.1 The schematic representations for two photoelectrolysis approaches

semiconductor–liquid interface. The ability of a semiconductor photoelectrode to drive either the oxidation of water to O_2, or the reduction of water to H_2, or the entire water splitting reaction is determined by its band gap and the position of the valence and conduction band edges relative to the water redox reactions [1]. Besides band edge positions, the semiconductor has to be active over a broad spectral range, and, upon light absorption, the material should efficiently separate the generated charges. Moreover, the immersed semiconductor has to be stable in the electrolyte.

In this chapter, we consider semiconductor liquid junctions in application to water splitting [1] with the photoelectrolysis cell consisting of TiO_2 nanotube arrays on a conducting substrate. Contact between the two electrodes through an aqueous electrolyte, where there is no illumination, results in charge transfer from the TiO_2 anode having a lower work function to the platinum cathode having a higher work function resulting in band bending. Under illumination, the photoanode surface potential and the (H^+/H_2) water reduction potential are each lowered, but the (H^+/H_2) water reduction potential still remains above the cathode Fermi level. Anodic bias is thus needed to elevate the Fermi level of cathode above the water reduction potential, making the water splitting process feasible. This applied bias provides overvoltage at the metal cathode necessary to sustain the current flow, and increases the semiconductor band bending to maintain the required electric field driven charge separation in the semiconductor. The bias can be provided either by an external voltage (power) source, or by immersing the anode in a basic solution and the cathode in acidic solution. In the first report on water photoelectrolysis by Fujishima and Honda [16] using an n-type TiO_2 anode and Pt cathode, it was observed that an additional bias voltage of 0.25–0.50 V was required to achieve simultaneous oxygen evolution at the TiO_2 electrode and hydrogen evolution at the Pt cathode.

Several oxide semiconductors have flat band potentials above the H^+/H_2 level; therefore, no external bias is needed to produce H_2 and O_2. Unfortunately, these oxide semiconductors have relatively large bandgap energies that result in low optical absorption and hence low visible spectrum photoconversion efficiencies.

The photoelectrochemical (PEC) reaction separates the oxidation and reduction processes into half-cell reactions; the half-cell reactions with their corresponding standard reduction potential E° with respect to the standard hydrogen electrode (SHE) are shown. The negative ΔE° indicates that water splitting is not a thermodynamically spontaneous process; for the reaction to proceed, 1.23 V must be externally provided.

$$\text{Oxidation}: 2H_2O \rightarrow O_2 + 4H^+ + 4e^- \quad E^\circ = 1.23\,\text{V vs. SHE} \tag{4.1}$$

$$\text{Reduction}: 2H^+ + 2e^- \rightarrow H_2 \quad E^\circ = 0.00\,\text{V vs. SHE} \tag{4.2}$$

$$\text{Overall}: 2H_2O \rightarrow O_2 + 2H_2 \quad \Delta E^\circ = -1.23\,\text{V} \tag{4.2}$$

The bandgap E_g of the material determines its light harvesting ability. There are inherent losses associated with any solar energy conversion process involving materials [17–20]. Consider, since any absorbed energy higher than the bandgap is lost as heat, only a partial fraction of the excited state energy can be converted into electrical energy or stored as chemical energy, and losses occur due to spontaneous emissions. Other losses include those due to electron transport within the electrode during charge separation, transport of electrons from photoanode to platinum electrode, transport of electrons/holes to the photoelectrode/electrolyte interface, Joule heating due to the electron flow through the external circuit, and the cathodic and anodic overpotentials. Overvoltage losses associated with electrodes are the functions of current density, electrode structure and surface properties, temperature, electrolyte composition, etc. [21]. In practical systems, energy losses can be greater than 0.8 eV. Considering this, a bandgap of ≈ 2.0 eV appears optimal for semiconductor photoelectrodes applied to water splitting.

A photoelectrolysis cell can have two illuminated semiconductor–liquid junctions, with an n-type semiconductor used for O_2 evolution through use of the photogenerated holes, and a p-type semiconductor for H_2 generation through use of the photogenerated electrons [1]. Two semiconductors with smaller band gaps can be utilized since each needs only to provide part of the water splitting potential. The smaller band gap means more absorption in the visible region of the solar spectrum. The majority carriers of each photoelectrode, electrons formed in the n-type semiconductor (photoanode) and holes formed in the p-type semiconductor (photocathode), combine at an ohmic contact. This is theoretically possible only if the valence band of the photocathode is positive (higher electrochemical potential) with respect to the conduction band of the photoanode [1]. Proper selection of both semiconductor electrodes ensures that the energy necessary for water photoelectrolysis is gathered entirely from the illumination.

4.2.1 *Water Splitting Efficiency*

The usefulness of a water photoelectrolysis cell is primarily determined by the efficiency with which it converts light energy into chemical energy, which is stored in the form of hydrogen. In the case of water splitting, the redox properties of the electrolyte are fixed, and hence the light harvesting and utilization properties of the photoelectrode decide cell efficiency. The overall photoconversion efficiency is defined as the ratio of the maximum energy output that can be obtained from the final products, hydrogen and oxygen, to the energy supplied in the form of light to produce them.

4.2.1.1 Two Electrode Configuration

Photoelectrolysis cells consist of two electrolyte immersed electrodes, with the bias voltage, if any, applied between them; the overall chemical reaction in such a cell is made of two independent half-reactions [22]. The photoconversion efficiency (solar conversion efficiency if sunlight is used) in these practical cells is calculated as:

$$\varepsilon_0 = \frac{\Delta G^0 R_{H_2}}{P_t} \tag{4.4}$$

where R_{H2} is the rate of production (mol/s) of hydrogen in its standard state per unit area of the photoelectrode. The standard Gibbs energy $\Delta G^0 = 237.2$ kJ/mol at 25°C and 1 bar, and P_t is the power density (W/m^2) of illumination.

Equation (4.4) is based on the assumption that the free energy ΔG^0 can be completely retrieved in an ideal fuel cell run by the products from the photoelectrolysis cell for which the relation is applied. Instead of the free energy ΔG^0, the enthalpy (heat) of water splitting ΔH_0 also has been used in some cases. Here, it is assumed that the heat of water splitting is completely retrieved by burning hydrogen. At 25°C and 1 bar, $\Delta H_0 = 285$ kJ/mol. However, the thermodynamic definition of "efficiency" involves work, and hence it is appropriate to use ΔG^0. Using ΔH gives exaggerated values for efficiency. If I is the current density responsible for the generation of hydrogen at the rate of R_{H2} in (4.4), then under 100% Faradaic conversion, $R_{H2} = I/nF$, where n, which is the number of moles of electrons used for generating 1 mol of hydrogen, is 2 and F, the Faraday constant, is 96,485 C. Therefore, the voltage corresponding to the Gibbs energy is $V_{rev} = \Delta G^0/nF = 1.229$ V.

$$\varepsilon_0 = 1.229 I_p / P_t \tag{4.5}$$

where the constant is expressed in volts (1.229 V), I_p in A/m^2 and P_t in W/m^2.

Spontaneous water splitting upon illumination requires semiconductors with appropriate electron affinity and flat band conditions. The flat band positions shift with electrolyte pH. Hence, an external bias needs to be applied between the electrodes in most cases to effect water splitting. The external bias can be either electrical or chemical. This external bias contribution should be subtracted from (4.4) or (4.5) to obtain overall photoconversion efficiency. In the case of an external electrical bias, the efficiency can be defined as:

$$\varepsilon_0\,(\mathit{efficiency}) = \frac{\text{Energy stored as hydrogen} - \text{Energy input from power supply}}{\text{Light energy input}}$$

$$\varepsilon_0 = \frac{\left(\Delta G^0 R_{H_2} - V_{\text{bias}} I_{\text{p}}\right)}{P_{\text{t}}} \tag{4.6a}$$

which simplifies to

$$\varepsilon_0 = \frac{(1.229V - V_{\text{bias}})I_{\text{p}}}{P_{\text{t}}} \tag{4.6b}$$

with all units in SI. The basis of this definition is that a fuel cell run by the products of the photoelectrolysis cell supplies a part of its output to the photoelectrolysis cell as electrical bias [23, 24].

4.2.1.2 Three-Electrode Configuration

In laboratory water photoelectrolysis experiments, a three-electrode geometry is commonly used to measure photocurrent. This geometry involves a working electrode (photocathode or photoanode), a counter electrode that is generally platinum, and a reference electrode. The internationally accepted primary reference is the standard hydrogen electrode (SHE) or normal hydrogen electrode (NHE) which has all components at unit activity [(P_t/H_2 (a = 1)/H^+ (a = 1, aqueous)]. However, using such an electrode is impractical, and hence other reference electrodes, such as silver–silver chloride (Ag/AgCl/KCl) and saturated calomel electrodes (SCE) (Hg/Hg_2Cl_2/KCl), are generally used. Their potentials may then be converted in terms of NHE potential. The potential of an Ag/AgCl electrode is 0.197 V versus NHE and that of saturated calomel electrode is 0.242 V versus NHE [25]. Equations (4.6a) and (4.6b) give a thermodynamical measure of efficiency and should be applied, in general, to all cases involving an electrical bias irrespective of the electrode configuration (two- or three-electrode) used. Note that in the case of a three-electrode configuration V_{bias} should be measured between the working and counter electrodes [26]. The reference electrode in the three-electrode geometry does not draw any current and the current flows between the working and counter electrodes [27–29]. Here, we will consider two common approaches used which avoid the use of a comparative noble metal electrode and neglect overpotential losses at the electrodes. In this method, the potential applied at the hydrogen (or oxygen) electrode (in a three-electrode configuration) is compared with the potential generated at an ideal fuel cell anode (or cathode). In the case of an n-type semiconductor photoanode:

$$\varepsilon_0 = \left(V^0_{\text{ox}} - V_{\text{app}}\right) I_{\text{P}} / P_{\text{t}} \text{ (Oxygen evolution using photoanodes)} \tag{4.7a}$$

OR

$$\varepsilon_0 = \left(V_{\text{app}} - V^0_{\text{H}_2}\right) I_{\text{P}} / P_{\text{t}} \text{ (Hydrogen evolution using photoanodes)} \tag{4.7b}$$

V^0_{ox} and $V^0_{\text{H}_2}$ represent, respectively, the standard potentials of oxygen and hydrogen electrodes. V^0_{ox} takes a value of +0.401 V versus NHE in alkaline electrolytes (pH = 14) and +1.229 V versus NHE in acidic electrolytes (pH = 0). The corresponding values of $V^0_{\text{H}_2}$ are −0.828 V versus NHE and 0 V versus NHE.

Another form of this definition has sparked much debate in the scientific community [30–35]. In this approach, V_{app} (or V_{bias}) is taken as the absolute value of the difference between the potential at the working electrode measured with respect to a reference electrode (V_{meas}) and the potential of the photoelectrode under open circuit conditions measured with respect to the same reference electrode in identical conditions (in the same electrolyte solution and under the same illumination). In the case of a semiconductor photoanode where oxygen evolution takes place, the efficiency is calculated as:

$$\varepsilon_0 = \frac{\left(V^0_{\text{rev}} - |V_{\text{app}}|\right) I_{\text{P}}}{P_{\text{t}}} = \frac{\left(V^0_{\text{rev}} - |V_{\text{meas}} - V_{\text{aoc}}|\right) I_{\text{P}}}{P_{\text{t}}} \tag{4.8}$$

$V^0_{\text{rev}} = 1.229$ V is the standard state reversible potential for the water splitting reaction and V_{aoc} is the anode potential measured with respect to the reference electrode at open circuit conditions. The term $V_{\text{meas}} - V_{\text{aoc}}$ arises from the fact that V_{oc} (the open circuit voltage of the cell which may be obtained from V_{aoc}) represents the contribution of light toward the minimum voltage needed for water splitting potential (1.229 V) and that the potential of the anode measured with respect to the reference electrode V_{meas} has contributions from the open circuit potential and the bias potential applied by the potentiostat (i.e., $V_{\text{meas}} = V_{\text{app}} + V_{\text{aoc}}$). The term $V_{\text{meas}} - V_{\text{aoc}}$ makes relation (4.8) independent of the electrolyte pH and the type of reference electrode used. Thus the use V^0_{rev} in (4.8) instead of V^0_{ox} or $V^0_{\text{H}_2}$ as in the case of relation (4.7a) or (4.7b) is justified. Although this approach has received wide attention and is commonly being used, it is suggested that the relation (4.8) may give exaggerated photoconversion efficiency values. For both two- or three-electrodes, the voltage measured between the working and the counter electrodes gives the actual applied bias voltage V_{app} [or V_{bias} in (4.6a) and (4.6b)]. This voltage multiplied by the cell current gives the electrical energy supplied by the electrical power supply. But in practice, where a potentiostat is used to apply an external bias to the photoelectrode, this actual voltage V_{app} (between the working and counter electrodes) may slightly exceed the bias voltage measured as $V_{app} = V_{meas} - V_{aoc}$ with respect to the reference electrode [36]. Thus, the use of the latter term in relation (4.8) can show a higher efficiency value than when the term V_{app}, measured between the counter and working electrodes, is directly used as in the case of relation (4.6a) and (4.6b).

4.2.1.3 Efficiency Comparison Determined Using Two- and Three-Electrode Configurations

Figure 4.2 is plotted to demonstrate the difference in the efficiency values calculated by different approaches represented by relations (4.6), (4.7a), and (4.8) [37]. A two electrode geometry was used for photocurrent measurements and efficiency calculation using relation (4.6) and a three-electrode geometry was employed for relations (4.7a) and (4.8). A 6-μm long TiO_2 nanotube array film (polycrystalline) on titanium foil was used as the photoanode. Platinum served as the counter electrode and the 1 M KOH solution as the electrolyte. In the three-electrode configuration, an Ag/AgCl reference electrode was used to measure the photoanode potentials. As the TiO_2 bandgap is 3.0–3.2 eV (depending upon whether the crystalline phase is rutile or anatase), a near UV light illumination (320–400 nm) from a 50 W metal-hydride lamp was used. For the two electrode geometry, a Keithley source meter and for the three-electrode geometry a potentiostat was used to apply the bias. The x-axis of the plot (Fig. 4.2) shows the voltage measured between the working (TiO_2 nanotube array) and counter electrodes for relation (4.6) and the potential at the photoanode anode measured with respect to Ag/AgCl reference electrode for relations (4.7a) and (4.8).

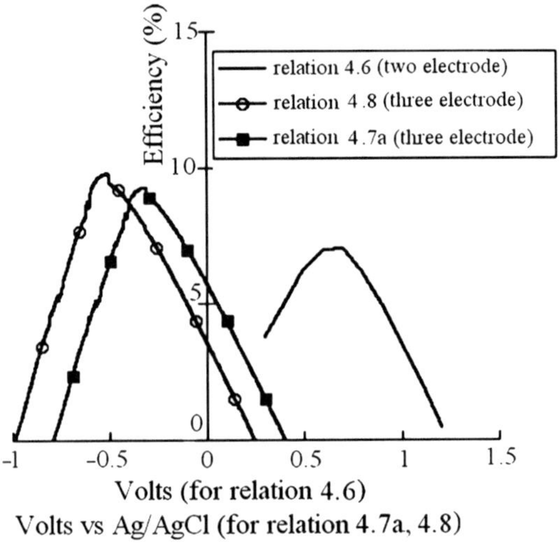

Fig. 4.2 Efficiency under near UV illumination of a photoelectrochemical cell comprised of a TiO_2 nanotube array photoanode and Pt counter electrode. For the calculation of efficiency using (4.6), a two-electrode geometry was used while for the calculation using (4.7a) and (4.8), a three-electrode geometry was used [37]

It can be seen from Fig. 4.2 that the relation (4.8) yielded the highest efficiency of about 9.5% whereas the efficiency calculated using relation (4.6) employing a two-electrode geometry has a maximum value of about 7%. Although the relative values may vary with respect to the experimental setup and measurement conditions, this exercise demonstrates that a certain degree of exaggeration could occur in the efficiency values calculated using (4.8) in a three-electrode geometry.

4.2.2 *Quantum Efficiency Calculation*

A very useful parameter for evaluating the performance of a photoelectrolysis cell is the incident photon to current conversion efficiency (IPCE). This is a measure of the effectiveness in converting photon's incident on the cell to photocurrent flowing between the working and counter electrodes. IPCE is also called the external quantum efficiency [38].

$$\mathrm{IPCE} = \frac{I_\mathrm{p}(\lambda)}{eF(\lambda)} \tag{4.9a}$$

$$\mathrm{IPCE} = \left(\frac{hc}{e}\right)\left(\frac{I_\mathrm{p}(\lambda)}{P(\lambda)\lambda}\right) \tag{4.9b}$$

$$\mathrm{IPCE} = 1240\frac{I_\mathrm{p}(\lambda)}{P(\lambda)\lambda} \tag{4.9c}$$

$I_\mathrm{p}(\lambda)$ is the photocurrent density at wavelength λ. The IPCE becomes 100% when all photons generate electron-hole pairs. However, in practical situations, the IPCE is always less than 100% due to the losses corresponding to the reflection of incident photons, their imperfect absorption by the semiconductor and recombination of charge carriers within the semiconductor, etc. IPCE is calculated by measuring the current in a cell when a particular wavelength or a small group of wavelengths (band pass, usually up to 12 nm) with a known power density $P(\lambda)$ incident on it. It is usually measured at a bias voltage corresponding to the maximum power point (voltage corresponding to the peak efficiency). The IPCE enables the estimation of the total photocurrent as well as efficiency of a photoelectrolysis cell under any type of illumination, say for example global sunlight [39]. If $I_\mathrm{p}(\lambda)$ is the current density corresponding to wavelength λ, the photocurrent spectrum corresponding to a particular energy distribution can be obtained by multiplying the IPCE with the photon flux density of that distribution.

$$I_\mathrm{p}(\lambda) = \mathrm{IPCE}\,F(\lambda)\mathrm{d}\lambda = \mathrm{IPCE}\,P(\lambda)\lambda\left(\frac{e}{hc}\right) \tag{4.10}$$

The total photocurrent that can be obtained from this electrode is:

$$I_\mathrm{p} = \int_{\lambda_\mathrm{min}}^{\infty} I_\mathrm{p}(\lambda)\mathrm{d}\lambda \tag{4.11}$$

Another parameter of interest, used mainly in photoelectrochemical solar cells, is the absorbed photon to current conversion (APCE) efficiency, also called the internal quantum efficiency. APCE is defined as the number of electrons (or holes) collected per absorbed photon. It is calculated after considering the losses in the incident photons like reflection, scattering, absorption, etc. APCE and IPCE are related by [40]:

$$\mathrm{IPCE}(\lambda) = \mathrm{LHE}(\lambda)\varphi_\mathrm{inj}\eta \tag{4.12}$$

LHE is the light harvesting efficiency or absorptance, defined as LHE $= 1 - 10^{-A}$ where A is the absorbance, φ_inj is the quantum yield of charge injection, and η is the efficiency of transporting injected electrons in to the external circuit. Equation (4.12) can be written as:

$$\mathrm{APCE} = \frac{\mathrm{IPCE}}{1 - 10^{-A}} \tag{4.13}$$

4.3 Photoelectrolysis Using Unmodified TiO_2 Nanotubes

Figure 4.3 depicts the experimental setup for water photoelectrolysis measurements with the nanotube arrays used as the photoanodes from which oxygen is evolved.

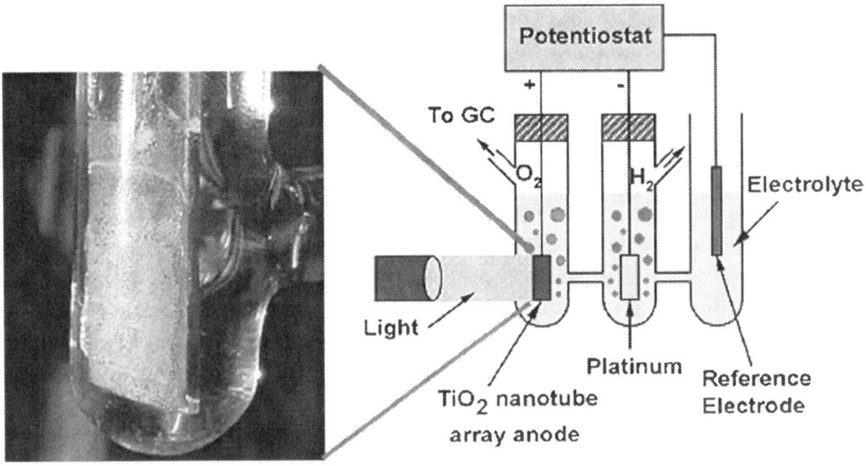

Fig. 4.3 Depiction of experimental set-up for hydrogen generation by water photoelectrolysis [58]

4.3.1 Short Nanotubes

TiO_2 nanotubes were grown by potentiostatic anodization of titanium foil at 10 V in an electrolyte of 0.5% HF + acetic acid mixed in a 7:1 ratio [41, 42]. Four different electrolyte bath temperatures, 5°C, 25°C, 35°C, and 50°C were tried to vary the nanotube wall thickness and length. The *I-V* characteristics of the TiO_2 nanotube-array electrodes, photocurrent density versus potential, measured in 1 M KOH electrolyte as a function of anodization bath temperature under UV (320–400 nm, 100 mW/cm^2) illumination are shown in Fig. 4.4. At 1.5 V, the photocurrent density of the 5°C anodized sample is more than three times the value for the sample anodized at 50°C. The lower anodization temperature also increases the slope of the photocurrent–potential characteristic. On seeing the photoresponse of a 10 V 5°C anodized sample to monochromatic 337 nm 2.7 mW/cm^2 illumination, it was found that at high anodic polarization, greater than 1 V, the quantum efficiency is larger than 90%.

The TiO_2 nanotube array architecture results in a large effective surface area in close proximity with the electrolyte, thus, enabling diffusive transport of photo-generated holes to oxidizable species in the electrolyte. Separation of photogenerated charges is assisted by action of the depletion region electric field [43, 44]. Minority carriers generated within a "retrieval" length from the material surface, that is, a distance from the surface equal to the sum of the depletion layer width and the diffusion length, escape recombination and reach the electrolyte [45]. The relevant structure sizes of the TiO_2 nanotube arrays, that is, half the wall thickness, are all smaller than 20 nm which is less than the retrieval length of crystalline TiO_2 [46]; hence, bulk recombination is greatly reduced and the quantum yield enhanced [46]. Because of the light scattering within a porous structure incident photons are more effectively absorbed than on a flat electrode [47]. However, while bulk recombination is reduced by the nanotube architecture, photogenerated minority carriers can be trapped by surface states [46, 48]. The photoconversion efficiency as a function of potential for the different photoanodes is shown in Fig. 4.5. A maximum conversion efficiency of 6.8% is obtained for nanotubes anodized

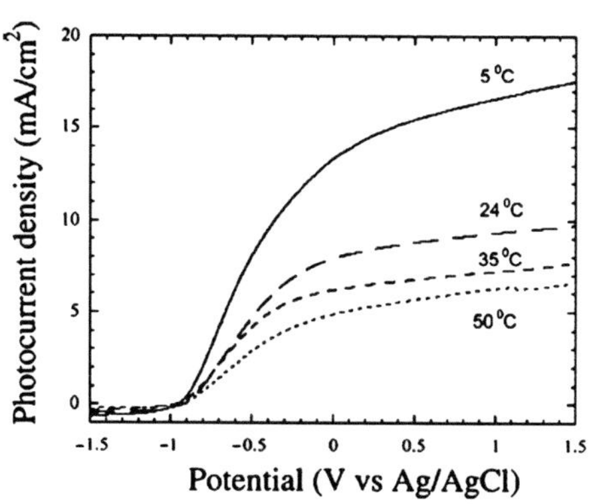

Fig. 4.4 The photocurrent density versus potential [vs. Ag/AgCl] for 10 V samples anodized at four anodization bath temperatures, that is, 5°C, 25°C, 35°C, and 50°C [41]

Fig. 4.5 Photoconversion efficiency as a function of measured potential [vs. Ag/AgCl] for 10 V samples shown in Fig. 4.4 [41]

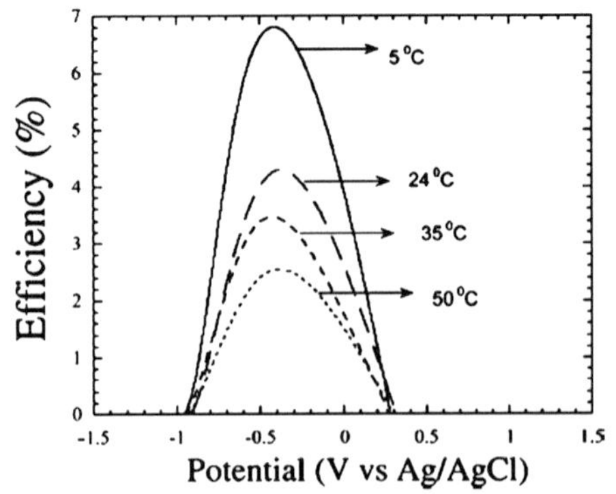

at 5°C. For this sample, gas chromatographic analysis verified that the volume ratio of the evolved hydrogen and oxygen was 2:1, which confirmed water splitting. With the nanotube array photoanodes held at constant voltage bias, determined by the peak position in the photoconversion efficiency curve with respect to the Ag/AgCl electrode, during 1,800 s of exposure, 48 μmol of hydrogen gas was generated. Normalizing this rate to time and incident power, we find a hydrogen generation rate of 960 μmol/hW, or 24 mL/hW. Oxygen bubbles evolving from the nanotube-array photoanode do not remain on the sample; hence, the output remains stable with time irrespective of the duration of hydrogen production.

Under UV (320–400 nm, 100 mW/cm^2) illumination, a maximum photoconversion efficiency of 7.9% was obtained for nanotube arrays, up to 600 nm in length, anodized in boric acid contained electrolyte [49], with a hydrogen generation rate of 42 mL/hW. Under full spectrum illumination (AM 1.5, 100 mW/cm^2), a photoconversion efficiency of 0.45% was obtained. The significant photoresponse of the boric acid anodized sample may be due to boron incorporation into the nanotube array structure. It is possible that boron, which is difficult to identify by XPS, remains inside the TiO_2 matrix and affects its charge transfer properties [50].

Figure 4.6 shows the photocurrent as a function of applied voltage for 500°C annealed TiO_2 nanotube electrodes prepared at 10, 18, or 20 V anodization potentials in aqueous or ethylene glycol (EG) solutions containing 0.5 HCl and 0.4 M H_2O_2, under AM 1.5 simulated sunlight in 1.0 M KOH [51]. The dark current was less than 5 $\mu A/cm^2$ for all samples over the displayed potential range [38, 52].

Nanotube arrays of 570 nm length and 70 nm pore size were synthesized on titanium foil in 0.15 M HF–0.5 M HNO_3 [53]. Samples were then annealed at 450°C for 2 h. The photocurrent and photoelectrochemical current response were measured in 0.01 M Na_2SO_4 under 0.68 mW/cm^2 UV illumination. The photocurrent density (measured at 0 V vs. SCE) increased from almost zero in the dark to 0.105 mA/cm^2 under UV illumination. The photoelectrochemical current of TiO_2 nanotube array films is compared with that of porous TiO_2 in Fig. 4.7. The higher electrochemical current density of the nanotube array samples was ascribed to the higher surface area.

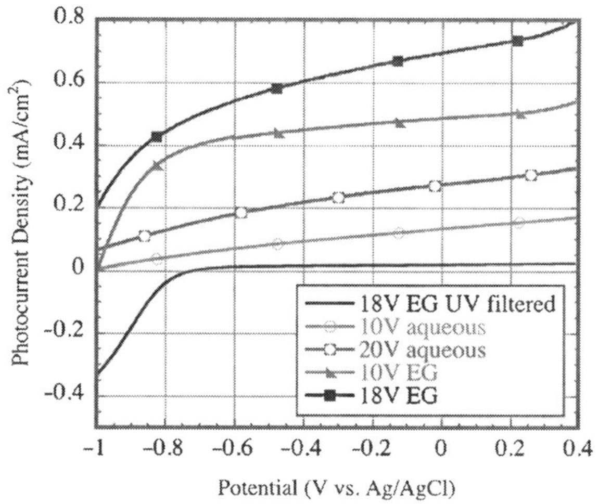

Fig. 4.6 Current–voltage characteristics (light less dark currents) tested under AM 1.5 sun condition for the annealed TiO_2 electrodes prepared in 0.5 M HCl + 0.4 M H_2O_2 solution [51]

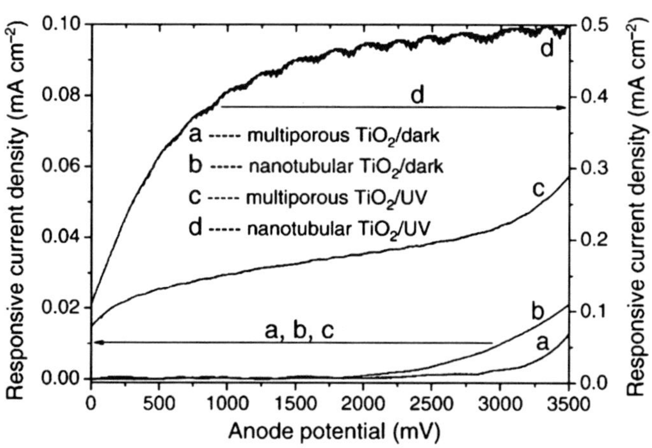

Fig. 4.7 Measured current of crystallized porous and nanotubular TiO_2/Ti electrodes under linearly swept potential from 0 to 3,500 mV vs. SEC [53]. Reprinted from *Materials Letters*, Vol. 60, Xie Y, Zhou LM, Huang H, Enhanced photoelectrochemical current response of titania nanotube array, 3558–3560, Copyright 2006, with permission from Elsevier

4.3.2 Medium Length Nanotubes

The photoelectrochemical properties of nanotube arrays $\approx 2\,\mu m$ in length, tube diameter 100 nm, and wall thickness of 10 nm anodically grown on titanium samples at 20 V Ag/AgCl in 1 M $(NH_4)_2SO_4$ + 0.5 wt% NH_4F, were investigated in 0.1 M Na_2SO_4 using a 1,000 W Xe lamp [54]. In the UV range, the IPCE of these annealed nanotubes (450°C for 3 h) was found to be about three to four times higher than that of compact annealed TiO_2 films, with an enhanced photocurrent over the entire potential range. However, at potentials close to the flat band (the optical flat-band potential E_{fb} is about at −0.1 to −0.2 VAg/AgCl), the absolute photocurrent response for the nanotubes found to be 10 to 20 times higher than for the compact oxide. At higher anodic potentials (>1 V), a saturation of the photoresponse was observed for the nanotubes, while the compact oxide still showed an increasing behavior.

Figure 4.8(a) shows the *I-V* characteristics of 6 μm nanotube arrays prepared in KF buffer bath [55] (on foil), annealed at different temperatures, under UV (320 nm to 400 nm) illumination with an intensity 100 mW/cm^2; the dark current in all cases is approximately 10^{-7} to 10^{-6} A/cm^2 [56]. The photocurrent increases with increasing annealing temperature to 675°C, after which it reduces with samples annealed at 700°C showing a low photocurrent (~10^{-4} A) due to the disruption of the nanotube-array architecture. The corresponding light energy to chemical energy conversion (photoconversion) efficiencies is shown in Fig. 4.8(b). The highest efficiency of about 12.25% was obtained for samples annealed in the range 580°C to 620°C. The increase in photocurrent and efficiency is due to the increased crystallinity of the nanotube-walls, with the reduction of the amorphous regions and grain boundaries in turn reducing the number of charge carrier recombination centers. However, at temperatures near 675°C, the densification of the bottom part of the nanotubes starts isolating the undestroyed nanotubes from the metal electrode reducing the number of charge carriers reaching the electrode.

The photoresponse increases with nanotube-array length, saturating when the light is fully absorbed [57]; in our experiments, we found this saturation with length to occur at approximately 20 μm; however, this is dependent on the nanotube array density and hence optical absorbance. On exposing a 6 μm nanotube-array samples annealed at 600°C to individual wavelengths of 337 nm (3.1 mW/cm^2) and 365 nm (89 mW/cm^2), the quantum efficiency was calculated as 81% and 80%,

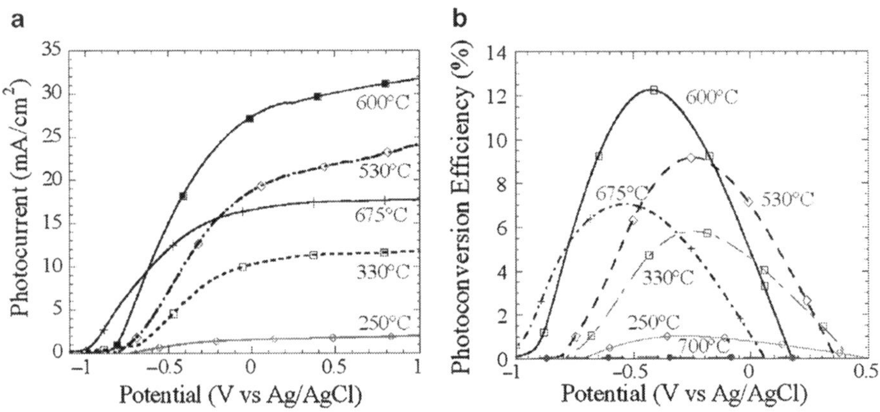

Fig. 4.8 (**a**) Photocurrent generated from 6 μm long nanotube-arrays (in 1 M KOH solution) with respect to annealing temperature, and (**b**) the corresponding photoconversion efficiencies [56]

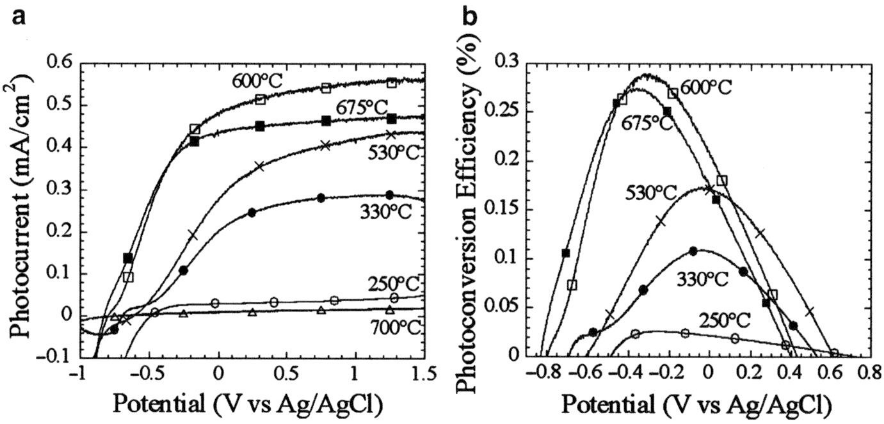

Fig. 4.9 (**a**) Visible light (AM 1.5) photocurrent generated from 6 μm long nanotube-arrays with respect to annealing temperature, and (**b**) the corresponding photoconversion efficiencies [58]

respectively. For a 6 μm nanotube array annealed at 600°C, the hydrogen evolution rate is approximately 76 mL/hW.

Figure 4.9 shows the *I-V* characteristics under UV filtered AM 1.5 illumination of 6-μm long nanotube arrays annealed at different temperatures [58]. The photocurrent increases with increasing annealing temperature to approximately 620°C, after which it reduces with samples annealed at 700°C, in which the nanotube array geometry is destroyed by the growth of the underlying oxide layer from the Ti foil substrate, showing a low photocurrent (~10^{-4} A). The highest AM 1.5 full spectrum efficiency of about 0.6% was obtained for samples annealed in the range 580°C to 620°C. For a 6 μm nanotube array annealed at 600°C, the hydrogen generation rate is 1.75 mL/Wh.

We note that the unique, highly ordered TiO_2 nanotube array structure enables the conductive electrolyte to permeate the entire internal and external surfaces; hence, there is a constant electrostatic potential along the length of the tubes. Therefore, long-range electron transport is dominated by diffusion rather than drift. In this case, for a nanotube array of length d, the diffusion driving force is nearly constant and approximately equal to $2k_BT/d$ [59]. The nanotube-array architecture, with a wall thickness of 20 nm, ensures that the holes are never generated far from the semiconductor-electrolyte interface. Furthermore, since half the wall thickness is significantly less than the minority carrier diffusion length $L_p \approx 20$ nm in TiO_2 [60], charge carrier separation takes place efficiently. The increased crystallinity of the samples annealed at elevated temperatures reduces the number of grain boundaries, improves connectivity between grains, and eliminates any amorphous regions that provide defects acting as carrier recombination centers. The width of the anatase crystallites in the walls is restricted by the wall thickness, to approximately 10 nm. The potential drop within the wall can be represented as:

$$\Delta\phi_0 = kTr_0^2/6eL_D^2 \tag{4.14}$$

where r_0 is half the width of the wall, T is the temperature, and L_D is the Debye length given by

$$L_D = \left[\varepsilon_0 \varepsilon kT / 2e^2 N_D\right]^{1/2} \tag{4.15}$$

where N_D is the number of ionized donors per cm^3 [61]. This potential drop across the wall thickness may not be enough to separate the photogenerated electrons and holes; however, because of the nanoscale dimensions of the walls, the holes can reach the surface through diffusion, which takes place on a scale of picoseconds [42–44, 47–61]. Minority carriers generated within a "retrieval" length from the material surface, that is a distance from the surface equal to the sum of the depletion layer width and the diffusion length, escape recombination and reach the electrolyte [43]. The relevant dimensional features of the TiO_2 nanotube arrays, that is, half the wall thickness, are all smaller than the crystalline TiO_2 retrieval length of 10 nm [44]; hence, bulk recombination is greatly reduced and the quantum yield enhanced [44, 60, 62, 63]. It is noted that van de Lagemaat et al. observed a substantial enhancement of the quantum yield in SiC made nanoporous by anodic etching in HF solution [43]. Furthermore, charge carriers near the electrolyte-nanotube interface region are readily accessible to the electrolyte species due to overlapping wave functions [61, 64].

4.3.3 Long Nanotubes

Figure 4.10(a) shows the measured photocurrent density of nanotube array samples fabricated using a formamide electrolyte, 0.56 g of ammonium fluoride in a solution mixture of 5 mL de-ionized water + 95 mL formamide, at 15 V, 20 V, and 25 V

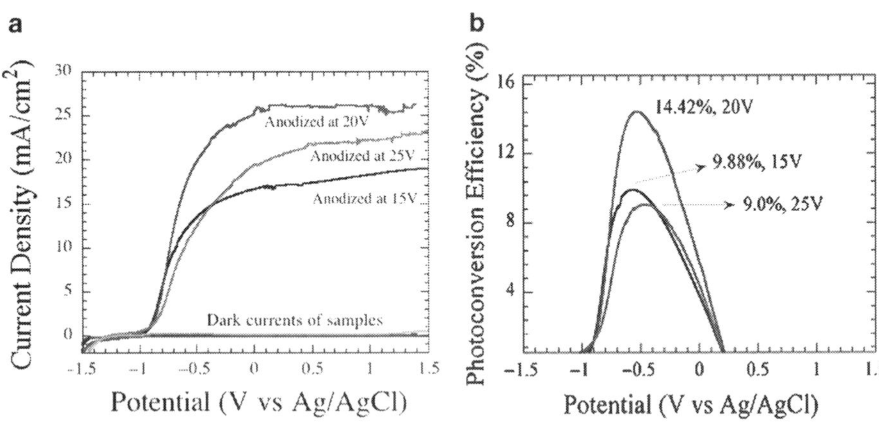

Fig. 4.10 (**a**) Photocurrent density and (**b**) corresponding photoconversion efficiency, of nanotube array samples fabricated using a formamide solvent (K1 bath). All samples were annealed at 525°C for 1 h in oxygen [65]

anodization potentials [65]. All samples were annealed at 525°C for 1 h in oxygen. Figure 4.10(b) shows the corresponding photoconversion efficiency of the three nanotube array samples. The use of different anodization voltages resulted in variation of nanotube array length and tube outer diameter (hence packing fraction): 15 V resulting in a nanotube array 8.2 μm long, 80 nm outer diameter; 20 V resulting in a nanotube array 14.4 μm long, 94 nm outer diameter; 25 V resulting in a nanotube array16 μm long, 140 nm outer diameter. The 20 V sample, 14.4 μm long with 94 nm outer diameter, achieves a photoconversion efficiency of 14.42%. A trade-off exists between nanotube length and internal surface area. Thus, the 15 V nanotubes have a relatively large internal surface area, but their optical absorption is limited by their shorter length while the 25 V nanotubes are longer but have less surface area. The maximum conversion efficiency is obtained for nanotubes anodically formed at 20 V. Figure 4.11(a) shows the *I-V* characteristics of 30 μm long 205 nm diameter nanotube arrays fabricated in an electrolyte of 1.2 g of NH_4F in a solution of 5 mL deionized water + 95 mL formamide at 35 V. The samples were annealed at 525°C and 580°C for 1 h in oxygen prior to measurement. The 580°C annealed sample had an open circuit voltage V_{OC} of −0.925 V (vs. Ag/AgCl); the 525°C annealed sample had an open circuit voltage V_{OC} of −0.80 V (vs. Ag/AgCl). Figure 4.11(b) shows the corresponding photoconversion efficiencies of the two samples.

Figure 4.12(a) shows the measured photocurrent density of nanotube array samples fabricated in an EG electrolyte, 0.25 wt% NH_4F and 1% H_2O at 60 V for 6 h [65]. The resulting nanotube array samples were approximately 24 μm in length, with an inner pore diameter of 110 nm and outer diameter of 165 nm. The samples were annealed at 625°C, 580°C, and 525°C for 1 h in oxygen prior to measurement. Figure 4.12(b) shows the corresponding photoconversion efficiencies of the two samples.

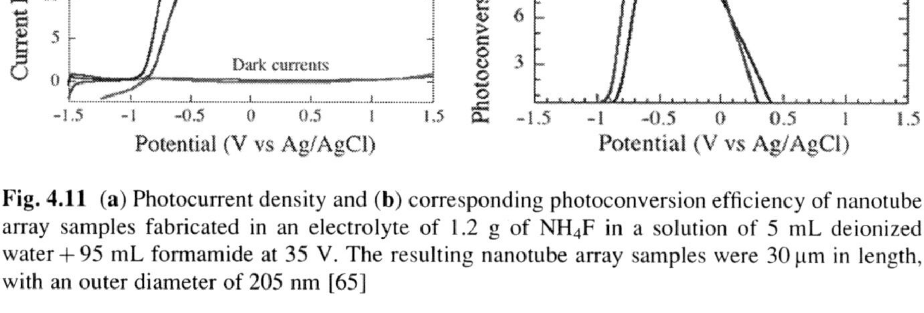

Fig. 4.11 (**a**) Photocurrent density and (**b**) corresponding photoconversion efficiency of nanotube array samples fabricated in an electrolyte of 1.2 g of NH_4F in a solution of 5 mL deionized water + 95 mL formamide at 35 V. The resulting nanotube array samples were 30 μm in length, with an outer diameter of 205 nm [65]

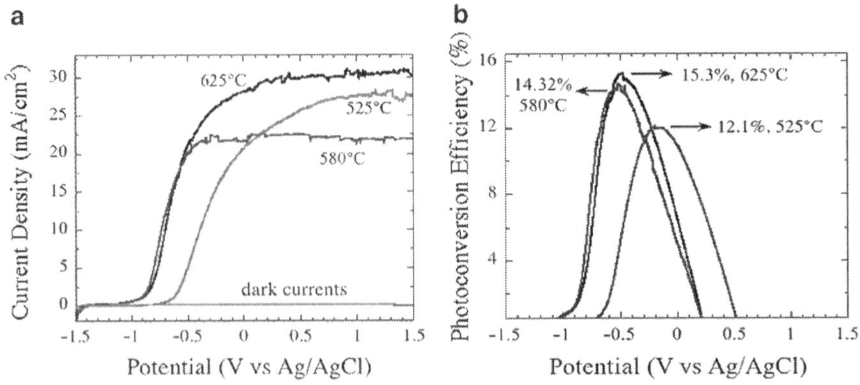

Fig. 4.12 (a) Photocurrent density and (b) corresponding photoconversion efficiency of nanotube array samples fabricated in an EG electrolyte, 0.25 wt% NH_4F and 1% H_2O at 60 V for 6 h. The samples were annealed at indicated temperatures for 1 h in oxygen prior to measurement. The resulting nanotube array samples were ~24 μm in length, with a pore size of 110 nm and wall thickness 25 nm [65]

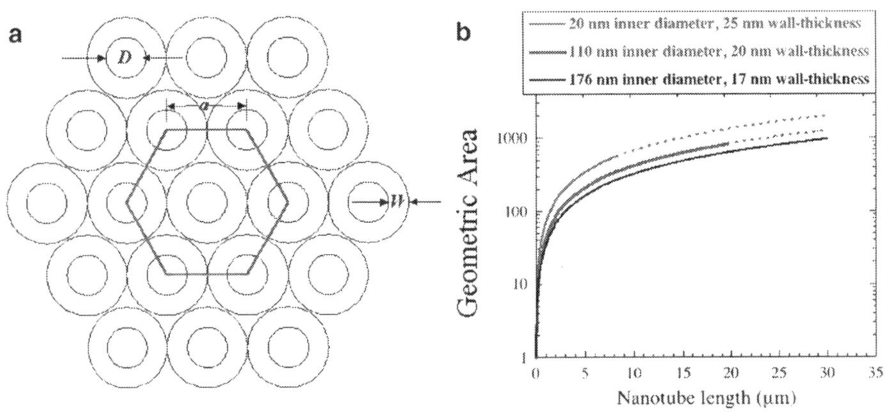

Fig. 4.13 (a) Idealized unit cell of TiO_2 nanotube array with inner diameter D, wall-thickness W, and $a = D + W$. (b) Calculated geometric roughness factor as a function of nanotube length. The solid, versus dashed lines, denote nanotube array lengths considered in this work [65]

4.3.4 Roughness Factor

The roughness factor, that is, the physical surface area of the film per unit projected area, measures the internal surface area of the electrode and is of crucial significance in applications such as sensing and catalysis. Assuming an idealized nanotubular structure shown in Fig. 4.13(a) with inner diameter D, wall-thickness W, and tube-length L, the purely geometric roughness factor is calculated as [65, 66]

$$G = \left[\frac{4\pi L(D+W)}{\sqrt{3}(D+2W)^2} \right] + 1 \tag{4.16}$$

Equation (4.3) assumes all surfaces of the nanotubes to be perfectly smooth. In reality, the surfaces are not smooth, and the actual roughness factor is higher due to the ridging effect associated with fluctuations in voltage, and/or local pH of the anodizing solution. Figure 4.13(b) is a plot of the calculated geometric roughness factor for nanotubes with three different geometries. The plot shows that higher surface area is more easily obtained with nanotubes of smaller pore size. Table 4.1 displays the roughness factor of nanotubes formed in different electrolytes, determined using dye desorption measurements. The table shows that nanotubes formed in formamide electrolytes have the highest roughness factors per unit length, which also correlates with the high photocurrents and conversion efficiencies obtained

Table 4.1 Surface area of nanotubes determined by dye desorption measurements

Electrolyte	Anodic potential (V)	Inner diameter (nm)	Length (μm)	Surface dye coverage (nmol/cm^2)	Effective roughness factor	Geometric roughness factor
Aqueous 0.5% HF	10	22	0.2	5.25	52	38
Aqueous 0.5% HF	10–23					
(Nanocones)	-	0.4	1.75	17	-	
Aqueous buffered 0.1 M KF (pH~5)	25	110	6.6	50	496	252
0.1 M NH_4F in 1:1 DMF:H_2O	15 Porous	-	-	30	296	136
0.15 M NH_4F + 5% H_2O in Formamide	10	12	3.6	55	546	325
0.15 M NH_4F + 5% H_2O in Formamide	15	29	5.4	79	784	390
0.15 M NH_4F + 5% H_2O in Formamide	20	70	14.4	181	1,797	778
0.22 M NH_4F + 5% H_2O in Formamide	35	180	30	277	2,750	928
0.27 M NH_4F + 5% H_2O in Formamide	20	65	19.6	304	3,019	1,224
0.3% NH_4F in EG + 1% H_2O	30	45	9	57	566	544
0.3% NH_4F in EG + 1% H_2O	40	70	12	66	655	609
0.3% NH_4F in EG + 1% H_2O	65	135	30	116	1,152	1,018

with nanotubes formed in this electrolyte. Anodization variables and annealing conditions impact the photoresponse of the resulting TiO_2 nanotubes [67–70].

4.3.5 Effect of Electrolyte Additives

Irrespective of the morphology and method of preparation, TiO_2 is known to contain defects [71] that act as charge carrier traps and recombination centers, which reduce the photoactivity of the material. Recombination losses can be reduced by utilizing the holes before they recombine by addition of sacrificial organic compounds, such as methanol, propanol and EG, to the electrolyte [72–77]. Further, these alcohols modify the energy levels of the electrolyte in comparison to the O_2/H_2O energy level [78, 79]. For metal oxide semiconductors, electron-hole recombination losses are high in alkaline aqueous solutions. Adding a sacrificial organic to the alkaline aqueous solution helps facilitate separation of the photogenerated electrons and holes. An organic additive reacts with the hole in the presence of water to generate protons and CO_2. Protons are then potentially driven to the cathode where they are reduced to evolve hydrogen. Mohapatra et al. investigated the effects of various additives on the photoactivity of TiO_2 nanotube array photoanodes [80]. TiO_2 nanotube samples were made in 0.5 M H_3PO_4 + 0.14 M NaF (pH 2.1) solution and crystallized at 500°C for 6 h in oxygen ambient. The photoelectrochemical measurements of these samples were carried out in various solutions such as 1 M KOH; 1 M KOH + 5 vol% methanol; 1 M KOH + 1–10 vol% EG; and 1 M KOH + 5 vol% glycerol. A 300 W solar simulator with AM 1.5 filter in the front was used as a light source. The intensity of the illumination on the surface of photoanode was set to 87 mW/cm^2.

Figure 4.14 shows the photocurrent generated by the TiO_2 nanotubes in 1 M KOH solution with variable EG concentration. The maximum photocurrent density obtained in 1 M KOH solution was 0.87 mA/cm^2. Upon adding 1 vol% EG, the photocurrent increased by more than 150%. The value of photocurrent density reached to 3.97 mA/cm^2 at 0.2 V versus Ag/AgCl for 10 vol% EG. In case of 1 and 2 vol% EG addition, the photocurrent saturated at potentials more anodic than −0.6 V versus Ag/AgCl. The increase in the photocurrent density could be correlated to the photo induced open circuit potential which became more negative with the addition of EG, see Table 4.2.

Figure 4.15 demonstrates the effect of other organic additives in 1 M KOH solution over the photocurrent values of TiO_2 nanotubes tested under solar simulated 1 sun condition. Addition of 5 vol% methanol moved the photocurrent onset potential of TiO_2 to −1.02 V versus Ag/AgCl and increased the photocurrent density to 2.43 mA/cm^2. Addition of 5 vol% of glycerol shifted the photocurrent onset potential to a more negative value (−1.26 V versus Ag/AgCl) and showed a photocurrent density of 2.55 mA/cm^2. It was found that all three alcohols acted effectively to enhance the photoactivity of the TiO_2 nanotubes. Other hydroxyl organic additives, such as ethanol (2.1 mA/cm^2), isobutyl alcohol (2.6 mA/cm^2),

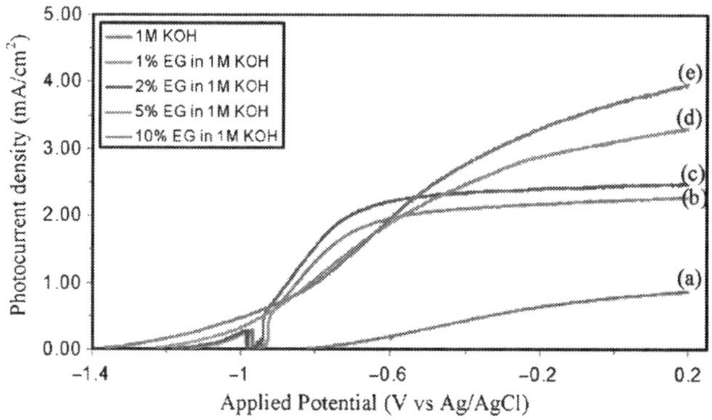

Fig. 4.14 Potentiodynamic plot of TiO_2 photoanode in 1 M KOH solution containing (*a*) without any organic additives, (*b*) 1% EG, (*c*) 2% EG, (*d*) 5% EG and (*e*) 10% EG as organic additive [80]. Reprinted with permission from *Journal of Physical Chemistry C* 2008, 112, 11007–11012. Copyright 2008 American Chemical Society

Table 4.2 Electrochemical measurements of TiO_2 with and without the addition of EG as organic additive [80]

conditions	N_D [a]	U_B [b] ($V_{Ag/AgCl}$)	OCP[c] ($V_{Ag/AgCl}$)	J_p (mA/cm^2)[d]
1 M KOH dark	8.91×10^{17}	−0.725	−0.04	0.001
1 M KOH bright	1.92×10^{19}	−0.725	−0.93	0.87
1 M KOH + 5 vol% EG dark	6.01×10^{20}	−1.34	−0.36	0.0012
1 M KOH + 5 vol% EG bright	1.54×10^{22}	−1.34	−1.38	3.40

Reprinted with permission from American Chemical Society, copyright 2008, ref. [80]
[a]Charge carrier density
[b]Flat band potential
[c]Open circuit potential
[d]Photocurrent density

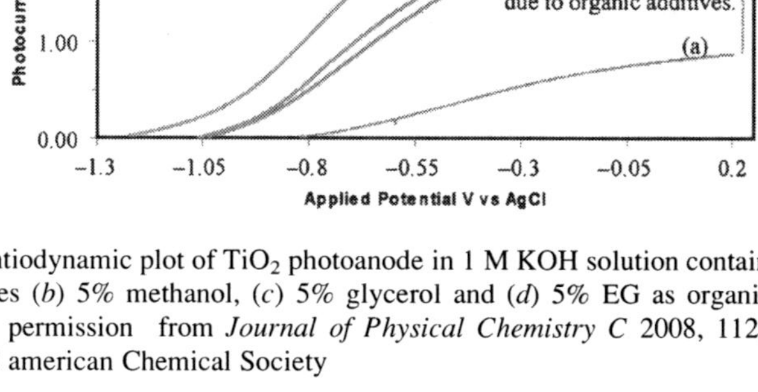

Fig. 4.15 Potentiodynamic plot of TiO_2 photoanode in 1 M KOH solution containing (*a*) without organic additives (*b*) 5% methanol, (*c*) 5% glycerol and (*d*) 5% EG as organic additive [80]. Reprinted with permission from *Journal of Physical Chemistry C* 2008, 112, 11007–11012. Copyright 2008 american Chemical Society

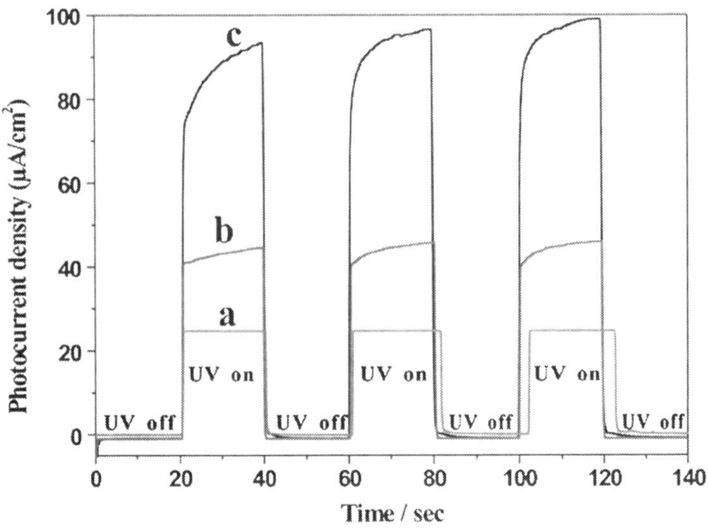

Fig. 4.16 Photocurrent-time characteristics of TiO_2 nanotubes electrode in (*a*) supporting electrolyte, (*b*) 100 mM methanol and (*c*) 1 M methanol. Conditions: 0.5 M Na_2SO_4 as supporting electrolyte, illumination with light of 253.7 nm, photocurrent collected at 0.6 V [85]. Reprinted from *Journal of Electroanalytical Chemistry*, Vol. 610, Zhang Z, Yuan Y, Fang Y, Liang L, Ding H, Shi G, Jin L, Photoelectrochemical oxidation behavior of methanol on highly ordered TiO_2 nanotube array electrodes, 179–185, Copyright 2007, with permission from Elsevier

and sucrose (2.25 mA/cm^2) showed photocurrent densities lower than EG at 0.2 VAg/AgCl. The effect of various organic additives on the reduction of the electrons and holes recombination losses depends on their redox potential and reactivity in a particular solution [81–84]. As seen in Fig. 4.16, the effect of methanol addition to 0.5 M Na_2SO_4 was also investigated as a water photoelectrolysis electrolyte, as a means to improve the photocurrent generated by illuminated TiO_2 nanotubes (253.7 nm illumination) [85].

4.4 Photoelectrolysis Using Anionic and Cationic Doped TiO_2 Nanotubes

4.4.1 N-Doped TiO_2 Nanotubes

Shankar et al. reported an electrochemical method to incorporate anionic dopants, specifically nitrogen, by an anodization process that allows simultaneous nanostructuring and doping of the anodic TiO_2 film [12]. Titanium foils were anodized in electrolytic solutions containing 0.07 M HF and varying concentrations of NH_4NO_3 from 0.2 to 2.5 M, with NH_4OH added to adjust the pH to 3.5. Such anodization chemistry restricted the electrolytic ions to nitrogen and fluorine bearing species to be incorporated into the anodic TiO_2 films. Sample A was removed after 17 s of anodization, while Sample B was anodized for 240 s in 0.4 M NH_4NO_3 and 0.07 M HF. Sample C was anodized for 6 h at 20 V in an electrolyte containing 2.5 M NH_4NO_3 and 0.07 M HF. XPS confirmed the presence of N doping in TiO_2 matrix, maximum concentration of nitrogen was found in Sample A. The doping of nitrogen was found to be inhomogeneous with the maximum nitrogen being incorporated close to the surface, then linearly decreasing with increasing depth inside the film. Diffuse reflectance measurements exhibited optical absorption in the wavelength range 400–510 nm for N-doped samples; however, the magnitude change was small due to small thickness of sample A and B. Figure 4.17 shows the photocurrent-potential characteristics of the nitrogen doped TiO_2 samples A, B, and C in 1 M KOH aqueous solution under 1 sun (AM 1.5, 100 mW/cm^2). Higher nitrogen doping levels resulted in superior electrode photoactivity. The photocurrent onset potentials are −0.85 V versus Ag/AgCl for Samples A and B and −0.95 V for Sample C.

A similar approach for nitrogen doping in TiO_2 was adopted by Li and coworkers where titanium foils were anodized at 20 V with a platinum cathode in an electrolyte of 0.5 wt% HF, 0.2 M $Na_2SO_4 + 0.5$ wt% NaF, and 1 M $H_2SO_4 + 0.15$ wt% HF, containing different concentrations of $C_6H_{12}N_4$ (0.3–0.6 M) and CH_3CN (0.01–1 M) as the nitrogen source [86–88]. The total concentration of N was found to be 2.7 at.% using XPS. Figure 4.18 shows a comparison of photocurrent spectra for annealed N-doped and undoped TiO_2 nanotube layers on Ti substrate in 0.2 M Na_2SO_4 solution. The N-doped sample (curves a, b, c) had a significantly enhanced absorption to the visible wavelength with a broad absorption maximum centered in 550 nm at bias potentials of 300 mV, 1 V, and 3 V, respectively. For the pure TiO_2 nanotube films (curve d), no photocurrent response to the visible light could be observed at the same given bias potentials.

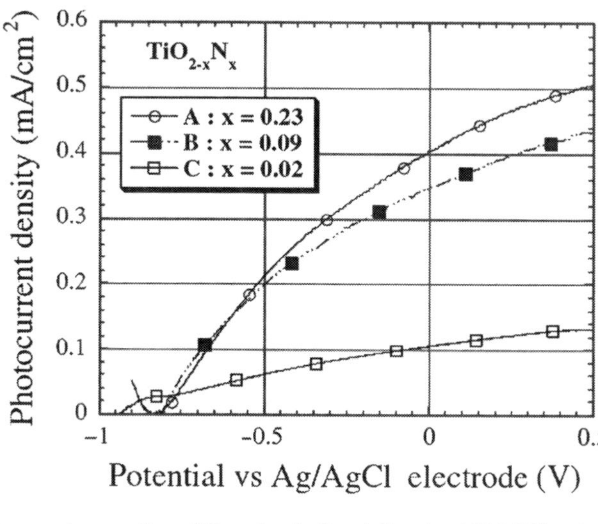

Fig. 4.17 Photocurrent generation of Samples *A*, *B* and *C* under AM 1.5 illumination [12]

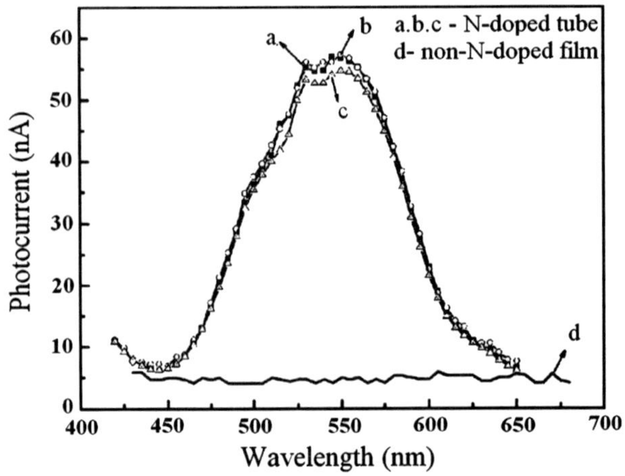

Fig. 4.18 Comparison of photocurrent spectra of N-doped (curves *a*, *b*, *c*), and those without N-doping (curve *d*), TiO_2 nanotube films recorded under the different bias voltages: (*a*) 1 V, (*b*) 3 V, (*c*) 300 mV, (*d*) 3 V [88]. Li Y, Yun H, Lin CJ, *Journal of The Electrochemical Society*, 2007, 154, C631–C636. Li J, Yun H, Lin CJ, *Journal of The Electrochemical Society*, 2007, 154, C631–C636. Reprinted by permission of ECS–The Electrochemical Society

Kim and co-workers formed N-doped TiO_2 nanotubes by anodizing nitrogen-containing titanium alloy slices (5 at.% of N) in a glycerol + water (50:50 vol %) + 0.27 M NH_4F mixture at 20 V for 2 h [89]. The amorphous nanotubes were converted to anatase by annealing them at 450°C for 1 h with a heating and cooling rate of 30°C/min. Photocurrent spectra of $TiO_{2-X}N_X$ sample acquired in 0.1 M Na_2SO_4 at 500 mV showed increased photoresponse in the visible light.

N-doping of TiO_2 nanotubes was also carried out using ion implantation [90, 91]. Here, nanotube arrays were formed in a H_2SO_4/HF electrolyte, annealed at 450°C for 3 h, implanted with nitrogen using ion bombardment at 60 keV accelerating energy at a nominal dose of 10^{16} ions/cm^2, and then again annealed to remove the defects. Photoelectrochemical measurements were carried out with a 150 W Xe arc lamp. Photocurrent spectra of the nanotube layers (a) after formation and annealing, (b) after nitrogen implantation, (c) and after the subsequent reannealing step were measured [90]. The photocurrent spectra were acquired in 0.1 M Na_2SO_4 at a potential of 500 mV (Ag/AgCl). The implantation step resulted in a significant drop in the photocurrent magnitude. It was also noted that thermal annealing of the sample after implantation leads to an enhancement of the photocurrent in the visible light and in the UV range, with this reannealing step increasing the UV response to values larger than those for the undoped samples.

In another method of nitrogen doping of TiO_2 nanotubes, thermal annealing was performed in pure ammonia gas [92, 93]. Samples were treated in a pipe furnace with a quartz glass tube with an ammonia flow of 32 ccm/h through the tube at temperatures of 300–700°C and durations of 10 min to 3 h. Temperatures higher than 600°C resulted in structural damage and collapse of the nanotube array structure. The photocurrent action spectra recorded at 2 V Ag/AgCl for an NH_3-treated (at 600°C) sample in 0.1 M Na_2SO_4 showed significantly higher photocurrents in the visible range than the undoped samples. It was evident that N-doping also enhanced the UV light photoresponse.

For making N, B, or F doped TiO_2 nanotubes [94–96], Su et al. first anodized Ti foil at 20 V for 1 h in 0.08 M $C_2H_2O_42H_2O$ containing 0.5 wt% NH_4F, dried in a stream of dry nitrogen at 300°C for 30 min. The sample was then annealed, 400°C or 600°C, while exposed to a compositional gas stream, nitrogen passing a (2.5 wt% NaF + 2.5 wt% H_3BO_3 solution heated at 80°C) at a rate of 3,300 mL/min. To make fluorine doped samples, a precursor solution containing 2.5 wt% NaF was used; for B-doped sample, a 2.5 wt% H_3BO_3 solution was used as the precursor. Experiments were carried out under visible light irradiation (300 W Xe lamp) using a UV filter [97, 98] in 0.1 M Na_2SO_4 electrolyte. N-F doped TiO_2 nanotubes made at 600°C showed modestly improved photocurrents. Chen et al. reported fabrication of P-F doped TiO_2 nanotubes by anodizing the Ti foil in 0.5 M H_3PO_4 containing a small amount of HF (0.2 wt%) [99].

Beranek and co-workers reported nitrogen and carbon surface-modified TiO_2 nanotubes by heating TiO_2 in a gaseous atmosphere of urea pyrolysis products at 400°C [100]. The TiO_2 nanotube array samples were made using 1 M H_2SO_4 + 0.16 M HF, glycerol/water (50:50 vol%) + 0.27 M NH_4F, glycerol + 0.14 M NH_4F, and EG + 0.27 M NH_4F electrolytes and then annealed at 450°C for 3 h in air. For surface modification [101, 102], the electrodes were placed into a 170 mL Schlenk tube connected via an adapter with a 70 mL round bottom flask containing 1 g of urea and heated in a muffle oven for 30 min at 400°C. Photocurrent measurements were carried out in a 0.1 M $LiClO_4$ + 0.1 M KI electrolyte. The photocurrent wavelength dependence was measured at a constant potential of 0.5 V versus Ag/AgCl using a tunable monochromatic light source provided with a 1,000 W Xenon lamp. The authors suggest that the nanotube walls are chemically modified by covalently bound nitrogen-containing species, having their occupied electronic states lying energetically above the valence band edge of neat TiO_2. The electronic coupling of these states with the electronic continuum of TiO_2 valence band states yields a band of surface states from which electrons can be optically excited into the conduction band also by low-energy visible light irradiation [101, 102]. These surface states act also as efficient trapping sites for photogenerated holes, so that most photogenerated holes relax deep into these states before they can react with hole-scavenging species in the electrolyte. Since the reacting holes are not positive enough to induce efficient one-electron oxidation of water ($E \sim 2.0$ V vs. NHE at pH = 7 for one electron process) enhanced recombination was observed in aqueous electrolytes unless a more readily oxidizable reducing agent (such as iodide, $E = 1.3$ V vs. NHE) [101–104] was present. Figure 4.20 shows IPCE action spectra recorded for both unmodified and modified TiO_2 nanotubes of various lengths in iodide-containing electrolytes. When compared with unmodified TiO_2 nanotubes (Fig. 4.19a), the IPCE of the UV wavelengths decreased significantly, a behavior which can be attributed to enhanced recombination.

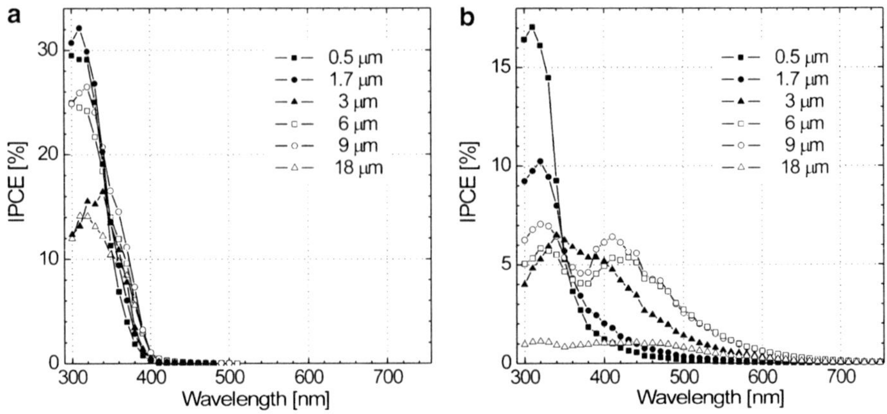

Fig. 4.19 Photoaction spectra recorded in 0.1 M $LiClO_4$ + 0.1 M KI electrolyte at 0.5 V vs. Ag/AgCl for electrodes with unmodified (**a**) and surface-modified (**b**) nanotubes of different lengths [100]

4.4.2 Carbon Doped TiO_2 Nanotubes

Bard and coworkers reported fabrication of C-doped TiO_2 nanotubes [10]. TiO_2 nanotube arrays with 3 μm length, 70 nm pore diameter, and 20 nm wall thickness were grown by anodization of titanium foil at 15 V in an electrolyte of 1 M $(NH_4)H_2PO_4$ + 0.5 wt% NH_4F [10]. Anodized samples were then annealed at 450°C in oxygen for 1 h with heating and cooling rates of 2°C/min. To introduce carbon, the nanotube arrays were subsequently annealed at high temperatures (500–800°C) under controlled CO gas flow with no significant morphological changes observed. It was reported that the doping concentration of carbon in the TiO_2 nanotube array could be controlled between 8% and 42% (as determined from the XPS peaks) depending on the CO annealing temperature [10]. Figure 4.20 shows the measured photocurrent density of a $TiO_{2-x}C_x$ nanotube array under visible light in 1 M KOH electrolyte. The proposed model for visible-light absorption in doped TiO_2 is due to substitution of oxygen by another atom, such as N or C, which results in a mix of N, C2*p*, and O2*p* states [15, 105]. The linear portion of the curve in their $(j_{ph}h\nu)^{1/2}$ versus $h\nu$ plot indicated a transition above an optical band-edge and an intercept of about 2.22 eV, corresponding to the bandgap energy of the $TiO_{2-x}C_x$ nanotube array. This energy is slightly larger than the earlier reported value of ~2 eV [106]. The gases collected at the TiO_2 and Pt counter electrode during the photoreaction showed a 2.1:1 volume ratio of hydrogen and oxygen.

Hu et al. incorporated carbon in TiO_2 by annealing as-formed nanotubes in a continuous flow of Ar and acetylene gases, then investigated the photocatalytic activity by evaluating the photodegradation of aqueous methyl blue under sunlight illumination [107]. Hahn et al. performed carbon doping by annealing the TiO_2 nanotubes in N_2 and acetylene gas mixture at 500°C [108]. Raja et al. prepared carbon modified TiO_2 nanotubes by annealing them in a carbonaceous gas mixture at

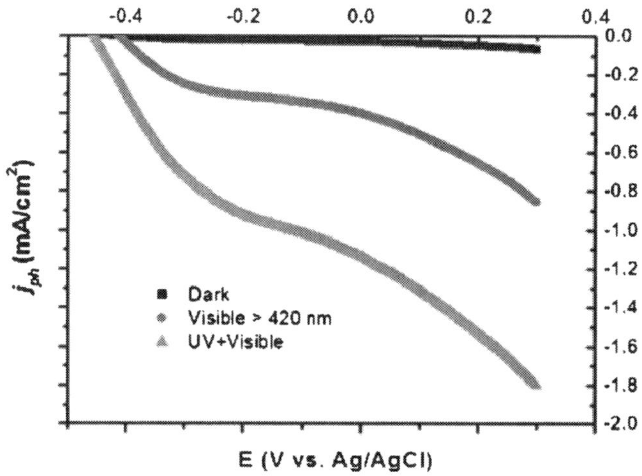

Fig. 4.20 Current-potential characteristics of the $TiO_{2-x}C_x$ nanotube array prepared at 600°C in the dark, and under both visible and white-light illumination [10]. Reprinted with permission from *Nano Letters*, 2006, 6, 24–28. Copyright 2006 American Chemical Society

650°C for 5–20 min [109]. Flame annealing was used for carbon doping, IPCE, and optical absorption measurements indicated that the impurity levels induced by carbon doping were located deep in the gap, rather than a modification of the band gap itself [110]. Xu et al. pursued carbon doping of TiO_2 nanotubes by annealing in air at 500°C for 1 h, and natural gas flame oxidation at 820°C for 18 min, with no significant change found in their quantum efficiency over the visible range [111]. Mohapatra et al. reported carbon-doped TiO_2 nanotubes by producing nanotubes from EG using a sonochemical method [112, 113]. Murphy has reported an excellent study revisiting the various reports on different methods of carbon doping in TiO_2 nanotubes, and their resulting photocurrents [114]. As Murphy indicates [114], clear demonstration of successful carbon doping requires accurate IPCE data.

4.4.3 *Sulfur-Doped TiO_2 Nanotubes*

Based on calculations using a two-unit supercell, Umebayashi et al. found that mixing of the S3*p* with O2*p* orbitals on the upper edge of the valence band helps promote visible light activity of the material [115]. Yamamoto et al. reported a theoretical study on S-doped rutile TiO_2, and Tian et al. on S-doped anatase TiO_2 [116, 117]. Since sulfur has a larger ionic radius compared to nitrogen and carbon, Tang and Li expected that sulfur could considerably modify the electronic structure of TiO_2 and, thus, influence its photocatalytic properties [118]. For fabricating S-doped nanotubes, first TiO_2 nanotubes were produced by anodization at 60 V for 12 h in EG electrolyte containing 0.25 wt% NH_4F and 2 vol% water. Then, the as-prepared titania nanotube arrays were annealed at 380°C for 14 h in a tube

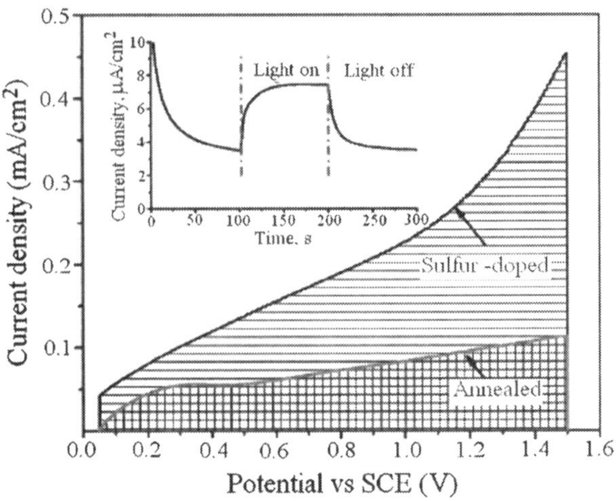

Fig. 4.21 Photocurrent versus applied potential of the annealed and the sulfur-doped TiO_2 nanotube arrays under visible light illumination in a 0.1 M Na_2SO_4 solution. The inset shows the transient photocurrent generated under pulse visible light illumination at a fixed bias potential of 0.1 V (SCE) for the sulfur-doped TiO_2 nanotube arrays [118]. Reprinted with Permission from *Journal of Physical Chemistry C*, 2008, 5405–5409. Copyright 2008 American Chemical Society

furnace, heating/cooling rate of 0.75°C/min, exposed to a H_2S flux of 10 mL/min. For comparison, an otherwise identical undoped TiO_2 nanotube array sample was annealed in air but otherwise identical thermal conditions. FESEM imaging indicated nanotube arrays 36 μm in length, with a roughness factor (the real surface area per unit geometric surface area) of 960 (tubular packing, 140 nm inner diameter, 10 nm wall thickness) and a length-to-width aspect ratio of 250. XPS spectra confirmed the presence of S in the TiO_2 nanotubes. XRD showed an increase in lattice parameters, implying that oxygen ions in the TiO_2 lattice were replaced by larger radii ions. Diffuse reflectance spectra exhibited absorption, of small magnitude, up to 650 nm. The photoelectrochemical properties of the sulfur-doped TiO_2 nanotubes were measured in 0.1 M Na_2SO_4 solution. A 50 W fiber optic illuminator (model 190, Dolan-Jenner Industries, Inc.) was used as the light source with a cutoff filter used to remove any radiation below 400 nm to ensure illumination only by visible light. The photocurrent transient was measured at fixed bias potential, 0.1 V versus SCE, with a light pulse of 100 s under visible light illumination. Figure 4.21 shows the photocurrent of the S-doped sample.

4.4.4 Boron-Doped TiO_2 Nanotubes

The preparation of B-doped TiO_2 nanotube arrays through use of a sol–gel technique was reported [119, 120]. Ruan et al. observed photocurrent enhancement for

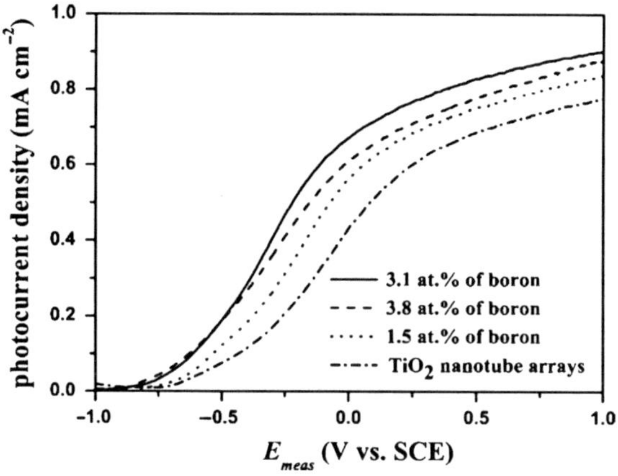

Fig. 4.22 Photocurrent density generation of non-doped TiO_2 nanotube arrays and B-doped TiO_2 nanotube arrays under simulated sunlight irradiation (80 mW/cm^2) [121]. Reprinted from *Separation and Purification Technology*, Vol. 62, Lu N, Zhao H, Li J, Quan X, Chen S, Characterization of boron-doped TiO_2 nanotube arrays prepared by electrochemical method and its visible light activity, 668–673, Copyright 2008, with permission from Elsevier

nanotube arrays fabricated in boric acid containing HNO_3/HF electrolyte [49]. Nanotube samples, approximately 1.4 μm long, were prepared by anodization in an electrolyte containing 0.625 to 1.25 mM $NaBF_4$ [121, 122], with XPS confirming the presence of boron. The amount of boron within the TiO_2 nanotube arrays was found to vary linearly as a function of $NaBF_4$ concentrations below 1.25 mM. All B-doped samples exhibited red shifts in the absorption edge, with a sample containing 3.1 at.% boron having the largest red shift of $\approx$30 nm. The photocurrent densities as a function of applied potential for B-doped and undoped TiO_2 nanotube array electrodes, in a 1 M KOH electrolyte, under simulated sunlight irradiation (80 mW/cm^2), are shown in Fig. 4.22. Clearly, the photocurrent densities of the B-doped samples are higher than the undoped samples. The photocurrent density of the sample with 3.1 at.% boron is the highest of the samples, approximately 1.17 times that of the undoped TiO_2 nanotube arrays.

4.4.5 Silicon-Doped TiO_2 Nanotubes

Embedding silicon in the TiO_2 matrix was found to induce new structural and physicochemical properties such as high thermal stability and good surface wettability [123, 124], with which Su et al. demonstrated an improved photoactivity [125]. TiO_2 nanotubes were anodized at 20 V for 2 h in 0.5 wt% $NH_4F + 1$ M $(NH_4)_2SO_4$ aqueous solution. The electrolyte pH was adjusted close to 6 by a solution of 2 M $H_2SO_4 + 0.5$ wt% NH_4F. After anodization, the sample was rinsed with deionized water, dried in a N_2 stream then annealed at 450°C in air for 3 h with heating and cooling rates of 2°C/min. For incorporating

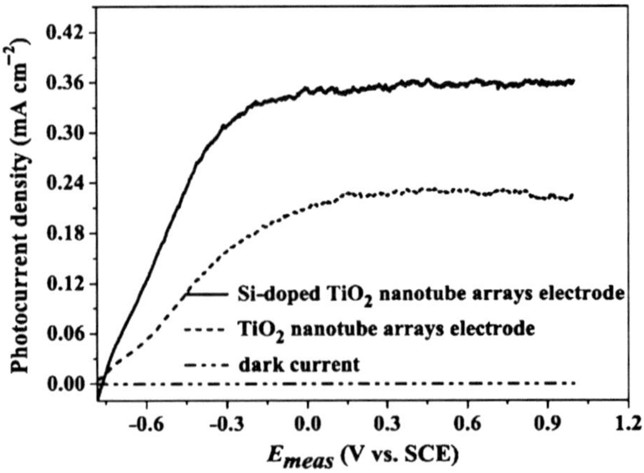

Fig. 4.23 *I-V* characteristics of Si-doped and undoped TiO_2 nanotube array electrodes in 1 M KOH solution under UV light irradiation (0.75 mW/cm^2) [125]. Reprinted from *Applied Surface Science*, Vol. 255, Su Y, Chen S, Quan X, Zhao H, Zhang Y, A silicon-doped TiO_2 nanotube arrays electrode with enhanced photoelectrocatalytic activity, 2167–2172, Copyright 2008, with permission from Elsevier

silicon, tetraethylorthosilicate (TEOS) was admitted to the tube furnace carried by argon gas, Si concentration 156 mg/L, with controlled gas flow at 500°C for 15 min, after which the sample was annealed at 650°C for 30 min. For comparison, undoped TiO_2 nanotubes were prepared under identical thermal conditions in air. The length of the resulting nanotubes was 1 μm and pore size of 50–65 nm. XRD patterns revealed that the rutile fraction in Si-doped TiO_2 nanotube arrays was lower than that of undoped TiO_2 nanotube arrays annealed at 650°C [124, 126, 127]. XPS results showed the presence of silicon within the anatase TiO_2 nanotubes and the formation of Si-O-Ti cross-linking bonds. Photocurrent densities were measured using 1 M KOH purged with N_2, and a 300 W high-pressure mercury lamp with a principal wavelength of 365 nm. Figure 4.23 shows *I-V* characteristics of Si-doped TiO_2 and undoped TiO_2 nanotube array electrodes. For both samples, the dark currents are negligibly small, in the range of 10^{-6}A.

4.5 Photoelectrolysis Using Surface-Sensitized TiO_2 Nanotubes

4.5.1 CdS Sensitized TiO_2 Nanotubes

The *I-V* characteristics of CdS-sensitized TiO_2 nanotube electrodes are presented in Fig. 4.24 for an illumination intensity of 1 sun (AM 1.5, 100 mW/cm^2, wavelengths below 400 nm removed by use of filter) [128]. Measurements were conducted in 1.0 M Na_2S solution, an efficient hole scavenger for CdS in which the electrodes

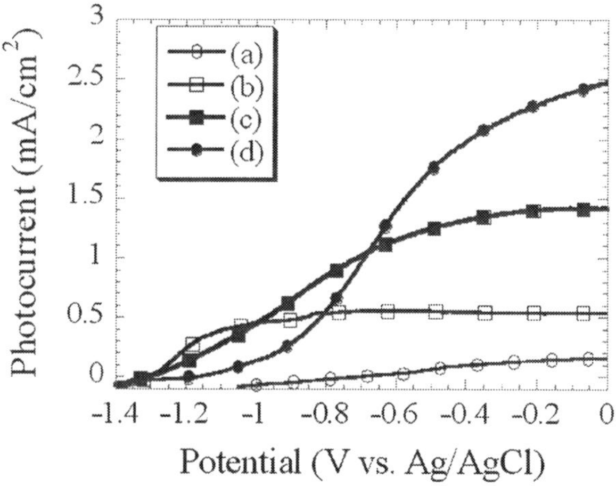

Fig. 4.24 Photocurrent versus voltage in 1 M Na_2S under AM 1.5 (1 sun), 100 mW/cm^2 illumination: (*a*) bare TiO_2 nanotube electrode. (*b*) As-prepared electrodeposited CdS film (−0.5 V, 30 min.) upon TiO_2 nanotube-array electrode. (*c*) CdS (−0.5 V, 30 min.)-TiO_2 electrode after annealing at 350°C in N_2 for 60 min. (*d*) CdS (−0.5 V, 30 min.)-TiO_2 electrode after annealing at 400°C in N_2 for 1 h [128]

are stable. For the as-prepared CdS-TiO_2 electrode, photocurrent onset occurs at −1.30 V versus Ag/AgCl, a −0.60 V negative shift compared to a TiO_2 nanotube-array electrode. In comparison to the plain TiO_2 nanotube-array electrode, addition of the CdS film increased the photocurrent from 0.16 to 0.55 mA/cm^2 [129–131]. The photocurrent response is sensitive to the annealing temperature, as discernable in Fig. 4.24. The sample annealed at 350°C, trace (c), reaches $\approx$ 1.42 mA/cm^2; the 400°C annealed sample, trace (d), reaches $\approx$ 2.51 mA/cm^2, respectively 9 and 16 times higher than that of bare TiO_2 nanotube array electrode of trace (a). The *I-V* curves of these samples gradually lose their well-defined photocurrent saturation as the annealing temperature increases. Higher temperature annealing may result in pore blockage due to sintering of the CdS nanoparticles, reducing the area accessible to the hole scavenging electrolyte solution.

Recently, Sun et al. sensitized TiO_2 nanotube array films with CdS quantum dots (QDs) ranging in size from 2 to 10 nm [132]. Highly ordered TiO_2 nanotube arrays prepared in NH_4F organic electrolyte were coated with CdS QDs by the use of sequential chemical bath deposition [133]. Photocurrents were measured in 1 M Na_2S under AM 1.5 at 138.4 mW/cm^2 illumination. The current-voltage characteristics measured from an as-synthesized nanotube array photoelectrode (tube length about 19.2 μm) showed that the open-circuit photovoltage V_{oc} is 0.94 V versus Ag/AgCl electrode for the plain TiO_2 nanotube film electrode, and for the as-prepared CdS deposited TiO_2 nanotube film electrode V_{oc} is about 1.27 V. The generated photocurrent increased from 0.22 for the plain TiO_2 nanotubes to 7.82 mA/cm^2 for the CdS modified. The cell efficiency was determined to be 4.15% with a fill factor of 0.578 for the as-modified CdS-TiO_2 nanotube film.

4.5.2 CdSe Sensitized TiO_2 Nanotubes

For sensitization of the nanotube arrays with CdSe QDs, TiO_2 nanotube array films were prepared by anodizing a 150 μm thick titanium foil in 0.5% HF for 30 min [134], followed by a 450°C anneal in air for 3 h. The nanotube samples were dipped into an acetonitrile solution of thioglycolic acid ($HOOCCH_2SH$) for 12 h, and then washed with both acetonitrile and toluene before being transferred to a glass vial containing a CdSe QD toluene solution. TiO_2 has a strong affinity for the carboxylate (-COOH) group [39], while the thiol group binds strongly to CdSe [135]. Therefore, $HOOCCH_2SH$ was used as a bifunctional linker to attach CdSe QDs onto the TiO_2 films. After being immersed in the CdSe solution for 12 h, the samples were washed with toluene and dried in vacuum. The CdSe QDs were from 3 to 4.5 nm in diameter as measured by TEM. Fluorescence emission spectra showed that the CdSe particles exhibit band edge emission (480 nm excitation) at 567 nm (3 nm) and 609 nm (4.5 nm), respectively.

Kongkanand and co-workers assembled CdSe quantum dots, 2.3 nm to 3.7 nm diameter, onto TiO_2 nanotube arrays using a bifunctional linker molecule [136]. Here, they used 8 μm long TiO_2 nanotube arrays with an average outer diameter of 90 nm and a wall thickness of 12 nm, prepared by anodization of Ti foil at 20 V in 0.27 M NH_4F in formamide (containing 5% water). After annealing (400°C for 1 h), the TiO_2 electrodes were immersed in a 1 M mercaptopropionic acid + 0.1 M sulfuric acid acetonitrile solution for 12 h. The electrodes were then rinsed with acetonitrile and toluene before being transferred to the CdSe QD solution (toluene) for 3 days to ensure saturated adsorption of the QD onto the TiO_2 electrodes. The photoaction spectra of TiO_2 nanotube array samples sensitized with different sized QDs exhibited a maximum IPCE of 45% in the visible range tested in 0.1 M Na_2S solution. Results indicate two major findings: (a) the ability to tune the photoelectrochemical response and photoconversion efficiency via size control of the CdSe quantum dots, and (b) improvement in the photoconversion efficiency by facilitating the charge transport through TiO_2 nanotube architecture [136].

4.5.3 CdTe Sensitized TiO_2 Nanotube Arrays [137]

Transparent nanotubes were formed on FTO glass by potentiostatic anodization of 500 nm thick titanium films at an applied bias of 14 V in a 0.5 vol% HF solution. The resultant amorphous tubes were crystallized to anatase TiO_2 by annealing at 450°C in oxygen ambient for 2 h. Nanotubes had an average inner pore diameter near 50 nm, wall thickness of approximately 13 nm, and were 250–300 nm in length. A conventional three-electrode set-up in an undivided cell was used for cathodic deposition of CdTe films ($3H^+ + Cd^{2+} + HTeO_2{}^+ + 6e^- \rightarrow CdTe + 2H_2O$) [138–140]. Counter electrodes were prepared by e-beam evaporation of

200 Å of Ti followed by 1,000 Å of Pt on cleaned glass slides. The reference electrode used was an Ag/AgCl electrode in 3.8 M KCl solution. Either FTO or TiO_2 nanotube arrays prepared on FTO substrates were used as the working electrodes. Deposition was carried out potentiostatically at –0.4 V versus the reference electrode. The aqueous deposition solutions consisted of 0.1 M $CdSO_4$ and 0.1 mM TeO_2 at pH 1.4 (H_2SO_4) and were maintained at 85°C throughout the deposition. Newly made films were rinsed with DI water and dried under a stream of nitrogen. Annealed films were heated to 300°C for 45 min, with a ramp of 2°C/min. Some films were treated with $CdCl_2$ before annealing. This treatment composed of a 2 min immersion in a saturated $CdCl_2$/methanol solution.

Photoelectrochemical properties of CdTe/TiO_2 electrodes were characterized by measuring short-circuit photocurrent using a 0.6 M Na_2S solution with pH adjusted to 12. A UV filter was used to expose the samples only to visible light. This effectively eliminated the photocurrent contribution from the TiO_2 nanotubes and allowed the examination of the photocurrent generated by only the CdTe layer. The best photocurrents were obtained from CdTe/TiO_2 electrodes that were subjected to $CdCl_2$-treatment and a 300°C thermal anneal. This is expected since the crystallinity of the CdTe layer is significantly improved by annealing the sample at 300°C for 45 min in air. Furthermore, the $CdCl_2$-treatment is known for promoting recrystallization and CdTe grain growth during the annealing process as well as passivating defects [141–144]. This process can effectively passivate defects and increase minority carrier lifetimes, thereby, improving device characteristics of the CdTe-based solar cells (e.g., open circuit voltage and fill factor) [144–147]. Consequently, the crystallinity of the CdTe films was further improved by dipping the CdTe layer in a methanol solution saturated with $CdCl_2$ before annealing. In order to make the best use of the nanotubular structures of TiO_2, it is critical that the internal area of each tube is sufficiently covered with CdTe before the entrances can become clogged. The challenge of depositing CdTe in the TiO_2 pores is that the bottom of the pores is no more conductive than the entrance of the pores, while diffusion makes deposition at the bottom of the pores less favorable than deposition at the entrance of the pores. This process can result in the deposition of a thick CdTe layer at the entrance of the pores before a significant portion of the internal area is covered with CdTe. To overcome this problem, the pores of the TiO_2 tubes were filled with an electrolyte containing Cd^{2+} and $HTeO_2^+$ ions, utilizing capillary forces by dipping TiO_2 tubes in the plating solution, and then moved the TiO_2 electrode into a new medium that contains only an inert supporting electrolyte without Cd^{2+} and $HTeO_2^+$ ions. Immediate electrodeposition in this medium resulted in deposition of CdTe only within the pores. This procedure was repeated until a desired amount of CdTe is deposited in the pores. The process is depicted in Fig. 4.25(a, b) [137]. For building a photoelectrochemical cell, incomplete filling of TiO_2 nanotubes with CdTe was desired so that the electrolyte could penetrate into the pores, increasing the electrolyte/electrode junction area. A CdTe/TiO_2 junction prepared with 20 cycles of the dipping and deposition method (sample I) was found optimal in achieving the highest photocurrent [137].

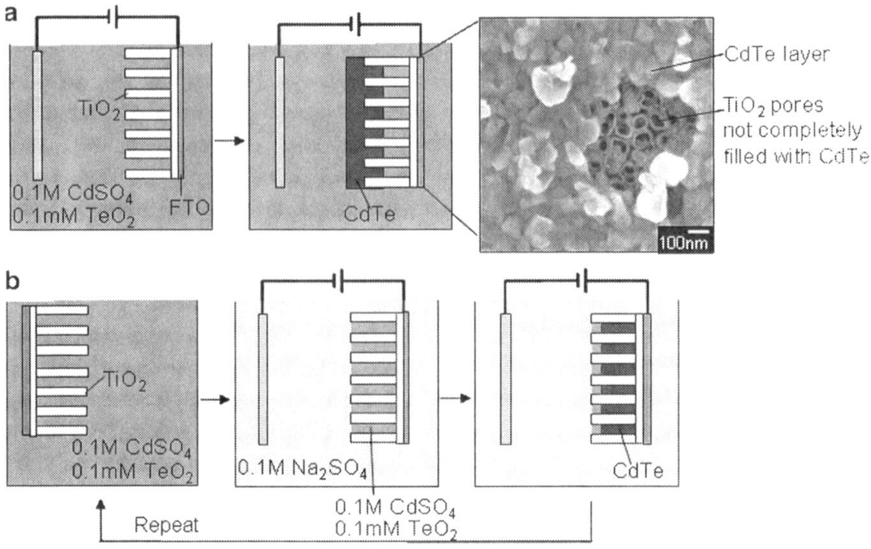

Fig. 4.25 Schematic illustration of: (a) a regular deposition method, and (b) a newly developed dipping & deposition technique [137]

All CdTe/TiO_2 electrodes (as-deposited, annealed, and annealed after $CdCl_2$-treatment) commonly generated anodic short-circuit photocurrent, indicating the presence of an upward band bending at the electrode/electrolyte interface. In this case, holes are consumed at the electrode/electrolyte surface to oxidize S^{2-} ions to polysulfide $(S_n)^{2-}$ ions (e.g., $S_2{}^{2-}$) while electrons are injected into the TiO_2conduction band, transferred to the counter electrode, and used to reduce water or polysulfide ions to hydrogen or sulfide ions, respectively [137]. When a CdTe film was deposited directly onto an FTO substrate, it generated a cathodic photocurrent [137]. This indicated that the CdTe layer prepared using our electrodeposition condition is *p*-type, forming a downward band bending at the electrolyte/CdTe junction [137, 140]. However, when *p*-CdTe is deposited on *n*-TiO_2, the upward band bending formed at the *p*-CdTe/*n*-TiO_2 junction is steeper than the downward band bending formed at the *p*-CdTe/electrolyte junction, making the CdTe/TiO_2 electrode exhibits the overall behavior of a photoanode.

An important difference between CdTe/TiO_2 and CdTe electrodes is the photostability of the CdTe layer [137]. The photocurrent generated by a CdTe layer deposited directly on an FTO substrate gradually diminishes over time. This is most likely due to photocorrosion, which involves the anodic formation of Te ($CdTe + 2OH^- + 2p^+ \rightarrow HCdO_2{}^- + H^+ + Te$) and/or the cathodic formation of Cd ($CdTe + 2e^- \rightarrow Cd + Te^{2-}$) [148]. In contrast, the steady state photocurrent observed by the CdTe/TiO_2 electrode was stable and no sign of photocorrosion was observed [137]. It was believed that this is because of the more efficient hole consumption at the electrolyte/CdTe junction and electron transfer at the CdTe/TiO_2 junction (from the CB of CdTe to CB of TiO_2) achieved by enhanced

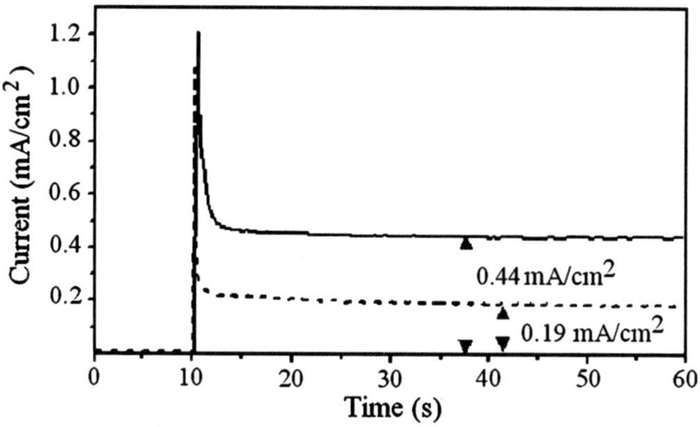

Fig. 4.26 Short-circuit photocurrent measured under visible light illumination of CdTe coated TiO_2 nanotube arrays using 20 cycles of dipping & deposition method (*solid line*) and a regular deposition method for 30 min (*dotted line*) [137]

electrolyte/CdTe/TiO_2 junction areas, which kinetically suppresses photocorrosion of the CdTe layer. Figure 4.26 [137] shows that although sample I showed a photon absorption slightly lower than sample II, the steady state photocurrent generated by sample I (0.44 mA/cm^2) was significantly higher than sample II (0.19 mA/cm^2). This difference in photocurrent appears due to the reduced electron-hole recombination achieved in sample I, resulting from the formation of a more intimate CdTe/TiO_2 junction by the new deposition technique. Thinner and more even CdTe coatings on the TiO_2 tubes allowed for more efficient electron transfer at the CdTe/TiO_2 junction and hole transfer at the CdTe/electrolyte junction, minimizing electron-hole recombination in the CdTe layer.

4.5.4 WO_3 Coated TiO_2 Nanotubes

WO_3 as a photoanode material for water splitting offers a relatively small band gap (2.5 eV) and corrosion stability in aqueous solutions [149, 150], although reported quantum yields are still low [151, 152]. TiO_2 nanotube arrays, 2 μm length, 80 nm pore diameter, and 20 nm wall thickness, were coated with WO_3 using electrochemical deposition [153]. Peroxotungstic acid sol ($WO_{3-x} \cdot nH_2O$) was synthesized by dissolving 1 g of tungsten powder in an ice-cooled beaker containing a 5 mL of 30% H_2O_2 aqueous solution and then diluting the solution using a water and 2-propanol mixture (100 mL, volume ratio of 5:2). WO_3 was deposited on nanotube array films by electrophoresis (−400 mV versus Ag/AgCl reference electrode, Pt mesh counter electrode) from the solution for 5 min and then annealed at 450°C for 30 min in air. Photoelectrochemical measurements were carried out in a three-electrode system by illuminating the xenon lamp (150 mW/cm^2). The electrolyte solution used for the water splitting reaction was 1 M $HClO_4$.

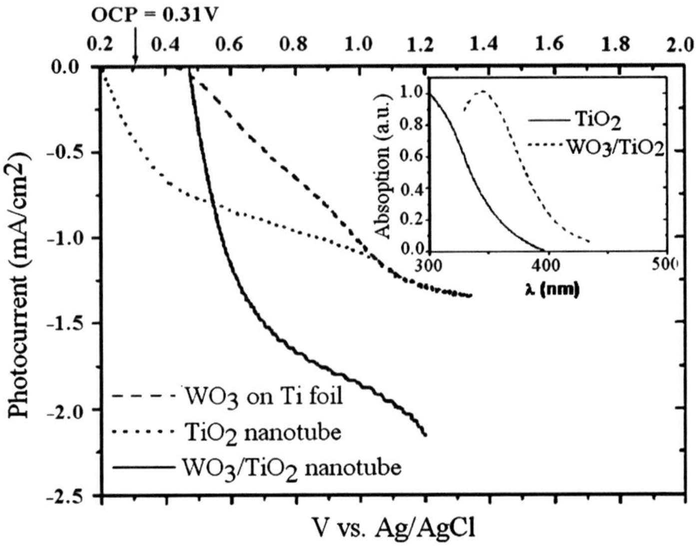

Fig. 4.27 Photocurrent density vs applied potential for the WO_3/Ti foil, TiO_2 nanotube arrays, and WO_3/TiO_2 nanotube illuminated by 150 mW/cm^2 xenon lamp. Open circuit potential of the WO_3/TiO_2 nanotube: 0.31 V. *Inset*: Normalized UV-visible diffuse reflectance spectra of the TiO_2 nanotube arrays and the WO_3/TiO_2 nanotube [153]. Reprinted with permission from Park JH, Park OO, Kim S, Photoelectrochemical water splitting at titanium dioxide nanotubes coated with tungsten trioxide, *Applied Physics Letters*, 89, 16, 163106. Copyright 2006, American Institute of Physics

Figure 4.27 shows photocurrent-potential plot for TiO_2 and the WO_3 coated TiO_2 nanotube array samples; the photocurrent of the WO_3/TiO_2 nanotube array sample is higher due to additional light absorption in the visible range. The photocurrent density of the WO_3/TiO_2 nanotube array sample was significantly higher than that of a WO_3/Ti planar foil photoanode of similar thickness. This suggests that the TiO_2 nanotube structure can harvest solar light more effectively than photoanodes with a nanocrystalline structure under the same illumination. Collection of evolved gases at the WO_3/TiO_2 nanotube photoanode and the Pt counter electrode during the photoreaction found a 2.2:1 volume ratio of hydrogen and oxygen.

4.5.5 Pt Sensitized TiO_2 Nanotubes

TiO_2 nanotubes with a length of 200–650 nm, an inner diameter of 50–180 nm and a wall thickness of about 15–35 nm were prepared by anodizing Ti foil in 0.5–5 vol% HF solution and subsequently annealed at 450°C for 3 h in air. Platinum was deposited on the nanotube films [154] by e-beam evaporation. Photocatalytic hydrogen evolution was performed in a small Pyrex glass closed vessel [154].

The Pt coated nanotubes were set on the bottom of the reactor and 0.01 M KOH solution used as the electrolyte with 1 mM KI as a redox mediator; the solution was purged with argon gas for 30 min. A 400 W metal halide lamp was used as the light source for water splitting. The evolved H_2 gas was analyzed by gas chromatography every 30 min. Without Pt sensitization very little hydrogen was evolved even for extended periods of UV irradiation. Activity increased with Pt content, reaching a maximum at 1.13 wt% Pt, and then decreased with any further increase in Pt content. The maximum hydrogen generation rate reported for 1.1 wt% Pt loaded TiO_2 was 70–72 μmol/cm^2 per 6 h of UV irradiation. It appears that a small amount of Pt facilitates the hydrogen evolution; however, as the amount of Pt on the TiO_2 surface increases, it acts to block the photosensitive TiO_2 surface, thus decreasing the surface concentration of electrons and holes available for the reaction. It has also been suggested that at higher metal loadings the deposited metal particles may act as recombination centers for photoexcited electrons and holes [155]. It was also reported that conformal Pt coated TiO_2 nanotubes exhibited higher O_2 reduction initiating potential, promising it as a cathodic oxygen reduction catalyst for fuel cell applications [156]

4.6 Other Approaches

4.6.1 Polyoxophosphotungstate Encapsulated in TiO_2 Nanotubes

Polyoxometalates (POMs) are photocatalytically active materials with unique physicochemical properties, in terms of molecular and electronic versatility, reactivity, and stability [157–161]. A variety of organic substrates can be oxidized photocatalytically by either POMs or TiO_2 under UV illumination [162, 163], with POMs acting in a manner analogous to photocatalytic TiO_2. A comparison of the photocatalytic oxidation of organic compounds by POMs (such as phosphotungstate, phosphomolybdate) and TiO_2 were reported [164–166]. Studies indicated that the immobilization of POMs on solid supports, such as silica and zeolite, can induce a faster photocatalytic reaction than POMs alone [167, 168]. As a kind of Keggin-type POM, phosphotungstic acid is used as an original chemical modifying reagent, as well as a coupled catalyst with TiO_2. Integrative polyoxophosphotungstate–TiO_2 /titanium (POW–TiO_2/Ti) hybrid electrodes were synthesized by a soft-chemical approach to encapsulate POW aggregates inside the TiO_2 nanotubular channels [169–171]. A galvanostatic–potentiodynamic anodization was initially performed in an aqueous 0.5 M $H_3PO_4 + 0.1$ M HF solution to form an ultra thin oxide layer at a constant current density of 1.2 mA/cm^2 with a continuous increase in the anode-to-cathode voltage to 20 V. The following potentiostatic–galvanodynamic anodization was maintained at 20 V, to a final current density of 2.5 mA/cm^2. Multiporous microstructure TiO_2 samples were made by high-voltage anodization (at 120 V) in aqueous 1.5 M $H_2SO_4 + 0.3$ M $H_3PO_4 + 0.3$ M $H_2O_2 + 0.03$ M HF solution.

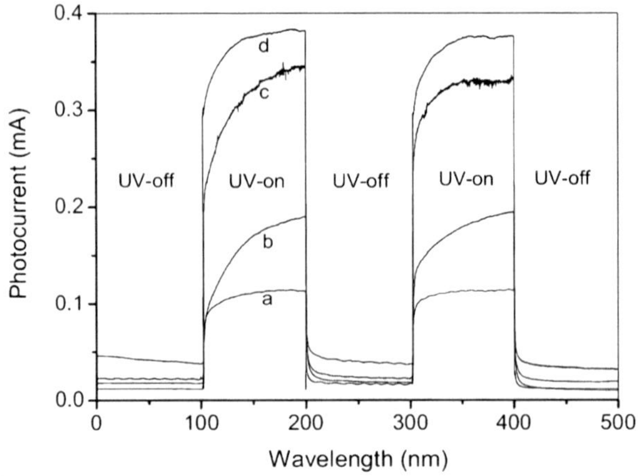

Fig. 4.28 Photocurrent responses of TiO_2/Ti and POW–TiO_2/Ti electrodes in 0.01 M Na_2SO_4 electrolyte solution under UV pulse illumination (the working electrode potential was constant at 3.0 V). (*a*) Microstructured TiO_2; (*b*) nanotubular TiO_2; (*c*) microstructured POW–TiO_2; (*d*) nanotubular POW–TiO_2 [169]. Xie YB: Photoelectrochemical Reactivity of a Hybrid Electrode Composed of Polyoxophosphotungstate Encapsulated in Titania Nanotubes. *Advanced Functional Materials*, 2006, 16, 1823–1831. Copyright Wiley-VCH Verlag GmbH & Co. KGaA. Reproduced with permission

Figure 4.28 shows the photocurrent–time profiles of TiO_2/Ti and POW–TiO_2/Ti electrodes under pulsed UV irradiation [169]. A medium-pressure mercury lamp acted as the UV-light source (main wavelength of 365 nm and irradiation intensity of 0.68 mW/cm^2). The measurements were performed at an applied constant potential of +3.0 V to TiO_2/Ti or POW–TiO_2/Ti electrodes, in 0.01 M Na_2SO_4 solution. The weak dark current (below 0.04 mA) at an anodic potential of 3.0 V resulted only from pure electrochemical reaction on TiO_2/Ti or POW–TiO_2/Ti electrodes. Two functionalized POW–TiO_2/Ti electrodes showed much higher photocurrent intensities than the pure TiO_2/Ti electrodes. The sharp increase in current intensity with light illumination was ascribed to the fast transfer of the photogenerated electrons on the surface of the functionalized electrodes. Under UV irradiation and applied positive potential, the photoelectrons from the TiO_2/Ti or POW–TiO_2/Ti photoanode moved to the counter electrode forming an external current circuit. Therefore, the overall photocurrent responsive is attributed to a UV-induced polarization current, with a minor anodic electrochemical reaction current.

4.6.2 Light Sensitized Enzymatic System with TiO_2 Nanotubes

The light sensitized enzymatic (LSE) system is a way of producing hydrogen by coupling an inorganic semiconductor with enzymes in a photoelectrochemical cell configuration [172–175]. The system uses the intrinsic proton reduction ability of hydrogenase enzyme in tandem with an anodic compartment where the electron donors, such as water, undergo oxidative reaction on a light sensitized photoanode. The generated electrons are separated and moved to the cathodic compartment through an external circuit and used to reduce protons into hydrogen on the active enzyme sites. Here, oxidized ions move to the cathodic compartment through an ion bridge. In the solar cell-applied enzymatic hydrogen production system, the light-sensitized anodized tubular TiO_2 was taken as the photoanode, and the anodized tubular TiO_2 electrode as a cathode. Nanotubes were grown by anodization of titanium foil at 20 V in 0.5 M $H_3PO_4 + 0.14$ M $NaF + 0.1$ M $NaNO_3$ [174]. A 0.05 M Tris–HCl buffer was used as a cathodic electrolyte. Purified hydrogenase (Pyrococcus furiosus, "*Pfu*"), whose activity assay (21,834 U/mL) was much higher than that of Clostridium butyricum (1,442 U/mL) and Thiocapsa roseopersicina (1,704 U/mL), was used. One unit of hydrogenase activity catalyzes the production of 1 μmol of H_2/min.

Experiments were conducted in a two-compartment reactor, with the compartments connected through a polyamide "Nanofiltration" (NF) membrane (SU-210, Toray, Japan), as shown in Fig. 4.29. The cylindrical shaped anodic compartment contained an aqueous solution of 1.0 M KOH for water oxidation. The cathodic compartment was a water jacketed, cylindrical-shaped cell with Tris–HCl buffer sealed with a silicone rubber gasket. Prior to reaction, the mixture was bubbled with argon gas for 20 min. In the LSE H_2 production system, replacing the Pt cathode with TiO_2 did not affect the rate of H_2 production. The light source used was a 1,000 W xenon lamp (Oriel, USA), filtered through a 10 cm IR water filter. The irradiated light intensity was measured as ca. 6873.4 mW/cm^2, further reduced by reflection and absorption at the chamber window and electrolyte.

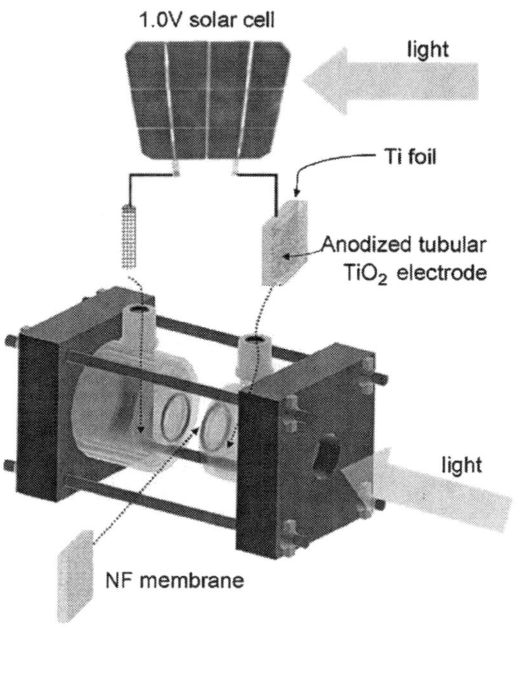

Fig. 4.29 Schematic view of the reactor with anodized tubular TiO_2 photoanode, solar cell, and NF membrane for hydrogen production [174]. Reprinted from *Solar Energy Materials & Solar Cells*, Vol. 92, Bae S, Shim E, Yoon J, Joo H, Enzymatic hydrogen production by light-sensitized anodized tubular TiO_2 photoanode, 402–409, Copyright 2008, with permission from Elsevier

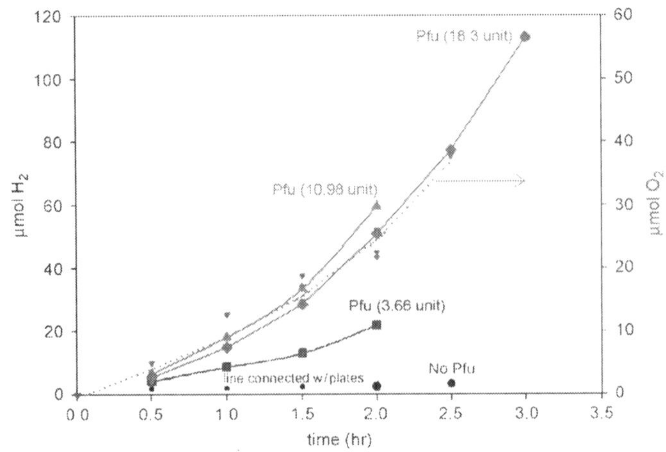

Fig. 4.30 Time-coursed profile of enzymatic hydrogen evolution with the anodized tubular TiO_2 photoanode (1.0 M KOH in anodic compartment, Tris–HCl buffer in cathodic compartment, irradiated photoanode area 1.0 cm^2, ~68 mW/cm^2 intensity with Xe lamp) [174]. Reprinted from *Solar Energy Materials & Solar Cells*, Vol. 92, Bae S, Shim E, Yoon J, Joo H, Enzymatic hydrogen production by light-sensitized anodized tubular TiO_2 photoanode, 402–409, Copyright 2008, with permission from Elsevier

Enzymatic hydrogen evolution with oxygen in the anodic compartment was performed with an external bias (~1.0 V) applied by a solar cell (100 cm^2, crystalline silicon) to the TiO_2 nanotube anode with respect to the platinum mesh cathode [174]. The solar cell and photoanode were both illuminated to increase the Fermi level and photogenerate charge carriers. With the *Pfu* amount optimized at 10.98 units (Fig. 4.30), this system configuration led to the stoichiometric evolution of H_2 and O_2 (H_2:O_2 = 2:1). Without any electron donor in the anodic compartment, oxygen was produced at a rate of about 20 μmol/h.cm^2 and hydrogen was produced in the cathodic compartment at a rate of about 43 μmol/h cm^2 when one of the TiO_2 nanotube (20 V/1 h/5°C in 0.5 wt% HF-650°C /5 h) was used. Without any KOH, the TiO_2 nanotube and *Pfu* system showed much lower activity in the H_2 and O_2 evolution [174]. If the rate of hydrogen production was about 40 μmol/h.cm^2, that is, ca. 9 l/h.m^2, a rough calculation suggests that with 6.5 h sunlight 585,000 L of H_2 could be evolved over an area of 10^4 m^2 (100 m × 100 m).

4.7 Self-Biased Photoelectrochemical Diodes Using Cu-Ti-O Ternary Oxide Nanotubes

CuO is a *p*-type semiconductor with a band gap of 1.2–1.4 eV commonly studied for its photoconductive properties while Cu_2O is a *p*-type semiconductor with a direct band gap of approximately 2 eV, which makes it a promising material for the conversion of solar energy [176–178]. CuO and Cu_2O films have traditionally been prepared by thermal oxidation of copper metal, offering limited control over the resulting material structure [179]. The abundance of copper and the non-toxicity of copper oxides offer added advantages [180]. While semiconductor materials with bandgaps suitable for capturing a significant fraction of incident solar spectrum energy typically suffer UV induced photocorrosion [181–183], the stability of CuO, Cu_2O and heterojunctions, thereof, are significantly altered by the material it is coupled with. For example, Cu_2O heterojunctions with In_2O_3, SnO_2, ZnO are reported to be unstable while Cu_2O-CdO junctions are relatively stable [184, 185]. Siripala and coworkers reported that depositing a thin layer of TiO_2 on *p*-type Cu_2O successfully protects the Cu_2O film against photocorrosion [186], suggesting that TiO_2 may be one of the best candidates to couple with CuO or Cu_2O to enhance photocorrosion stability.

Morphology also affects copper oxide stability, with rapid removal of the photon-generated electrons and holes from the copper oxide significantly improving its stability [187, 188]. Since the properties of nanoarchitectured materials are greatly influenced by their shape and dimensional constraints, the synthesis of the copper oxide into ordered one-dimensional structures, such as nanotubes or nanowires, could enhance its intrinsic characteristics. Many recent efforts have been directed toward the fabrication of nanostructural CuO to enhance its performance in currently existing applications. To date, well-defined CuO nanostructures with different dimensionalities such as nanoparticles, nanowires, nanowhiskers, nanoleaves, nanorods, nanotubes, and nanoribbons have been obtained successfully by a series of wet chemical routes, electrochemical deposition techniques and vapor-phase processes [189–191]. Of these different synthesis techniques, anodic oxidation is distinct in providing a strategy for the top-down growth of self-organized structures with a robust interface to the underlying substrate [65, 66, 192, 193].

Despite the excellent light harvesting potential of low band-gap CuO, its performance as a photoelectrode for water photoelectrolysis is rather poor due to the presence of recombination centers in the band gap and its chemical instability [194, 195]. By incorporating copper oxide inside a TiO_2 nanotube array matrix, we attempted to extend the unique benefits of the nanotube array architecture to the ternary Cu-Ti-O semiconductor system in the fabrication of photocorrosion stable *p*-type Cu-Ti-O nanotube array semiconductor electrodes by anodization of Cu-Ti films [196]. Unlike the easily anodized valve metals, the anodization of copper to simultaneously achieve oxidation with morphological control has proven quite difficult. In this regard, Ti is used as a supporting matrix to provide chemical resistance and mechanical support during the anodic oxidation of copper.

4.7.1 Fabrication of p-Type Copper Rich Cu-Ti-O Nanotubes

Ti-Cu films of various compositions were deposited on FTO coated glass substrates, maintained at 300°C, by simultaneous cosputtering from titanium and copper targets. The composition of the films was determined by Energy Dispersive X-ray Spectroscopy. To form the nanotube array structure, the Ti-Cu films were potentiostatically anodized at 30 V in EG containing 0.25 wt% NH_4F and 1.0 vol% deionized water. The real time variation of the anodization current is similar to that observed for pure Ti films [197]; however, the final saturation current value is quite high. An initial drop in anodization current due to oxide formation is followed by an increase in the current due to pitting of the oxide layer by the fluoride ions. The current then gradually decreases to a steady state value, at which stage the nanotube structure has been fully formed. From this point forward, maintaining the samples in the anodization bath results in an increase in the nanotube length. Anodization of the Cu-Ti metal film is stopped just prior to reaching the FTO substrate, hence in the as-anodized films a thin discontinuous film of Cu-Ti remains between the Cu-Ti-O nanotube array film and substrate. A summary of film parameters is given in Table 4.3. FESEM views of as-anodized films are shown in Fig. 4.31.

Figure 4.32 shows the GAXRD spectrum of a sample 60 and sample 74 oxygen annealed at 450°C for 2 h with a ramp up/down rate of 2°C/min. The predominant peaks observed in the samples at $2\theta = 36.4°$ and $2\theta 2 = 38.8°$ correspond to the (111) and (002) reflections of CuO. The residual presence of a Cu_4Ti_3 alloy phase is detected. The formation of Cu-Ti-O films with superior photoelectrochemical properties require a higher degree of crystallization, however, two parameters

Table 4.3 Summary of film properties used in photoelectrochemical diode [196]

Sample name	Film thickness (nm)	Composition (%) determined using EDS	Tube length, pore, & wall thickness (nm)	Semiconducting behavior of crystallized films	Light absorption range (nm)
74	1,500	74% Cu–26% Ti	1,000, 65,35	*p*-type	350–885
60	1,000	60% Cu–40% Ti	850, 65, 35	*p*-type	350–885
TiO_2	600	Pure Ti	600, 30, 7	*n*-type	300–400

Sample name denotes the copper content of the start Cu-Ti metal film

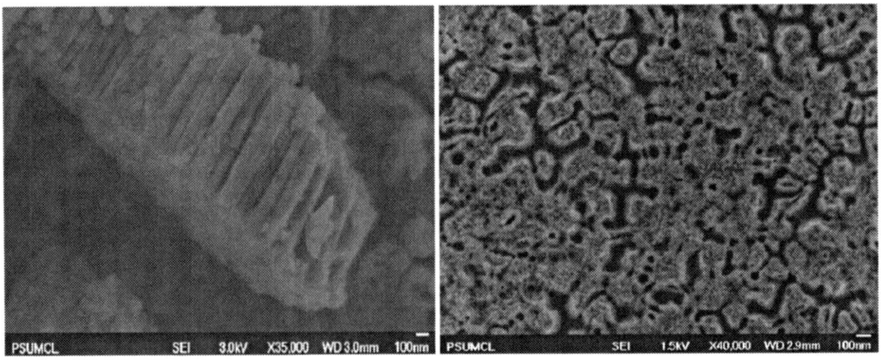

Fig. 4.31 SEM images of lateral and top views of Cu-Ti-O nanotube array films [196]

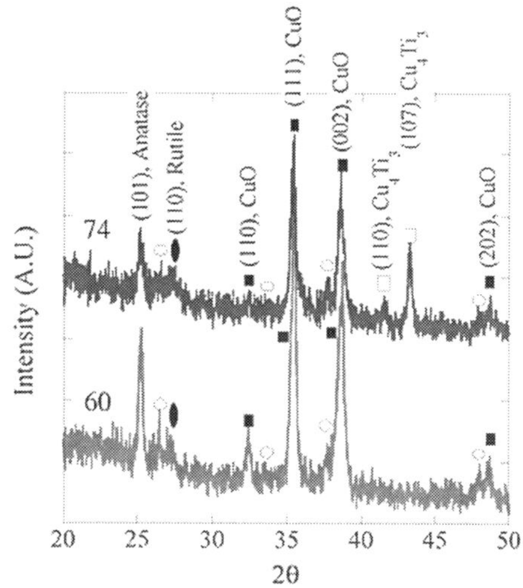

Fig. 4.32 XRD patterns of 450°C, 2 h O_2 annealed Cu-Ti-O samples 60 and 74. *Open circles* denote SnO_2 peaks

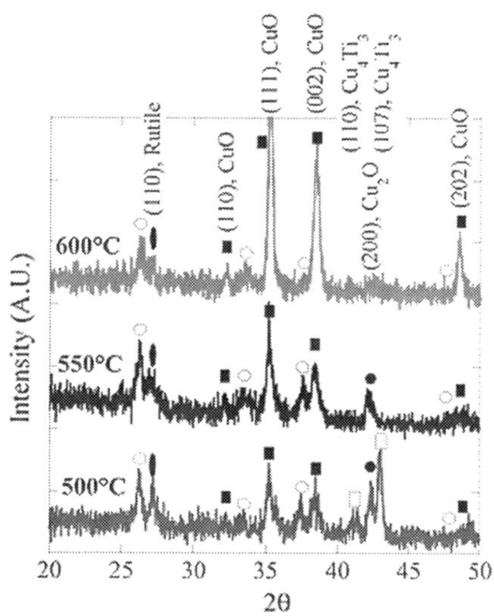

Fig. 4.33 XRD patterns of Cu-Ti-O samples 74 vacuum annealed at 500°C, 550°C, and 600°C for 2 h. *Open circles* denote SnO_2 peaks [196]

change with annealing, the film crystallization and the resistivity of the 15 Ω/□ FTO layer underlying the Cu-Ti-O films. At elevated temperatures, the FTO layer will diffuse into the Cu-Ti-O film, with a corresponding decrease in FTO conductivity, hence optimal sample performance is a trade-off between crystallization and sheet resistance of the underlying FTO film. Oxygen annealed temperatures greater than 450°C significantly increase the sheet resistance of the FTO layer, a behavior not as pronounced with vacuum (20 Torr with oxygen background) annealing. Figure 4.33 shows the GAXRD spectrum of samples 74 annealed in vacuum at 500°C, 550°C, and 600°C. All samples were annealed for 2 h. The predominant peaks observed in the samples at $2\theta = 36.4°$ and $2\theta = 38.8°$ correspond to the (111) and (002) reflections of CuO. The samples annealed by both methods do not indicate the formation of Cu_2O. Samples vacuum annealed at 500°C and 550°C show the reflections of Cu_4Ti_3, most likely present in the barrier layer separating the tubes from the FTO substrate; as the annealing temperature is increased to 600°C oxygen diffuses deeper into the barrier layer converting Cu_4Ti_3 to CuO.

4.7.2 *Photoelectrochemical Properties*

The optimum band gap of an effective semiconductor for water photoelectrolysis should be in the range of 1.7–2.0 eV, higher than a theoretical minimum water splitting energy of 1.23 eV per photon to compensate for overpotential losses [198]. Since the energy needed to split methanol into H_2 and CO_2 is 0.7 eV [199], and CuO is unstable in aqueous alkaline electrolytes, we initially used a 1:1 water and methanol solution to study the photoelectrochemical properties of the samples, with a 5:5:2 methanol:water:diethylamine solution providing the largest photocurrent amplitudes due to the relatively higher solubility of diethylamine compared with other amines in the methanol-water mixture and increased electron transfer from CuO to the added molecules. The global AM 1.5 current versus voltage behavior of a sample 74 annealed in vacuum at 500°C (2 h) and 550°C (1 h, 2 h), tested in 5:5:2 methanol/water/DEA solution (pH ≈ 8) is shown in Fig. 4.34. No significant degradation in the photoresponse of these *p*-Cu-Ti-O samples was observed in this electrolyte during a continuous 24 h monitoring and under periodic testing over several weeks thereafter. The FTO sheet resistance of the 500°C annealed sample is 15 Ω/□. For the 550°C sample, increasing the anneal duration from 1 to 2 h increased the measured FTO film sheet resistance from 60 to 1.2kΩ/□, with a significant drop in the measured photocurrent. For 600°C vacuum annealed samples the measured sheet resistance is approximately 32kΩ/□, with a measured

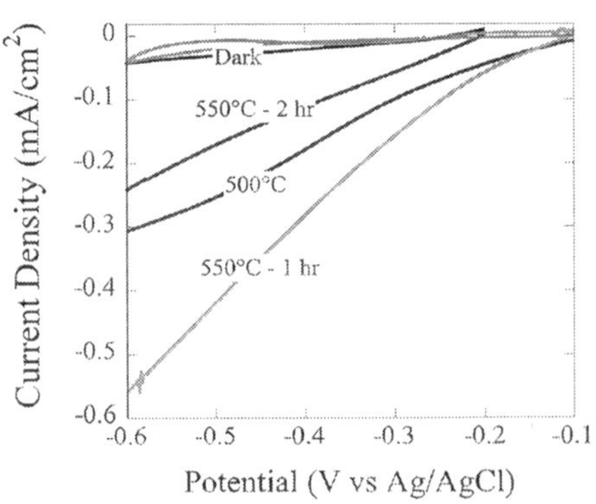

Fig. 4.34 Photocurrent density versus potential for 500°C and 550°C vacuum annealed samples 74 in 5:5:2 methanol:water:diethylamine solution, under dark and global AM 1.5 illumination [196]

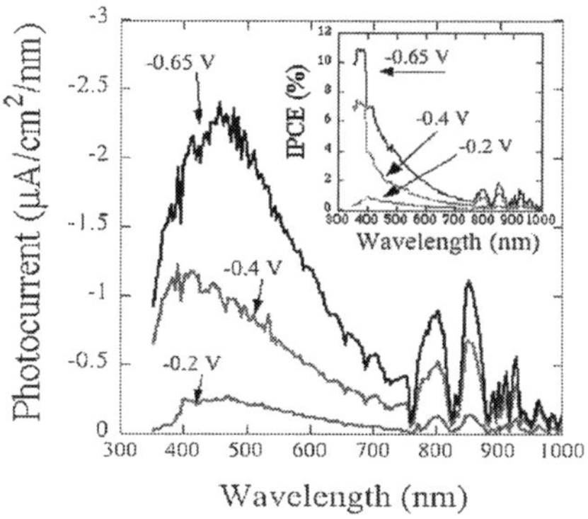

Fig. 4.35 Photoelectron flux dependence of a 550°C (1 h) vacuum annealed sample 74 at applied bias amplitudes (measured with respect to Ag/AgCl reference electrode) of: −0.2 V, −0.4 V and –0.65 V, tested in a 5:5:2 methanol: water:diethylamine solution under AM 1.5 illumination. Dark currents were nulled before starting the measurements. *Inset* shows IPCE curves for each applied bias [196]

photocurrent of only a few μA/cm^2. Photocathode bias assists in separation of the electron-hole pairs, thereby, enhancing photocurrent values. The photocurrent values obtained for this sample in a potential scan are in close agreement with the values obtained by integration of the measured photon flux values (μA/cm^2/nm) at three applied bias values of −0.2, −0.4 and −0.65 V versus Ag/AgCl. The IPCE is shown in the inset of Fig. 4.35. For an applied potential of –0.65 V Ag/AgCl, we typically see IPCE values of 10–11% in the UV range and 1–7% in the visible range.

4.7.3 *Self-Biased Heterojunction Photoelectrochemical Diodes*

The use of a self-biased *pn* junction, a so called photoelectrochemical diode, for solar water-splitting was first reported by A. J. Nozik in 1977 [35], motivating subsequent efforts [200–204]. In a manner analogous to photosynthesis, photoelectrochemical diodes require as input only sunlight for the creation of a portable chemical fuel is this case hydrogen by splitting of the water molecule, with hydrogen evolved from the *p*-side and oxygen from the *n*-side of the photoelectrochemical diode. That useful photoelectrochemical diodes have not been realized is largely due to the lack of a suitable *p*-type semiconductor resistant to photocorrosion.

A depiction of the *n*-TiO_2:*p*-Cu-Ti-O coupled photoelectrochemical diode system is shown in Fig. 4.36. The photoresponse of this system under global AM 1.5 illumination is shown in Fig. 4.37. The *n*-TiO_2 nanotube array film was made by sputtering a Ti film upon a FTO coated glass substrate, then performing an anodization [197] to achieve a TiO_2 nanotube array structure approximately 600 nm long, 30 nm pore size, with wall thickness of about 7 nm; the initially amorphous sample was oxygen annealed at 450°C for 1 h. The *p*-Cu-Ti-O sample is that of sample 74, 550°C vacuum annealed for 1 h, nanotube array length 1 μm, pore size ≈ 65 nm, wall thickness ≈ 35 nm, see Table 4.3. The films, atop FTO coated glass substrates,

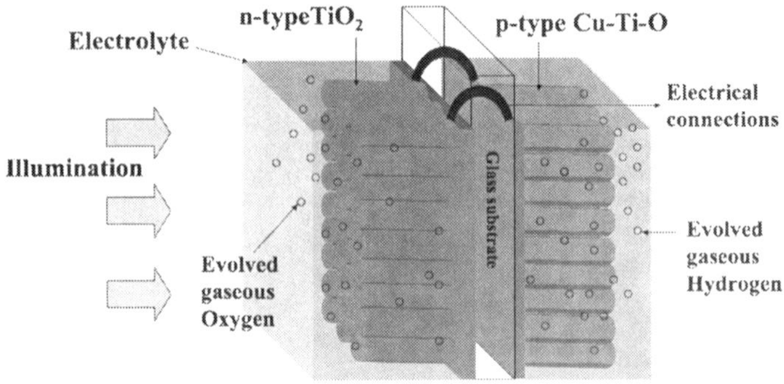

Fig. 4.36 Illustration of photoelectrochemical diode for water splitting comprised of *n*-type TiO_2 and *p*-type Cu-Ti-O nanotube array films, with their substrates connected through an ohmic contact. The oxygen evolving TiO_2 side of the diode absorbs UV light, passing the visible light to the hydrogen evolving Cu-Ti-O side [196]

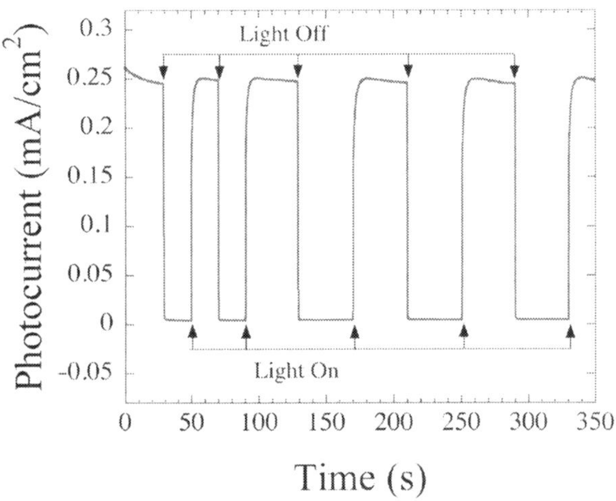

Fig. 4.37 Photocurrent from self-biased photochemical diode comprised of an *n*-type TiO_2 nanotube array – *p*-type Cu-Ti-O nanotube array under global AM 1.5 illumination (geometry shown in Fig. 4.36). Light is incident upon the oxygen evolving TiO_2 side of the diode, with the visible portion of the spectrum passing to the Cu-Ti-O side. The *n*-TiO_2 side of the diode is kept immersed in a 1 M KOH aqueous solution, the *p*-Cu-Ti-O side is kept in 0.1 M Na_2HPO_4 with a salt bridge linking the two sides solution [196]

were mounted by use of epoxy into a glass sheet that allowed separate collection of the evolved gases. The TiO_2 nanotube array side of the diode was kept in 1 M KOH, and the Cu-Ti-O side kept in 0.1 M Na_2HPO_4 with a salt bridge linking the two sides. Keeping the Cu-Ti-O side in KOH solution resulted in significantly greater photocurrents, $\approx 20\times$; however, the performance decayed with time as the CuO turned into copper. The AM 1.5 illumination was incident upon the TiO_2 side of the diode, which absorbs Ultraviolet photons while passing visible light to the Cu-Ti-O side, thus minimizing any UV-assisted photocorrosion of the low bandgap material. A photocurrent density of approximately 0.25 mA/cm^2 is achieved corresponding to a photoconversion efficiency of 0.30% not taking into account potential effects of chemical bias. An overall energy conversion efficiency of 0.48% was determined for sample 74 vacuum annealed at 550°C for 1 h using a two-electrode photoelectrochemical measurement. After 4 to 5 h of operation, only a minimal amount of photocathode photocorrosion was seen for such samples in 0.1 M Na_2HPO_4 electrolyte.

4.8 Visible light responsive Ti-Fe-O ternary oxide nanotubes

4.8.1 Benefits of nanostructuring hematite

Hematite, α-Fe_2O_3 or iron(III) oxide, is a *n*-type semiconductor with a ~2.2 eV bandgap [205] that is sufficiently large for water splitting but small enough to collect a significant fraction, $\approx$40%, of solar spectrum energy. However, iron oxide suffers from low electron mobility, generally in the range of 0.01 cm^2/Vs [206] to 0.1 cm^2/Vs [207] resulting in a rapid electron-hole recombination, and an insufficiently negative flat band potential as needed for spontaneous water splitting [208]. Therefore, α-Fe_2O_3 solar spectrum photoconversion efficiencies reported to date are far from the theoretical maximum value of 12.9% for a material of this band-gap [26]. Because of its visible spectrum bandgap, widespread availability, and non-toxicity iron (III) oxide nanostructures are of considerable interest for achieving efficient, cost effective and durable photoelectrochemical solar cells. Various techniques have been investigated for the synthesis of Iron (III) oxide nanostructures as it has proven to be a promising route to overcome low charge carrier mobility and high bulk recombination losses while offering large surface areas [209]. For example, Duret [210] demonstrated photocurrents in the mA range under 1.5 AM solar illumination from mesoporous hematite thin films consisting of 5–10 nm thick nanoleaflets synthesized via ultrasonic spray pyrolysis. Recently, silicon doped α-Fe_2O_3 photoanodes were reported by Gratzel and coworkers [211], while Ingler and Khan recently reported an α-Fe_2O_3 based self-biased photoelectrochemical cell [212].

We first review the synthesis of self-aligned nanoporous iron(III) oxide formed by electrochemical oxidation of Fe foil samples to achieve an amorphous nanoporous film and subsequent conversion to crystalline hematite by annealing [213]. Then, we review our efforts to achieve high solar energy photoelectrochemical properties by combining the charge transport and photocorrosion properties of the TiO_2 nanotubes with the bandgap of α-Fe_2O_3 [52]. Since both Ti and Fe can be anodized in a fluoride ion containing EG solution, the possibility exists for obtaining highly-ordered, vertically oriented Ti-Fe oxide nanotube arrays, the material architecture we desire for application to water photoelectrolysis, by anodization of Ti-Fe metal films.

4.8.2 Self-Aligned Nanoporous Iron (III) Oxide

Pure iron foil (99.9%) of 0.25 mm thickness was anodized in a two-electrode cell with iron foil as the anode and platinum foil as the counter electrode. The anodization bath temperature was controlled by placing the electrolytic assembly within a recirculating water cooler. After the removal from the anodization bath the samples were thoroughly rinsed with deionized water and isopropyl alcohol, and then blow-dried with 99.99% pure nitrogen. Electrolytes having lower diffusion coefficients are known to prevent local acidification at the oxide–electrolyte interface [214]. Glycerol is a highly viscous solvent, where ions diffuse slowly. Hence, when we began our studies of nanoporous hematite formation, we used glycerol as the electrolytic medium. Random self-organized nanopores were obtained in electrolytes under a narrow window of processing conditions. The electrolytes consisted of 1 vol% HF + 0.2–0.6 wt% NH_4F + 0.1–0.2% 0.1 M HNO_3 in glycerol (pH 3) at temperatures ranging from 5(C to 15(C at anodization voltages of 40–90 V for anodization durations ranging from 30 min to 2 h. It was found that the overall thickness of the oxide layer was greatly dependent on the amount of HF and HNO_3 added while the anodization bath temperature and NH_4F concentration were the primary parameters in controlling the sample morphology. Prakasam et al. provide a more extensive discussion of the processing window in glycerol necessary to obtain nanoporous Fe_2O_3 and of the resulting morphologies [213]. EG is a viscous organic solvent containing two alcoholic hydroxyl groups as opposed to glycerol's three hydroxyl groups. When an electrolyte comprised of 0.3% wt NH_4F (99.8% purity, Sigma Aldrich) in EG was used to anodize Fe foil at a potential of 40 V at 10°C, a highly-ordered nanoporous structure was obtained, with clearly defined channels similar to those seen in the nanoporous alumina [215]. Figure 4.38 shows the corresponding FESEM images. Using an electrolyte held at 10°C, the pore depth was found to be around 2.5 μm in 15 min and ~ 3 μm in 1 h, with 180-nm thick oxide barrier layer underneath the nanoporous layer. At 10°C, the structure remained nanoporous for an anodization duration up to ≈ 70 min beyond which the pores collapsed. Room temperature anodization in the same electrolyte exhibited similar nanoporous surface, depth ~ 2 μm in 10 min and 2.2 μm in 15 min, with the nanoporous structure collapsing for anodization of 30 min duration or beyond. Independent of the anodization bath temperature, the resulting pore diameters are ≈ 30 nm at 40 V and ≈ 60 nm at 60 V.

Owing to the fact that iron forms wires, rods, flakes and ribbons [216, 217] in the process of thermal oxidation, the samples were subjected to relatively brief anneals in a nitrogen atmosphere. The extent of pore filling due to oxide growth from the barrier layer is a function of annealing time and temperature, which limited the extent of sample crystallization. Samples were annealed at temperatures ranging from 150°C to 600°C. We observed that longer annealing periods and higher temperatures resulted in a greater fraction of magnetite and pore filling. Thus, a balance between structural stability and crystallization is achieved by annealing the samples at 400°C for 30 min, with a ramp rate of 10°C/min, in nitrogen atmosphere. Figure 4.39

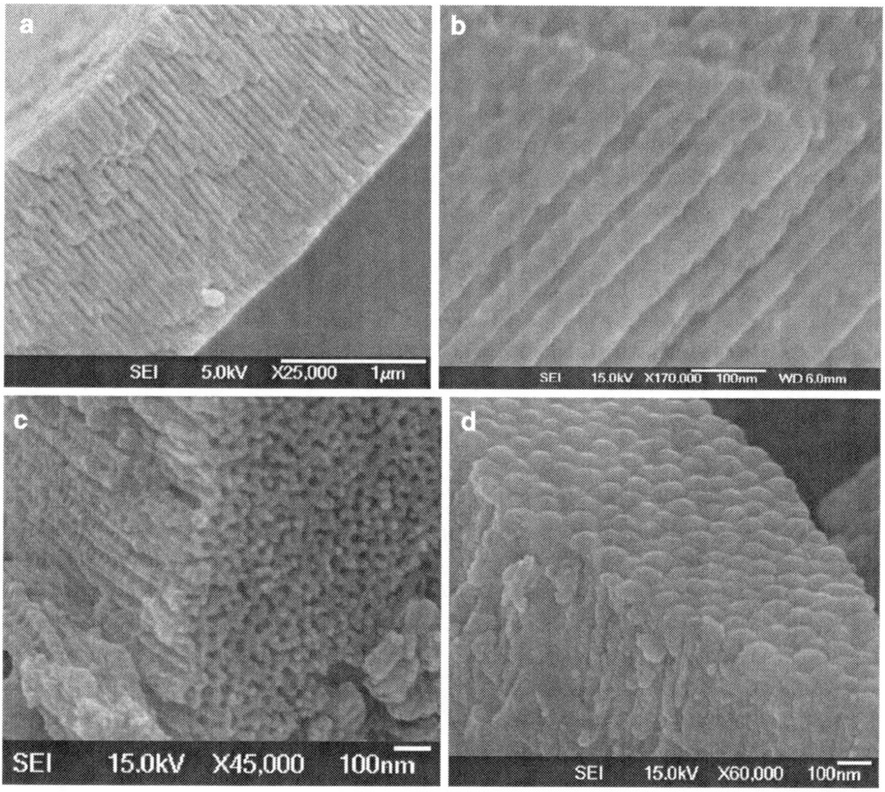

Fig. 4.38 FESEM images of Fe sample anodized in 0.3 wt% NH_4F in EG at 40 V, 10°C for 15 min. (**a–c**) show cross-sectional images of a mechanically fractured sample, while (**d**) shows the underside (*barrier layer*) of the highly-ordered nanoporous Fe_2O_3 [213]

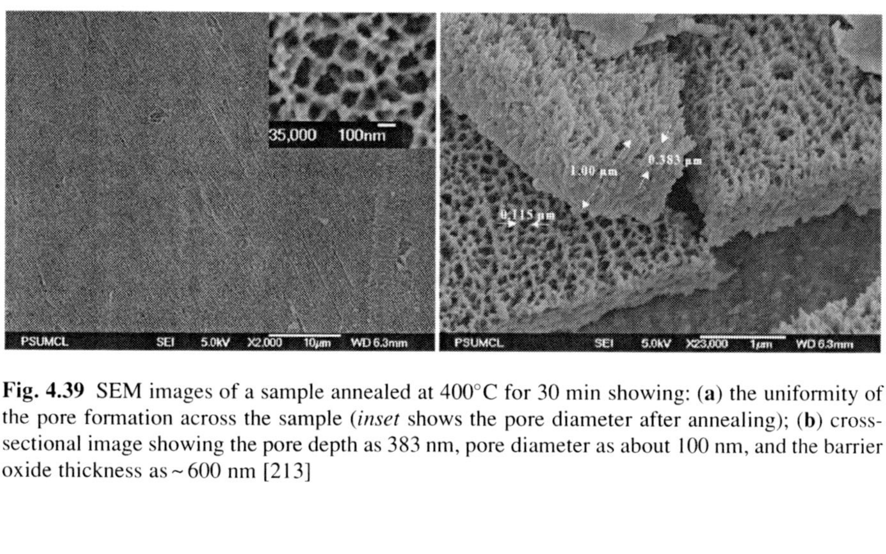

Fig. 4.39 SEM images of a sample annealed at 400°C for 30 min showing: (**a**) the uniformity of the pore formation across the sample (*inset* shows the pore diameter after annealing); (**b**) cross-sectional image showing the pore depth as 383 nm, pore diameter as about 100 nm, and the barrier oxide thickness as ~600 nm [213]

shows an FESEM image of a sample anodized at 90 V in an electrolyte containing 1% HF + 0.3 wt% NH_4F + 0.2% 0.1 M HNO_3 in glycerol (pH 3) at 10°C and annealed at 400°C for 30 min. From Fig. 4.39(a) we see that a crystalline nanoporous structure has been achieved with excellent uniformity, while the insert to Fig. 4.39(a) demonstrates that the pore geometry has been retained. The cross-sectional image of Fig. 4.39(b) shows the extent of the pore depth (~ 380 nm) and the barrier layer thickness (~ 600 nm) of the annealed sample.

4.8.3 *Photoelectrochemical Properties of Self–Aligned Nanoporous Iron (III) Oxide*

In photoelectrochemical measurements, the unanodized part of the iron foil underneath the nanoporous array and the barrier layer serves as the conducting substrate to transport the photogenerated electrons. Figure 4.40 shows the current-potential response of an illustrative annealed 90 V sample, 115 nm pore diameter, 400 nm pore depth, with potential measured relative to Ag/AgCl standard electrode under dark and 1.5 A.M simulated solar illumination. The sample was anodized in a HNO_3, HF, and NH_4F bath at 10°C. Figure 4.40(a) shows the photoelectrochemical response in an electrolyte containing 1 M NaOH; the onset potential is $\approx$ 0.24 mV. The dark current increases from zero at 0.46 V–0.020 mA at 0.6 V. After the measurement of Fig. 4.40(a), the sample was rinsed in DI water then measured in an electrolyte containing 0.5 M H_2O_2 (50%) and 1 M NaOH (50%), with the measured photoresponse plot given in Fig. 4.40(b). The onset potential is –0.37 mV, several hundred mV more negative than in the 1 M NaOH solution [210]. The addition of H_2O_2 enhances the reaction kinetics as the photo oxidation rate of H_2O_2 is much larger than that of water [218]. Thus, by introducing 50% of 0.5 M H_2O_2, the nanoporous structure demonstrated a net, light illuminated less dark, photocurrent of 0.51 mA at 0.6 V versus Ag/AgCl.

4.8.4 *Fabrication and Structural Characterization of Ti-Fe-O Nanotubes*

Ti-Fe metal films were deposited on fluorine doped tin oxide (FTO) coated glass substrates by simultaneous cosputtering from titanium and iron targets. The substrate temperature was maintained at 350°C–400°C. Details of sputtering conditions, the resulting thickness of the Ti-Fe metal films and their crystallinity are given in Table 4.4. Distinct nanotube arrays were formed from samples having iron concentrations less than about 70%. Films of higher Fe content, up to 100% Fe, were fabricated and characterized; however, since they showed significantly

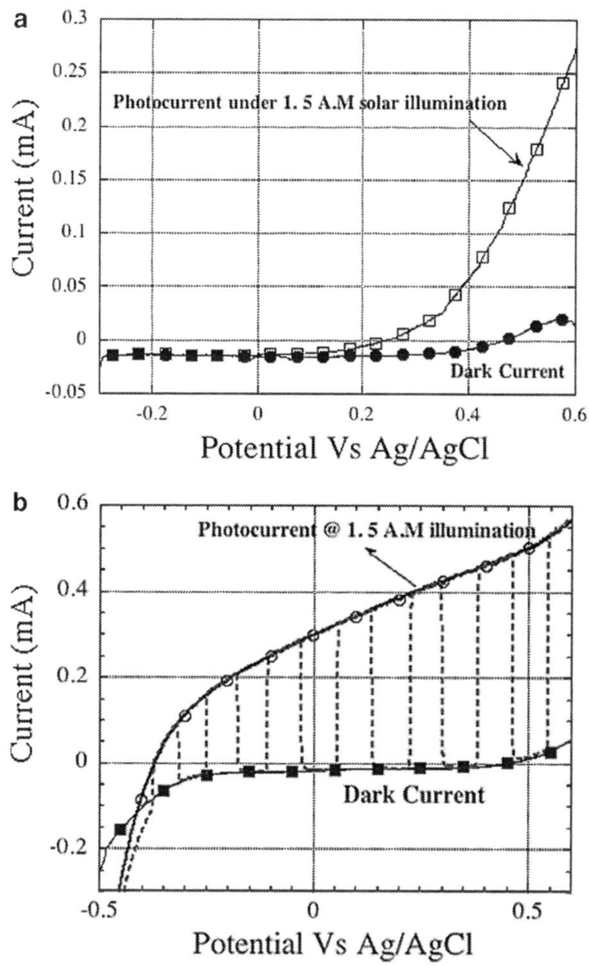

Fig. 4.40 Photocurrent as a function of measured potential for the Iron (III) Oxide photoanode (anodized in 0.2 % 0.1 M HNO_3, 1% HF, and 0.3 wt% NH_4F in glycerol at 10°C) recorded in: (**a**) 1 M NaOH solution, and (**b**) 0.5 M H_2O_2 + 1 M NaOH solution under 1.5 A.M simulated solar illumination [213]

diminished photoelectrochemical properties, they are not considered herein. We consider Ti:Fe metal films in the ratio 31:69, 56:44, 63:37, 74:26, 80:20, 93.4:6.6, and 96.5:3.5 as determined from the relative sputtering rates, with identifying sample numbers given by the Fe content. Anodization of the Ti-Fe films was performed at a constant voltage of 30 V in EG containing 0.3 wt% NH_4F and 2.0 vol% de-ionized water. Illustrative cross-sectional and bottom side FESEM images of Sample 20 are shown in Fig. 4.41.

As-anodized Ti-Fe-O nanotube films are amorphous, partially crystallized by annealing in oxygen atmosphere at 500°C for 2 h with a ramp up and down rate of

Table 4.4 Summary of Fe-Ti-O film formation parameters [52]

Sample	Fe target power (W)/ sputter rate (nm/s)	Ti target power (W)/ sputter rate (nm/s)	Substrate temp °C	Film thickness (nm)	Film composition	Crystallinity of metal starting film
69	300/5.5	100/2.5	350	400	69% Fe–31% Ti	Crystalline
44	300/5.5	300/7.0	350	1,000	44% Fe–56% Ti	Crystalline
37	250/4.1	300/7.0	350	500	37% Fe–63% Ti	Amorphous
26	200/2.5	300/7.0	350	400	26% Fe–74% Ti	Amorphous
20	150/1.3	300/7.0	400	750	20% Fe–80% Ti	Amorphous
6.6a	100/0.5	300/7.0	400	750	6.6% Fe–93.4% Ti	Crystalline
6.6b	100/0.5	300/7.0	400	1,500	6.6% Fe–93.4% Ti	Crystalline
3.5	50/0.25	300/7.0	400	750	3.5% Fe–96.5% Ti	Crystalline

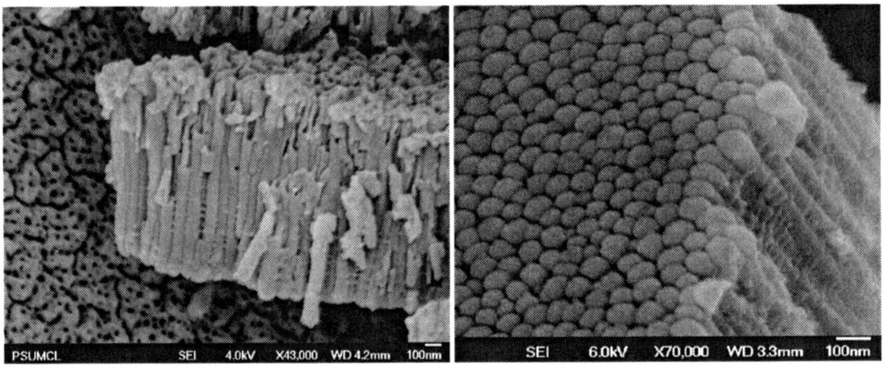

Fig. 4.41 Illustrative FESEM images of Ti-Fe-O nanotube array sample 20, image of *lateral view* (**a**) and *bottom view* (**b**) [52]

1°C /min. GAXRD patterns of the annealed films are seen in Fig. 4.42. All studies on TiO_2 nanotube array films made by anodization have indicated the tubes to be anatase, fixed atop a rutile base [219]. The Ti-Fe-O samples with low iron concentrations show the presence of anatase phase; Sample 3.5 is largely anatase with a dominant (110) plane, and a small amount of rutile. The anatase phase disappears at higher iron concentrations (see the patterns of Samples 20 and 37). The presence of the rutile peak is evident in Samples 6.6 and 20, which is consistent with the fact that low amounts of iron in TiO_2 aid rutile formation [220]. The predominant (104) plane of α-Fe_2O_3 coincides with the second intense plane of FTO, used as the conductive layer on the glass substrate, and its subsequent two less intense planes, (110) and (116), appear at $2\theta = 35.6$ and $2\theta = 54.1$, respectively. The XRD pattern of Sample 26 is omitted from Fig. 4.42 because of its relatively weak peaks. In Sample 37, the proportion of rutile increases slightly, and anatase is almost negligible. In general, Samples 20 and 37 showed poor crystallinity, Sample 44 showed pseudobrookite phase, while Sample 69 showed hematite phase. In general, it appears that the presence of iron inside the TiO_2 matrix degrades the crystallization

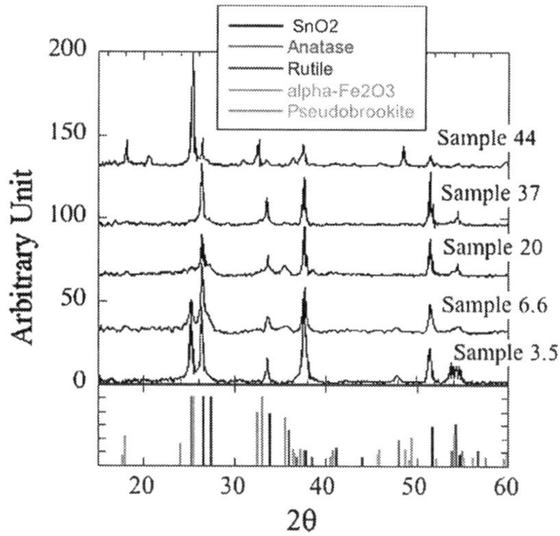

Fig. 4.42 GAXRD patterns of Ti-Fe-O nanotube array samples annealed at 500°C in dry oxygen for 2 hr, see Table 4.4 for fabrication details. Standard patterns for anatase, rutile, pseudobrookite, α-Fe_2O_3 (hematite), tin oxide (substrate) are shown for phase identification [52]

process since the proportion of rutile does not increase with decreasing amounts of anatase, an outcome possibly due to the fixed 500°C annealing temperature. However, higher annealing temperatures spoil the FTO–TiFeO interface, resulting in samples of high series resistance demonstrating essentially nil photoelectrochemical properties.

The crystalline nature of the nanotube walls is critical to applications involving light absorption, electrical carrier generation, and carrier transport, therefore, the crystalline nature of the nanotube walls were studied using HRTEM. Sample 20 with moderate iron doping was used for the study. TEM studies confirmed the relatively poor crystallinity of samples annealed at 500°C; hence, further studies were conducted on a Sample 20 annealed at 600°C for 2 h in oxygen. As seen in Fig. 4.41, and as described in our earlier articles on TiO_2 nanotube arrays formed in HF aqueous electrolytes [197, 221], the nanotube morphology appears as a column of stacked torus (doughnut) shaped rings as seen in the TEM image of Fig. 4.43(a). A selected area electron diffraction (SAED) pattern from the corresponding region is shown in Fig. 4.43(b). Although the pattern shows mainly rutile phase, reflections from hematite phase also can be seen. Figure 4.44(a) shows a HRTEM image of a nanotube wall, with the polycrystalline nature evident from the figure. The crystallites have a wide size distribution. The crystallite marked R has a rutile structure the lattice image of which is given in the upper left inset. A Fast Fourier Transform (FFT) of regions R and H are given in Fig. 4.44(b) and 4.44(c) indicating the presence of rutile and hematite phases in the respective regions. To find the distribution of iron within the nanotube walls, an energy dispersive x-ray spectroscopy (EDS) line scan was performed via scanning transmission electron microscopy (STEM). The intensity of both the TiK_α and FeK_α lines was found maximum at the center of the wall due to its torus shape. Despite the presence of isolated hematite crystallites, a more or less uniform distribution of iron relative to the titanium can be seen across the wall. STEM line scans were performed across a

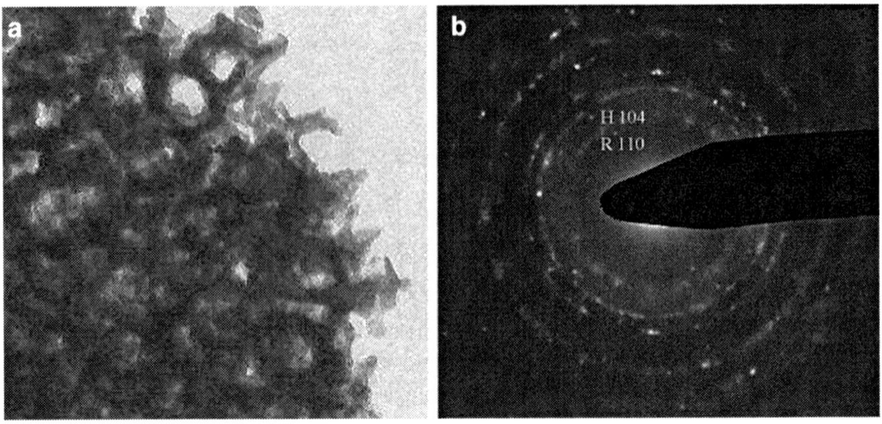

Fig. 4.43 (**a**) TEM image of Sample 20 annealed at 600°C, and (**b**) selected area diffraction pattern (*H* hematite and *R* rutile) [52]

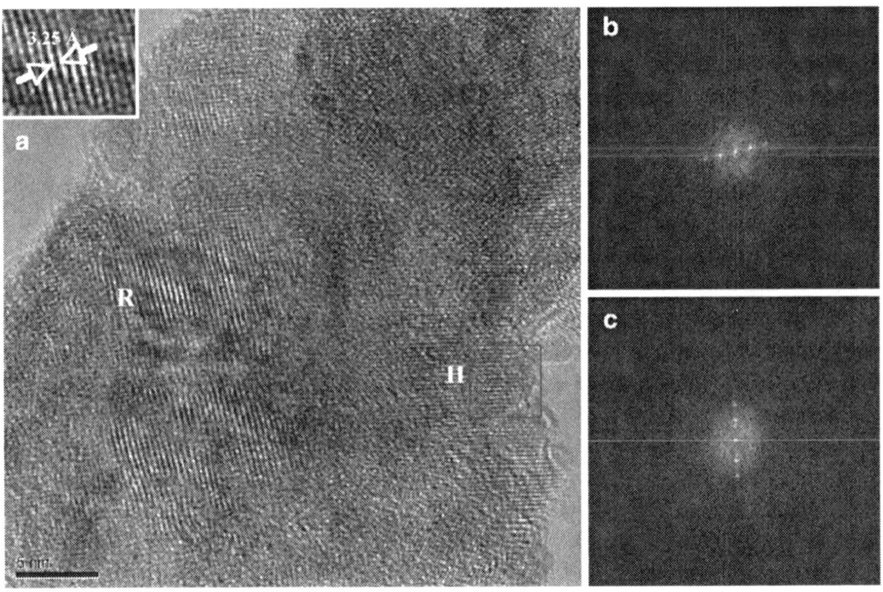

Fig. 4.44 (**a**) HRTEM image of a nanotube wall (Sample 20 annealed at 600°C) showing rutile (*region R*) and hematite (*region H*) crystallites with the lattice image of region R given in the *upper-left inset*; (**b**) FFT of region R showing a predominate 110 plane, and (**c**) FFT of region H showing a predominate 104 plane [52]

number of walls, and while the average relative intensity of the TiK_α and FeK_α lines varied from wall to wall, the relative distribution across a single wall remained uniform. It appears that some of the iron goes into the titanium lattice substituting titanium ions, and the rest either forms hematite crystallites or remains in the amorphous state.

Crystallization of the as-anodized amorphous nanotubes takes place through nucleation and growth [220]. Hence, a sufficiently high temperature and appropriate duration are needed for the complete transformation of the amorphous phase into crystalline. The removal of the amorphous phase is critical for effective device application; however, even after annealing at 600°C, Sample 20 showed the presence of amorphous phase. HRTEM studies indicate that rutile and hematite phases are formed in separate events depending upon the temperature and nature of the nucleation sites. In general, the nanotubes of moderate iron concentration consist mainly of rutile phase with hematite crystallites randomly distributed.

4.8.5 Photoelectrochemical Properties of Ti–Fe–O Nanotubes

Figure 4.45 shows the absorbance of the annealed samples; as expected the absorbance edge shows a red shift with increasing Fe content. Figure 4.46 shows current versus voltage characteristics for the nanotubular Ti-Fe-O electrodes under dark and simulated sunlight in 1.0 M NaOH. The electrodes show *n*-type behavior, that is, positive photocurrents at anodic potentials. A metal-coated glass substrate was placed adjacent to translucent Sample 6.6b, reflecting the transmitted light back into the sample. This method increased the photocurrent from 1.50 to 2.0 mA/cm^2 at 0.65 V (vs. Ag/AgCl). For comparison, the photocurrent of a pure nanoporous α-Fe_2O_3 film 1.5 μm thick on FTO glass is also shown in Fig. 4.46, prepared by anodic oxidation of an iron film at 30 V in EG containing 0.3 wt% NH_4F and 0.5% de-ionized water. The poor electron mobility results in a maximum photocurrent of 20–25 μA/cm^2 at 0.4 V (vs. Ag/AgCl). The highest photocurrent is exhibited by Sample 6.6, while the lowest photocurrent is demonstrated by Sample 3.5. It appears that minimal levels of Fe^{3+} ions act as trap sites, while significant amounts of Fe^{3+} ions increase the recombination rate of the photogenerated electron-hole

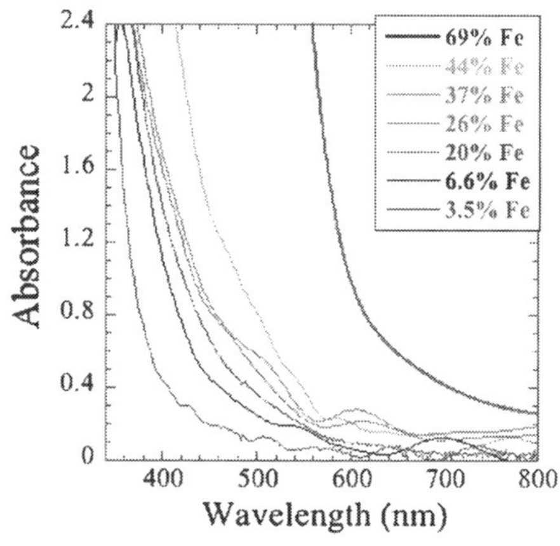

Fig. 4.45 Absorbance spectra of 500°C annealed Ti-Fe-O nanotube array films, see Table 4.4

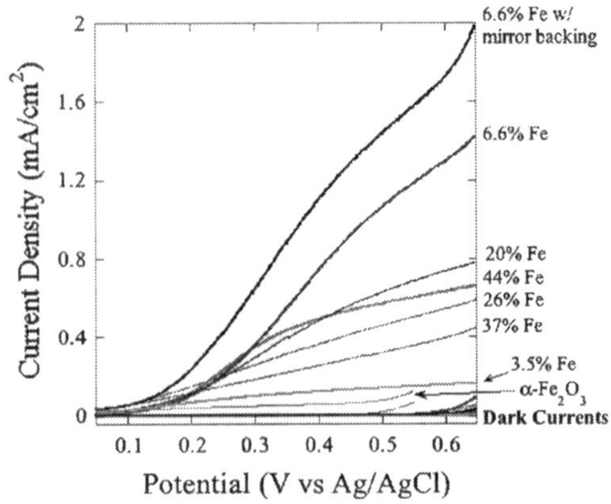

Fig. 4.46 Photocurrent density versus potential in 1 M NaOH solution for annealed Ti-Fe-O nanotube array samples, and α-Fe_2O_3 nanoporous film, under AM 1.5 (100 mW/cm^2) illumination. Dark currents are also shown for each sample [52]

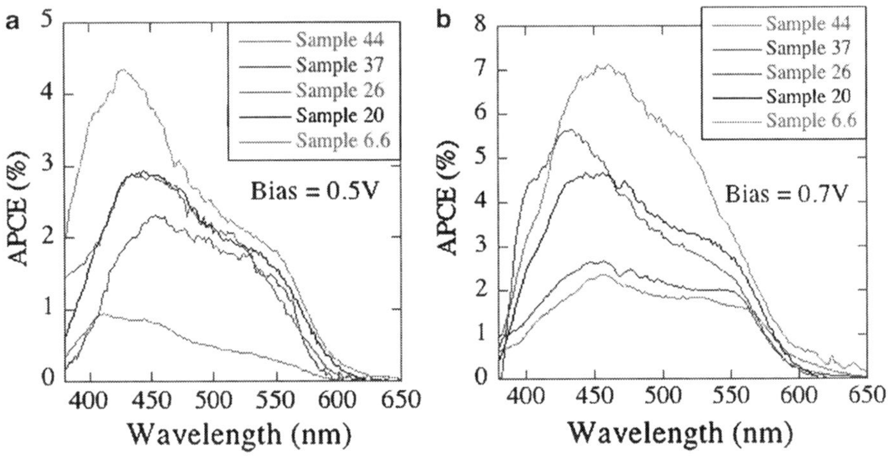

Fig. 4.47 Absorbed-photon-to-current-efficiency (*APCE*) of Ti-Fe-O nanotube array samples at: (**a**) 0.5 V bias, and (**b**) 0.7 V bias [52]

pairs. We note that Sample 44 does not comply with this trend, possibly due to the presence of the pseudobrookite phase, $FeTiO_5$.

Figure 4.47a and b show APCE values corresponding to two applied voltages, 0.5 V and 0.7 V; APCE values are appreciable only in the visible range, from 400 to 600 nm. Energy level band diagrams indicate electrons photogenerated in α-Fe_2O_3 cannot transfer to the TiO_2 conduction band without an applied bias. Since the redox potential for the $H_2O/\bullet OH$ reaction for oxygen evolution is above the valence band of α-Fe_2O_3 in 1 M NaOH electrolyte (pH = 14), and the thickness of the α-Fe_2O_3 crystallites is comparable to the hole diffusion length, application of a small positive bias can be used to separate the photogenerated electrons and holes. As shown in Fig. 4.47, APCE values of the films decrease as we move below 400 nm toward the UV region, possibly due to the Fe^{3+} ions of the α-Fe_2O_3 and iron incorporated TiO_2 crystallites becoming Fe^{2+} as associated with oxygen vacancy defects. Because of this, the position of the valence band at solid-electrolyte shifts upward, crossing the redox potential of the H_2O/•OH reaction inhibiting hole transfer and leading to increased recombination.

4.9 Compositionally Graded Ternary Oxide Nanotube Arrays

Compositionally graded ternary oxide nanotube arrays, depicted in Fig. 4.48, enable the same photoelectrode to efficiently absorb light of different wavelength-regions of the solar spectrum. An analogous approach has been proposed in the field of solar energy conversion, namely the so-called "rainbow solar cells" [136], where by varying the size of the semiconductor quantum dots the solar cells can be tuned to absorb light of certain wavelengths. Additionally, as suggested for the case of TiO_2/PbS [128], size quantization may result in more favorable energy levels sufficient to initiate charge injection from excited PbS into TiO_2. In Fig. 4.48, we show a hypothetical self-biased photoelectrochemical diode consisting of a compositionally graded *n*-type nanotube array photoanode and a *p*-type cathode. The larger bandgap stoichiometries, closer to the illumination source, absorb high energy photons while allowing optical transmission of lower energy photons. A broad swath of light energy ranging from the absorption edge of the large band-gap oxide to the absorption-edge of the smaller band-gap oxide would be more efficiently harvested than by the smaller band-gap semiconductor working in isolation. Ideally, photogenerated holes quickly oxidize solution ions while photogenerated electrons percolate down through the nanotube walls toward the substrate where they are collected. A key design criterion is the appropriate matching of energy levels such that photogenerated electrons easily travel down the conduction band of the compositionally graded nanotube array. Practical realization of such architecture would require cosputtering a bimetallic film with graded composition, followed by anodization in a suitable electrolyte to form the nanotube arrays and a subsequent thermal anneal to induce crystallinity.

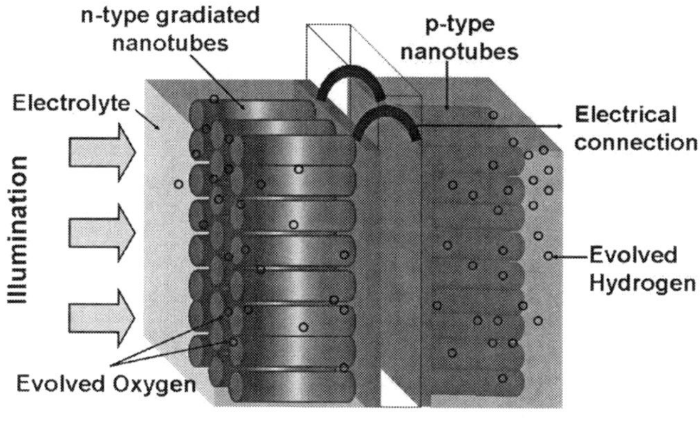

Fig. 4.48 Illustrative self-biased photoelectrochemical diode comprised of a *n*-type compositionally graded nanotube array photoanode and a *p*-type nanotube array cathode [222]

References

1. Grimes CA, Varghese OK, Ranjan S (2008) Light, water, hydrogen: the solar generation of hydrogen by water photoelectrolysis. Springer, New York
2. Paulose M, Shankar K, Yoriya S, Prakasam HE, Varghese OK, Mor GK, Latempa TA, Fitzgerald A, Grimes CA (2006) Anodic growth of highly ordered TiO_2 nanotube arrays to 134 μm in Length. J Phys Chem B 110:16179–16184
3. Prakasam HE, Shankar K, Paulose M, Varghese OK, Grimes CA (2007) A new benchmark for TiO_2 nanotube array growth by anodization. J Phys Chem C 111:7235–7241
4. Wilke K, Breuer HD (1999) The influence of transition metal doping on the physical and photocatalytic properties of titania. J Photochem Photobiol A 121:49–53
5. Wilke K, Breuer HD (1999) Transition metal doped titania: Physical properties and photocatalytic behaviour. Z Phys Chem - Int J Res Phys Chem Chem Phys 213:135–140
6. Burda C, Lou YB, Chen XB, Samia ACS, Stout J, Gole JL (2003) Enhanced nitrogen doping in TiO_2 nanoparticles. Nano Lett 3:1049–1051
7. Ihara T, Miyoshi M, Iriyama Y, Matsumoto O, Sugihara S (2003) Visible-light-active titanium oxide photocatalyst realized by an oxygen-deficient structure and by nitrogen doping. Appl Catal B-Environ 42:403–409
8. Sakthivel S, Kisch H (2003) Photocatalytic and photoelectrochemical properties of nitrogen-doped titanium dioxide. Chem phys chem 4:487–490
9. Irie H, Watanabe Y, Hashimoto K (2003) Carbon-doped anatase TiO_2 powders as a visible-light sensitive photocatalyst. Chem Lett 32:772–773
10. Park JH, Kim S, Bard AJ (2006) Novel carbon-doped TiO_2 nanotube arrays with high aspect ratios for efficient solar water splitting. Nano Lett 6:24–28
11. Yamaki T, Umebayashi T, Sumita T, Yamamoto S, Maekawa M, Kawasuso A, Itoh H (2003) Fluorine-doping in titanium dioxide by ion implantation technique. Nucl Instrum Methods Phys Res B 206:254–258
12. Shankar K, Tep KC, Mor GK, Grimes CA (2006) An electrochemical strategy to incorporate nitrogen in nanostructured TiO_2 thin films: modification of bandgap and photoelectrochemical properties. J Phys D Appl Phys 39:2361–2366
13. Lindgren T, Mwabora JM, Avendano E, Jonsson J, Hoel A, Granqvist CG, Lindquist SE (2003) Photoelectrochemical and optical properties of nitrogen doped titanium dioxide films prepared by reactive DC magnetron sputtering. J Phys Chem B 107:5709–5716
14. Nakamura R, Tanaka T, Nakato Y (2004) Mechanism for visible light responses in anodic photocurrents at N-doped TiO_2 film electrodes. J Phys Chem B 108:10617–10620
15. Torres GR, Lindgren T, Lu J, Granqvist CG, Lindquist SE (2004) Photoelectrochemical study of nitrogen-doped titanium dioxide for water oxidation. J Phys Chem B 108:5995–6003
16. Fujishima A, Honda K (1972) Electrochemical photolysis of water at semiconductor electrode. Nature 238:37–38
17. Bolton JR (1996) Solar photoproduction of hydrogen: a review. Sol Energy 57:37–50
18. Bolton JR, Strickler SJ, Connolly JS (1985) Limiting and realizable efficiencies of solar photolysis of water. Nature 316:495–500
19. Bolton JR (1978) Solar fuels. Science 202:705–711
20. Archer MD, Bolton JR (1990) Requirements for ideal performance of photochemical and photovoltaic solar energy converters. J Phys Chem 94:8028–8036
21. Grimes DM, Grimes CA (2006) A unique electromagnetic photon field using Feynman's electron characteristics and Maxwell's equations. J Comput Theor Nanosci 3:649–663
22. Fujishima A, Kohayakawa K, Honda K (1975) Hydrogen production under sunlight with an electrochemical photocell. J Electrochem Soc 122:1487–1489
23. Parkinson B (1984) On the efficiency and stability of photoelectrochemical devices. Acc Chem Res 17:431–437

24. Murphy OJ, Bockris JOM (1982) On the efficiency of conversion in photo-electrochemical cells. J Electrochem Soc 129:C332–C332
25. Bard AJ, Faulkner LR (2001) Electrochemical methods: fundamentals & applications. Wiley, New Jersey
26. Murphy AB, Barnes PRF, Randeniya LK, Plumb IC, Grey IE, Horne MD, Glasscock JA (2006) Efficiency of solar water splitting using semiconductor electrodes. Int J Hydrogen Energy 31:1999–2017
27. Khan SUM, Al-Shahry M, Ingler WB (2002) Efficient photochemical water splitting by a chemically modified n- TiO_2. Science 297:2243–2245
28. Ang PGP, Sammells AF (1984) Hydrogen evolution at P-InP photocathodes in alkaline electrolyte. J Electrochem Soc 131:1462–1464
29. Dohrmann JK, Schaaf NS (1992) Energy-conversion by photoelectrolysis of water – determination of efficiency by in situ photocalorimetry. J Phys Chem 96:4558–4563
30. Wrighton MS, Ellis AB, Wolczanski PT, Morse DL, Abrahamson HB, Ginley DS (1976) Strontium titanate photoelectrodes. Efficient photoassisted electrolysis of water at zero applied potential. J Am Chem Soc 98:2774–2779
31. Nozik AJ (1976) p-n Photoelectrolysis cell. Appl Phys Lett 29:150–153
32. Ohashi K, McCann J, Bockris JOM (1977) Stable photoelectrochemical cell for splitting of water. Nature 266:610–611
33. Lee J, Fujishima A, Honda K, Kumashiro Y (1985) Photoelectrochemical behaviour of p-type boron phosphide photoelectrode in acidic solution. Bull Chem Soc Jpn 58:2634–2637
34. Kainthala RC, Zelenay B, Bockris JOM (1987) Significant efficiency increase in self-driven photoelectrochemical cell for water photoelectrolysis. J Electrochem Soc 134:841–8450
35. Nozik AJ (1977) Photochemical diodes. Appl Phys Lett 30:567–569
36. Raja KS, Mahajan VK, Misra M (2006) Determination of photo conversion efficiency of nanotubular titanium oxide photo-electrochemical cell for solar hydrogen generation. J Power Sources 159:1258–1265
37. Varghese OK, Grimes CA (2008) Appropriate strategies for determining the photoconversion efficiency of water photo electrolysis cells: a review with examples using titania nanotube array photoanodes. Sol Energy Mater Sol Cells 92:374–384
38. Shankar K, Mor GK, Fitzgerald A, Grimes CA (2007) Cation effect on the electrochemical formation of very high aspect ratio TiO_2 nanotube arrays in formamide-water mixtures. J Phys Chem C 111:21–26
39. Kay A, Cesar I, Gratzel M (2006) New Benchmark for water photooxidation by nanostructured α-Fe_2O_3 films. J Am Chem Soc 128:15714–15721
40. Nazeeruddin MK, Kay A, Rodicio I, Humphry-Baker R, Muller E, Liska P, Vlachopoulos N, Gratzel M (1993) Conversion of light to electricity by cis-X2Bis(2, 2(-bipyridyl- 4, 4 dicarboxylate) ruthenium(II) charge-transfer sensitizers (X=Cl-, Br-, I-, CN-, and SCN-) on nanocrystalline TiO2 electrodes. J Am Chem Soc 115:6382–6390
41. Mor GK, Shankar K, Paulose M, Varghese OK, Grimes CA (2005) Enhanced photocleavage of water using titania nanotube arrays. Nano Lett 5:191–195
42. Mor GK, Shankar K, Varghese OK, Grimes CA (2004) Photoelectrochemical properties of titania nanotubes. J Mater Res 19:2989–2996
43. Sukamto JPH, Mcmillan CS, Smyrl W (1993) Photoelectrochemical investigations of thin metal-oxide films – TiO_2, Al_2O_3, and HfO_2 on the parent metals. Electrochim Acta 38:15–27
44. Sukamto JPH, Smyrl WH, Mcmillan CS, Kozlowski MR (1992) Photoelectrochemical measurements of thin oxide-films - multiple internal-reflection effects. J Electrochem Soc 139:1033–1043
45. van de Lagemaat J, Plakman M, Vanmaekelbergh D, Kelly JJ (1996) Enhancement of the light to-current conversion efficiency in an n-SiC/solution diode by porous etching. Appl Phys Lett 69:2246–2248

46. Lubberhuizen WH, Vanmaekelbergh D, Van Faassen E (2000) Recombination of photogenerated charge carriers in nanoporous gallium phosphide. J Porous Mater 7:147–152
47. Marin FI, Hamstra MA, Vanmaekelbergh D (1996) Greatly enhanced sub-bandgap photocurrent in porous GaP photoanodes. J Electrochem Soc 143:1137–1142
48. Gratzel M (2001) Photoelectrochemical cells. Nature 414:338–344
49. Ruan CM, Paulose M, Varghese OK, Grimes CA (2006) Enhanced photo electrochemical-response in highly ordered TiO_2 nanotube-arrays anodized in boric acid containing electrolyte. Sol Energy Mater Sol Cells 90:1283–1295
50. Allam NK, Grimes CA (2008) Effect of cathode material on the morphology and photoelectrochemical properties of vertically oriented TiO_2 nanotube arrays. Sol Energy Mater Sol Cells 92:1468–1475
51. Allam NK, Shankar K, Grimes CA (2008) Photoelectrochemical and water photoelectrolysis properties of ordered TiO_2 nanotubes fabricated by Ti anodization in fluoride-free HCl electrolytes. J Mater Chem 18:2341–2348
52. Mor GK, Prakasam HE, Varghese OK, Shankar K, Grimes CA (2007) Vertically oriented Ti-Fe-O nanotube array films: Toward a useful material architecture for solar spectrum water photoelectrolysis. Nano Lett 7:2356–2364
53. Xie Y, Zhou LM, Huang H (2006) Enhanced photoelectrochemical current response of titania nanotube array. Mater Lett 60:3558–3560
54. Beranek R, Tsuchiya H, Sugishima T, Macak JM, Taveira L, Fujimoto S, Kisch H, Schmuki P (2005) Enhancement and limits of the photoelectrochemical response from anodic TiO_2 nanotubes. Appl Phys Lett 87:243114 3 pp
55. Cai Q, Paulose M, Varghese OK, Grimes CA (2005) The effect of electrolyte composition on the fabrication of self-organized titanium oxide nanotube arrays by anodic oxidation. J Mater Res 20:230–236
56. Varghese OK, Paulose M, Shankar K, Mor GK, Grimes CA (2005) Water-photolysis properties of micron-length highly-ordered titania nanotube-arrays. J Nanosci Nanotechnol 5:1158–1165
57. Cao F, Oskam G, Meyer GJ, Searson PC (1996) Electron transport in porous nanocrystalline TiO_2 photoelectrochemical cells. J Phys Chem 100:17021–17027
58. Paulose M, Mor GK, Varghese OK, Shankar K, Grimes CA (2006) Visible light photoelectrochemical and water-photoelectrolysis properties of titania nanotube arrays. J Photochem Photobiol A Chem 178:8–15
59. Vanmaekelbergh D, de Jongh PE (1999) Driving force for electron transport in porous nanostructured photoelectrodes. J Phys Chem B 103:747–750
60. Butterfield IM, Christensen PA, Hamnett A, Shaw KE, Walker GM, Walker SA (1997) Applied studies on immobilized titanium dioxide films as catalysts for the photoelectrochemical detoxification of water. J Appl Electrochem 27:385–395
61. Hagfeldt A, Gratzel M (1995) Light-induced redox reactions in nanocrystalline systems. Chem Rev 95:49–68
62. Kopidakis N, Benkstein K, van de Lagemaat J, Frank AJ (2003) Transport-limited recombination of photocarriers in dye-sensitized nanocrystalline TiO_2 solar cells. J Phys Chem B 107:11307–11315
63. Benkstein KD, Kopidakis N, van de Lagemaat J, Frank AJ (2003) Influence of the percolation network geometry on electron transport in dye-sensitized titanium dioxide solar cells. J Phys Chem B 107:7759–7767
64. Ruan C, Paulose M, Varghese OK, Mor GK, Grimes CA (2005) Fabrication of highly ordered TiO_2 nanotube arrays using an organic electrolyte. J Phys Chem B 109:15754–15759
65. Shankar K, Mor GK, Prakasam HE, Yoriya S, Paulose M, Varghese OK, Grimes CA (2007) Highly-ordered TiO_2 nanotube-arrays up to 220 μm in length: use in water photoelectrolysis and dye-sensitized solar cells. Nanotechnol 18:065707 11 pp

66. Zhu K, Neale NR, Miedaner A, Frank AJ (2007) Enhanced charge-collection efficiencies and light scattering in dye-sensitized solar cells using oriented TiO_2 nanotubes arrays. Nano Lett 7:69–74
67. Watcharenwong A, Chanmanee W, de Tacconi NR, Chenthamarakshan CR, Kajitvichyanukul P, Rajeshwar K (2007) Self-organized TiO_2 nanotube arrays by anodization of Ti substrate: effect of anodization time, voltage and medium composition on oxide morphology and photoelectrochemical response. J Mater Res 22:3186–3195
68. Ghicov A, Tsuchiya H, Macak JM, Schmuki P (2006) Annealing effects on the photoresponse of TiO_2 nanotubes. Phys Stat Sol 203:R28–R30
69. Xiao P, Liu D, Garcia BB, Sepehri S, Zhang Y, Cao G (2008) Electrochemical and photoelectrical properties of titania nanotube arrays annealed in different gases. Sens Actuators B Chem 134:367–372
70. Mahajan VK, Mohapatra SK, Misra M (2008) Stability of TiO_2 nanotube arrays in photoelectrochemical studies. Int J Hydrogen Energy 33:5369–5374
71. Nowotny J, Bak T, Nowotny MK, Sheppard LR (2007) Titanium dioxide for solar-hydrogen II. Defect chemistry. Int J Hydrogen Energy 32:2630–2643
72. Moser J, Gratzel M (1982) Photoelectrochemistry with colloidal semiconductors – laser studies of halide oxidation in colloidal dispersions of TiO_2 and alpha-Fe_2O_3. Helv Chim Acta 65:1436–1444
73. Morand R, Lopez C, Koudelka-Hep H, Kedzierzawaski P, Augustynski J (2002) Photoelectrochemical behavior in low-conductivity media of nanostructured TiO_2 films deposited on interdigitated microelectrode arrays. J Phys Chem B 106:7218–7224
74. Yamakata A, Ishibashi T, Onishi H (2002) Electron- and hole-capture reactions on Pt/ TiO_2 photocatalyst exposed to methanol vapor studied with time-resolved infrared absorption spectroscopy. J Phys Chem B 106:9122–9125
75. Yoshihara T, Katoh R, Furube A, Tamaki Y, Murai M, Hara K, Murata S, Arakawa H, Tachiya M (2004) Identification of reactive species in photoexcited nanocrystalline TiO2 films by wide wavelength range (400–2500 nm) transient absorption spectroscopy. J Phys Chem B 108:3817–3823
76. Shkrob IA, Sauer MC Jr, Gosztola D (2004) Efficient, rapid photooxidation of chemisorbed polyhydroxyl alcohols and carbohydrates by TiO_2 nanoparticles in an aqueous solution. J Phys Chem B 108:12512–12517
77. Ni M, Leung MKH, Leung DYC, Sumathy K (2007) A review and recent developments in photocatalytic water-splitting using TiO_2 for hydrogen production. Renew Sustainable Energy Rev 11:401–425
78. Poznyak SK, Kokorin AI, Kulak AI (1998) Effect of electron and hole acceptors on the photoelectrochemical behaviour of nanocrystalline microporous TiO_2 electrodes. J Electroanal Chem 442:99–105
79. Khan SUM, Sultana T (2003) Photoresponse of n- TiO_2 thin film and anowire electrodes. Sol Energy Mater Sol Cells 76:211–221
80. Mohapatra SK, Raja KS, Mahajan VK, Misra M (2008) Efficient photoelectrolysis of water using TiO_2 nanotube arrays by minimizing recombination losses with organic Additives. J Phys Chem C 112:11007–11012
81. Harvey PR, Rudham R, Ward S (1983) Photocatalytic oxidation of liquid alcohols and binary alcohol mixtures by rutile. J Chem Soc Faraday Trans 79:2975–2981
82. Elshafei AA, Elmaksoud SAA, Fouda AS (1995) Noble-metal-modified glassy-carbon ‘electrodes for ethylene-glycol oxidation in alkaline-medium. J Electroanal Chem 395:181–187
83. Schell M, Xu Y, Zdraveski Z (1996) Mechanism for the electrocatalyzed oxidation of glycerol deduced from an analysis of chemical instabilities. J Phys Chem 100:18962–18969
84. Matsuoka K, Iriyama Y, Abe T, Matsuoka M, Ogumi Z (2005) Electro-oxidation of methanol and ethylene glycol on platinum in alkaline solution: poisoning effects and product analysis. Electrochim Acta 51:1085–1090

85. Zhang Z, Yuan Y, Fang Y, Liang L, Ding H, Shi G, Jin L (2007) Photoelectrochemical oxidation behavior of methanol on highly ordered TiO_2 nanotube array electrodes. J Electroanal Chem 610:179–185
86. Yin S, Yamaki H, Komatsu M, Zhang Q, Wang J, Tang Q, Saito F, Sato T (2003) Preparation of nitrogen-doped titania with high visible light induced photocatalytic activity by mechanochemical reaction of titania and hexamethylenetetramine. J Mater Chem 13:2996–3001
87. Chuang CC, Wu WC, Lee MX, Lin JL (2000) Adsorption and photochemistry of CH_3CN and CH_3CONH_2 on powdered TiO_2. Phys Chem Chem Phys 2:3877–3880
88. Li J, Yun H, Lin C-J (2007) A photoelectrochemical study of N-doped TiO_2 nanotube arrays as the photoanodes for cathodic protection of SS. J Electrochem Soc 154:C631–C636
89. Kim D, Fujimoto S, Schmuki P, Tsuchiya H (2008) Nitrogen doped anodic TiO_2 nanotubes grown from nitrogen-containing Ti alloys. Electrochem Commun 10:910–913
90. Ghicov A, Macak JM, Tsuchiya H, Kunze J, Haeublein V, Frey L, Schmuki P (2006) Ion implantation and annealing for an efficient N-doping of TiO_2 nanotubes. Nano Lett 6:1080–1082
91. Ghicov A, Macak JM, Tsuchiya H, Kunze J, Haeublein V, Kleber S, Schmuki P (2006) TiO_2 nanotube layers: dose effects during nitrogen doping by ion implantation. Chem Phys Lett 419:426–429
92. Vitiello RP, Macak JM, Ghicov A, Tsuchiya H, Dick LFP, Schmuki P (2006) N-Doping of anodic TiO_2 nanotubes using heat treatment in ammonia. Electrochem Commun 8:544–548
93. Macak JM, Ghicov A, Hahn R, Tsuchiya H, Schmuki P (2006) Photoelectrochemical properties of N-doped self-organized titania nanotube layers with different thicknesses. J Mater Res 21:2824–2828
94. Lei L, Su Y, Zhou M, Zhang X, Chen X (2007) Fabrication of multi-non-metal-doped TiO_2 nanotubes by anodization in mixed acid electrolyte. Mater Res Bull 42:2230–2236
95. Su Y, Zhang X, Zhou M, Han S, Lei L (2008) Preparation of high efficient photoelectrode of N–F-codoped TiO_2 nanotubes. J Photochem Photobiol A Chem 194:152–160
96. Su Y, Zhang X, Han S, Chen X, Lei L (2007) F–B-codoping of anodized TiO_2 nanotubes using chemical vapor deposition. Electrochem Commun 9:2291–2298
97. Geng J, Yang D, Zhu J, Chen D, Jiang Z (2009) Nitrogen-doped TiO_2 nanotubes with enhanced photocatalytic activity synthesized by a facile wet chemistry method. Mater Res Bull 44:146–150
98. Yu JC, Ho WK, Yu JG, Hark SK, Iu K (2003) Effects of trifluoroacetic acid modification on the surface microstructures and photocatalytic activity of mesoporous TiO_2 thin films. Langmuir 19:3889–3896
99. Chen X, Zhang X, Su Y, Lei L (2008) Preparation of visible-light responsive P–F-codoped TiO_2 nanotubes. Appl Sur Sci 254:6693–6696
100. Beranek R, Macak JM, Gartner M, Meyer K, Schmuki P (2009) Enhanced visible light photocurrent generation at surface-modified TiO2 nanotubes. Electrochim Acta 54:2640–2646
101. Beranek R, Kisch H (2007) Surface modified anodic TiO_2 films for visible photocurrent response. Electrochem Commun 9:761–766
102. Beranek R, Kisch H (2008) Tuning the optical and photoelectrochemical properties of surface-modified TiO_2. Photochem Photobiol Sci 7:40–48
103. Beranek R, Neumann B, Sakthivel S, Janczarek M, Dittrich T, Tributsch H, Kisch H (2007) Exploring the electronic structure of nitrogen-modified TiO_2 photocatalysts through photocurrent and surface photovoltage studies. Chem Phys 339:11–19
104. Wardman P (1989) Reduction potentials of one-electron couples involving free-radicals in aqueous-solution. J Phys Chem Ref Data 18:1637–1755
105. Asahi R, Morikawa T, Ohwaki T, Aoki K, Taga Y (2001) Visible-light photocatalysis in nitrogen-doped titanium oxides. Science 293:269–271

106. Barborini E, Conti AM, Kholmanov I, Piseri P, Podesta A, Milani P, Cepek C, Sakho O, Macovez R, Sancrotti M (2005) Nanostructured TiO_2 films with 2 eV optical gap. Adv Mater 17:1842–1846
107. Hu X, Zhang T, Jin Z, Zhang J, Xu W, Yan J, Zhang J, Zhang L, Wu Y (2008) Fabrication of carbon-modified TiO_2 nanotube arrays and their photocatalytic activity. Mater Lett 62:4579–4581
108. Hahn R, Ghicov A, Salonen J, Lehto V-P, Schmuki P (2007) Carbon doping of self-organized TiO_2 nanotube layers by thermal acetylene treatment. Nanotechnol 18:105604–105608
109. Raja KS, Misra M, Mahajan VK, Gandhi T, Pillai P, Mohapatra SK (2006) Photo-electrochemical hydrogen generation using band-gap modified nanotubular titanium oxide in solar light. J Power Sources 161:1450–1457
110. Shankar K, Paulose M, Mor GK, Varghese OK, Grimes CA (2005) A study on the spectral photoresponse and photoelectrochemical properties of flame-annealed titania nanotube-arrays. J Phys D Appl Phys 38:3543–3549
111. Xu C, Shaban YA, Ingler WB, Khan SUM (2007) Nanotube enhanced photoresponse of carbon modified (CM)-n-TiO2 for efficient water splitting. Sol Energy Mater Sol Cells 91:938–943
112. Mohapatra SK, Misra M, Mahajan VK, Raja KS (2007) Design of a highly efficient photoelectrolytic cell for hydrogen generation by water splitting: Application of TiO_{2-x} C_x nanotubes as a photoanode and Pt/ TiO_2 nanotubes as a cathode. J Phys Chem C 111:8677–8685
113. Mohapatra SK, Misra M, Mahajan VK, Raja KS (2007) A novel method for the synthesis of titania nanotubes using sonoelectrochemical method and its application for photoelectrochemical splitting of water. J Catal 246:362–369
114. Murphy AB (2008) Does carbon doping of TiO_2 allow water splitting in visible light? Comments on "Nanotube enhanced photoresponse of carbon modified (CM)-n- TiO_2 for efficient water splitting". Sol Energy Mater Sol Cells 92:363–367
115. Umebayashi T, Yamaki T, Itoh H, Asai K (2002) Band gap narrowing of titanium dioxide by sulfur doping. Appl Phys Lett 81:454–456
116. Yamamoto T, Yamashita F, Tanaka I, Matsubara E, Muramatsu A (2004) Electronic states of sulfur doped TiO_2 by first principles calculations. Mater Trans 45:1987–1990
117. Tian FH, Liu CB (2006) DFT Description on electronic structure and optical absorption properties of anionic S-doped anatase TiO_2. J Phys Chem B 110:17866–17871
118. Tang X, Li D (2008) Sulfur-doped highly ordered TiO_2 nanotubular arrays with visible light response. J Phys Chem C 112:5405–5409
119. Zhao W, Ma W, Chen C, Zhao J, Shuai Z (2004) Efficient degradation of toxic organic pollutants with $Ni_2O_3/TiO_{2-x}B_x$ under visible irradiation. J Am Chem Soc 126:4782–4783
120. Chen D, Yang D, Wang Q, Jiang Z (2006) Effects of boron doping on photocatalytic activity and microstructure of titanium dioxide nanoparticles. Ind Eng Chem Res 45:4110–4116
121. Lu N, Zhao H, Li J, Quan X, Chen S (2008) Characterization of boron-doped TiO_2 nanotube arrays prepared by electrochemical method and its visible light activity. Sep Purif Technol 62:668–673
122. Li J, Lu N, Quan X, Chen S, Zhao H (2008) Facile method for fabricating boron-doped TiO_2 nanotube array with enhanced photoelectrocatalytic properties. Ind Eng Chem Res 47:3804–3808
123. Fu XZ, Clark LA, Yang Q, Anderson MA (1996) Enhanced photocatalytic performance of titania-based binary metal oxides: TiO_2 /SiO_2 and TiO_2 /ZrO_2. Environ Sci Technol 30:647–653
124. Jung KY, Park SB (1999) Anatase-phase titania: preparation by embedding silica and photocatalytic activity for the decomposition of trichloroethylene. J Photochem Photobiol A Chem 127:117–122
125. Su Y, Chen S, Quan X, Zhao H, Zhang Y (2008) A silicon-doped TiO_2 nanotube arrays electrode with enhanced photoelectrocatalytic activity. Appl Sur Sci 255:2167–2172

126. Zhang HM, Quan X, Chen X, Zhao HM (2006) Fabrication and characterization of silica/titania nanotubes composite membrane with photocatalytic capability. Environ Sci Technol 40:6104–6109
127. Hirano M, Ota K, Iwata H (2004) Direct formation of anatase (TiO_2)/silica (SiO_2) composite nanoparticles with high phase stability of 1300 degrees C from acidic solution by hydrolysis under hydrothermal condition. Chem Mater 16:3725–3732
128. Chen SG, Paulose M, Ruan C, Mor GK, Varghese OK, Grimes CA (2006) Electrochemically synthesized CdS nanoparticle-modified TiO_2 nanotube-array photoelectrodes: Preparation, characterization, and application to photoelectrochemical cells. J Photochem Photobiol A Chem 177:177–184
129. Vogel R, Hoyer P, Weller H (1994) Quantum-sized PbS, CdS, Ag_2S, Sb_2S_3, and Bi_2S_3 particles as sensitizers for various nanoporous wide-bandgap semiconductors. J Phys Chem 98:3183–3188
130. Sant PA, Kamat PV (2002) Interparticle electron transfer between size-quantized CdS and TiO_2 semiconductor nanoclusters. Phys Chem Chem Phys 4:198–203
131. Yin Y, Jin Z, Hou F (2007) Enhanced solar water-splitting efficiency using core/sheath heterostructure CdS/TiO_2 nanotube arrays. Nanotechnol 18:495608 6 pp
132. Sun W-T, Yu Y, Pan H-Y, Gao X-F, Chen Q, Peng L-M (2008) CdS quantum dots sensitized TiO_2 nanotube-array photoelectrodes. J Am Chem Soc 130:1124–1125
133. Larramona G, Chone C, Jacob A, Sakakura D, Delatouche B, Pere D, Cieren X, Nagino M, Bayon R (2006) Nanostructured photovoltaic cell of the type titanium dioxide, cadmium sulfide thin coating, and copper thiocyanate showing high quantum efficiency. Chem Mater 18:1688–1696
134. Si H-Y, Sun Z-H, Zhang H-L (2008) Photoelectrochemical response from CdSe-sensitized anodic oxidation TiO_2 nanotubes. Colloids Surf A 313–314:604–607
135. Cassagneau T, Mallouk TE, Fendler JH (1998) Layer-by-layer assembly of thin film zener diodes from conducting polymers and CdSe nanoparticles. J Am Chem Soc 120:7848–7859
136. Kongkanand A, Tvrdy K, Takechi K, Kuno M, Kamat PV (2008) Quantum dot solar cells. Tuning photoresponse through size and shape control of CdSe-TiO_2 architecture. J Am Chem Soc 130:4007–4015
137. Seabold JA, Shankar K, Wilke RHT, Paulose M, Varghese OK, Grimes CA, Choi K-S (2008) Photoelectrochemical properties of heterojunction CdTe/ TiO_2 electrodes constructed using highly ordered TiO_2 nanotube arrays. Chem Mater 20:5266–5273
138. Panicker MPR, Knaster M, Kroger FA (1978) Cathodic deposition of CdTe from aqueous-electrolytes. J Electrochem Soc 125:566–572
139. Lepiller C, Lincot D (2004) New facets of CdTe electrodeposition in acidic solutions with higher tellurium concentrations. J Electrochem Soc 151:C348–C357
140. Guo YP, Deng XN (1993) Electrodeposition of CdTe thin-films and their photoelectrochemical behavior. Sol Energy Mater Sol Cells 29:115–122
141. Qi B, Kim DW, Williamson DL, Trefny JU (1996) Effects of post deposition heat-treatment on morphology and microstructure of CdTe grown by electrodeposition. J Electrochem Soc 143:517–523
142. Saraie J, Kitagawa M, Ishida M, Tanaka T (1978) Liquid-phase epitaxial-growth of CdTe in CdTe-$CdCl_2$ system. J Cryst Growth 43:13–16
143. Levi DH, Moutinho HR, Hasoon FS, Keyes BM, Ahrenkiel RK, AlJassim M, Kazmerski LL, Birkmire RW (1996) Micro through nanostructure investigations of polycrystalline CdTe: correlations with processing and electronic structures. Sol Energy Mater Sol Cells 41–42:381–393
144. Durose K, Edwards PR, Halliday DP (1999) Materials aspects of CdTe/CdS solar cells. J Cryst Growth 197:733–742
145. Beach JD, McCandless BE (2007) Materials challenges for CdTe and $CuInSe_2$ photovoltaics. MRS Bull 32:225–229

146. Kampmann A, Lincot D (1996) Photoelectrochemical study of thin film semiconductor heterostructures: junction formation processes in CdS vertical bar CdTe solar cells. J Electroanal Chem 418:73–81
147. Moutinho HR, Al-Jassim MM, Levi DH, Dippo PC, Kazmerski LL (1998) Effects of $CdCl_2$ treatment on the recrystallization and electro-optical properties of CdTe thin films. J Vac Sci Technol A 16:1251–1257
148. Elsirafy AA, Eldessouki MS, Elbasiouny MS (1987) CdTe semiconductor electrochemical-behavior – scope of application as an analytical sensor. J Electrochem Soc 134:221–226
149. Santato C, Ulmann M, Augustynski J (2001) Photoelectrochemical properties of nanostructured tungsten trioxide films. J Phys Chem B 105:936–940
150. Wang HL, Lindgren T, He J, Hagfeldt A, Lindquist SE (2000) Photolelectrochemistry of nanostructured WO_3 thin film electrodes for water oxidation: Mechanism of electron transport. J Phys Chem B 104:5686–5696
151. Erbs W, Desilvestro J, Borgarello E, Gratzel M (1984) Visible-light-induced O_2 generation from aqueous dispersions of WO_3. J Phys Chem 88:4001–4006
152. Darwent JR, Mills A (1982) Photooxidation of water sensitized by WO_3 powder. J Chem Soc-Faraday Trans 78:359–367
153. Park JH, Park OO, Kim S (2006) Photoelectrochemical water splitting at titanium dioxide nanotubes coated with tungsten trioxide. Appl Phys Lett 89:163106 3 pp
154. Nam W, Han GY (2007) Preparation and characterization of anodized Pt- TiO_2 nanotube arrays for water splitting. J Chem Eng Jpn 40:266–269
155. Thompson GE (1997) Porous anodic alumina: fabrication, characterization and applications. Thin Solid Films 297:192–201
156. Lee W-J, Alhosan M, Yohe SL, Macy NL, Smyrlz WH (2008) Synthesis of Pt/TiO_2 nanotube catalysts for cathodic oxygen reduction. J Electrochem Soc 155:B915–B920
157. Kim EY, Park JH, Han GY (2008) Design of TiO_2 nanotube array-based water-splitting reactor for hydrogen generation. J Power Sources 184:284–287
158. Gong Y, Hu CW, Liang H (2005) Research progress in synthesis and catalysis of polyoxometalates. Prog Nat Sci 15:385–394
159. Kang ZH, Wang YB, Wang EB, Lian SY, Gao L, You WS, Hu CW, Xu L (2004) Polyoxometalates nanoparticles: synthesis, characterization and carbon nanotube modification. Solid State Commun 129:559–564
160. Yamase T (2003) Photoredox chemistry of polyoxometalates as a photocatalyst. Catal Surv Asia 7:203–217
161. Gkika E, Troupis A, Hiskia A, Papaconstantinou E (2005) Photocatalytic reduction and recovery of mercury by polyoxometalates. Environ Sci Technol 39:4242–4248
162. Mylonas A, Hiskia A, Papaconstantinou E (1996) Contribution to water purification using polyoxometalates. Aromatic derivatives, chloroacetic acids. J Mol Catal A 114:191–200
163. Ozer RR, Ferry JL (2000) Kinetic probes of the mechanism of polyoxometalate-mediated photocatalytic oxidation of chlorinated organic. J Phys Chem B 104:9444–9448
164. Kim S, Park H, Choi W (2004) Comparative study of homogeneous and heterogeneous photocatalytic redox reactions: $PW_{12}O_{403}$- vs TiO_2. J Phys Chem B 108:6402–6411
165. Hiskia A, Mylonas A, Papaconstantinou E (2001) Comparison of the photoredox properties of polyoxometallates and semiconducting particles. Chem Soc Rev 30:62–69
166. Kormali P, Dimoticali D, Tsipi D, Hiskia A, Papaconstantinou E (2004) Photolytic and photocatalytic decomposition of fenitrothion by $PW_{12}O_{403}$- and TiO_2: a comparative study. Appl Catal B 48:175–183
167. Ozer RR, Ferry JL (2002) Photocatalytic oxidation of aqueous 1, 2-dichlorobenzene by polyoxometalates supported on the NaY zeolite. J Phys Chem B 106:4336–4342
168. Song XC, Yue LH, Liu B, Han G, Chen WX, Xu ZD (2003) Preparation of Fe-doped TiO_2 nanotubes with small diameter by hydrothermal method. Chin J Inorg Chem 19:899–901
169. Xie Y (2006) Photoelectrochemical reactivity of a hybrid electrode composed of polyoxophosphotungstate encapsulated in Titania nanotubes. Adv Funct Mater 16:1823–1831

170. Xie Y (2006) Photoelectrochemical reactivity of polyoxophosphotungstates embedded in titania tubules. Nanotechnol 17:3340–3346
171. Xie Y, Zhou L, Huang H (2007) Enhanced photoelectrocatalytic performance of polyoxometalate-titania nanocomposite photoanode. Appl Catal B 76:15–23
172. Bae S, Kang J, Shim E, Yoon J, Joo H (2008) Correlation of electrical and physical properties of photoanode with hydrogen evolution in enzymatic photo-electrochemical cell. J Power Sources 179:863–869
173. Yoon J, Joo H (2007) Photobiocatalytic hydrogen production in a photoelectrochemical cell. Korean J Chem Engg 24:742–748
174. Bae S, Shim E, Yoon J, Joo H (2008) Enzymatic hydrogen production by light-sensitized anodized tubular TiO_2 photoanode. Sol Energy Mater Sol Cells 92:402–409
175. Bae S, Shim E, Yoon J, Joo H (2008) Photoanodic and cathodic role of anodized tubular titania in light-sensitized enzymatic hydrogen production. J Power Sources 185:439–444
176. Wang WZ, Varghese OK, Ruan CM, Paulose M, Grimes CA (2003) Synthesis of CuO and Cu_2O crystalline nanowires using $Cu(OH)_2$ nanowire templates. J Mater Res 18:2756–2759
177. Ghijsen J, Tjeng LH, van Elp J, Eskes H, Westerink J, Sawatzky GA, Czyzyk MT (1988) Electronic structure of Cu_2O and CuO. Phys Rev B 38:11322–11330
178. Briskman RN (1992) A Study of electrodeposited cuprous-oxide photovoltaic cells. Sol Energy Mater Sol Cells 27:361–368
179. Musa AO, Akomolafe T, Carter MJ (1998) Production of cuprous oxide, a solar cell material, by thermal oxidation and a study of its physical and electrical properties. Sol Energy Mater Sol Cells 51:305–316
180. Georgieva V, Ristov M (2002) Electrodeposited cuprous oxide on indium tin oxide for solar applications. Sol Energy Mater Sol Cells 73:67–73
181. de Jongh PE, Vanmaekelbergh D, Kelly JJ (1999) Cu_2O: a catalyst for the photochemical decomposition of water? Chem Commun 12:1069–1070
182. Hardee KL, Bard AJ (1977) Semiconductor electrodes.10. Photoelectrochemical behavior of several polycrystalline metal-oxide electrodes in aqueous-solutions. J Electrochem Soc 124:215–224
183. Hara M, Kondo T, Komoda M, Ikeda S, Shinohara K, Tanaka A, Kondo JN, Domen K (1998) Cu_2O as a photocatalyst for overall water splitting under visible light irradiation. Chem Commun 3:357–358
184. Papadimitriou L, Economou NA, Trivich D (1981) Heterojunction solar-cells on cuprous-oxide. Sol Cells 3:73–80
185. Herion J, Niekisch EA, Scharl G (1980) Investigation of metal-oxide cuprous-oxide heterojunction solar-cells. Sol Energy Mater 4:101–112
186. Siripala W, Ivanovskaya A, Jaramillo TF, Baeck SH, McFarland EW (2003) A Cu_2O/ TiO_2 heterojunction thin film cathode for photoelectrocatalysis. Sol Energy Mater Sol Cells 77:229–237
187. Rakhshani AE (1986) Preparation, characteristics and photovoltaic properties of cuprous-oxide – a review. Solid-State Electron 29:7–17
188. Rai BP (1988) Cu_2O solar-cells – a review. Sol Cells 25:265–272
189. Liu J, Huang X, Li Y, Sulieman KM, He X, Sun F (2006) Self-assembled CuO monocrystalline nanoarchitectures with controlled dimensionality and morphology. Cryst Growth Des 6:1690–1696
190. Oh J, Tak Y, Lee Y (2004) Electrodeposition of Cu_2O nanowires using nanoporous alumina template. Electrochem Solid State Lett 7:C27–C30
191. Jiang XC, Herricks T, Xia YN (2002) CuO nanowires can be synthesized by heating copper substrates in air. Nano Lett 2:1333–1338
192. Mor GK, Shankar K, Paulose M, Varghese OK, Grimes OK, Grimes CA (2006) Use of highly-ordered TiO_2 nanotube arrays in dye-sensitized solar cells. Nano Lett 6:215–218

193. Shankar K, Bandara J, Paulose M, Wietasch H, Varghese OK, Mor GK, LaTempa TJ, Thelakkat M, Grimes CA (2008) Highly efficient solar cells using TiO_2 nanotube arrays sensitized with a donor-antenna dye. Nano Lett 8:1654–1659
194. Koffyberg FP, Benko FA (1982) A Photo-electrochemical determination of the position of the conduction and valence band edges of p-Type CuO. J Appl Phys 53:1173–1177
195. Sumikura S, Mori S, Shimizu S, Usami H, Suzuki E (2008) Photoelectrochemical characteristics of cells with dyed and undyed nanoporous p-type semiconductor CuO electrodes. J Photochem Photobiol A Chem 194:143–147
196. Mor GK, Varghese OK, Wilkie RHT, Sharma S, Shankar K, Latempa TJ, Choi KS, Grimes CA (2008) p-Type Cu-Ti-O nanotube arrays and their use in self-biased heterojunction photoelectrochemical diodes for hydrogen generation. Nano Lett 8:1906–1911
197. Mor GK, Varghese OK, Paulose M, Grimes CA (2005) Transparent highly ordered TiO_2 nanotube arrays via anodization of titanium thin films. Adv Funct Mater 15:1291–1296
198. Tomkiewicz M, Fay H (1979) Photoelectrolysis of water with semiconductors. Appl Phys 18:1–28
199. Choi HJ, Kang M (2007) Hydrogen production from methanol/water decomposition in a liquid photosystem using the anatase structure of Cu loaded TiO_2. Int J Hydrogen Energy 32:3841–3848
200. Nozik AJ, Memming R (1996) Physical chemistry of semiconductor-liquid interfaces. J Phys Chem 100:13061–13078
201. de Tacconi NR, Chenthamarakshan CR, Rajeshwar K, Tacconi EJ (2005) Selenium-modified titanium dioxide photochemical diode/electrolyte junctions: photocatalytic and electrochemical preparation, characterization, and model simulations. J Phys Chem B 109:11953–11960
202. Frank AJ, Honda K (1983) Oxygen and hydrogen generation from water on polymer-protected CdS photo-anodes. J Electroanal Chem 150:673–678
203. Frank AJ, Honda K (1982) Visible-light induced water cleavage and stabilization of N-type CdS to photocorrosion with surface attached polypyrrole catalyst coating. J Phys Chem 86:1933–1935
204. Honda K, Frank AJ (1984) Polymer-catalyst-modified cadmium-sulfide photochemical diodes in the photolysis of water. J Phys Chem 88:5577–5582
205. Bjorksten U, Moser J, Gratzel M (1994) Photoelectrochemical studies on nanocrystalline hematite films. Chem Mater 6:858–863
206. Morin FJ (1954) Electrical properties of alpha-Fe_2O_3. Phys Rev 93:1195–1199
207. Gardner RFG, Tanner DW, Sweett F (1963) Electrical properties of alpha ferric oxide.2. Ferric oxide of high purity. J Phys Chem Solids 24:1183–1186
208. Sato N (1998) Electrochemistry at metal and semiconductor electrodes. Elsevier, Amsterdam
209. Beermann N, Vayssieres L, Lindquist SE, Hagfeldt A (2000) Photoelectrochemical studies of oriented nanorod thin films of hematite. J Electrochem Soc 147:2456–2461
210. Duret A, Gratzel M (2005) Visible light-induced water oxidation on mesoscopic alpha-Fe_2O_3 films made by ultrasonic spray pyrolysis. J Phys Chem B 109:17184–17191
211. Cesar I, Kay A, Martinez JAG, Gratzel M (2006) Translucent thin film Fe_2O_3 photoanodes for efficient water splitting by sunlight: nanostructure-directing effect of Si-doping. J Am Chem Soc 128:4582–4583
212. Ingler WB, Khan SUM (2006) A self-driven p/n-Fe_2O_3 tandem photoelectrochemical cell for water splitting. Electrochem Solid State Lett 9:G144–G146
213. Prakasam HE, Varghese OK, Paulose M, Mor GK, Grimes CA (2006) Synthesis and photoelectrochemical properties of nanoporous iron (III) oxide by potentiostatic anodization. Nanotechnol 17:4285–4291
214. Macak JM, Tsuchiya H, Taveira L, Aldabergerova S, Schmuki P (2005) Smooth anodic TiO_2 nanotubes. Angew Chem Int Ed Engl 44:7463–7465
215. Gong DW, Yadavalli V, Paulose M, Pishko M, Grimes CA (2003) Controlled molecular release using nanoporous alumina capsules. Biomed Microdevices 5:75–80

216. Zong BY, Wu YH, Han GC, Yang BJ, Luo P, Wang L, Qiu JJ, Li KB (2005) Synthesis of iron oxide nanostructures by annealing electrodeposited Fe-based films. Chem Mater 17:1515–1520
217. Wen XG, Wang SH, Ding Y, Wang ZL, Yang SH (2005) Controlled growth of large-area, uniform, vertically aligned arrays of alpha-Fe_2O_3 nanobelts and nanowires. J Phys Chem B 109:215–220
218. Itoh K, Bockris JO (1984) Stacked thin-film photoelectrode using iron-oxide. J Appl Phys 56:874–876
219. Varghese OK, Gong DW, Paulose M, Grimes CA, Dickey EC (2003) Crystallization and high-temperature structural stability of titanium oxide nanotube arrays. J Mater Res 18:156–165
220. Gennari FC, Pasquevich DM (1998) Kinetics of the anatase rutile transformation in TiO_2 in the presence of Fe_2O_3. J Mater Sci 33:1571–1578
221. Gong D, Grimes CA, Varghese OK, Hu WC, Singh RS, Chen Z, Dickey EC (2001) Titanium oxide nanotube arrays prepared by anodic oxidation. J Mater Res 16:3331–3334
222. Shankar K, Basham JI, Allam NK, Varghese OK, Mor GK, Feng X, Paulose M, Seabold JA, Choi KS, Grimes CA (2009) A review of recent advances in the use of TiO_2 nanotube and nanowire arrays for oxidative photoelectrochemistry. J Phys Chem C 113:6327–6359

Chapter 5
Dye-Sensitized and Bulk-Heterojunctions Solar Cells: TiO_2 Nanotube Arrays as a Base Material

5.1 Introduction

A dye-sensitized solar cell (DSC) is a relatively low cost photovoltaic device using a photosensitized anode and hole transporting electrolyte [1], where charge separation occurs at the dye layer between the semiconductor and electrolyte. DSC efficiencies have reached over 11% using nanocrystalline TiO_2 films. In such DSCs, the slow percolation of electrons through the random polycrystalline network and the poor absorption of low energy photons by the available dyes are two of the major factors limiting further improvement in photoconversion efficiencies. Here, we review the progress with respect to the use of anodically grown and crystallized TiO_2 nanotube arrays as the base electron transporting material for DSC use, offering large surface areas with vectorial charge transport along the length of the nanotubes.

Polymer-based solar cells offer the prospect of lightweight, flexible photovoltaics as well as potentially low manufacturing costs [2–7]. Heeger and co-workers have reported a maximum 6.5% efficiency using tandem solar cells, with each layer processed from solution with the use of bulk heterojunction materials comprising semiconducting polymers and fullerene derivatives [8]. Inorganic–organic hybrid photovoltaic devices based on conjugated polymers combine the high electron mobility of inorganic semiconductors with the solution processing capabilities of polymers [9–11]. Inorganic n-type nanocrystals [12–17] are commonly dispersed in conjugated p-type polymer matrices to provide intimate bulk heterojunctions for exciton splitting, and to provide distinct paths for electron and hole transport. However, in this type of heterojunction device, the organic and inorganic phases form disordered interpenetrating networks, and a continuous transport path for photogenerated electrons is lacking. Consequently, both types of charge carriers are transported through the isotropic host organic material for significant periods of time. This, along with the random charge carrier paths, increases the probability of recombination resulting in a decrease of the overall photoconversion efficiency [18]. To overcome this charge transport limitation of isotropic bulk heterojunction

C.A. Grimes and G.K. Mor, *TiO_2 Nanotube Arrays*,
DOI 10.1007/978-1-4419-0068-5_5, © Springer Science+Business Media, LLC 2009

solar cells, it has been proposed that the ideal heterojunction geometry of an inorganic-organic device should consist of a vertical array of highly crystallized nanostructures, such as nanotubes or nanorods, encased in a polymer film with size and spacing of the nanostructures controlled to be on the order of the exciton diffusion length [19, 20]. Such geometry provides, respectively, direct and spatially separated charge transport channels for electron and holes. In response, many efforts have been devoted to create films with ordered one-dimensional architectures, such as TiO_2 nanopores and nanotubes [21–24], ZnO and SnO_2 nanorods [25–31], InP nanowires [32], etc. Herein, we consider the working principles of bulk heterojunction devices and the development of polymeric solar cells using TiO_2 nanotube arrays as the base material.

5.2 Dye Sensitized Solar Cells: Operating Principles

The operating principle of a DSC is shown schematically in Fig. 5.1, comprised of a mesoporous oxide layer of anatase TiO_2, the surface of which is sensitized with a monolayer of, typically, a ruthenium complex dye, for example, N719, N3, Black dye, anchored to the TiO_2 surface by a carboxylated bipyridyl ligand. The visible light absorption of these types of complexes is based on a metal-to-ligand charge transfer. The carboxylate groups are directly coordinated to the surface titanium

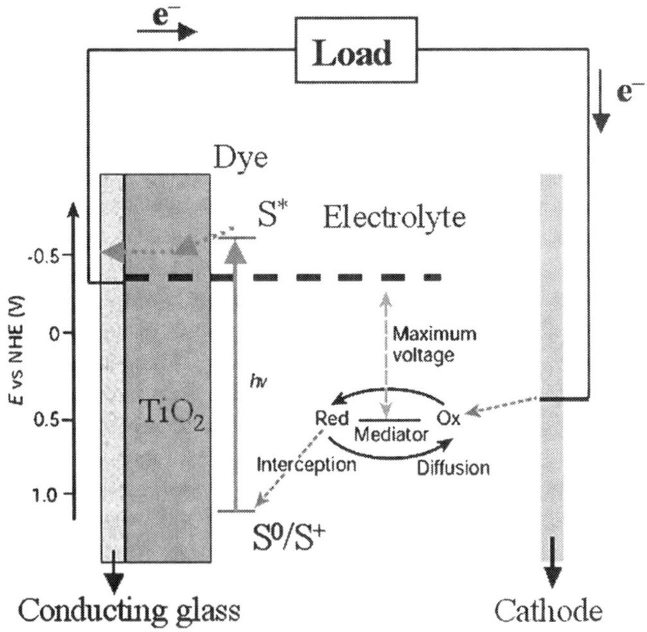

Fig. 5.1 Operation principles and energy levels of nanocrystalline dye-sensitized solar cell

ions producing intimate electronic contact between the sensitizer and the semiconductor. Photo-excited dye molecules inject electrons into the TiO_2 conduction band, and redox species such as the iodide/triiodide couple in the electrolyte reduce the oxidized dye molecules back to their original state. Dye regeneration by the iodide happens through the thiocyanate group and, thereby, intercepts the recapture of the conduction band electron by the oxidized dye. In turn, the iodide is regenerated by the reduction of triiodide at the counterelectrode, with the circuit completed via electron migration through the external load. The value of the redox potential (0.15 V vs. SCE) is the formal potential in the solution, whereas the value of the oxidation potential of the dye is the standard potential. The absorption threshold for Ru dye adsorbed on TiO_2 is about 750 nm, corresponding to a potential of 1.7 V, which added to the dye oxidation potential gives a reduction potential of −0.85 V vs. SCE. The voltage generated under illumination corresponds to the difference between the Fermi level of the electron in the solid and the redox potential of the electrolyte. About 0.4 V is lost in the reduction of the photoexcited dye by iodide, but so far no other candidates have been as efficient as the I^-/I_3^- couple when it comes to the kinetics of dye interception and electron uptake at the counter electrode.

5.2.1 Key DSC Processes

An important kinetic parameter for characterization of the charge recombination rate is the rate constant k_b, see Fig. 5.2. It is of great interest to develop sensitizer systems for which the value of the electron injection, k_{in}, is high and that of k_b low.

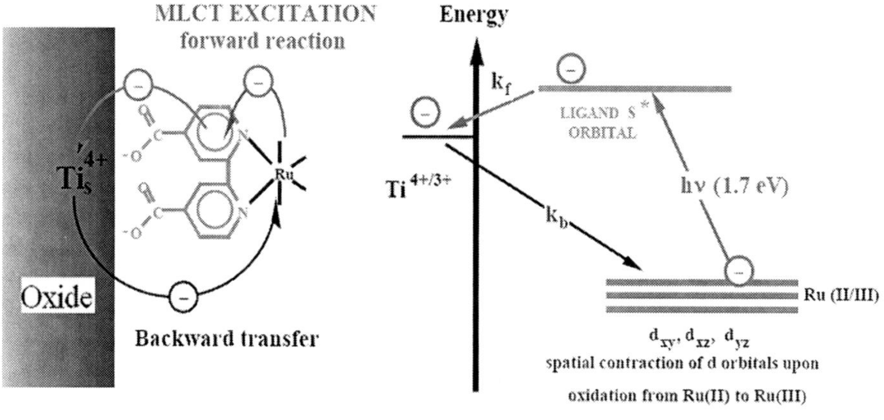

Fig. 5.2 Molecular orbital diagram for ruthenium complexes anchored to the TiO_2 surface by a carboxylated bipyridyl ligand. The visible light absorption of these types of complexes is a metal-to-ligand charge transfer (MLCT) transition [61]. Reprinted with permission from *Chemical Reviews*, 1995, 95, 49–68. Copyright 1995 American Chemical Society

Fortunately, for the transition metal complexes, the ratio of injection over recapture rate k_{in}/k_b is greater than 10^3. The carbxylate groups interact directly with the surface Ti(IV) ions resulting in good electronic coupling of the π^* wave function of the carboxylated bipyridyl ligand with the 3d orbital manifold of the conduction band of the TiO_2. As a result, electron injection from the excited sensitizer into the semiconductor is an extremely rapid process occurring on a femtosecond scale [33, 34]. By contrast, electron back-reactions with the oxidized ruthenium complex involves a *d* orbital localized on the ruthenium metal whose electronic overlap with the TiO_2 conduction band is small. The spatial contraction of the wave function upon oxidation of the Ru(II) to the Ru(III) state further reduces this electronic coupling causing relatively slow backward electron transfer which, typically, is on the order of microseconds. The oxidized dye is restored by electron donation from the organic solvent-based electrolyte containing iodide/tri-iodide redox couple. The rate constant is about 10^7 per second. As a consequence, the presence of a local electrostatic field is not required to achieve good efficiencies for the process.

The reciprocal absorption length α is given as

$$\alpha = \sigma C \tag{5.1}$$

where σ and C are the optical absorption cross-section of the sensitizer and its concentration within the mesoporous film, respectively. Values of σ for N3 and N719 are in the range of 1 to $2 \times 10^7 \text{cm}^2/\text{mol}$. The dye concentration in the film depends on the surface area of the mesoporous film with a monolayer dye coverage. The absorption length $1/\alpha$ of the N3 and N719 is about 2–5 μm [35].

5.2.2 *Factors Influencing Conversion Efficiencies*

Using the triiodide/iodide redox couple under full sunlight, an open-circuit DSC voltage of 0.7–0.9 V can be measured. Under a 1,000-fold lower intensity, the cell voltage is about 200 mV lower, a relative change of cell voltage of only 20–30%. For the conventional silicon photovoltaic cell, the cell voltage decreases by a factor of 3 for a comparable change of light intensity. Measurements in nonaqueous solutions have revealed that V_{cb}, the potential of the conduction band edge, is directly related to the proton generating ability of the solvent [36]. A distinction between protic and aprotic solvents seems, therefore, to be more appropriate than distinctions between aqueous and nonaqueous electrolytes [36]. In aprotic solvents, V_{cb} may depend on the cation, and Li^+, Na^+, and Mg^{2+} have been found to be potential determining ions [36]. Lower photovoltages have resulted from adding adsorptive cations, such as Li^+, to the electrolyte, which are presumed to charge the surface positively, causing the band edge to move downward, toward positive potentials [37].

Charge carriers in nanocrystalline films can be trapped in localized energy levels in the band gap region affecting film properties. Trapped electrons in nanocrystalline semiconductor electrodes have been observed in terms of a tail of localized energy levels below the conduction band edge [38]. The chemical nature of a localized energy level just below the conduction band edge is well established as a Ti^{3+} (3d) state [39, 40]. Trapping of electrons in the bulk of the TiO_2 particle leads to a slow photocurrent response but not to recombination losses. It will, however, lead to a lower position of the quasi-Fermi level for the electrons during illumination and, thus, to a reduced photovoltage. If electrons are trapped at the TiO_2 surface, it may be a recombination pathway, that is, it acts as an intermediate in the reduction of triiodide by the conduction band electrons. This may lead to a lowering of the photocurrent, and decrease in the photovoltage [41]; however, since the photocurrent quantum yield is practically unity in the device [1, 41], it largely affects the photovoltage. The key issue to improve the efficiencies of TiO_2 dye-sensitized solar cells is to reduce the rate for triiodide reduction by blocking the surface states below the conduction band edge.

Changing either the rate of recombination or the band-edge position of TiO_2 can strongly affect V_{OC} [42]. A substantial effort has been made in devising surface treatment strategies to suppress the recombination kinetics or to shift the band edges upward, toward negative potentials, as a means of improving the photovoltage. Several studies have reported that chemically treating the TiO_2 surface with certain organic molecules influences the photovoltage and/or the photocurrent [42–53]. Passivation or shielding of recombination centers by the organic molecules is often given as the explanation for the observed effect [44–52]. There is also evidence that adsorbed molecules improve the cell performance by inducing band-edge movement [42, 48–50, 53, 54]. Treatment of the TiO_2 surface with 4-tert-butylpyridine increases the photovoltage by about 0.3 V. It was also shown that treating the TiO_2 surface with ammonia causes a significant negative potential shift of the band edges resulting in an increased open-circuit voltage at a given photoinjected charge density. The band edge shift was attributed to the amines negatively charging the TiO_2 surface by deprotonating it [42]. Surface treatment of TiO_2 with amphiphilic molecules, consisting of a carboxylic or phosphonic moiety on one end and an alkyl group on the other end, improves V_{OC} and J_{SC}. The carboxylic or phosphonic group of the molecule is believed to bind to coordinatively unsaturated Ti species on the TiO_2 surface, leaving the hydrophobic end as a buffer between the TiO_2 and the electrolyte. The higher photovoltage and photocurrent have been attributed to the shielding of recombination centers by the hydrophobic layer created by the adsorbed molecule; coadsorbents grafted with the sensitizer include hexadecylmalonic acid [44] and decylphosphonic acid [46].

The gain in open circuit voltage that can be achieved by minimization of the dark current can be calculated from the diode equation:

$$V_{OC} = \frac{(nRT)}{F} \ln\left(\frac{I_{Ph}}{I_o}\right) \tag{5.2}$$

where n is the ideality factor whose value is between 1 and 2 for the DSC, I_{ph} is the photocurrent, and I_o is the reverse saturation current. Thus, for each order of magnitude decrease in the dark current, the V_{OC} would increase 59 mV at room temperature. Clearly, the ability to control the dark current is indispensable in the goal of raising device photoconversion efficiency.

It has been shown in other systems that surface modification can reproducibly protect semiconductor surfaces and increase overall efficiencies. This was accomplished by preventing the formation of electronic surface states lying below the conduction band edge which mediate recombination of the injected conduction band electrons by chemical modification of the surfaces [55–57]. In particular, some suppression of back electron transfer in TiO_2 and other semiconductor systems has been achieved by passivating recombination centers with a variety of pyridine derivatives [49, 58, 59]. It is difficult to directly measure the energetics of the surface state or to isolate the states that are directly participating in electron transfer with the redox couple in solution. The kinetics of electron transfer, however, can be indirectly measured by monitoring the dark current. An understanding of the effect of surface modification on the relative rates of this electron transfer process will open the door to a better understanding of the mechanism behind the recombination reaction and may ultimately result in increases in overall cell efficiency.

As a factory product, it is considered that a photovoltaic device must remain serviceable for 20 years of continuous operation in natural sunlight without noticeable loss of performance, which corresponds to $5\text{–}10 \times 10^7$ dye redox cycles. A recent 12,000 h test under continuous full intensity light exposure has confirmed inherent DSC stability [60]. Ruthenium polypyridyl complexes such as the cis-Ru $(SCN)_2L_2$ (L = 2.2′-bipyridyl-4,4′-dicarboxylate (N3) or the black dye Ru-$(SCN)_3$L (L = 4,4′,4″-tricarboxy 2,2′:6′,2″-terpyridine) and cis-di(thiocyanato)-N,N′-bis (2,2′-bipyridyl-4-carboxylic acid-4′-tetrabutylammonium carboxy late) ruthenium (II) (N-719) have shown the best performance as sensitizers, to date, to their advantageous spectral properties and high stability.

Improvement of the electron diffusion coefficient is important to further increase DSC efficiencies. A crucial parameter is the electron diffusion length

$$L_n = \sqrt{(D_n \tau_t)} \tag{5.3}$$

where D_n and τ_t are the diffusion coefficient and lifetime of the electron, respectively [61, 62]. Collection of charge carriers is achieved only if the electron diffusion length is of a magnitude comparable to or greater than the film thickness d. The thickness of the film should satisfy the following condition for full harvesting of light in the spectral absorption range of the dye and for efficient removal of the injected electron from dye into the TiO_2:

$$L_n > d > 1/\alpha$$

5.2.3 Nanocrystalline DSCs

In 2001, nanocrystalline DSCs making use of "black dye" achieved 10.4% (air mass 1.5) solar to power conversion efficiency in full sunlight [63]. Recently, this record has been extended to over 11% by using the N3 dye in conjunction with guanidinium thiocyanate, a self-assembly facilitating additive allowing to increase the open circuit voltage (V_{OC}) of the solar cell as shown in Fig. 5.3. The guanidinium cations are adsorbed along with the N3 anions at the interface, screening the lateral Coulombic repulsion of the sensitizer and facilitating in this fashion the self-assembly of a compact dye monolayer. Figure 5.4 compares the spectral response of the photocurrent observed with the two sensitizers, N3 and black dye. With the AM 1.5 solar emission, the short circuit photocurrents of the N3 and black dye-sensitized cells were found to be 16.5 and 20.5 mA/cm^2 [64]. The DSC based on N719 dye currently reaches 11.2% energy conversion efficiencies under standard reporting conditions (AM 1.5 global sunlight at 100 mW/cm^2 intensity, 298 K temperature) in liquid junction devices [65, 66]. Figure 5.5 shows the *J–V* characteristics and spectral response of liquid based DSCs sensitized with N719 and the molecular structure of N3, black dye and N719 dyes.

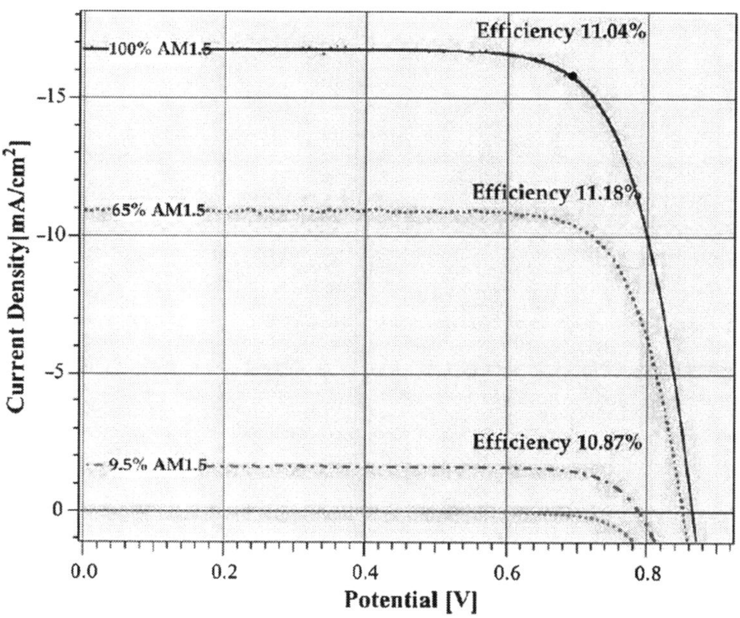

Fig. 5.3 *J–V* curve of a solar cell (N3 dye) based on an electrolyte containing guanidinium thiocyanate as self-assembly facilitating agent. The cell was equipped with an anti-reflecting coating ([64]; Copyright 2004. Reprinted with permission from Elsevier)

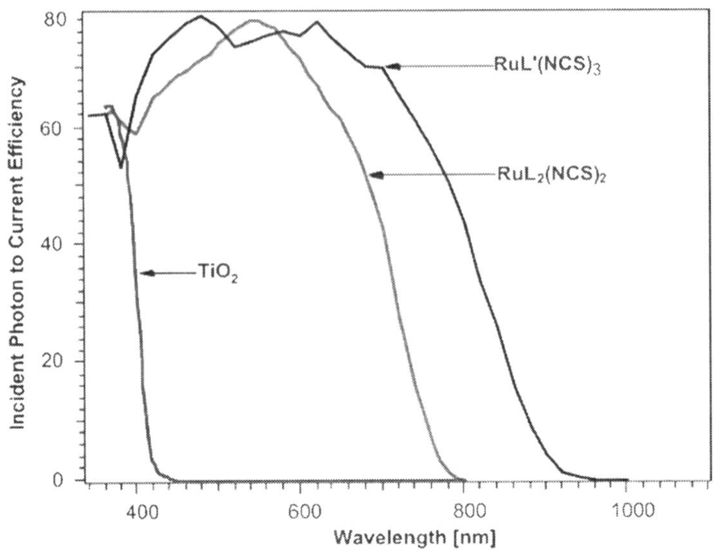

Fig. 5.4 Photocurrent action spectra obtained with the N3 (ligand L) and the black dye (ligand L′) as sensitizer. The photocurrent response of a bare TiO_2 films is also shown for comparison ([37, 64]; Copyright 2004. Reprinted with permission from Elsevier)

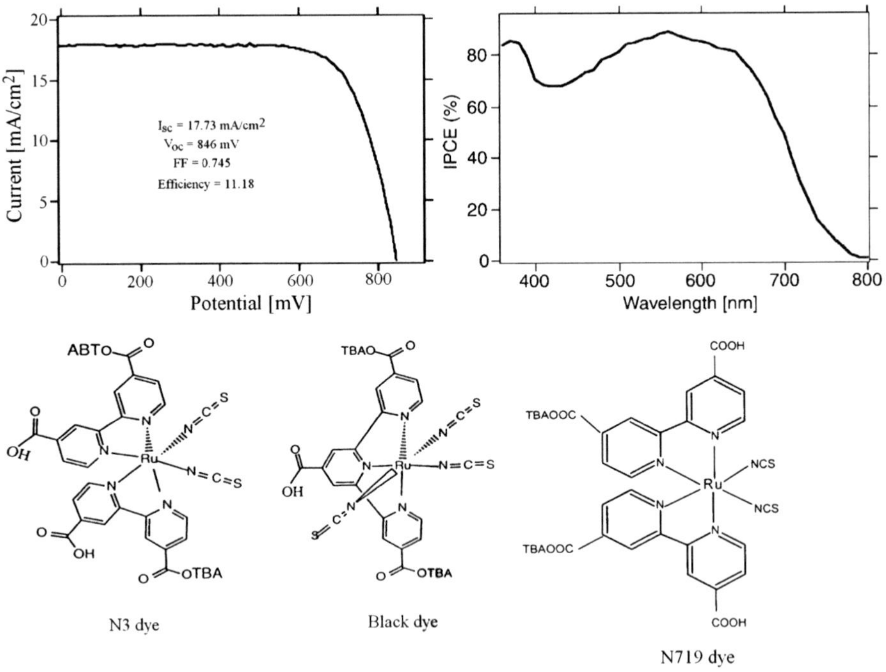

Fig. 5.5 *J–V* curve and IPCE as function of light wavelength of DSC (N719 dye) tested under AM 1.5 conditions and molecular structure of N3, black dye and N719 dye ([66]; Copyright Wiley-VCH Verlag GmbH & Co. KGaA. Reproduced with permission)

5.3 Solar Cell Parameters

Overall device photoconversion efficiency (η_{global}) is calculated from the short circuit photocurrent density (I_{sc}), the open circuit photovoltage (V_{OC}), the fill factor of the cell (*ff*), and the intensity of the incident light ($I_p = 100$ mW/cm^2)

$$\eta_{global} = \frac{I_{sc} V_{OC} FF}{I_p} \tag{5.4}$$

The fill factor is calculated as

$$FF = \frac{I_{max} V_{max}}{I_{sc} V_{OC}} \tag{5.5}$$

where I_{max}, V_{max} are the coordinates of the power point curve maximum. The fill factor indicates the deviation of the *I–V* characteristic from a square like curve and is, therefore, dependent on series and shunt resistance. Figure 5.6 illustrates the influence of increasing series resistance R_s and decreasing shunt resistance R_p on the *I–V* characteristic. To obtain high fill factors, R_s has to be as small as possible, while R_p needs to be as high as possible.

The incident photon to electron conversion efficiency, IPCE, is defined by the ratio of the number of electrons generated in the solar cell to the number of photons incident on the photoactive surface of the device at a given wavelength. This key parameter can be expressed as

$$IPCE(\lambda) = LHE(\lambda)\varphi_{inj}\eta_{cell} \tag{5.6}$$

Here, LHE(λ) is the light harvesting efficiency of photons of wavelength λ, φ_{inj} is the quantum yield for electron injection from the excited sensitizer into the conduction band of the semiconductor oxide, and η_{cell} is the elecron collection efficiency of the electrode [67]. LHE(λ) is derived from the reciprocal absorption length α via

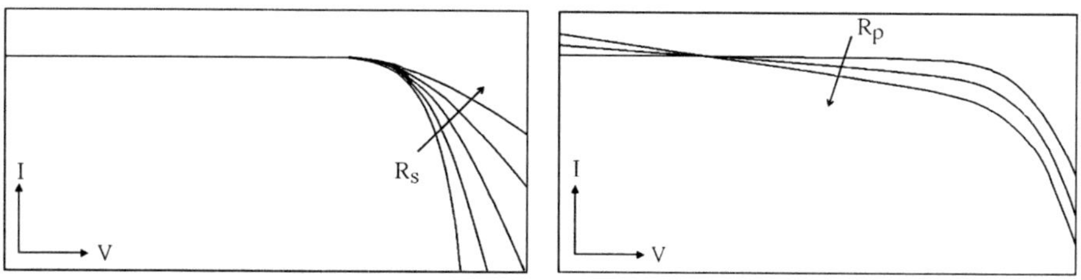

Fig. 5.6 Illustration of the influence of increasing series resistance R_s and decreasing shunt resistance R_p on the current-voltage characteristic of a solar cell

$$LHE(\lambda) = 1 - 10^{\alpha d} \tag{5.7}$$

where d is the thickness of the film and α is determined from (5.1). IPCE and Photon flux measured in terms of current $I_P(\lambda)$ can be calculated by the following two simplified expressions:

$$I_P(\lambda) = \frac{I_S(\lambda) \times P_D(\lambda)}{P_L(\lambda)} \tag{5.8}$$

$$IPCE(\%) = \frac{1240 \times I_S(\lambda)}{P_L(\lambda) \times \lambda \times 10} \tag{5.9}$$

where $I_S(\lambda)$ is sample photocurrent in $\mu A/cm^2$, $P_D(\lambda)$ is direct AM 1.5 solar irradiance in $mW/cm^2/nm$, and $P_L(\lambda)$ the light intensity in mW/cm^2 (light from xenon lamp after passing through direct AM 1.5 filter) observed at wavelength λ. $I_P(\lambda)$ units are $\mu A/cm^2/nm$, with IPCE measured in percentage. The $P_L(\lambda)$ values were obtained from NREL [36].

5.4 *J–V* Characterization Under Standard Conditions

The complete procedure for accurate determination of cell efficiencies according to standard reporting conditions (AM 1.5 Global, 1,000 W/m^2, 25°C) by JV (or IV, there seems to be a immodest lack of consistency with the symbols) characterization is as follows [68]: (1) Determination of the spectral response of the test cell and the reference cell; (2) Calculation of the spectral mismatch factor, using spectral response, lamp spectrum and standard AM 1.5 global spectrum; (3) Adjustment of the lamp output to correct for spectral mismatch; (4) Measurement of the *I–V* characteristics; and (5) Calculation of the V_{OC} I_{sc}, *FF* and efficiency.

5.4.1 *Calibrating the Solar Simulator for DSC and Polymeric Solar Cells*

Calibration errors can result from a mismatch between the spectrum of a solar simulator and that of AM 1.5 and from a mismatch between the action spectrum of the reference cell and that of the test cell. In evaluating DSCs and polymer solar cells, photocurrents under natural sunlight are measured using a crystalline Si solar cell (c-Si) capped by an IR filter (KG-3 filter for DSC and KG-5 filter for polymer solar cell, Schott) as a reference cell. The solar simulators are adjusted until currents obtained under simulator illumination match those of the reference cells obtained under outdoor sunlight. This calibration procedure can lower the spectral mismatch, since the action spectra of IR-filtered c-Si is similar to that of DSC and P3HT/PCBM based polymer solar cell.

The spectral mismatch correction factor (M) [69–74] is calculated by the expression:

$$M = \frac{\int_{\lambda_1}^{\lambda_2} E_{\mathrm{Ref}}(\lambda) S_{\mathrm{R}}(\lambda) \mathrm{d}\lambda \int_{\lambda_1}^{\lambda_2} E_{\mathrm{S}}(\lambda) S_{\mathrm{T}}(\lambda) \mathrm{d}\lambda}{\int_{\lambda_1}^{\lambda_2} E_{\mathrm{Ref}}(\lambda) S_{\mathrm{T}}(\lambda) \mathrm{d}\lambda \int_{\lambda_1}^{\lambda_2} E_{\mathrm{S}}(\lambda) S_{\mathrm{R}}(\lambda) \mathrm{d}\lambda} \tag{5.10}$$

where $S_{\mathrm{R}}(\lambda)$ and $S_{\mathrm{T}}(\lambda)$ are the spectral responses of reference cell and test cell, respectively. The spectral irradiance of the solar simulator and the global AM 1.5 illumination denoted, respectively, as $E_{\mathrm{S}}(\lambda)$ and $E_{\mathrm{Ref}}(\lambda)$. To avoid errors, the limits of integration λ_1 and λ_2 should encompass the range of the reference cell and the test-device spectral responses, and the simulator and reference spectra should encompass λ_1 and λ_2 [72]. Here, the spectral response is defined as:

$$S_{\mathrm{RorT}}(\lambda) = \frac{I(\lambda)}{P(\lambda)} \tag{5.11}$$

where "I" is the photocurrent density (A/m^2) of the reference or test cell and "P" the monochromatic light power in W/m^2. The spectral response is measured relative to the known spectral response of a reference cell.

5.4.2 Experimental Setup

Since the dye sensitized solar cells and polymeric solar cells absorb light between 350 and 750 nm, it is necessary to closely match the spectral irradiance of the light source to that portion of the global AM 1.5 spectrum. The authors employ the following technique to ensure spectral matching: Photovoltaic measurements employ an AM 1.5 solar simulator (Newport) equipped with a 500 W xenon lamp. The power of the simulated light was calibrated to 100 mW/cm^2 using a NREL calibrated and certified crystalline Si solar cell (Burdick Technologies Unlimited, LLC). Spectral irradiance of the light source was measured using UV-VIS-NIR Spectrometer (OSM2-400DUV-U, Newport) to determine the mismatch of the measured irradiance with respect to standard global AM 1.5 spectral irradiance. This was found to be less than 5% over wavelength range 350–800 nm. *J–V* measurements are done using linear sweep voltammetry (CHI 600B, CH Instruments) with the applied voltage scan at the rate of 10 mV/s. For IPCE measurements, light of a 300 W Xenon lamp (Newport), integrated with a parabolic reflector, is spectrally narrowed to monochromatic light using a motor driven monochromator (Oriel). The spectral response of the test cell as a function of wavelength is determined relative to the spectral response of a calibrated silicon reference cell (Oriel). The photocurrent measurement scan is done from low to higher wavelengths. No corrections for surface- and interface reflections losses are made for standard characterizations. Other authors report the same general method for light source calibration [63, 75–77].

5.5 Benefits of Vertically Oriented Uniformly Aligned TiO_2 Nanotube Arrays in DSCs

The working electrodes in typical DSCs consist of a 10 μm thick film comprised of a three-dimensional network of interconnected nanocrystalline semiconductor nanoparticles, typically 15 to 20 nm in diameter [1, 64]. The large surface area associated with such films enables efficient light harvesting that in turn maximizes the amount of photogenerated charge. However the structural disorder at the contact between two crystalline nanoparticles leads to enhanced scattering of free electrons, thus, reducing electron mobility [78]. From the light intensity dependence of the electron diffusion coefficient in randomly packed TiO_2 nanoparticle films, it has been inferred that transport is limited by the residence time of electrons in traps [79–83] and the morphology associated with both the particle network (e.g., distribution of the number of interparticle connections) [84–86] and the interparticle contact area [87]. Electron transport is a limiting factor in the performance of these nanoporous nanocrystalline electrodes, hindering the progress in achieving higher efficiencies.

The slow percolation of electrons through a random polycrystalline network and the poor absorption of low energy photons by available dyes are two of the major factors limiting the further improvement in the photoconversion efficiencies achievable using nanocrystalline dye-sensitized solar cells [88]. A desirable film morphology would have the mesoporous channels aligned parallel to each other, vertically oriented with respect to the TCO glass current collector. The arrangement of the highlyordered TiO_2 nanotube arrays perpendicular to the surface permits directed charge transfer along the length of the nanotubes from the solution to the conductive substrate, thereby reducing the losses incurred by charge-hopping across the nanoparticle grain boundaries [89]. Easier access to the nanotube-array surface, as well as better control of the interface makes this morphology desirable for DSCs [64, 90]. The enhancement in charge transport allows for improved light harvesting as thicker films can be used to increase the optical density, thus, improving the absorption of low energy photons in the red and infrared regions without losing the additionally harvested charge carriers to recombination [88]. As discussed in Chapter 1, TiO_2 nanotube arrays can be self-organized on Ti foil as well as on conducting FTO glass. Use of nanotube arrays on Ti foil substrates in DSCs requires backside illumination, a less optimal approach for light to electrical energy conversion because the platinized counter electrode partially reflects light, while iodine in the electrolyte absorbs photons.

5.5.1 Finite Difference Time Domain Application to DSCs

The light absorbing behavior of dye sensitized solar cells (DSCs) having photo-electrodes comprised of TiO_2 nanotube-arrays were examined using Finite-Difference Time-Domain (FDTD) [91, 92]. The numerical simulation model is comprised of nanotube-arrays on a transparent conducting glass substrate under front-side illumination. In the FDTD analysis, a transverse electromagnetic (TEM) wave is incident onto a N719 dye-coated nanotube array initially passing through the barrier layer; light that emerges from the nanotubes is reflected by a perfectly electrical conducting layer (PEC) boundary that simulates the effect of the DSC platinum counter electrode. An observation plane placed between the electromagnetic source and DSC detects the intensity of both the incident wave and the wave returning back from the DSC structure. The absorbance and transmittance spectra are determined in the wavelength range 300–700 nm as a function of nanotube-array dimensions including length, pore size, barrier layer thickness, and surface roughness while keeping the wall thickness constant at 12 nm. A significant increase in the light absorption by the dye-coated nanotubes was observed for increasing nanotube length, smaller pore sizes, and increased surface roughness. Changes in the barrier layer thickness had a negligible effect on the absorbance spectrum. Our efforts demonstrate FDTD to be a broadly applicable technique capable of guiding design of an optimal DSC architecture.

With FDTD, the space containing the object of interest is divided into two or three-dimensional grids. On the basis of the Maxwell's equations, the FDTD operator updates the electric and magnetic fields in all grids at increasing time steps, allowing determination of the electromagnetic wave in real time. We apply FDTD to study the propagation of light through dye-covered TiO_2 nanotube arrays fabricated from titanium thin films on FTO-coated glass in an attempt to find the optimum nanotube dimensions and device architecture [93] for the desired light-material interactions. Figure 5.7 shows a schematic of the front-side illuminated DSC structure with a TiO_2 nanotube array cathode. Ti films were sputtered onto FTO coated glass substrates having a sheet resistance of $8\,\Omega$ per square followed by anodization at a constant voltage in an appropriate electrolyte to yield nanotube arrays with a semitransparent metal film underneath. The films were then annealed at about 420°C in oxygen ambient for crystallization and also for oxidizing the metallic region. The resulting transparent nanotube array films were immersed in $TiCl_4$ for a day, and then annealed at 450°C for 30 min. The treated samples were dipped in a 0.3 mM solution of N719 dye ($RuL_2(NCS)_2$:2 TBA) in acetonitrile overnight. The cells were fabricated with this structure as the negative electrode and a platinized FTO coated glass as the positive electrode. The electrodes were separated using a $25\,\mu m$ thick SX1170 hot melt film. An electrolyte consisting of

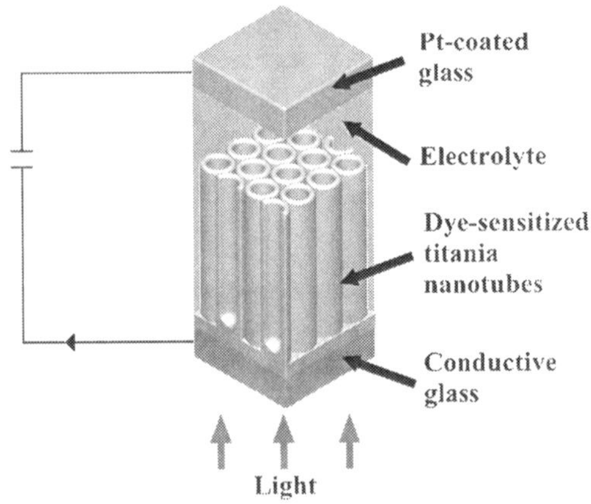

Fig. 5.7 An illustrative drawing of the TiO_2 nanotube dye-sensitized solar cell [91]

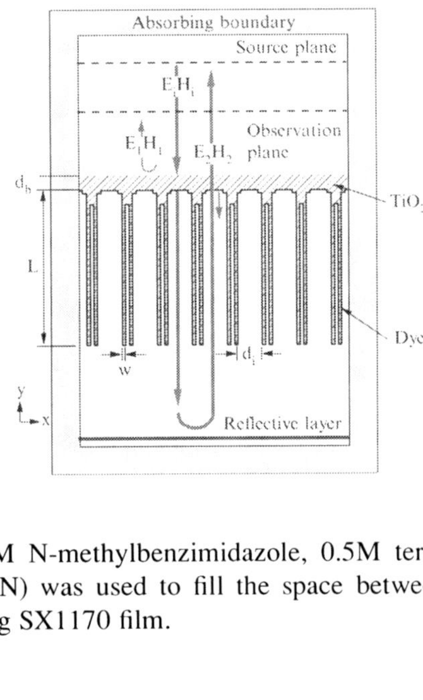

Fig. 5.8 Two dimensional FDTD model used for determining the propagation of a TEM wave through a TiO_2 nanotube DSC

0.5M LiI, 0.05M I_2, 0.6M N-methylbenzimidazole, 0.5M tert-butylpyridine in methoxypropionitrile (MPN) was used to fill the space between the electrodes. The cells were sealed using SX1170 film.

5.5.1.1 FDTD Simulations

The numerical model used in this work is shown in Fig. 5.8. The nanotube array dimensions are defined by the tube length L, inner diameter of the tube d_i, wall thickness w, and barrier layer thickness d_b. The separation distance between two adjacent tubes was kept constant at 10 nm, and wall thickness was set at 12 nm. The transparent conductive substrate supporting the nanotube array was omitted in the model (100% light transmission), while the Pt-coated positive electrode was represented by a perfect electric conductor (PEC) with 100% light reflection. To avoid complexity, we did not consider the effect of the electrolyte; light losses between the top of the nanotube array and PEC are not taken into account. In the model, the nanotubes were covered with a dye layer 2 nm thick. The FDTD space was terminated with an absorbing boundary condition made of uniaxial perfect matching layer [94] to absorb any stray fields. The grid size of the FDTD model was set to 2 nm × 2 nm, the lowest limit based on the 2 GHz processing speed of the computer used for the simulations. The excitation wave, generated at the Source Plane located above the barrier layer of the nanotube array, is a transverse electromagnetic wave propagating along the y-axis of a rectangular coordinate (TEM_y) system. The excitation source is chosen as a modulated Gaussian electric field, which is a sinusoidal wave windowed by a Gaussian pulse, as shown in Fig. 5.9. The modulated Gaussian source can be expressed in the time domain as:

$$E_i(t) = E_0 e^{\left(-(t-t_0)^2/t_w^2\right)} \sin(2\pi f_m t)\hat{x} \tag{5.12}$$

where E_0 is the field strength, t_0 the time delay before the beginning of the Gaussian pulse, t_w the Gaussian pulse width, f_m the frequency of the sinusoidal component, and $\hat{x}$ indicates the direction of the excitation wave. In this work, $t_w = 2/f_m$ and $t_0 = 1/f_m$. Note that the nature of the TEM_y wave and the choice of the excitation electric field result in the existence of only an x-directed electric field and z-directed magnetic field, that is, E_x and H_z.

Figure 5.9 shows the electric field captured at the Observation Plane. The excitation center frequency is 5.55×10^{14} Hz (wavelength = 540 nm), and the nanotube length, inner tube diameter, and barrier layer thickness are 1,000, 22, and 100 nm, respectively. As shown in the figure, the observation plane captures the excitation field E_i, as well as two packets of fields reflected from the barrier layer of the nanotubes E_1 and from the PEC layer E_2. We ignore the effect of the reflected field E_1 from the barrier layer since it forms only a small part of the excitation field due to the low refractive index of titanium dioxide, 1.5–2.5 for a dense film, and it

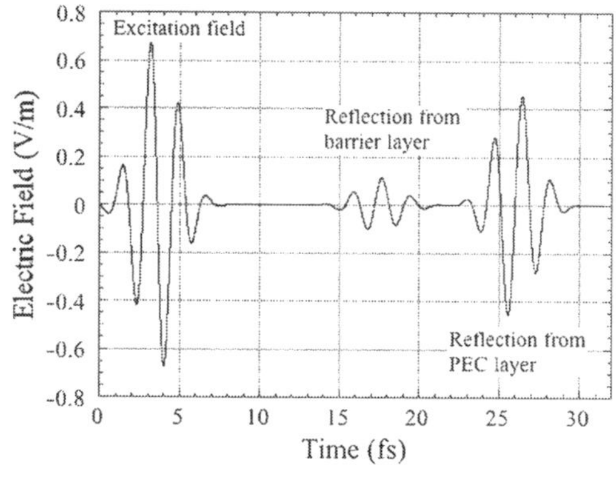

Fig. 5.9 The observation plane, see Fig. 5.8, captures the excitation electric field, as well as the electric field reflected from the barrier layer and reflected from the perfect electrical conductor (PEC) boundary [91]

does not interact with the dye-coated nanotubes. The transmitted wave propagates through the barrier layer and dye-coated nanotubes before being reflected from the PEC. This reflected wave travels back through nanotube array and barrier layer and reaches the observation plane carrying information about light-material interactions; this field (E_2) is, hence, referred to as the "transmitted field."

By performing a Fast Fourier Transformation (FFT) on E_i and E_2, as well as the corresponding H_i and H_2, the total incident and "transmitted" field, at f_m, is calculated using [95]:

$$P_i = E_i \times H_i^* \tag{5.13a}$$

$$P_t = E_2 \times H_2^* \tag{5.13b}$$

By definition, the transmittance T and absorbance A are calculated from P_i and P_t as:

$$T = P_t/P_i \times 100\% \tag{5.14a}$$

$$A = \log(P_i/P_t) \tag{5.14b}$$

The complex permittivities of TiO_2 and N719 dye vary with wavelength as shown in Fig. 5.10. As described fully in [93], the permittivity of the dye was determined by first obtaining the optical absorbance A of 19 μM N719 dye solution in ethanol as a function of wavelength. The molar absorption coefficient ε (units L/M cm) was then determined using Beer–Lambert law [96]:

$$\varepsilon = A/lC \tag{5.15}$$

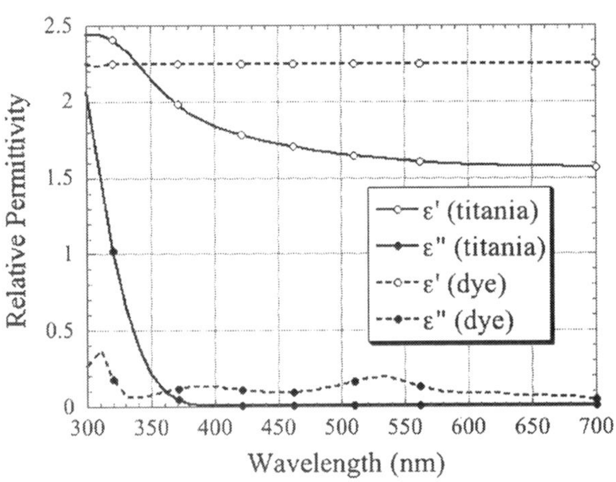

Fig. 5.10 The permittivity spectra of N719 dye and TiO_2 nanotubes [91]

where l is the path length of the light beam and C is the concentration of the dye. The absorption cross-section σ was obtained by multiplying ε with 1,000 cm^3/l [96]. Taking the area of a dye molecule as 1.65 nm^2, the number of moles of dye molecules present per cm^2 of the surface area τ was calculated. Hence, the absorption coefficient of a monolayer of dye was obtained from:

$$\alpha = \sigma\tau/l \tag{5.16}$$

where l, the thickness of the dye monolayer, is taken as 1 nm. Next, the extinction coefficient k of the dye monolayer was calculated from:

$$k = \alpha\lambda/4\pi \tag{5.17}$$

where λ is the wavelength of light. It is evident from Fig. 5.10 that in the visible region the dye has maximum absorption at wavelengths near 530 nm. We assigned a constant value of 1.5 for the dye refractive index in the wavelength range 300–700 nm used for the study. The real and imaginary parts of the complex permittivity were determined using the relations:

$$\varepsilon' = n^2 - k^2 \tag{5.18a}$$

$$\varepsilon'' = 2nk \tag{5.18b}$$

5.5.1.2 Validation of Computational Model

The FDTD simulation results were compared to experimentally measured results. Figure 5.11a plots the experimentally measured and FDTD simulated transmittance

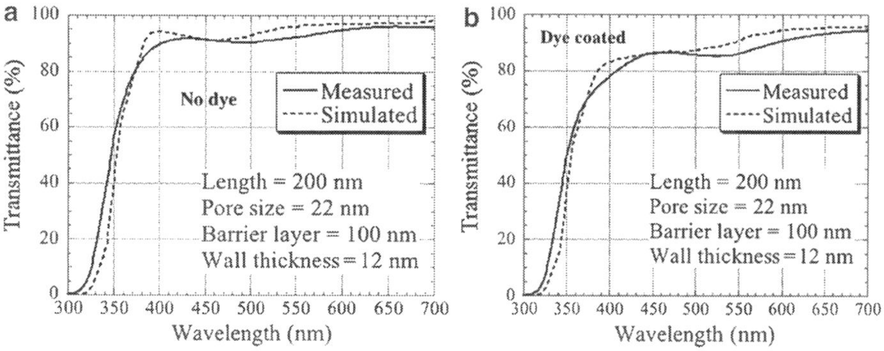

Fig. 5.11 The experimentally measured (solid line) and FDTD simulated (dashed line) transmittance of a TiO_2 nanotube array sample of 200 nm length and 22 nm pore diameter with: (**a**) no dye, and (**b**) dye layer [91]

of a TiO_2 nanotube sample with $L=200$ nm, $d_i=22$ nm, $w=13$ nm and $d_b=100$ nm without a dye layer. Experimental values were obtained using a UV-visible spectrophotometer (Cary, model 100). To match the conditions used in the spectrophotometer where light passes from one side and detected at the other side, we changed the arrangement shown in Fig. 5.8 for the FDTD simulations by replacing the PEC with the observation plane. Figure 5.11b shows the experimentally measured and FDTD simulated transmittance of an identical TiO_2 nanotube array ($L=200$ nm, $d_i=22$ nm, $w=12$ nm, $d_b=100$ nm) coated with a N719 dye layer. Figure 5.11a, b indicate that both FDTD simulated and experimentally measured results are similar with minor differences in the transmittance magnitude. These differences are believed to be largely caused by (1) errors in the determination of the permittivity values of TiO_2 nanotubes and the N719 dye, (2) the assumption that refractive index "n" of the dye takes the value 1.5 and is frequency independent and (3) the difference in various geometrical features, including value of dye monolayer thickness, as well as variation in surface and wall thicknesses.

Figure 5.12 shows the FDTD simulated propagation of a TEM derivative Gaussian wave [92] (center frequency $=8.3\times10^{14}$ Hz, bandwidth $=8.3\times10^{15}$ Hz) through the dye-sensitized TiO_2 nanotube array. The tube length, inner tube diameter, wall thickness, barrier layer thickness, and dye thickness are 200, 22, 12, 100, and 2 nm, respectively. Figure 5.12a shows the wave originating from the source and moving toward the nanotubes. When the wave reaches the barrier layer most of its energy is transmitted into the nanotubes with only a small portion reflected back. The reflected wave can be seen in Fig. 5.12b as a faint horizontal line above the barrier layer. As the wave passes outward from the nanotube array, another reflected wave, much smaller than the first reflected wave, is generated and can be seen in Fig. 5.12c. As shown in Fig. 5.12d, the main wave passes through the nanotube array reaching the reflective PEC layer. It then moves back toward the nanotube array, Fig. 5.12e–h. Small reflections can be observed during the backward journey of the wave front. However, the magnitudes of these subsequent reflections are very small and do not have a significant impact on the overall light propagation characteristic.

The light transmittance and absorbance through the TiO_2 nanotube DSC model are calculated as functions of tube length L, pore size d_i, barrier layer thickness d_b, and surface roughness, while keeping the average wall thickness at 12 nm and tube-to-tube separation 10 nm. Figure 5.13a, b plot the transmittance and absorbance, respectively, of DSCs that consist of 1,000 and 200 nm long TiO_2 nanotubes ($w=$ 12 nm, $d_b=100$ nm, $d_i=22$ nm), with and without the dye coating. As the band gap of the TiO_2 nanotube is approximately 3.2 eV, the absorption of light by the nanotube arrays dominates at lower wavelengths and, hence, the effect of dye cannot be discerned. Therefore, only wavelengths above 450 nm were considered in the simulation. The influence of the dye monolayer is evident in the case of higher length nanotubes. The 1,000 nm long nanotubes have a geometric surface area five times that of 200 nm nanotubes (with other dimensions same), and hence, it contains that much higher amount of dye leading to greater light absorption.

Figure 5.14 is plotted to show the effect of length more clearly on the transmittance, Fig. 5.14a, and absorbance, Fig. 5.14b, of dye coated samples. As the

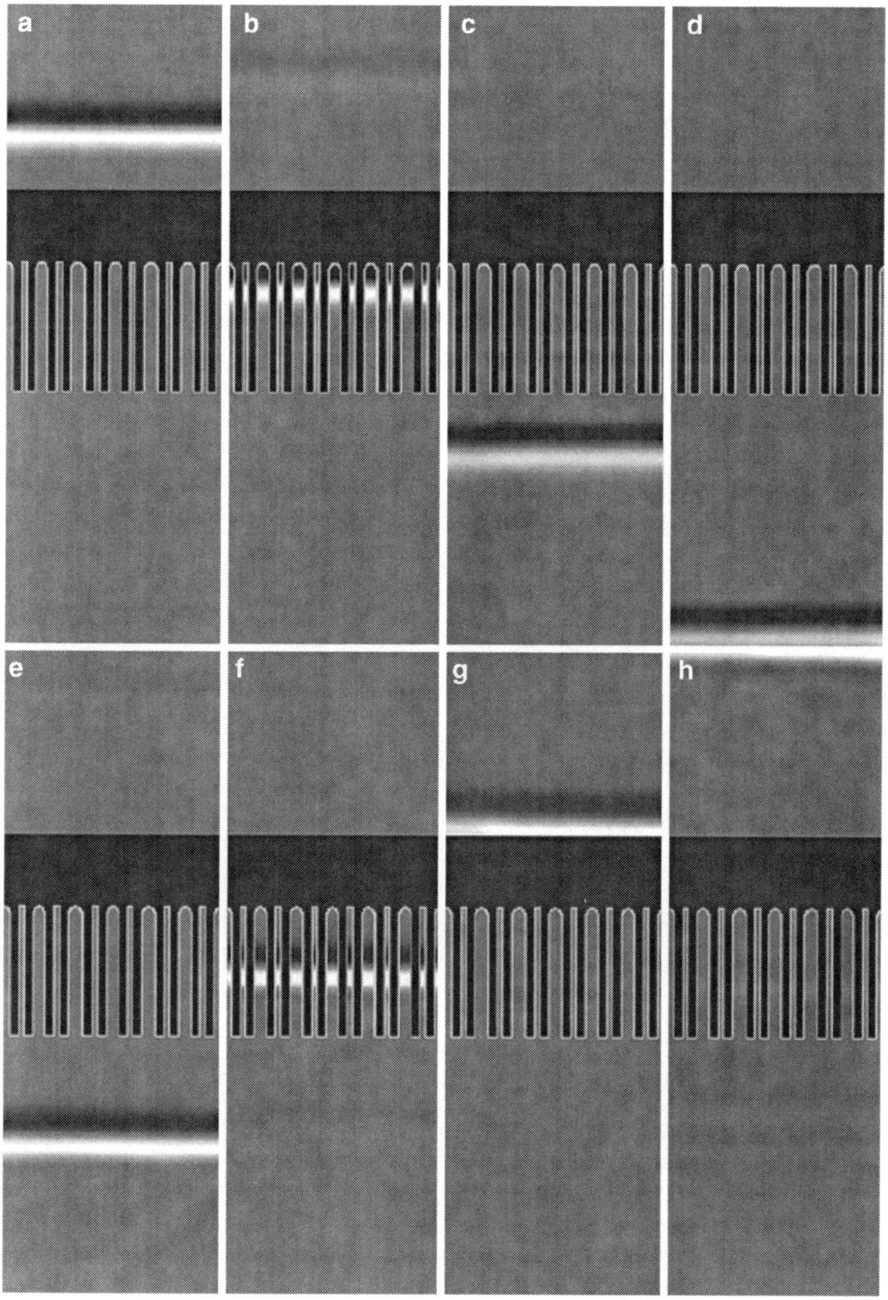

Fig. 5.12 FDTD simulated propagation of a TEM wave through the nanotube-arrays at (**a**) 1.4 fs, (**b**) 2.3 fs, (**c**) 3.3 fs, (**d**) 4.2 fs, (**e**) 5.1 fs, (**f**) 6.1 fs, (**g**) 7.0 fs, and (**h**) 7.9 fs [91]

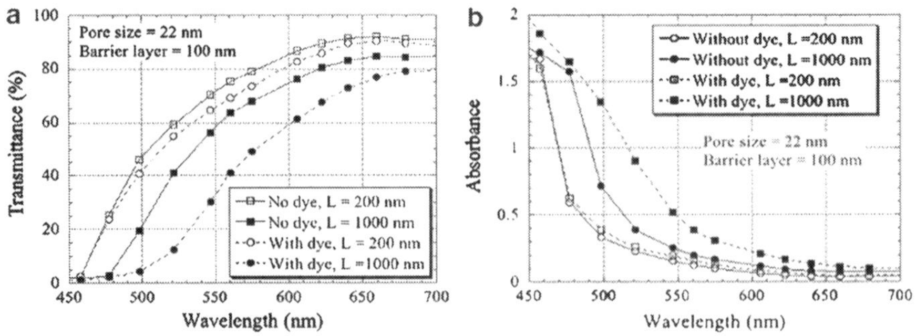

Fig. 5.13 FDTD simulated (**a**) transmittance, and (**b**) absorbance of TiO_2-based DSC of 1,000 and 200 nm in length, with and without dye coating [91]

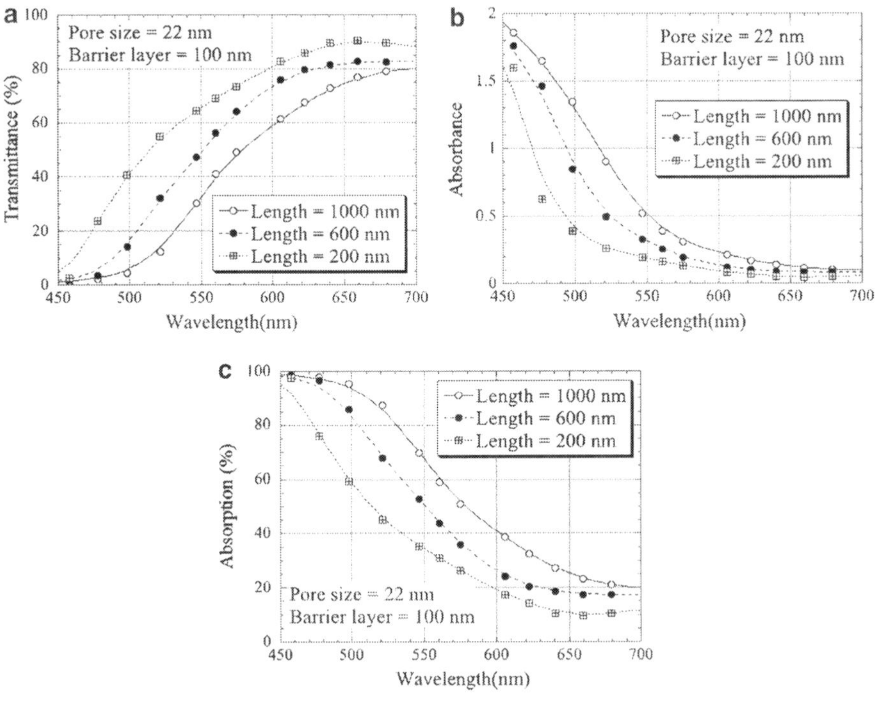

Fig. 5.14 FDTD simulated (**a**) transmittance, and (**b**) absorbance of TiO_2-based DSC as a function of tube length. The percentage absorption, 100-transmittance (%), is shown in (**c**) [91]

absorption of visible light by the dye peaks near 530 nm, Fig. 5.10, the shift in absorbance/transmittance is the largest in the wavelength range 500–550 nm, Figs. 5.13 and 5.14. Since photon induced electron-hole generation in the dye is a key mechanism responsible for the operation of dye-sensitized solar cells, higher absorption of the light by the dye is a necessary, though not sufficient, condition for better solar cell performance. Although we have limited our efforts to a nanotube length of 1000 nm due to technical limitations such as computer processing speed, Fig. 5.14 clearly demonstrates that on extending the FDTD simulations to greater nanotube array lengths it should be possible to find an optimum length for the dye-coated nanotubes for maximum light absorption. Figure 5.14c shows the percentage light absorption with nanotube array length, determined by 100 less the (%) transmittance.

Figure 5.15a, and b show respectively the light transmittance and absorbance of the dye-coated TiO_2 nanotubes at varying inner diameters, while the length of the tube, the wall thickness, and the barrier layer thickness are fixed. As shown in the figures, for a given nanotube length, wall thickness, and barrier layer thickness, the light absorption increases with smaller pore size indicating greater dye coverage for the array with narrower tubes.

Figure 5.16a and b plots, respectively, the transmittance and absorbance of nanotube arrays with two different barrier layer thicknesses. Varying the barrier layer thickness has negligible effect on the transmittance and the absorbance, as well as the initial reflection of the incident wave. This is consistent with the fact that increasing the barrier layer thickness does not increase the dye coverage. However, a small increase in absorbance, Fig. 5.16b, can be seen in the spectrum at wavelengths less than 530 nm, which we believe is due to the absorption associated with the passage of light through the thick TiO_2 barrier layer.

The effect of surface roughness on the transmittance and absorbance of the TiO_2 nanotube-array DSC was also investigated. TiO_2 nanotubes fabricated through anodization generally have fluctuations in wall, or surface, roughness the magnitude of which are dependent upon the electrolyte used and power supply stability. Dye absorption measurements show up to a factor of three increase in surface area beyond the geometrical surface area due to the surface roughness. By manipulating the electrolyte chemistry, the nanotube walls can be made smooth, Fig. 5.17a, or with periodic fluctuations in thickness, that is, "rough," as indicated by Fig. 5.17b. We note that the height of Region 2 in Fig. 5.17b is, in real nanotube arrays, actually much smaller than that of Region 1; however, a comparable size is given to this

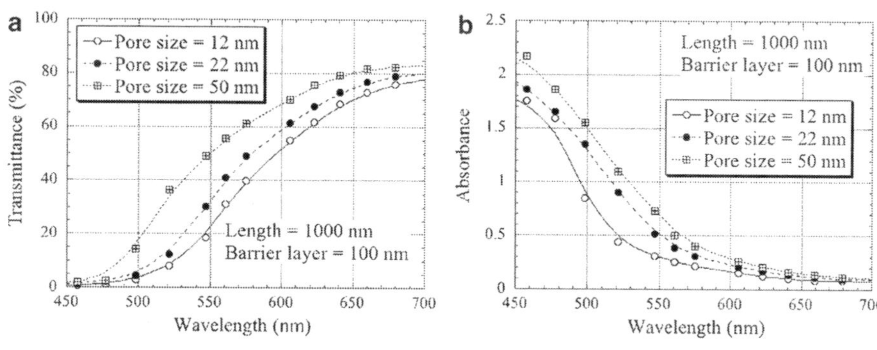

Fig. 5.15 FDTD simulated (**a**) transmittance, and (**b**) absorbance of TiO_2 -based DSC as a function of pore inner diameter (pore size) [91]

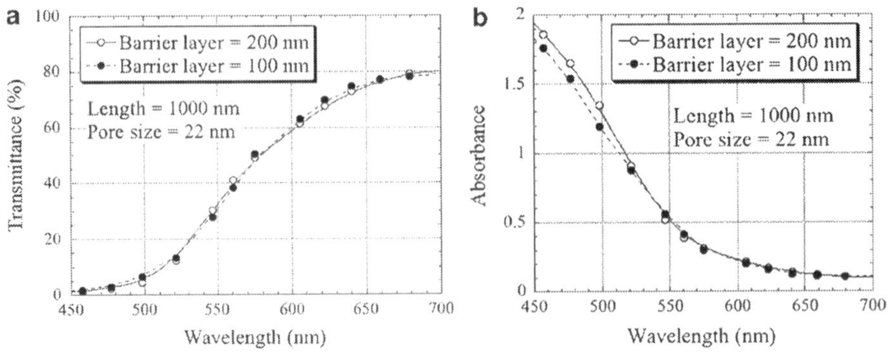

Fig. 5.16 FDTD simulated (**a**) transmittance, and (**b**) absorbance of TiO_2-based DSC as a function of barrier layer thickness [91]

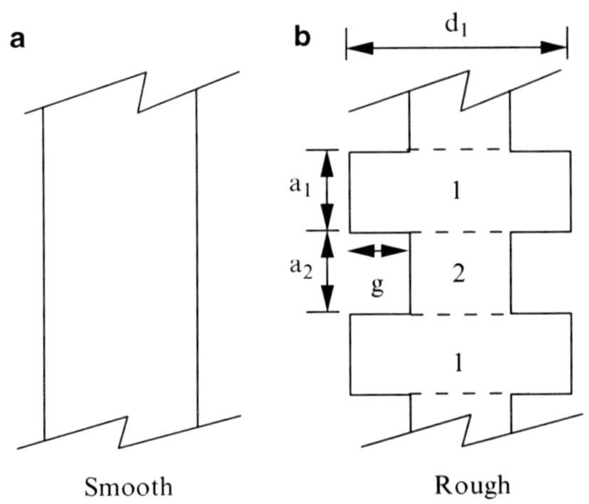

Fig. 5.17 The FDTD models for a: (**a**) "smooth" TiO_2 nanotube, and (**b**) "rough" TiO_2 nanotube [91]

region for the ease in performing the simulations. We define the surface roughness by term g that gives the difference in the thickness of the two regions.

Figure 5.18a and b plots, respectively, the change in transmittance of a cell on changing the value of surface roughness factor g from 0 (smooth) to 4 nm with $L=$ 1,000 nm, $w=22$ nm, $d_b=100$ nm. For the sample with $g=2$ nm, $a_1=a_2=8$ nm and $d_1=14$ nm; for the sample with $g=4$ nm, $a_1=12$ nm, $a_2=4$ nm and $d_1=14$ nm. Note that by adjusting a_1, a_2, and d_1, the total volume of the "rough" TiO_2 nanotubes was made identical to that of the "smooth" nanotubes regardless of the surface roughness. By calculating the dye surface in each FDTD model, it was found that increasing g from zero to 2 nm increased the dye coverage by 12%, while increasing it to 4 nm increased the dye coverage by 36%. From Fig. 5.18, it is evident that the enhancement of light absorption corresponds to the increase in dye coverage at higher surface roughness.

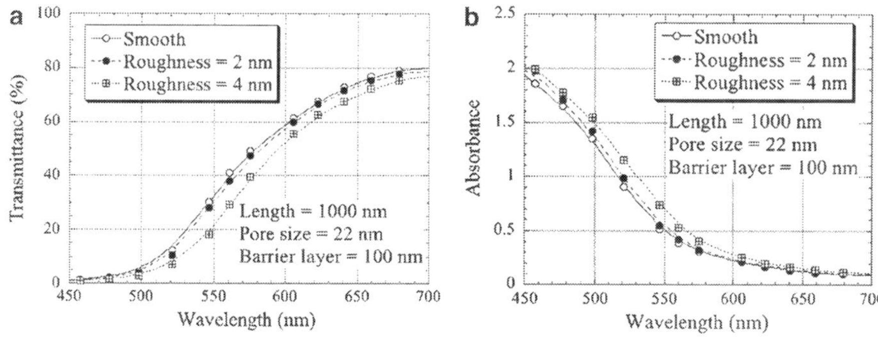

Fig. 5.18 FDTD simulated (**a**) transmittance, and (**b**) absorbance of TiO_2 -based DSC as a function of surface roughness [91]

5.6 Liquid Junction DSCs

5.6.1 Transparent TiO_2 Nanotube Arrays on FTO Coated Glass: Front Side Illumination

Figure 5.19 illustrates the DSC front sided illumination geometry in cross-section. Anodically grown TiO_2 nanotube array films 360 nm to 4.0 μm thick were formed by anodizing Ti films sputter deposited on FTO coated glass [97, 98]. Short nanotubes were synthesized at 10 V in an electrolyte of 0.5% HF+acetic acid mixed in a 7:1 ratio while the long nanotubes were made in fluoride ion containing ethylene glycol organic solvent. These nanotubes were crystallized by a 450°C anneal in oxygen for 3 h. Prior to their use, the annealed short tubes were placed in a 0.2M $TiCl_4$ solution for 60 min at room temperature to 100°C within airtight bottles, then rinsed in water and ethanol and finally annealed in air at 450°C for 30 min. Electrode spacing was ensured by the use of 25 μm thick sheets of SX1170 hot melt film (Solaronix). The nanotube samples were immersed overnight in 0.3 mM N719 (commercially available ruthenium based dye $RuL_2(NCS)_2$:2TBA, from Solaronix) in ethanol or ethanol+acetonitrile (1:1) solution. The electrolyte contained 0.5M LiI, 0.05M I_2, 0.6M N-methylbenzimidazole, 0.10M guanidinium thiocyanate, and 0.5M tert-butylpyridine in methoxypropionitrile (MPN). A conductive glass slide sputter-coated with 25 nm of Pt was used at the counter electrode in the fabricated DSCs. A drop of electrolyte was placed on dye-sensitized region before placing the counter electrode and clamping. Excess electrolyte coming out from the cell is wiped with a cotton swab before cell testing. Alternatively, the electrolyte can be introduced into the assembled DSC, by vacuum backfilling through a subsequently filled hole.

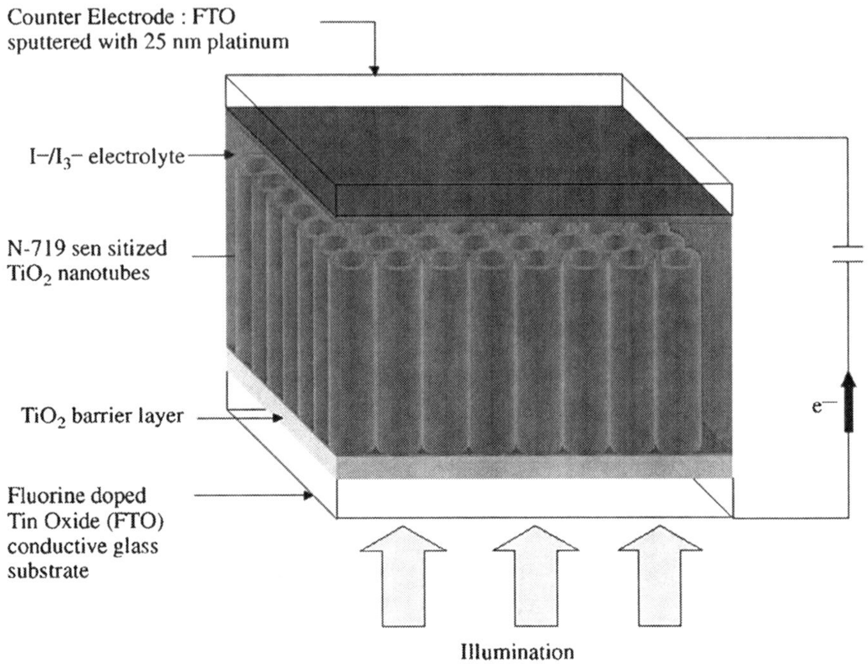

Fig. 5.19 Integration of transparent nanotube array architecture into front-side illuminated dye solar cell architecture [97]

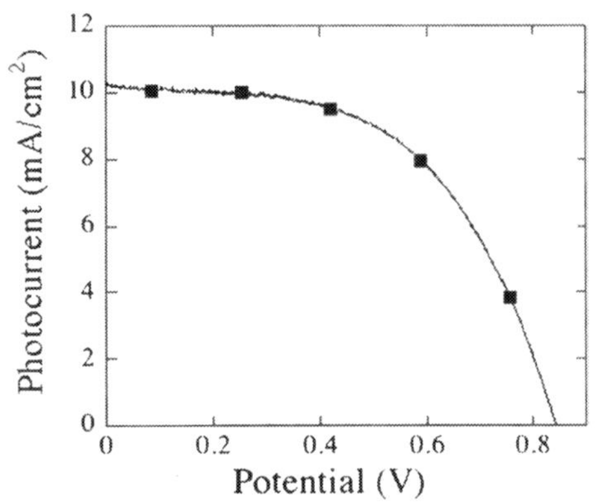

Fig. 5.20 Current–voltage characteristics of transparent, front-side illuminated 3.6 μm TiO_2 nanotube array DSC under AM 1.5 illumination when sensitized by a 0.3 mM solution of N-719 dye in acetonitrile. The active surface area is 0.4 cm^2 [97]

Figure 5.20 shows the current-voltage characteristic of an illustrative 0.4 cm^2 device under AM-1.5 illumination (150 W or 400 W Oriel Solar Simulator) [97]. At 100% sun, 3,600 nm thick nanotube array DSCs exhibit a J_{sc} of 10.3 mA/cm^2, a V_{oc} of 0.84 V and a fill factor (FF) of 0.54, with an overall conversion efficiency of 4.7%.

We note that the relatively short nanotube array films result in considerably less photoabsorption than, for example, the current DSC standard of a 10 μm thick layer of TiO_2 nanoparticles. A factor for improvement is the fill factor, which is reduced with increasing series resistance. The series resistance will be increased, and *ff* reduced, with increasing barrier layer thickness. The barrier layer thickness can be reduced using a step-wise reduction in the anodization voltage [99], followed as needed by an acid rinse to further thin the barrier layer. Yang and co-workers reported low photoconversion efficiency in $TiCl_4$ treated 680 nm long TiO_2 nanotubes/FTO glass tested under 100 mV/cm^2 (AM 1.5) illumination [100], due to the barrier layer thickness and poor annealing conditions.

Figure 5.21 shows the performance of a solar cell constructed by sensitizing a $TiCl_4$-treated 1 μm long transparent nanotube array with N-719 [101]. These nanotubes were crystallized at 400°C for 4 h in oxygen ambient and then at 450°C for 30 min, with a furnace temperature ramp rate of 1°C/min. The crystallized tubes were treated with 0.05M $TiCl_4$ solution for 15 h, then baked at 70°C for 30 min in airtight bottles. Subsequently, these samples were rinsed with water and ethanol and annealed in air at 450°C for 30 min.

Donor antenna dyes provide an exciting route to improving the efficiency of dye sensitized solar cells owing to their high molar extinction coefficients and the effective spatial separation of charges which decelerates the recombination of photogenerated charges. The donor-antenna dye cis-di(thiocyanato)(2,2′-bipyridyl-4,4′-dicarboxylic acid)-(2,2′-bipyridyl-4,4′-bis(vinyltriphenylamine) ruthenium(II) (from here on referred to as Ru-TPANCS) contains the electron rich donor triphenylamine linked to the bpy by a conjugated vinyl spacer. The molecular structures of the Ru-TPA-NCS donor-antenna dye and the commercially available N719 dye (Solaronix, Switzerland) are shown in Figs. 5.5 and 5.22, respectively.

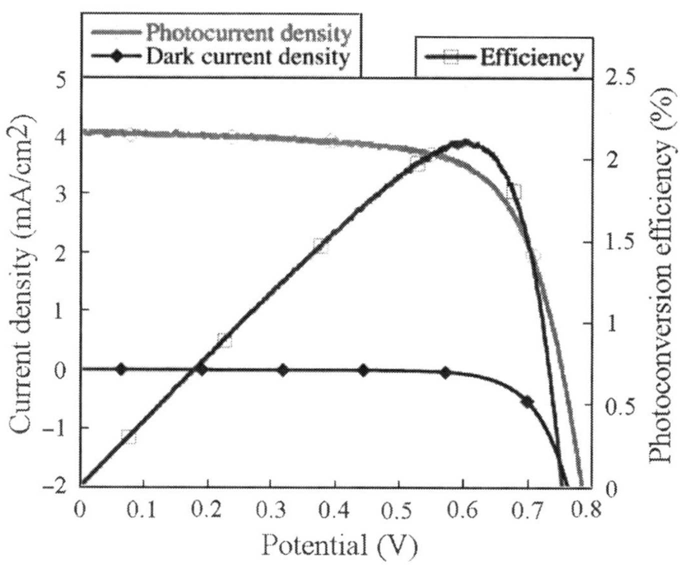

Fig. 5.21 Photocurrent-voltage characteristics of a $TiCl_4$-treated 1 μm long transparent nanotube array DSC using N-719 dye under AM 1.5 illumination [101]

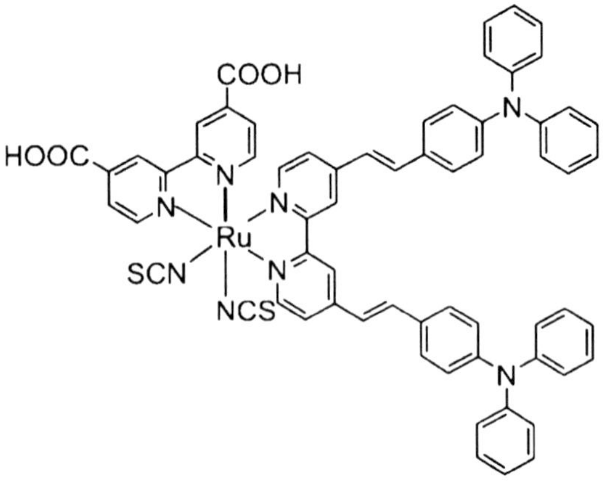

Fig. 5.22 Molecular structure of Ru-TPA-NCS [101]

Table 5.1 Absorption peaks and peak molar extinction coefficients of N719 and Ru-TPA-NCS dyes measured in solution [101]

Dye	λ_{max1}	ε_1	λ_{max2}	ε_2	λ_{max3}	ε_3
N-719	309	46400	378	11500	514	11700
Ru-TPA-NCS	307	58300	423	54500	526	24500

The extended π electron delocalization in the bpy ligand that results enables the donor-antenna dye molecules to have molar extinction coefficients more than twice that of the commonly used N719 dye [102]. Furthermore, unlike in N719, the highest occupied molecular orbitals (HOMO) of Ru-TPA-NCS are spread over the triphenyl amine moieties [103, 104]. The increased separation of the HOMO orbitals from the TiO_2 surface has been shown to retard the recombination process at the TiO_2–dye interface and at the TiO_2–hole conductor interface in solid state solar cells [105]. The procedures for the synthesis of the donor-antenna dye have been detailed elsewhere [105].

In films sensitized by molecular dyes, the thickness of the nanostructured TiO_2 film needs to be at least 10 μm to harvest the maximal amount of incident photons [96]. The absorption characteristics and molar extinction coefficient of a dye play a crucial role in determining the performance of a dye-sensitized solar cell. Table 5.1 gives the absorption peaks and the corresponding molar extinction coefficients of both dyes, Ru-TPA-NCS and N719. The unprecedented optical density values of Ru-TPA-NCS are the highest reported for (bipyridyl)Ru(II)-dyes, significantly greater that of the N719 dye. The donor-antenna dye Ru-TPA-NCS exhibits 4–5 times higher molar extinction coefficient than N719 at 400 nm [102]. Because of the much higher molar extinction coefficient of the Ru-TPA-NCS dye, a much lower film thickness should suffice to harvest a similar amount of incident photons. We employed vertically oriented TiO_2 nanotube arrays in conjunction with the Ru-TPA-NCS donor-antenna dye to fabricate liquid junction DSCs. TiO_2 nanotube array electrodes sensitized with Ru-TPA-NCS dye had significantly higher optical absorption than the N719 coated sample across the entire solar spectrum with an absorbance more than twice that of the N719 dye at 470 nm, which corresponds to the wavelength of maximum irradiance in the solar spectrum.

Various TiO_2 nanotube array samples were sensitized with N719and Ru-TPA-NCS by immersing them overnight in their respective dye solutions. The dyes were then desorbed from their respective dye-coated electrodes into methanolic KOH containing 50 vol% N,N′-dimethylformamide (DMF). From the amount of desorbed dye, the dye surface coverage was determined with the results shown in Table 5.2. The similar values of the surface coverage confirm that the increased optical density of the Ru-TPA-NCS coated samples arises from the higher extinction coefficient of the dye. Table 5.2 also indicates that the surface coverage of the dyes on 1 μm long transparent nanotube arrays is relatively poor, increasing by nearly a factor of 4 with the $TiCl_4$ treatment. Frontside illumination was employed using 1 μm long transparent nanotube arrays on fluorine-doped tin oxide (FTO) coated glass with an average pore diameter of 100 nm. The $TiCl_4$ treatment appears to improve bonding between the TiO_2 and the dye, resulting in better charge injection [106]. The $TiCl_4$ treated transparent nanotube arrays were sensitized with Ru-TPA-NCS dye and used to fabricate solar cells. With transparent nanotube arrays, the commercially available redox electrolyte MPN-100 (Solaronix) containing 100 mM of tri-iodide in methoxypropionitrile was used. A conductive glass slide sputter-coated with 80 nm of platinum was used as the counter-electrode for the transparent nanotube array-based solar cells, which were subjected to front-side illumination, where light is incident on the TiO_2 nanotube array after passing through the FTO coated glass. In Fig. 5.23, we see that a 1 μm long TiO_2 nanotube array yields a short circuit photocurrent density of 10.1 mA/cm^2, an open circuit potential of 743 mV, and a fill factor of 0.55 resulting in an overall conversion

Table 5.2 Surface coverage on TiO_2 nanotube arrays of N719 and Ru-TPA-NCS determined by dye desorption [101]

Nanotube type	N-719 (nmol cm^{-2})	Ru-TPA-NCS (nmol cm^{-2})
1 μm transparent	3.68	5.76
1 μm transparent ($TiCl_4$ treated)	17.3	20.5

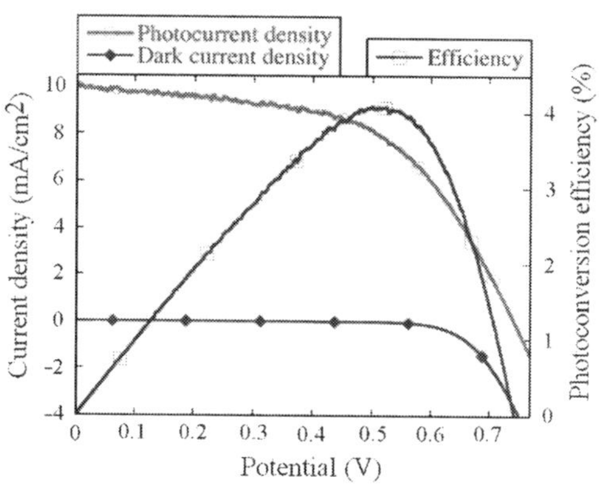

Fig. 5.23 Current-voltage characteristics of frontside illuminated solar cells with Ru-TPANCS sensitized 1 μm long $TiCl_4$ treated TiO_2 nanotube array under 1 sun AM 1.5 illumination [101]

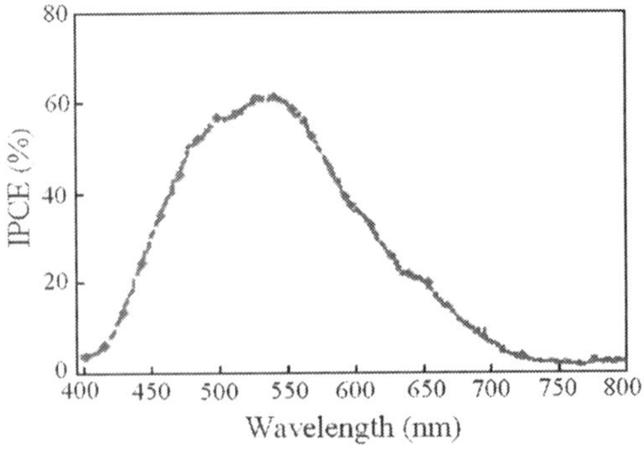

Fig. 5.24 Photocurrent action spectra of Ru-TPA-NCS sensitized front side illuminated 1 μm long transparent TiO_2 NT solar cells [101]

efficiency of 4.1%. While the use of N719 results in slightly larger values of the open circuit potential, the short circuit photocurrent density J_{sc} is nearly half that obtained with Ru-TPA-NCS sensitization because of the smaller amount of light harvested.

The action spectra of the transparent nanotube array-based solar cells are shown in Fig. 5.24; overlap integrals of the solar cell AM 1.5 global solar irradiance action spectra provided a good match to the values of short circuit photocurrent shown in Fig. 5.23. Examination of the action spectrum of the front side illuminated solar cell reveals how a 1 μm thick film of TiO_2 nanotube arrays is able to generate 10 mA/cm^2 of photocurrent under AM 1.5 illumination. At 423 nm, close to the observed IPCE maximum for this solar cell, the molar extinction coefficient of Ru-TPA-NCS is high 54500M^{-1}, more than 5 times the value for N719, thus enabling the achievement of 80% conversion of incident photons at this wavelength into electrons.

Park and co-workers demonstrated the fabrication of DSCs, of very small active areas 0.03–0.15 cm^2, using freestanding TiO_2 nanotube arrays [107]. This method of nanotube fabrication is composed of four steps: (i) growth of TiO_2 nanotube arrays on Ti foil, (ii) detachment of the nanotube arrays, (iii) transfer and fixation of nanotube arrays on FTO glass, and (iv) thermal annealing. For the first step, TiO_2 nanotube arrays with a length from 7 to 35 μm, pore size 130 nm, and wall thickness 15 nm were made by the anodization of Ti foil at 60 V in 0.25 wt% NH_4F and 2 vol% H_2O containing ethylene glycol for different durations. Separation of the nanotube array films from the metallic Ti foil was achieved by immersing the sample in 0.1M aqueous HCl for 1 h. In the next step, films were transferred onto FTO glass (or 0.04M $TiCl_4$ solution treated FTO glass), and two drops of 100 mM titanium isopropoxide were subsequently placed on the films to form interconnections between FTO glass and nanotube arrays film. The whole assembly was then annealed at 500°C for 30 min in air, with heating and cooling rates of 1°C/min. For dye sensitization, the samples were immersed in a 0.3 mM N719 ethanol solution for 4 days. The redox electrolyte used for solar cell testing was composed of 0.6M butylmethylimidazolium iodide, 0.03M I_2, 0.1M guanidinium thiocyanate and 0.5M 4-tert-butylpyridine in a mixture of acetonitrile and valeronitrile (v/v, 85:15). The photocurrent and photovoltage of the DSCs (35 μm tube length on untreated

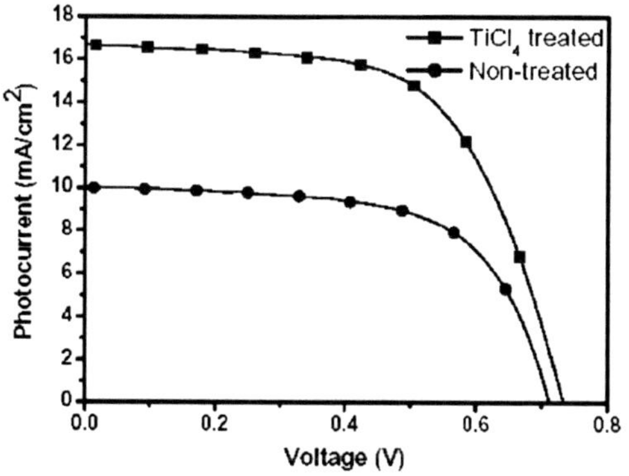

Fig. 5.25 Photocurrent–photovoltage characteristics of transparent nanotube array-based DSCs ([107]; Reproduced by permission of The Royal Society of Chemistry)

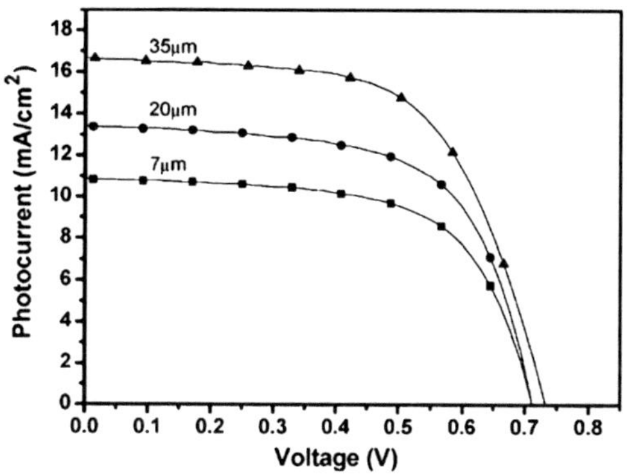

Fig. 5.26 Photocurrent–photovoltage characteristics of transparent nanotube array-based DSCs as a function of tube length ([107]; Reproduced by permission of The Royal Society of Chemistry)

and $TiCl_4$ treated FTO glass) were measured under solar simulated AM 1.5 light produced by a 1,000 W xenon lamp, see Fig. 5.25. The plot exhibits a short-circuit current density of 16.8 mA/cm^2, an open-circuit voltage of 0.73 V and a fill factor of 0.62, with an overall conversion efficiency of 7.6% for the treated substrates. Figure 5.26 compares the current-voltage properties of the TiO_2 nanotube based DSCs as a function of film thickness. While Park reports high photoconversion efficiencies, the small sample size of the DSCs must be kept in mind.

5.6.2 TiO_2 Nanotube Arrays on Ti Foil: Back Side Illumination

Figure 5.27 is an illustrative drawing of the nanotube array backside illuminated DSC structure, using nanotube arrays on Ti foil substrates.

Highly ordered TiO_2 nanotube arrays (~6 μm length, ~110 nm pore size, ~20 nm wall thickness) were grown by anodic oxidation of titanium in KF-based electrolyte (25 V, pH 5.0) and then crystallized by oxygen anneal at 580°C for 6 h. Unlike nanocrystalline electrodes where pore interconnections ensure an escape route for

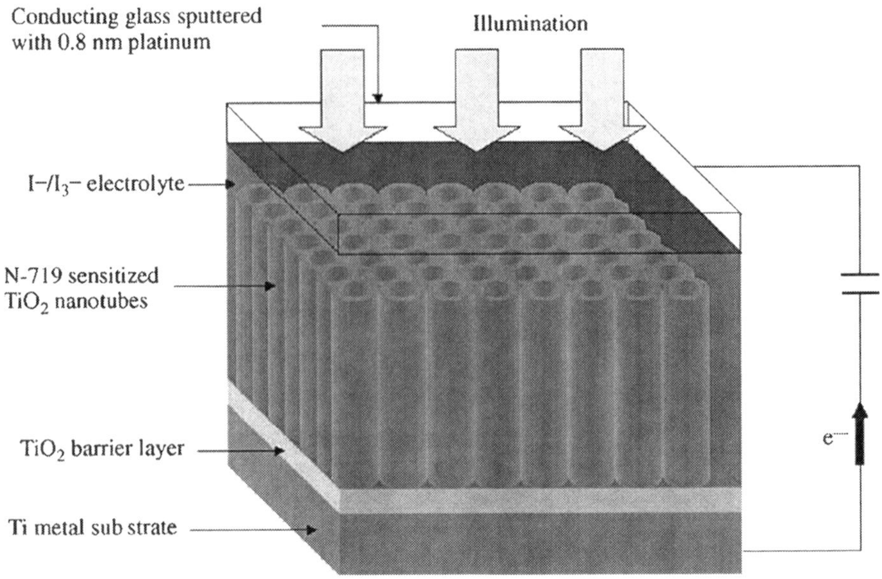

Fig. 5.27 Schematic diagram of back side illuminated nanotube array dye solar cell [108]

trapped air, the nanotube array geometry consists of nanotubes sealed at one end by the underlying barrier layer. This raises the concern that air trapped inside the nanotube may not pass out of the tube due to surface tension of the electrolyte. The nanotube samples were immersed overnight in 0.3 mM solutions of N-719 in acetonitrile. The electrolyte contained 0.5M LiI, 0.05M I_2, 0.6M N-methylbenzimidazole, 0.10M guanidinium thiocyanate, and 0.5M tert-butylpyridine in methoxypropionitrile (MPN). A conductive glass slide sputter coated with 1.3 nm of DC sputtered Pt was used at the counter electrode in the fabricated cells. The electrodes were spaced and sealed by 25 μm thick sheets of SX1170 hot melt film. The electrolyte was introduced with the help of two holes in the sealed film by capillary action, after which the holes were sealed using SX1170 film from Solaronix. The photocurrent density photovoltage performance of an illustrative device, 0.4 cm^2 active surface area, under simulated sunlight at AM 1.5 (100 mW/cm^2) is shown in Fig. 5.28 [108]. The $TiCl_4$ treated nanotube array DSC, infiltrated with N719 in acetonitrile, exhibits a J_{sc} of 15 mA/cm^2, a V_{oc} of 0.842 V and a *ff* of 0.43 to produce an overall conversion efficiency of 5.44%. Maximum power of 5.44 mW/cm^2 is delivered at $\approx I_{max} = 10.4$ mA/cm^2 and $V_{max} = 0.53$ V. Clearly, the device suffers from a low *ff*, which we believe is largely due to a relatively thick barrier layer. The annealing step used to crystallize the nanotube arrays can significantly increase the barrier layer thickness, up to ≈1 μm [109]. Compared with the use of acetonitrile, using ethanol to infiltrate the N719 resulted in a decrease of short-circuit current density and open-circuit voltage by about 25%. This is directly attributed to the differential adsorption of dye molecules [110].

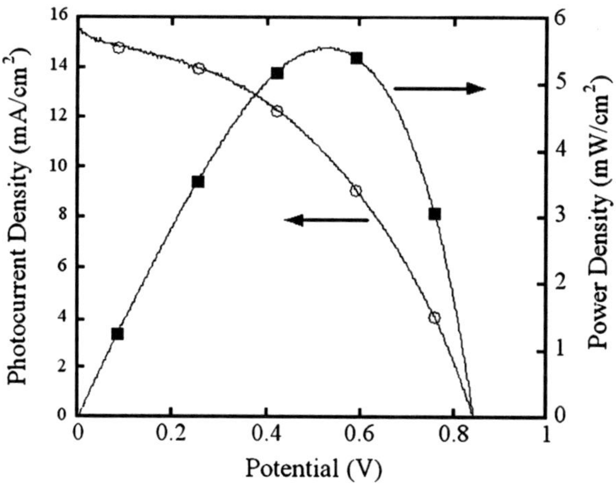

Fig. 5.28 Current–voltage characteristics of back-side illuminated 6.0μm TiO_2 nanotube array solar cell under 1 sun AM 1.5 illumination when sensitized by a 0.3 mM solution of N719 dye in acetonitrile. The active surface area is 0.4 cm^2 [108]

From dye desorption measurements, the surface coverage of dye in the back side illuminated 6 μm long nanotube samples was determined to be 50 nmol/cm^2. Light harvesting efficiency (LHE) is a measure of the fraction of light absorbed by the dye sensitized photoelectrode and is calculated using (5.7). The number of moles of sensitizer per square centimeter of projected surface area of the film was found to be 5×10^{-8}mol/cm^2. At 535 nm, the absorption maximum of the N719 dye, the dye has an absorption cross-sections equal to 1.42×10^7cm^2/mol. This results in a LHE at 535 nm of 80%, and 56% at the wavelength of half-maximum height [96].

Titania nanotube arrays were grown by anodic oxidation of titanium foil at 20 V for 24 h in an electrolyte comprised of 8 wt% t-butyl ammonium fluoride in formamide. The resulting nanotube arrays were thoroughly washed with isopropyl alcohol, then immersed in a dilute HCl (0.1M) solution for 1 h to remove any unwanted deposits and surface impurities introduced during the anodization, then washed with deionized water and dried in an oven at 100°C. The length of the resulting nanotube array was approximately 20μm. The nanotube array samples were crystallized by annealing in oxygen at 580°C for 3 h, with heating and cooling rates of 1°C/min. The crystallized nanotubes were immersed in 0.1M $TiCl_4$ solution for 24 h at room temperature, then rinsed in deionized water and annealed at 450°C for 30 min. After cooling to 80°C, the TiO_2 electrodes were immersed into N719 dye solution, 0.3 mM cis-bis(isothiocyanato)bis(2,2'-bipyridyl-4,4'-dicarboxylato)-ruthenium(II) bis-tetrabutylammonium in a mixture of acetonitrile and ethanol (v/v,1:1), and kept at room temperature for 48 h under partial vacuum. The electrolyte contained 0.6M butyl methyl imidazolium iodide (BMII), 0.01M I_2, 0.5M N-methylbenzimidazole, 0.1M guanidinium thiocynate in methoxypropionitrile (MPN). Counter electrodes were prepared by sputter coating a platinum layer of 0.6 nm thickness on FTO coated glass (TEC 15 Ω/□, 2.3 mm thick, Hartford

Glass Company). A 25 μm thick sheet of SX1170 hot melt film was used as the electrode spacer. The electrolyte was placed on the dye coated region, then the counter electrode clamped over it. Figure 5.29 shows the resulting photocurrent and power densities generated by the solar cell under AM 1.5 illumination. An overall photoconversion efficiency of 6.89% is achieved, with an open circuit voltage of 0.817 V, short circuit current density of 12.72 mA/cm^2, and fill factor of 0.663, representing a typical device performance out of perhaps 30 solar cells fabricated using nanotube array electrodes of approximately 20 μm length [111]. Current-voltage characteristics of a back side illuminated solar cell fabricated using 14.4 μm long nanotube arrays sensitized with Ru-TPANCS is shown in Fig. 5.30. The cell showed a short circuit photocurrent (J_{sc}) of 13.44 mA/cm^2, an open circuit potential (V_{oc}) of 723 mV, and a fill-factor of 0.63 resulting in an overall conversion efficiency of 6.1% [101].

Kuang et al. report the use of TiO_2 nanotube array films as the photoanode for ionic liquid electrolyte-based rigid and flexible dye-sensitized solar cells [112, 113]. Here, nanotube arrays of three different lengths 5, 7, and 14 μm were prepared at 35 V in a solution of 0.25 wt% NH_4F and 0.75 wt% H_2O in ethylene glycol. The as-prepared nanotube samples were soaked in 0.04M aqueous $TiCl_4$ solution for 30 min at 70°C, rinsed with water and ethanol and then annealed in air at 500°C

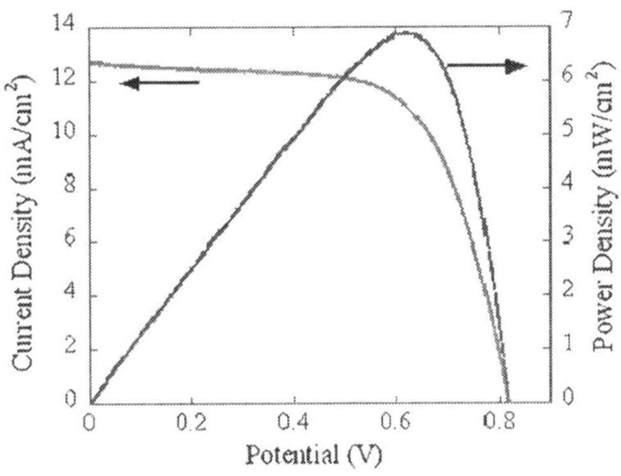

Fig. 5.29 Photocurrent and power density of backside illuminated dye sensitized solar cell under AM 1.5 illumination, fabricated using a 20 μm long nanotube array upon Ti foil [111]

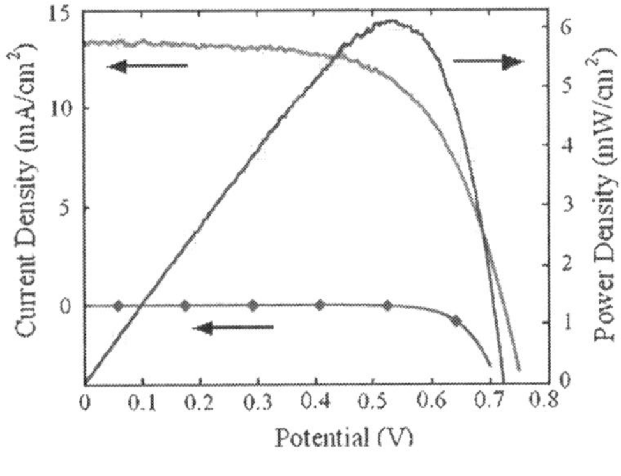

Fig. 5.30 Current-voltage characteristics of backside illuminated solar cells with Ru-TPA-NCS sensitized 14.4 μm long TiO_2 nanotube array under 1 sun AM 1.5 illumination [101]

Table 5.3 DSCs performance parameters on different TiO_2 nanotube length with Pt/FTO-Glass or Pt/ITO-PEN as the counter electrode. Cells were tested under AM 1.5 Full Sunlight Illumination (100 mW/cm^2) and the active area of the devices with a metal mask is 0.158 cm^2 ([112]; Copyright 2008 American Chemical Society. Reprinted with permission)

TiO_2 nanotube length (μm)	Counterelectrode	J_{sc} (mA cm^{-2})	V_{oc} (mV)	FF	η (%)
5	Pt/FTO-glass	4.95	763	0.666	2.52
7	Pt/FTO-glass	5.58	759	0.675	2.86
14	Pt/FTO-glass	6.11	743	0.725	3.29
14[a]	Pt/FTO-PEN	8.99	709	0.561	3.58
14	Pt/FTO-PEN	7.80	705	0.58	3.19

[a]denotes open cell

for 3 h. The sintered nanotubes were sensitized in 0.3 mM solution of a novel indoline organic dye denoted as D205 [114] in acetonitrile and tert-butyl alcohol (volume ratio 1:1) for 16 h. The binary ionic liquid electrolyte for the DSC contained 0.05M I_2, 0.5M N-butylbenzimidazole (NBB), 0.1M guanidinium thiocyanate (GuNCS) in a mixture of 65 vol% PMII and 35 vol% EMIB(CN)$_4$ [115, 116]. Pt/FTO-glass or Pt/ITO-polyethylene naphthalate were used as the counter electrode for rigid or flexible solar cells, respectively. For rigid cells, Surlyn was used as the spacer layer. In the fabrication of flexible DSCs, a drop of electrolyte was placed on top of dye sensitized TiO_2 nanotube array, and then the flexible counter electrode (Pt/ITO-PEN) was gently placed on top of the electrode. DSC performance using the different counter electrodes is shown in Table 5.3.

DSCs prepared with Pt/FTO-glass and Pt/ITO-PEN as the counter electrode were characterized using electrochemical impedance spectroscopy, in the dark at an applied potential bias of −0.7 V [117]. The reaction resistance of the counter electrode was analyzed by equivalent circuit software. The series resistance (R_s) was approximately 8.9 Ω for rigid solar cells (Pt/FTO-glass), and 49.6 Ω for flexible solar cells (Pt/ITO-PEN). Further, the interface resistance of Pt/electrolyte (R_{Pt}) was 4.6 Ω and 62.4 Ω for Pt/FTO-glass and Pt/ITO-PEN, respectively. Higher series resistance lead to lower fill factors in the flexible solar cells.

Lin et al. used Pt coated ITO/PET as counter electrodes in the assembly of flexible backside illuminated solar cells, using a redox electrolyte containing 0.5M LiI, 0.05M I_2, and 0.5M tert-butylpyridine in dry acetonitrile [118]. Their 24μm thick TiO_2 nanotube/Ti foil based DSCs exhibits V_{oc}=0.751 V, I_{sc}=11.12 mA/cm^2, FF=0.51, for an overall photoconversion efficiency of 4.3%. Wang and co-workers fabricated back side illuminated DSCs using 1 and 2μm long TiO_2 nanotubes [119]. These nanotubes were prepared at 25 V in KF based buffered solution (pH 2–4) and then annealed at 450°C for 3 h in air or ammonia. Annealed samples were soaked in N3 dye [cis-bis(isothiocyanato) bis(2,2′-bipyridyl-4,4′-dicarboxylato)-ruthenium (II)] solution (20 mg N3 in 100 ml ethanol) for 2 h at 80°C and then 15 h at room temperature. Platinum coated glass was used as the counter electrode. PMI-50 electrolyte (Solaronix) was used, with spacer layers of 25 and 60μm (SX1170-25, SX1170-60) for electrode separation. The device area was 0.785 cm^2. In their cell performance, it was observed that the short circuit current density increases with tube length due to the higher quantity of the dye adsorbed on the longer tubes leading to improved photon absorption and carrier generation. It was also observed that the spacer layer thickness significantly affects the device performance. Thicker spacer layers increase the series resistance [120] and, hence, resulted in reduced short circuit current density and reduced fill factor.

Stergiopoulos et al. reported 0.85% efficiency under AM 1.5 illumination in N719 dye sensitized 20μm TiO_2 nanotubes using a composite polymer electrolyte [121]. This electrolyte consisted of the usual I^-/I_3^- redox couple dissolved in acetonitrile, which solidifies (after evaporation of the organic solvent) by the addition of a polymer (polyethylene oxide: PEO; $MW=2\times10^6$) and a filler (TiO_2 nanoparticles, Degussa P25).

In an interesting report by Yip and co-workers, an interconnected network of TiO_2 was formed inside the TiO_2 nanotubes [122]. Here, nanotubes were formed at 30 V in anhydrous ethylene glycol containing 0.25 wt% NH_4F and were annealed at 450°C for 3 h first in air and then in ammonia. A TiO_2 network structure was formed within the tubes by the immersion of the film in a precursor solution, described below, at 80°C for 30 h. After rinsing with toluene, the sample was annealed at 400°C for 5 h in air with a 1°C/min ramp. The precursor solution prepared by mixing 1% w/v toluene solution of poly(styrene-b-ethylene oxide) (PS-b-PEO, with polydispersity 1.05 and average molecular weight of 62,500 g/mol and weight ratio of 80:20 PS:PEO) aged overnight, and a solution of 2.5% w/v titanium tetraisopropoxide (TTIP) in toluene. The PS-b-PEO and TTIP solutions were mixed in a 40:60 volume ratio to obtain a porous layer [123, 124]. This was followed by immersing the nanotube substrates into a 0.2M $TiCl_4$ solution at room temperature for 1 h. Samples were then washed in water and ethanol and finally annealed at 450°C for 30 min. These nanotubes were then sensitized with N719 solution in acetonitrile at 80°C for 15 h and then for 30 min at room temperature. The result was back side illuminated DSC photoconversion efficiencies in the range of 1.6–2.2% under AM 1.5 illumination. Yang and co-workers made nanotube array films in N,N-dimethyl formamide electrolyte containing 0.5 wt% HF and sensitized them after crystallization (anneal at 400°C for 3 h) by immersion in a 0.3 mM ethanol solution of Indoline dye D102 ($C_{37}H_{30}N_2O_3S_2$, Shanghai Green Technology Co. Ltd) for 15 h [117]. The electrolyte was comprises 0.3M LiI, 0.03M I_2, and 0.5M TBP in an acetonitrile solution. With an active area of 0.16 cm^2, under AM 1.5 solar condition, they observed a short-circuit current density of 2.49 mA/cm^2 and open circuit voltage of 0.595 V.

Xie et al. performed a systematic study on the effects of anodization parameters on the photovoltaic performance of the resulting DSCs [125]. Ti foil samples were anodized in a mixture of ethylene glycol and DI water (100:2) containing 0.3 wt% NH_4F, in the voltage range 10–60 V for 4 h. After anodization, the samples were rinsed in DI water, ultrasonicated in 0.1M HCl for 30 s and dried by nitrogen gas and then annealed at 450°C for 3 h in air. After cooling the samples to 80°C, nanotubes were dropped in 0.5 mM N3 in acetonitrile/ethanol (1:1) for 20 h. A 25 μm thick SX1170 hot melt film was used as a spacer. Platinum coated transparent glass (Dyesol) was used as a counter electrode, and a highly stable electrolyte (HSE, Dyesol) based on 3-methoxypropionitrile was used for the back side illuminated DSCs. The variation of pore diameter and tube length as a function of anodization voltages is shown in Fig. 5.31. The AM 1.5 100 mW/cm^2 photocurrent–voltage characteristics of the nanotube array based DSCs are shown in Fig. 5.32, based on nanotubes anodized at 20–60 V for 4 h (device area 0.5 cm^2). The DSC short-circuit current densities using 20 and 30 V anodized nanotubes were found to be smaller than the nanoparticle films, due to a considerably smaller surface area as confirmed by desorption experiments (20 V: 1.75×10^{-8}mol/cm^2; 30 V: 5.9×10^{-8}mol/cm^2). For a fixed anodization time of 4 h, the short-circuit current density and photoconversion efficiency increased with the anodization voltage.

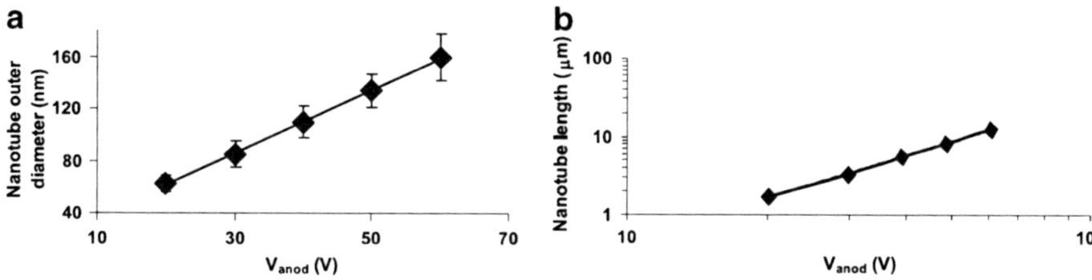

Fig 5.31 Nanotube pore diameter (**a**), and nanotube length (**b**) as a function of anodization voltage. The anodization time was fixed at 4 h. The markers are the experimental data ([125]; Copyright 2008 IOP Publishing. Reprinted with permission)

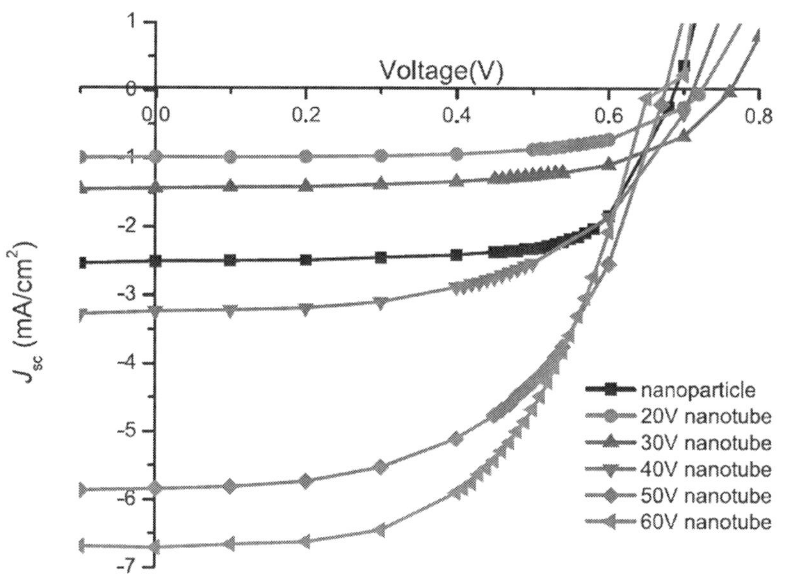

Fig. 5.32 *J–V* characteristics of the DSCs based on nanotubes anodized at different voltages for 4 h and nanoparticle paste under AM1.5, 100 mW/cm^2 illumination ([125]; Copyright 2008 IOP Publishing. Reprinted with permission)

5.6.3 Charge Collection Properties

Open-circuit photovoltage decay (OCVD) measurements were employed to investigate the charge collection properties of various DSCs. OCVD measurement consists of turning off the illumination at steady state and then monitoring the subsequent decay of the open circuit photovoltage V_{oc} [126]. In this case, the dye coated solar cells were exposed to AM 1.5 illumination for 100 s in a dark room before turning off the illumination. Transient values of the open circuit potential were measured as a function of time for 400 s after the illumination was turned off. Assuming electron recombination occurs only with the electrolyte, the potential dependent electron lifetime $\tau_n(V_{oc})$ (response time) is given by the reciprocal of the derivative of the decay curve normalized by the thermal voltage [127]:

$$\tau_n = \frac{-K_B T}{e}\left[\frac{dV_{oc}}{dt}\right]^{-1} \tag{5.19}$$

The thermal energy is given by $K_B T$, e is the positive elementary charge, and dV_{oc}/dt is the derivative of the open circuit voltage transient. Appropriate use of (5.19) assumes that recombination occurs only with the electrolyte [128]. The V_{OC} transient was recorded during relaxation from an illuminated quasi-equilibrium state to the dark equilibrium for the back-side and front-side illuminated DSCs; these data, and for comparison the response time data for TiO_2 nanoparticles (front-side illuminated) from [126], are shown in Fig. 5.33. In comparison with reported

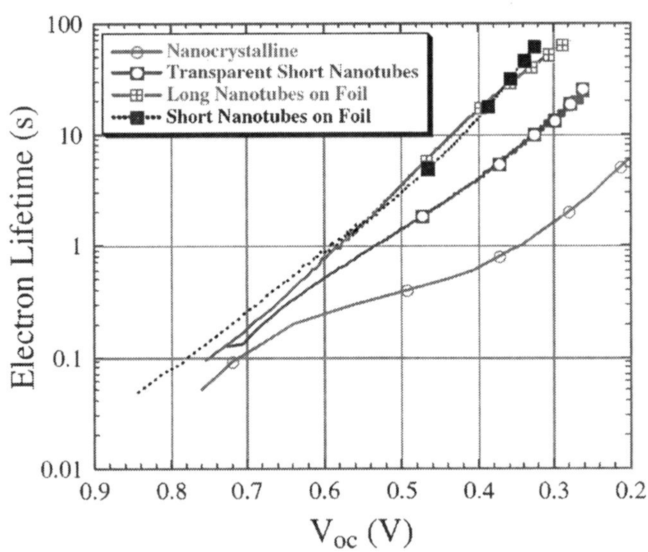

Fig. 5.33 Electron lifetimes determined by open circuit photovoltage decay measurements [97] for front and back side illuminated TiO_2 nanotube DSCs, as well as response times for TiO_2 nanoparticle DSC replotted from [126]

open circuit photovoltage decay measurements of nanoparticulate TiO_2 based DSCs [126, 128], the nanotube-arrays exhibit superior recombination characteristics. The longer lifetimes seen in the nanotube-array films indicate relatively fewer recombination centers. If the recombination rate is non-linear, its dependence on the free electron concentration may be expressed as $R=-k_r n^{\beta}$, where k_r is the recombination rate constant, and n is the free electron concentration [129]. The effective recombination order β, given by $1+\mathrm{d}\tau_n/\mathrm{d}t$ where τn is the electron lifetime, has been found to be nearly constant and is, hence, a convenient way to describe the lifetime dependence in DSCs [126, 127]. The recombination parameter β, $1+\mathrm{d}\tau_n/\mathrm{d}t$ of the nanotube-array samples, is constant at about 1.2, vs. 1.3–1.6 as reported for the nanoparticulate DSCs, indicating a smaller open circuit voltage dependence of the electron recombination rate in the nanotube-arrays.

The data of Fig. 5.33 indicate that backside illuminated nanotube arrays resting atop a Ti thick film have superior charge transfer properties than their FTO connected counterparts [97]. It was believed that this difference in charge transfer properties is due to differences in the back contacts, TiO_2 to FTO vs. TiO_2 to Ti, and crystallinity, since the transparent nanotube array films are limited to lower temperature anneals to retain the integrity of the FTO layer and, hence, are not as well crystallized as their foil counterparts [130].

Using OCVD measurements Shankar and co-workers found that the transition probability for electron transfer from Ru-TPA-NCS coated TiO_2 nanotubes to the electrolyte ionic species is lower than that for N-719 [101]. At first glance, this is somewhat counter-intuitive; since the size of the Ru-TPANCS molecule is larger than N-719, the self-assembled monolayer forms on the TiO_2 surface would be expected to be less well packed than N-179 because of the steric hindrance from the additional bulky triphenylamine groups. The resulting gaps in the Ru-TPA-NCS monolayer would provide less blocking behavior and allow easy access for the electrolyte ions to the TiO_2 surface. Yet, the work of Shankar and co-workers indicates the slope of the response time plot to be much higher for Ru-TPA-NCS with lifetimes approaching 100 s at low V_{OC} values [101]. One possibility is that the Ru-TPA-NCS forms a more closely packed layer than N-719. Alternatively, we surmise that (a) even though the Ru-TPA-NCS layer is less closely packed, the nonpolar TPA end groups may provide similar or better blocking action against electrolyte ionic species, and (b) the consequent larger spatial separation between the TiO_2 surface and the electrolyte ions due the larger size of the donor antenna molecule results in a lower transition probability. Additionally, dipole effects at the interface caused by the donor antenna dye may also play a role here in determining the charge transfer and recombination dynamics.

5.6.4 Electron Transport and Recombination Properties

Frank and co-workers studied the dynamics of electron transport and recombination in dye-sensitized solar cells incorporating vertically oriented TiO_2 nanotube arrays using frequency-resolved modulated photocurrent/photovoltage spectroscopies [131]. The transport and recombination properties of the nanotube and nanoparticle films used in the DSCs were determined and compared. While both morphologies had comparable transport times, recombination was slower in the nanotube films implying significantly higher charge-collection efficiencies in nanotube based DSCs than their nanoparticle based counterparts [131].

In the work of Frank [131], TiO_2 nanotube arrays 1.9 to 5.7 μm in length were prepared by anodizing Ti foils in 0.5 wt% NH_4F in glycerol [132] at 20 V for 16–70 h. Nanotubes were crystallized by annealing in air at a relatively low temperature of 400°C for 3 h. Dye sensitization of nanotubes with N719 were performed in the same method as for nanoparticle films, described in [133]; the amount of adsorbed N719 was measured by optical absorption of the desorbed dye. Anatase TiO_2 nanoparticle films were made on top of Ti foils via doctor blading, and the resulting films sintered and stained using the same procedure described for the nanotubular films. The average particle size in the resulting film was 24 nm with film thicknesses ranging from 2.5 to 7.3 μm, and a porosity of 63% [133]. The redox electrolyte contained 0.8M 1-hexyl-2,3-dimethylimidazolium iodide and 50 mM iodine in methoxypropionitrile. Transport and recombination properties were measured by intensity modulated photocurrent spectroscopy (IMPS) and intensity modulated photovoltage spectroscopy (IMVS) as described in [42, 134]. For these measurements, the back side illuminated DSCs were probed with a modulated beam of 680 nm light superimposed on a relatively large background (bias) illumination also at 680 nm. The probe and bias light entered the cell from the counter electrode side [133].

The surface roughness factor rf (μm^{-1}) and film porosity P of a hexagonal nanotube array can be determined from nanotube parameters including pore diameter d_p, wall thickness w and center to center nanotube distance l using the following relation [111]

$$rf = \frac{2(1-P)}{w} = \frac{4\pi}{\sqrt{3}} \frac{d_p + w}{l^2} \tag{5.20}$$

$$P = 1 - \frac{2\pi}{\sqrt{3}} \frac{d_p w + w^2}{l^2} \tag{5.21}$$

Using nanotube dimensions determined from FESEM images, Frank estimated a roughness factor of 94.5 μm^{-1}, a porosity of 62%, and a pore density of 3.7×10^{10} pores/cm^2 [131]. Nanoparticle and nanotube films were found to have nearly same roughness factor and dye loading, suggesting that dye coverage occurs on both the interior and exterior nanotube walls.

Figure 5.34 shows typical IMPS and IMVS response curves for the TiO_2 nanotube arrays, which display a semicircle in the complex plane, similar to that observed for nanoparticle cells [62, 83, 135]. The collection time can be estimated from the expression $\tau_c = 1/2\pi f_c$, where f_c is the characteristic frequency minimum of the IMPS imaginary component in Fig. 5.34a. Similarly, the recombination time can be determined from $\tau_r = 1/2\pi f_r$, where f_r is the characteristic frequency minimum of the IMVS imaginary component in Fig. 5.34b.

Figure 5.35 compares transport and recombination times for nanotube and nanoparticle based DSCs as a function of the incident photon flux (light intensity) I_0. The thicknesses of the nanotube and nanoparticle films (4.3 and 4.2 μm, respectively) in the cells were about the same. For both film morphologies, τ_c, Fig. 5.35a, exhibits a power-law dependence on the light intensity. It was found surprising that the transport times τ_c for both film morphologies are comparable [131] instead of expecting that transport would have been much faster in oriented nanotube films than in the randomly packed nanoparticle films for a given film thickness, albeit that the nanotube arrays are not well crystallized. Because the transport times τ_c and film thicknesses L for the nanotube and nanoparticle films are similar, it follows that the electron diffusion coefficients D for both morphologies are also similar as determined from the expression: $D = L^2/2.35\tau_c$ [84]. Figure 5.35b

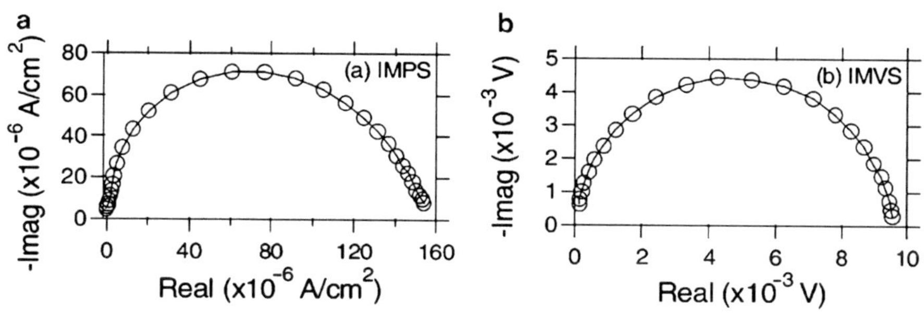

Fig. 5.34 Typical (**a**) IMPS and (**b**) IMVS responses in the complex plane for nanotube based DSCs ([131]; Copyright 2007 American Chemical Society)

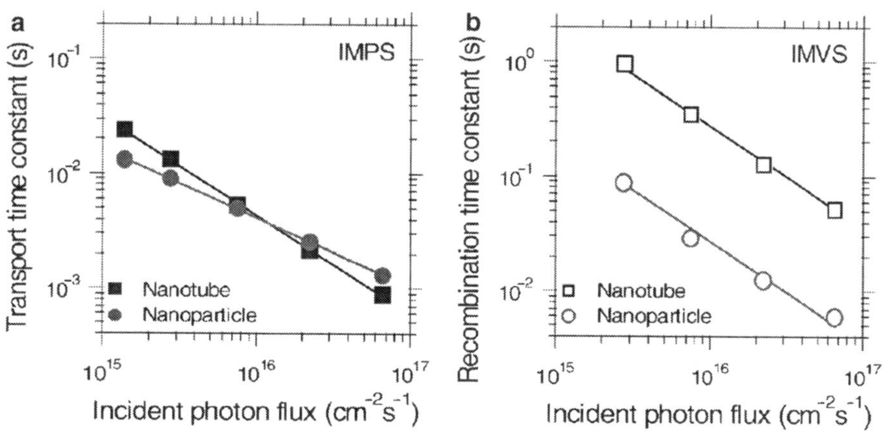

Fig. 5.35 Comparison of (**a**) transport and (**b**) recombination time constants for nanotube and nanoparticle based DSCs as a function of the incident photon flux for 680 nm laser illumination [131]. Reprinted with permission from *Nano Letters*, 2007, 7, 69–74. Copyright 2007 American Chemical Society

shows that the recombination time constants τ_r for the nanotube arrays are an order of magnitude larger than those for the nanoparticle films over the light-intensity range investigated [131]. The slower recombination could indicate that fewer potential surface recombination sites exist in nanotube arrays than in nanoparticle films. Given that recombination in the nanotube films is 10 times slower than that in the nanoparticle films and that the transport times for both film morphologies are about the same, the charge collection efficiency η_{cc} for nanotube electrodes, where [83] $\eta_{cc} = 1 - (\tau_c/\tau_r)$, must be significantly greater. A calculation of the charge-collection efficiency at the highest light intensity (6.6×10^{16} per cm^{2}s) in Fig. 5.35 shows that η_{cc} for the nanotube films is 25% larger than that for the nanoparticle films, which implies nanotube films can be much thicker than nanoparticle films for a given recombination loss. Therefore, nanotube films can be made thicker than their nanoparticle films, which would allow for a higher light-harvesting efficiency, especially at the long-wavelength end of the visible and in the near-infrared region.

Recombination times τ_r shows little dependence on the film thickness from 1.9 to 5.7 μm, which indicates that photocarriers recombine uniformly throughout film and, consequently, that recombination losses at the TCO substrate are negligible. This latter result was not surprising given that anodization is known to generate a thin TiO_2 barrier layer [132, 136] shielding the Ti substrate from direct contact with the electrolyte. On the other hand, transport time constant τ_c unexpectedly showed a minimal dependence on film thickness [131].

Figure 5.36 shows the short circuit photocurrent density J_{sc} dependences of the nanotube and nanoparticle films on film thickness at 680 nm illumination at fixed photon flux I_0 of 6.6×10^{16} per cm^{2}s. The photocurrent density was seen to increase linearly with the thickness of both types of films. However, the rate of increase of J_{sc} with L is larger for the nanotube films than for the nanoparticle films. In general, J_{sc} can be approximated by the expression

$$J_{sc} = e\eta_{lh}\eta_{inj}\eta_{cc}I_o \tag{5.22}$$

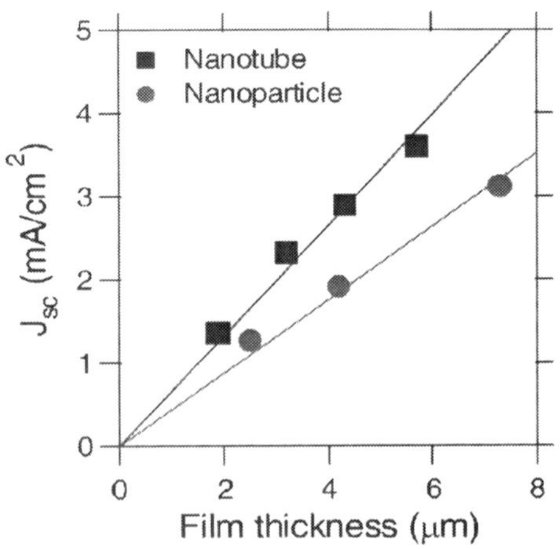

Fig. 5.36 Comparison of the short-circuit photocurrent densities of dye-sensitized nanotube and nanoparticle cells as a function of film thickness. The cells were illuminated by a 680 nm laser with an incident photon flux of 6.6×10^{16} per cm^{2}s. Lines represent linear fits of data ([131]; Copyright 2007 American Chemical Society. Reprinted with permission)

where e is the elementary charge, η_{lh} is the light-harvesting efficiency of a cell, and η_{inj} is the charge-injection efficiency. Of the three parameters, η_{lh} is determined by the amount of adsorbed dye, which is proportional to the film thickness for weakly absorbed light (as is the case for 680 nm illumination), the light-scattering properties of the film, the concentration of redox species, and other factors [43]; η_{cc} is largely determined by the competition between recombination and charge collection [83]; η_{inj} values of adsorbed N719 on the TiO_2 nanotubes and nanoparticles were assumed to be the same. The slopes of the plots of J_{sc} vs. film thickness are primarily determined by η_{cc} and the light scattering properties of the films. The ratio of slopes of the plots for the respective nanotube and nanoparticle films is about 1.5, which implies that the product of $\eta_{lh}\eta_{cc}$ for the nanotube films is about 50% larger than that for the nanoparticle films. From the respective 50% and 25% values of $\eta_{lh}\eta_{cc}$ and η_{cc}, η_{lh} of nanotube films was estimated to be at least 20% larger than that of nanoparticle films at short circuit. Considering the comparable dye coverage and redox electrolyte composition, the increased η_{lh} was attributed to enhanced internal light scattering in the nanotube films with respect to that in the nanoparticle films.

Table 5.4 compares the photocurrent density-voltage properties of nanotube and nanoparticle based DSCs as a function of film thicknesses under simulated AM 1.5 light [131]. At comparable film thicknesses, the photocurrent densities of the nanotube based DSCs are larger than those of nanoparticle based DSCs while there are essentially no substantial differences in the photovoltages. On the basis of the longer photoelectron lifetime τ_r (Fig. 5.35b) and the greater J_{sc} (Table 5.4) observed for nanotube based DSCs, one might expect that the larger photoelectron densities n ($n \propto J_{sc}\tau_r$) [42] would give rise to a higher V_{oc} [43]. The fill factors (FF) of the nanotube based DSCs are slightly lower than those of nanoparticle based DSCs for comparable film thicknesses, which may be attributed to both the relatively poor crystallization arising from the low temperature annealing step, and, possibly, the series (sheet) resistance due to thermally grown oxide formed at the nanotube/substrate interface during annealing process.

Zhu et al. observed the effect of post growth cleaning on the transport and recombination time constants of nanotube based, back side illuminated DSCs [137]. TiO_2 Nanotubes up to 6.1 μm length were prepared in a solution of 0.5 wt% NH_4F in glycerol. The as-anodized nanotube films were first cleaned with water then

Table 5.4 *J–V* characteristics of Dye-Sensitized nanotube and nanoparticle Solar Cells under Simulated AM 1.5 Light ([131]; Copyright 2007 American Chemical Society. Reprinted with permission)

Film type	L (μm)	J_{sc} (mA/cm^2)	V_{oc} (V)	FF	η (%)
Nanotube	1.9	4.4	0.64	0.60	1.7
	3.2	6.6	0.63	0.58	2.4
	4.3	7.8	0.61	0.57	2.7
	5.7	9.0	0.61	0.55	3.0
Nanoparticle	2.5	4.6	0.65	0.69	2.1
	4.2	6.5	0.63	0.63	2.6
	7.3	7.9	0.61	0.66	3.2

soaked sequentially in the bath of 20/80, 40/60, 60/40, 80/20, and 100/0 vol% ethanol/water and then dried using supercritical CO_2 ($scCO_2$, SPI-DRY CPD). The nanotubes were then annealed at 400°C in air for 1 h. The resulting films were then immersed in 0.3 mM N719 dye in ethanol for 24 h and then assembled into DSCs. From the dye desorption measurements, the average surface area of the ethanol/$scCO_2$ dried nanotube film was found to be 23% larger than that of the water/air dried film [137]. At any given light intensity ($I_0 = 6.5 \times 10^{16}$ per cm^2s), the transport times was found to be 43% shorter in the ethanol/$scCO_2$ dried film than in the H_2O/air dried film where as recombination times also decreased by about 23% in ethanol/$scCO_2$ dried nanotubes than in water/air dried nanotubes [137]. From the transport and recombination measurements, the values of η_{cc} at constant light intensity for H_2O/air and ethanol/$scCO_2$ dried nanotube based DSCs were calculated to be 93% and 95%, respectively that were 3% higher for ethanol/$scCO_2$ dried DSC than that for the H_2O/air dried DSCs [137].

Jennings and co-workers studied the electron transport, trapping, and back transfer behaviors of back side illuminated DSCs using nanotube arrays on Ti foil. Electron diffusion lengths were estimated using a method that takes into account that the occupancies of electron traps under open circuit and short circuit conditions were substantially different. Their experimental results showed that the electron diffusion length was of the order of 100 μm in TiO_2 nanotube array cells [138].

5.7 Polymer Based Bulk Heterojunction Solar Cells

For fabrication of an organic solar cell, the hole conducting donor-type semiconducting polymer is commonly blended with an electron accepting Buckminsterfullerene-type molecule, which acts as an active layer for light absorption, exciton generation and their separation into charge carriers. Various well known donor-type polymers are: (i) derivatives of phenylene vinylene backbones such as poly[2-methoxy-5-(3,7-dimethyloctyloxy)]-1,4-phenylenevinylene) (MDMOPPV); (ii) derivatives of thiophene chains such as poly(3-hexylthiophene) (P3HT); and (iii) derivatives of fluorine backbones such as (poly(9,9′-dioctylfluorene-co-bis-N,N′-(4-butylphenyl-1,4-phenylenediamine) (PFB). Electron accepting molecules commonly comprise $PC_{60}BM$ (1-(3-methoxycarbonyl) propyl-1-phenyl [6,6]C_{61}) and $PC_{70}BM$ (1-(3-methoxycarbonyl) propyl-1-phenyl[6,6]C_{71}). Most common organic materials used for TiO_2 based bulk heterojunction solar cell are $PC_{60}BM$, $PC_{70}BM$, and P3HT, with chemical structures shown in Fig. 5.37. The general structure used for organic solar cells is ITO/PEDOT:PSS/active layer/Al, as shown in Fig. 5.38. Coating the ITO glass with an initial layer, via spin coating, with poly(ethylene-dioxythiophene) doped with polystyrenesulfonic acid (PEDOT:PSS) reduces the probability of shorting and facilitates hole injection/extraction.

As illustrated in Fig. 5.39, the conversion of light into electricity by an organic solar cell is described by four steps [139]: (1) absorption of a photon leading to the

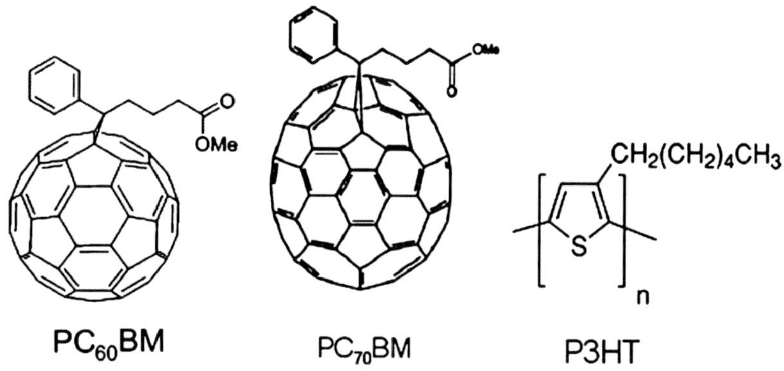

Fig. 5.37 Chemical structures of $PC_{60}BM$, $PC_{70}BM$ and P3HT

Fig. 5.38 Schematic structure showing the bulk heterojunction polymer solar cell

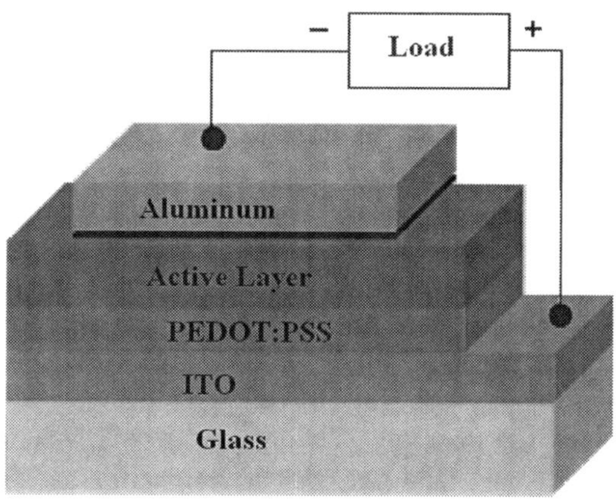

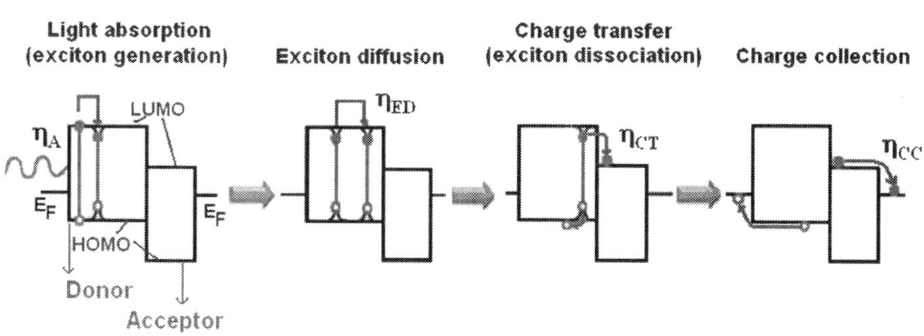

Fig. 5.39 Schematic representations of four steps involved in bulk heterojunction solar cell

formation of an excited state, which is the bound electron-hole pair (exciton) creation; (2) exciton diffusion toward interface; (3) exciton dissociation (separation of electron and hole) at the interface of polymer and PCBM; and (4) electron and holes collection at their respective electrodes. Only a small portion of the incident solar spectrum is absorbed by organic materials. Yet because the absorption coefficients of organic materials are as high as 10^5cm, only 100 nm layers are enough to absorb most of the photons when a reflective back contact is used [139].

Photoexcitation of organic materials leads to the formation of bound electron-hole pairs, called excitons, with an energy between 0.1 and 1.4 eV. This strong binding occurs because electronic wave functions in organic molecules are more localized, and electrostatic attraction can, thus, keep the electron and hole together as an exciton. It was estimated that only 10% of the photoexcitations could separate into free charge carriers in conjugated polymers [140], with strong electric fields necessary for efficient exciton dissociation. Such local fields can be supplied via externally applied electrical fields, as well as via interfaces where photoinduced charge transfer can occur if reached by the exciton within its lifetime. Therefore, the bilayer film thickness is limited by the exciton diffusion length [141] which needs to be of the same order of magnitude as the donor acceptor phase separation length. Otherwise, excitons decay via radiative or nonradiative pathways before reaching the interface, and their energy is lost for power conversion. Exciton diffusion lengths in polymers and in organic semiconductors are usually around 10 to 20 nm [139].

Blending conjugated polymers with electron acceptors, such as fullerenes, is an efficient way to break apart photoexcited excitons into free charge carriers and the photoinduced charge transfer in such blends happens on a time scale of 45 fs, which is much faster than other competing relaxation processes [142]. For efficient photovoltaic devices, the created charges need to be transported to the appropriate electrodes within their lifetime. A gradient in the chemical potentials of the electrons and holes is built up in a donor-acceptor junction as determined by the difference between the highest occupied molecular (HOMO) level of the donor and the lowest unoccupied molecular orbital (LUMO) level of the acceptor. This internal electrical field determines the maximum open circuit voltage (V_{oc}) and contributes to a field-induced drift of charge carriers. Another driving force can be the concentration gradients of the respective charges, which leads to a diffusion current; charge transport is affected by recombination during the journey to the electrodes, particularly if the same material serves as transport medium for both electrons and holes [141].

The use of asymmetrical contacts (low work-function metal for the collection of electrons, and a high work-function metal for the collection of the holes) has been put forth to create an external field at short circuit within a metal-insulator-metal (MIM) structure [143]. Transparent indium tin oxide (ITO) matches the HOMO levels of most of the conjugated polymers (hole contact). An evaporated aluminum metal contact with a work function of around 4.3 eV matches the LUMO of acceptor PCBM (electron contact) on the other side.

The difference between the HOMO level of the donor and LUMO acceptor level leads to the maximum photovoltage achieved in bulk heterojuction based devices. There are charge carrier losses at the electrodes lowering V_{oc}, as shown in Fig. 5.40 considering P3HT as the donor and PCBM as the acceptor. Open circuit voltages are also affected by nanoscale morphological changes in the active layer of the polymer fullerene bulk heterojunction solar cells [144]. Small changes at the metal

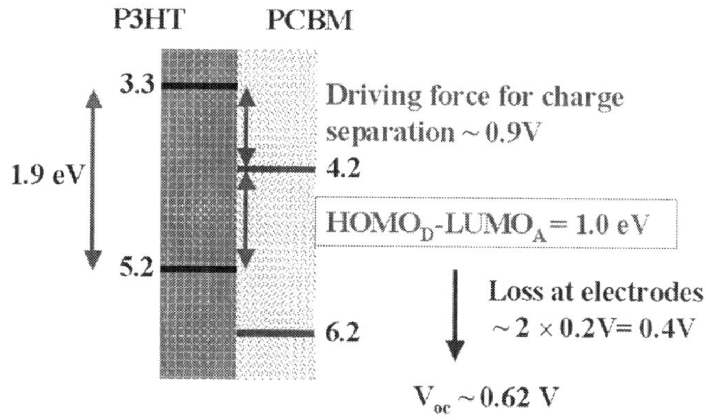

Fig. 5.40 Schematic illustration of photovoltage generated in P3HT and PCBM blend, determined using the HOMO level of the donor and LUMO level of PCBM

to organic semiconductor interface affect the electrode work functions and, thereby, influences the open circuit potential [145, 146].

In the ideal case, the product of the photoinduced charge carrier density and the charge carrier mobility within the organic semiconductors determines short circuit current, I_{sc}, given by:

$$I_{sc} = ne\mu E \tag{5.23}$$

where n is the density of charge carriers, e is the elementary charge, μ is the mobility, and E is the electric field. Here mobility is a device parameter, related to the morphology of the organic semiconductor thin films [147–151]. Film processing parameters such as the solvent type, the solvent evaporation (crystallization) time, the temperature of the substrate, and the deposition method can change the morphology [152, 153]. The fill factor depends on the mobility μ of charge carriers and their lifetime τ. Carrier drift length L_d, defined as $L_d = \mu\tau E$, must be longer than the active layer thickness to prevent significant recombination loss [154, 155]. Hence, a higher mobility, or a thin film geometry, is necessary for the efficient carrier extraction and improved fill factor. The series resistance should be minimized to maximize the shunt resistance improving the fill factor.

Ma et al. report bulk heterojunction solar cells using P3HT and $PC_{60}BM$ showing an optimistic 5% (I_{sc}=9.5 mA/cm^2, FF=0.68, V_{oc}=0.63 V) power conversion efficiency under AM 1.5 illumination at 80 mW/cm^2 [156]. The device structure used was ITO/PEDOT/P3HT:PCBM/Al. PEDOT:PSS was spin coated from an aqueous solution on an ITO substrate at 5,000 rpm, with a resulting thickness of approximately 40 nm. The substrate was dried for 10 min at 140°C in air and then moved to a glove box for spin coating of the photoactive layer. A chlorobenzene solution, 1 wt% P3HT (10 mg/mL) plus PCBM (8 mg/mL) was spin coated at 700 rpm on top of the PEDOT:PSS layer. A 100 nm thick aluminum film was deposited on top of the active layer under high vacuum, followed by a 150°C anneal for 30 min in a glove box filled with nitrogen gas.

Using regioregular (RR) P3HT as the donor and PCBM as the acceptor [2, 157], bulk heterojunction solar cells have shown external quantum efficiencies of around 75%, and power conversion efficiencies toward 5% [158]. It was suggested that the high efficiency of these devices was due to microcrystalline lamellar stacking [159] resulting in reduced recombination [160]. RR-P3HT gives a hole mobility of up to ~0.1 cm^2/Vs [161, 162], with a red shift of the RR-P3HT optical absorption due to this lamellar stacking (PCBM has an electron mobility of $2 \times 10^{-3} cm^2/Vs$ [163]). Further, we note that stability is a concern for organic solar cells, with protection against oxygen and humidity required to achieve extended device lifetimes. However photodegradation of the conjugated polymer was greatly reduced in films blended with PCBM due to efficient electron extraction.

The ideal structure of a bulk heterojunction solar cell is the one where two phases of donor and acceptor are interspaced with an average length scale of around 10–20 nm, equal to or less than the exciton diffusion length to ensure high mobility charge carrier transport with reduced recombination, provided a pure donor phase is at the hole collecting electrode and a pure acceptor phase at the electron collecting electrodes [164]. Such a well-organized nanostructure within polymer mixtures is difficult to achieve. However, self-organization of the organic semiconducting polymers (molecules) can be introduced by di-block copolymers [165–167], liquid crystalline self-organizing columns of donor acceptor phases [168], and inorganic oxide nanotubes filled with organic semiconductors [21–24, 169]. Furthermore, there are several nuances to the design of polymeric solar cells, such as a bulk heterojunction consisting of fullerenes [170], inorganic semiconductor nanorods [9], or nanocrystals [171] embedded in a polymer matrix, or an ordered heterojunction where a hole transporting polymeric film is infiltrated into a mesoporous architecture [172] or n-type inorganic semiconductor nanowire/nanorod arrays. Here, we emphasize the use of TiO_2 nanotube array films, synthesized by anodic oxidation, for construction of polymeric solar cells.

5.7.1 TiO_2 Nanotubes on FTO Glass: Polymeric Bulk Heterojunction Solar Cells

In a bulk heterojunction device, the ease of electron transfer from the semiconducting polymer to the fullerene and the intimate contact between the two materials allows the photogenerated excitons to be efficiently separated. While this configuration has provided the best reported efficiencies of up to 5.5% [173], it still suffers from poor charge collection efficiency resulting from the inherently random interpenetrating network morphology and the segregation of the donor and acceptor type phases. The poor collection efficiency also limits light harvesting by placing constraints on the thickness of the active layer. The alternative configuration, namely, the ordered heterojunction, utilizes an ordered mesoporous or nanoporous inorganic architecture but is still limited by the poor hole mobility in the polymer, the ability to infiltrate the polymer into the mesoporous structure, and non-ideal polymer crystallization resulting in reported efficiencies close to 1% [19, 20]. Since the typical range of exciton diffusion lengths in polymers is 5–20 nm, the pore size of the mesoporous architecture is typically chosen to be of comparable dimensions. However, at these small pore sizes, the structure of the polymer inside the pores is radically different from and inferior to bulk polymer chain packing [172]. The high degree of π–π stacking between the polymer chains in the bulk, which enables the delocalization of excitons over multiple chains, does not occur to a significant extent inside small pores [172].

A schematic of our device configuration is shown in Fig. 5.41; transparent TiO_2 nanotube array films on FTO coated glass are used for collecting electrons from PCBM molecules generated upon exciton dissociation at the P3HT/PCBM interface, as well as bulk heterojunctions present in polymer/organic molecule blend [22]. Charge separation occurs at the TiO_2/P3HT and PCBM/P3HT interfaces. Figure 5.42 indicates the favorable HOMO and LUMO energies of the component materials (P3HT/PCBM blend, PEDOT:PSS) with respect to the TiO_2 conduction and valence bands and electrode work functions (FTO and gold). The dimensions of the transparent TiO_2 nanotube arrays, 270 nm length and 50 nm pore size, were chosen to allow infiltration of the active polymeric layer and allow the polymer to form a crystal structure similar to that in the bulk. The polymer used is a blend of an electron donor P3HT and an electron acceptor $PC_{70}BM$; a 10 mg/ml solution of P3HT in chlorobenzene and 8 mg/ml solution of PCBM in chlorobenzene were mixed in a 1:1 ratio using the procedure outlined by Kim et al. [158] and infiltrated into the nanotubes. Subsequently, a PEDOT:PSS layer was deposited by spin coating an aqueous suspension of the polymer at 5,000 rpm. The samples were then placed on a hot plate at 170°C for 15 min to evaporate residual water and allow diffusion of the polymer into the TiO_2 nanotube arrays. All spin coating and baking steps were carried out in a nitrogen glovebox. A 50 nm film of gold was evaporated

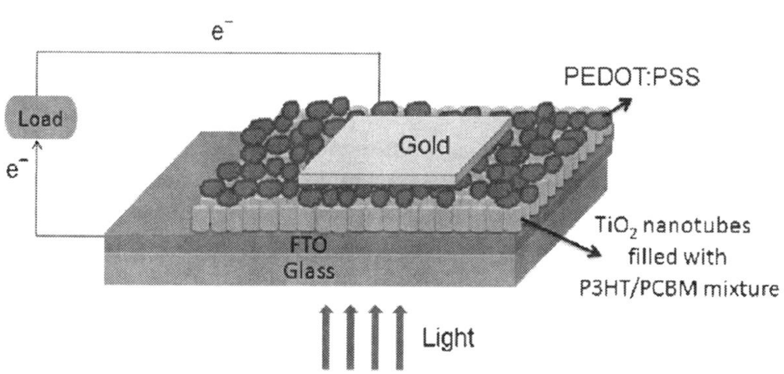

Fig. 5.41 Depiction of polymer TiO_2 nanotube array based solid state solar cell geometry

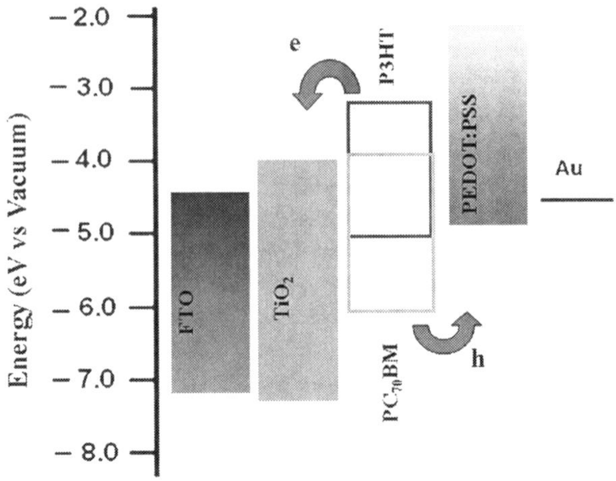

Fig. 5.42 HOMO and LUMO energies of the component materials of polymer TiO_2 nanotube array based solar cell

onto the devices though a shadow mask to form the top contact. The barrier layer of compact TiO_2 at the bottom of the nanotube array prevents the polymer from reaching the transparent conductive oxide contact.

A key advantage of the TiO_2 nanotube array electron accepting network is that it prevents the electron donating material (P3HT) and electron accepting materials (TiO_2 and PCBM) from being in contact with both anode and cathode of the solar cell at the same time, as is the case with the conventional bulk heterojunction device geometry. The direction of illumination and current flow in our devices is opposite to that in conventional bulk heterojunction cells. The poor hole mobility of the p-type semiconducting polymer and the large distance charge carriers need to travel before being collected are the limiting factors in the performance of disordered bulk heterojunction solar cells. To minimize the distance, charge carriers need to travel in the hole conducting polymer to reach the hole collection contact, the thickness of the hole conducting polymer in conventional bulk heterojunction devices is generally kept between 100 and 150 nm. However, the hole transporting polymer is responsible for much of the charge generation due its higher absorption coefficient with respect to the n-type material (e.g. PCBM derivative), therefore reducing the thickness of the hole conducting polymer decreases the amount of harvested incident light. The ordered geometry constituted by TiO_2 nanotube arrays affords the charge carriers direct paths to their respective contacts. Facile hole transport in the polymer is contingent on suitable chain aggregation to ensure delocalization of the generated excitons, and alignment of the chains in the desired direction of hole propagation. For regioregular P3HT infiltrated into an anodic alumina template with vertical channels similar to the TiO_2 nanotube array architecture, Coakley et al. [174] found that the mobility of the polymers in the alumina pores was enhanced by as much as a factor of 20 due to the alignment induced by the vertical

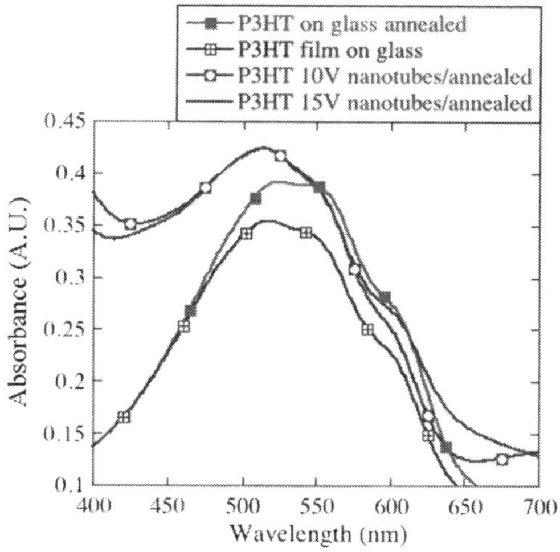

Fig. 5.43 UV-Visible absorption spectra of neat film of P3HT spin coated on glass, neat film of P3HT after annealing at 170°C, P3HT film inside nanotube arrays fabricated at 15 V after annealing at 170°C, and P3HT film inside nanotube arrays fabricated at 10 V after annealing at 170°C [22]

pore-channel confinement of the polymer chains perpendicular to the substrate. However, in very small pores, the chains of the polymer MEH-PPV were found to be isolated with a low polaron yield and little evidence of interchain contact [175].

The absorption spectra of TiO_2 nanotube arrays of two different pore sizes infiltrated with pristine P3HT are shown along with the absorption spectra of neat P3HT films in Fig. 5.43. The layer of P3HT on top of the nanotubes was removed so that only the P3HT inside the nanotubes contributed to the absorption spectrum. The absorption peak for the neat P3HT film in Fig. 5.43 occurred at 526 nm and shifted slightly to 518 nm upon annealing. The peak position for the nanotube arrays infiltrated with P3HT occurred at 514–515 nm indicating a small blue shift from the neat film. The blue shift relative to the rod-like chain in an unannealed neat film is indicative of a coiled configuration since the conjugation length is shorter on a coiled chain [176]. Roux and Leclerc [176] found that [poly-3-(octyloxy)-4-methylthiophene] showed a blue shift in its optical absorption maximum, which was related to a transition from a coplanar (rod) conformation to a nonplanar (coil) conformation by annealing at 150°C. However, the magnitude of the blue shift noticeable in Fig. 5.43 is smaller than that reported for the infiltration of P3HT into mesoporous structures of smaller pore sizes ($<$10 nm) [172].

The *J–V* characteristics of the fabricated TiO_2 nanotube array-polymer solar cells under AM 1.5 global illumination from a calibrated solar simulator are shown in Fig. 5.44. The short circuit density is 12.43 mA/cm^2, the open circuit potential 0.641 V, and the fill factor was 0.511, resulting in an overall conversion efficiency of 4.07%. Figure 5.45 shows the external quantum efficiency (EQE) plot of the nanotube array-polymer double heterojunction solar cells. A maximum EQE of 80% was seen at 538 nm. When adjusted for the photonic losses, which include an $\approx$15% absorption loss due to the FTO coating and glass substrate, the reflection losses at the FTO-TiO_2 interface, and the FTO-glass interface, the conversion of

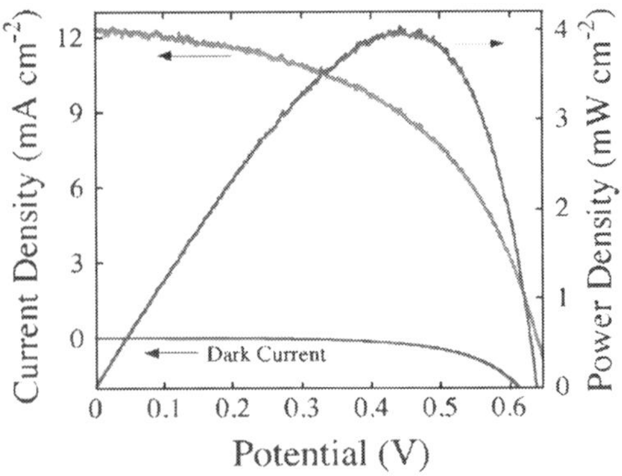

Fig. 5.44 *J–V* characteristics of TiO_2 nanotube array-polymer bulk heterojunction solar cells showing the dark current density, photocurrent density, and power density under 1 sun AM 1.5 illumination [22]

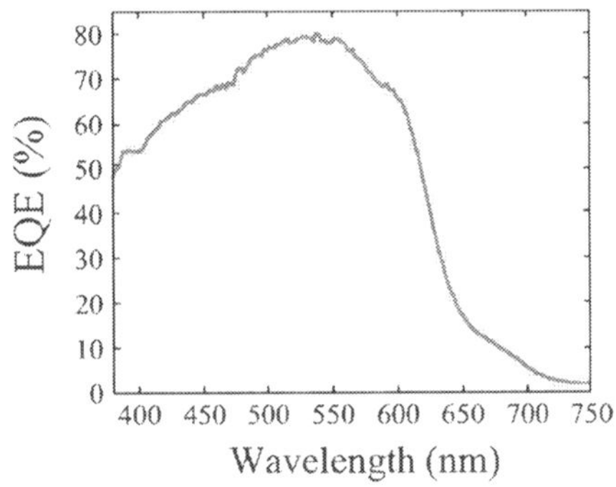

Fig. 5.45 EQE of the nanotube array-polymer bulk heterojunction solar cells [22]

incident photons at the absorption peak wavelength into collected electrons is nearly quantitative. The position of the EQE maximum at 538 nm is different from conventional bulk heterojunction cells where it occurs close to 500 nm [173], a difference we believed attributable to the device geometry and stacking order of the absorbing layers.

5.7.2 Solar Cell Fabrication and Performance

In the conventional bulk heterojunction solar cell, the device structure is: ITO/PEDOT:PSS/P3HT-PCBM blend/LiF or TiO_2 nanoparticles films/Al, with light incident upon the ITO side of the device. The layer arrangement in our so-called "inverted" configuration, Fig. 5.41, is FTO/TiO_2 nanotubes/P3HT-PCBM/PEDOT:PSS/gold, with illumination from the FTO side. The active layer thickness in these devices should be in the range of 150–250 nm due to the limited electron and hole mobilities of the active layer. The presence of PCBM at the interface with P3HT helps in photogenerated exciton dissociation, which is excellent compared to other n-type inorganic and organic semiconductors currently known. The inverted configuration has numerous benefits, including: (1) The conduction band position of TiO_2 is just below the LUMO energy level of PCBM for efficient extraction of electrons from PCBM received from the interface of P3HT/PCBM. (2) The nanotube array architecture enhances the interfacial contact area between TiO_2 and P3HT and/or PCBM of the active blend layer, which implies higher electron collecting surface area and improvement in electron extraction property. (3) A thin, conformal coating on the nanotubes could be infiltrated, at its core, with a hole extracting/injecting material such as PEDOT:PSS. In this way the nanotube lengths can be increased beyond the critical thickness of active layer increasing light absorption. (4) The inverted configuration has PEDOT:PSS on top of the active blended layer of P3HT/PCBM, which acts as a protective cover slowing down oxidation and humidity related degradation of the polymers. (5) With the inverted geometry light passes through the UV absorbing TiO_2 before reaching the active layer region. (6) Compared with a conventional bulk heterojunction device design, since the incident light does not have to pass through a PEDOT:PSS layer with the inverted geometry light absorption is increased in the 600 to 750 nm region.The solubility of $PC_{70}BM$ in O-DCB is quite high, about 70 mg/mL, whereas in comparison only 40 mg/mL of $PC_{70}BM$ can be dissolved in chlorobenzene. To infiltrate the polymer inside the nanotubes, we wet the nanotube surface with O-DCB such that the nanotubes are filled with solvent removing any of the air bubbles within the nanotubes. The nanotubes are then coated with concentrated P3HT/$PC_{70}BM$ solution, then the sample slowly spun to low rpm with the intention to mix the O-DCB solvent, within the nanotubes, with the concentrated blended solution. The polymer blend diffuses into the O-DCB wet nanotubes due to the concentration gradient. The nanotube dimensions, blended solution concentration, and spin speed are critical details and require rigorous process optimization to achieve conformal coating. Baking the samples on hot plate at 150°C further assists diffusion of the polymer blend into the nanotubes and polymer chain alignment. As the solvent evaporates, the volume of polymer blend inside the nanotubes shrinks resulting in a thin conformal polymer coating with space in the center of the nanotube.

High quality transparent TiO_2 nanotube array films were synthesized by anodizing Ti films, on FTO coated glass, in an ethylene glycol organic solvent containing NH_4F and water. Anodization duration was optimized such that the resulting as-anodized sample had minimal residual Ti underneath the titania barrier layer. The length of the nanotubes used for polymer based solar cells was 600–700 nm with a pore size of approximately 40 nm. Amorphous as-anodized nanotubes were crystallized by first annealing the sample at 400°C for 4 h and then at 450°C for 30 min, with temperatures and dwell times chosen to avoid degradation of FTO layer. These crystallized samples were wet with O-DCB with the excess solvent spun-off. P3HT/$PC_{70}BM$ was mixed in a ratio of 1:0.8 and dissolved in DCB:CB organic solvent mixture (from 2:1 to 4:1 and then from 1:2 to 1:4); this solution was placed on the wet sample and spun at 300 rpm. The sample was then enclosed within a petri dish, heated at 150°C for 30 min, and allowed to cool. A PEDOT:PSS layer was then spin coated from aqueous solution at 4,000 rpm for 2 min, then baked at 120°C for 8–10 min in a covered petri dish. The sample was cooled to room temperature, and taken from glove box for DC sputter deposition of Au back contacts. Cross-sectional image Fig. 5.46a shows the TiO_2 nanotube walls and tube opening coated with polymer. Figure 5.46b shows a cross-sectional view of the full device covered with PEDOT:PSS layer. Figure 5.47a, b shows another device fabricated under spin

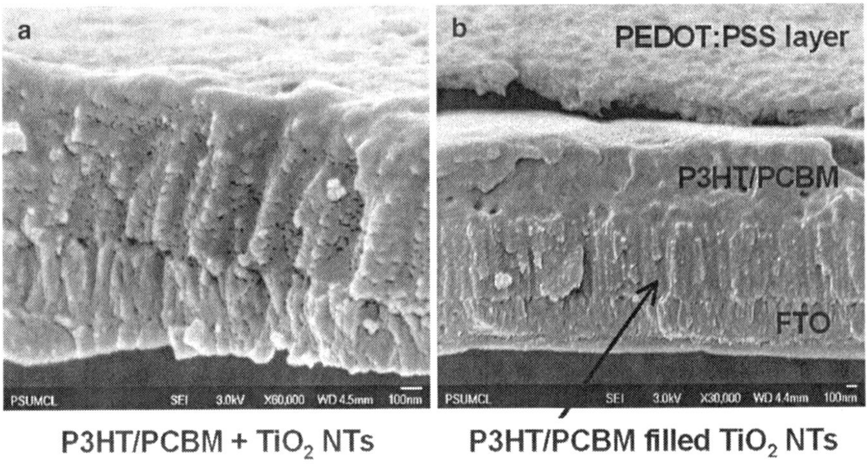

Fig. 5.46 FESEM cross-sectional view of: (**a**) polymer blend filled nanotubes; (**b**) full device, also showing the presence of PEDOT:PSS layer on top

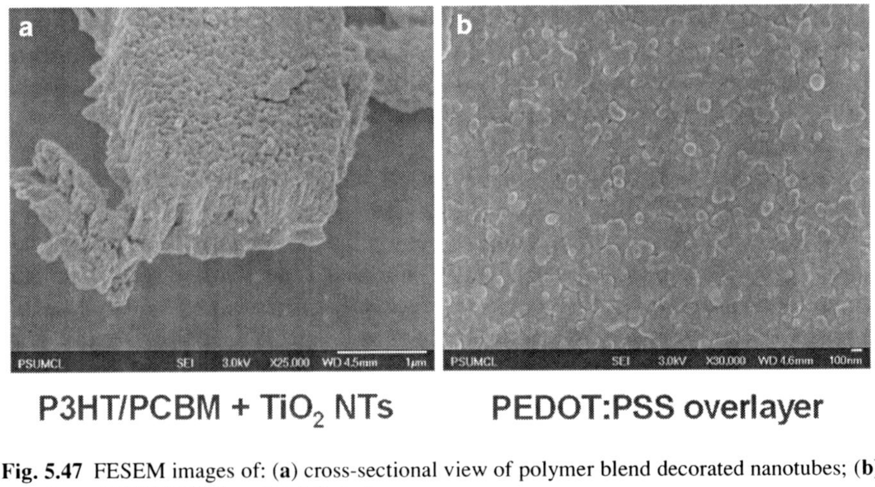

Fig. 5.47 FESEM images of: (**a**) cross-sectional view of polymer blend decorated nanotubes; (**b**) top view of PEDOT:PSS coated full device

conditions such that polymer decorated nanotubes are achieved without the blocking of the nanotube pore allowing for PEDOT:PSS layer infiltration.

Photocurrent–voltage characteristics and the overall conversion efficiency of the photovoltaic devices made using P3HT/PCBM coated TiO_2 nanotubes are shown in Fig. 5.48. There are five curves in the *J–V* plot, with each curve arising from the use of a different organic solvent to wet the nanotube surface before coating with a concentrated P3HT/PCBM solution. Curve 1 represents the device performance of nanotubes treated with t-butyl pyridine (TBP). TBP is a polar solvent, which easily wets the hydrophilic nanotube surface but does not allow the polymer film to penetrate deep within the nanotube pores with the active layer remaining close to the tube opening. Low polymer coverage of the TiO_2 nanotubes results in small current magnitudes. Curve 2 shows the effect of chlorobenzene (CB); although the photocurrent is low, the photovoltage is close to that expected for a conventional bulk heterojunction device. It was observed that the CB treated surface required a relatively large amount of blended solution to coat the entire film surface. Curve 3 and Curve 4 are due to O-dichlorobenzene (O-DCB), with Curve 4 arising from a sample immersed in O-DCB rather than simply wetted. O-DCB has a polarity higher than that of CB but lower than TBP.

Curves 1 to 4 denote samples that were baked at 150°C for 30 min. Use of a higher baking temperature, 165°C, for the device design of Curve 4 resulted in a reduced photocurrent of about 4 mA/cm^2 but improved photovoltage of 0.61 V as shown in Curve 5. Curve 4 shows the best conversion efficiency of ≈3.5%; the lower voltage of Curve 4 is due to high dark currents, which require further attention to improve.

The presence of well-stacked polymer chains, which is required for optimal device performance, can be verified by the device photoaction spectra. Figure 5.49

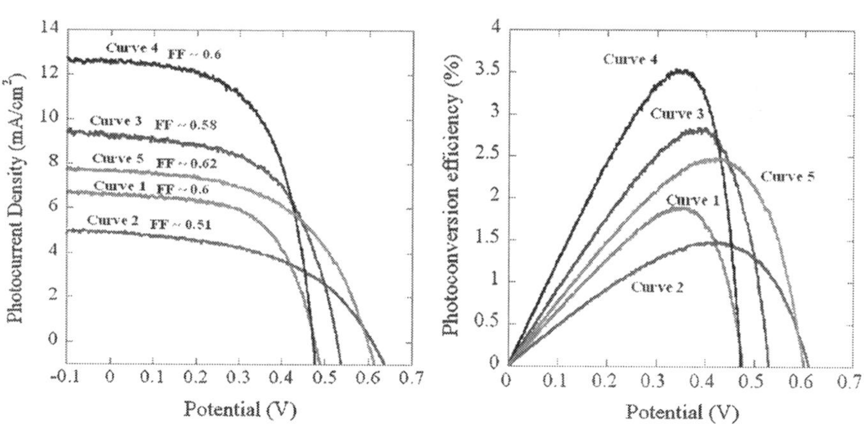

Fig. 5.48 Photocurrent–voltage characteristics and the overall conversion efficiency of photovoltaic devices made using P3HT/PCBM coated TiO_2 nanotubes. Five curves, 1 to 5, represent five different TiO_2 surface treatments before polymer blend spin coating. Treatment of nanotubes for (1) Curve 1 – TBP, (2) Curve 2 – CB, (3) Curve 3 – O-DCB, (4) Curve 4 – soaked in O-DCB, and (5) Curve 5 – soaked in O-DCB and baked at 165°C. Curves 1 to 4 were baked at 150°C

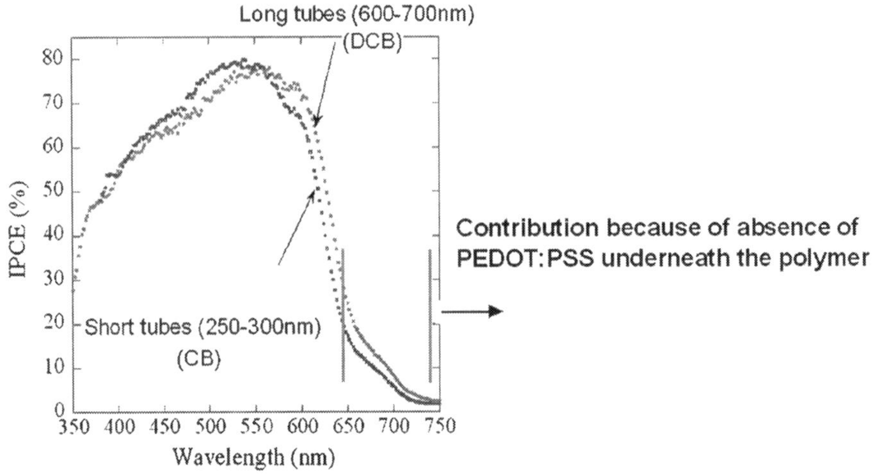

Fig. 5.49 IPCE as a function of wavelength for: "long" nanotube arrays of 600–700 nm length used in the fabrication of a polymer solar cell prepared from P3HT (20 mg/mL) plus PCBM (16 mg/mL) mixed in o-dichlorobenzene+CB solvent (4:1); "short" nanotube arrays of 250 to 300 nm used in the fabrication of a polymer solar cell prepared from P3HT (10 mg/mL) plus PCBM (8 mg/mL) mixed in chlorobenzene solution

presents the IPCE of the device giving rise to Curve 4 of Fig. 5.48, made with TiO_2 nanotubes ≈650 nm long using O-DCB as the solvent. Also shown in Fig. 5.49 is the IPCE of a solar cell synthesized with a CB based polymer blend solution in combination with nanotube arrays ≈275 nm long. In the IPCE spectra of the conventional bulk heterojunction device, the peak position occurs at 500 nm with a sharp drop to zero around 675 nm due to light absorption by the PEDOT:PSS layer. In contrast, the CB polymer solution device, prepared on short nanotubes, exhibits a red shifted peak position to 525 nm, and significant contribution from 650 to 735 nm. Solar cells prepared using O-DCB polymer solution show a further red shift in the peak position to 550 nm, which suggests ordered stacking of the polymer chains along the nanotube length.

5.7.3 *TiO_2-Polymer Based Solar Cells: Back Side Illumination Geometry*

Poly(3-hexylthiophene) (P3HT) is a conjugated polymer that strongly absorbs in the wavelength region 450–600 nm and possesses a hole mobility of 0.1 cm^2/Vs in its regioregular form [177]. P3HT has side chains that make it soluble in a variety of common organic solvents. Solution processing of the polymer is attractive because of its applicability in scalable large area devices. A carboxylated P3HT derivative can be used to self-assemble on the surface of the TiO_2 nanotube upon overnight immersion in a solution of the polymer [24]. Yanagida et al. obtained impressive photocurrents of up to 9.75 mA/cm^2 by using the carboxylated P3HT derivative poly-(3-thiophene acetic acid) in conjunction with 7 μm thick nanocrystalline TiO_2 films [163]. However, the values of the open circuit voltages obtained were low, close to 0.4 V. Shankar et al. [24] used the carboxylated P3HT derivative regioregular poly[3-(5-carboxypentyl)thiophene-2,5-diyl] with 4 μm long TiO_2 nanotube arrays on Ti foil (pore size ~70 nm, wall thickness ~22 nm; prepared in 0.3 wt% NH_4F and 2% deionized H_2O dissolved in ethylene glycol; and annealed in oxygen at 580°C for 4 h) and obtained open circuit potentials of 0.7–0.76 V upon AM 1.5 1 sun illumination with single hetrojunction devices. Mwaura and co-workers used the same carboxylated P3HT derivative and obtained open circuit potentials of 0.54 V when employed with 4 μm thick nanocrystalline TiO_2 films [178]. While non-polymeric, all-organic, dye sensitized TiO_2 nanoparticulate film based solid state devices showed efficiencies upward of 4% [179].

Shankar and co-workers prepared three kinds of photovoltaic devices with the polymer self-assembled on the TiO_2 nanotube array electrode (atop Ti foil) [24]. A liquid junction solar cell was prepared by infiltrating the polymer coated TiO_2 electrode with I_3^-/I_2 redox electrolyte where the polymer functioned primarily as a photosensitizer. A single heterojunction solid state solar cell using the self-assembled carboxylated P3HT derivative was fabricated by replacing the redox electrolyte with PEDOT:PSS. In third device configuration, P3HT-PCBM blend was infiltrated into the pores of self-assembled P3HT derivative coated TiO_2 nanotube arrays. Crystallized nanotubes were coated with polymer by immersion in a 7 mg mL^{-1} solution of regioregular poly[3-(5-carboxypentyl)thiophene-2,5-diyl] in N,N-dimethylformamide (DMF) overnight. For the liquid junction solar cells, the redox electrolyte used contained 0.5M LiI, 0.02M I_2, 0.6M N-methylbenzimidazole, 0.10M guanidinium thiocyanate, and 0.5M tert-butylpyridine in methoxypropionitrile (MPN). A FTO glass substrate sputter coated with 1 nm of Pt was used as the counter electrode. For the single heterojunction solid state solar cells, a layer of conducting polymer PEDOT:PSS was formed on the carboxylated P3HT coated nanotube array electrodes by spin coating (5,000 rpm). The device was subsequently placed on a hot plate for 5 min at 160°C to eliminate residual water and annealed at 200°C in nitrogen ambient for 1 min to cause the PEDOT/PSS film to penetrate into the pores of the nanotubular structure. PEDOT:PSS coated FTO glass was used as the counter electrode and pressed onto the PEDOT/PSS coated polymer sensitized nanotube array electrode. In the third device, P3HT(10 mg/mL)-PCBM (8 mg/mL) mixed in chlorobenzene solution was spin coated at 700 rpm on carboxylated P3HT coated nanotube array electrodes. The subsequent steps were identical to those for the single heterojunction cell.

The photocurrent density–voltage (*J*–*V*) characteristics of liquid junction based polymer sensitized nanotube array as shown in Fig. 5.50, exhibiting a J_{sc} of 5.5 mA/cm^2, a V_{oc} value of 0.7 V, and a fill factor (*FF*) of 0.55 to produce an overall conversion efficiency of 2.1%. It was noted that the open circuit photovoltage obtained using the carboxylated P3HT as the sensitizer (0.7–0.76 V) is similar to those obtained by us using ruthenium based dyes as the sensitizer (0.76–0.84 V).

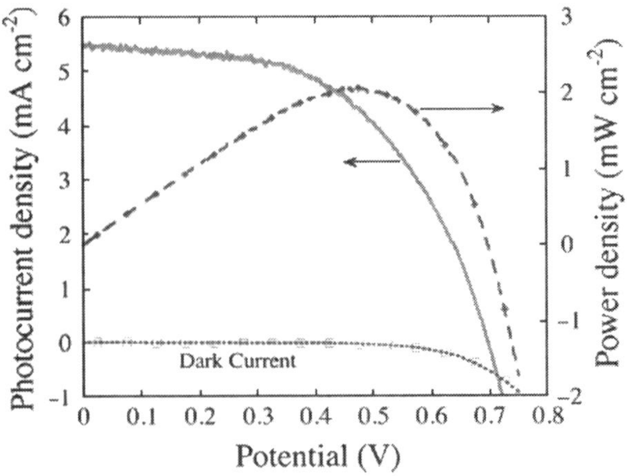

Fig. 5.50 *J–V* characteristics of backside illuminated liquid junction polymer based solar cells under 1 sun AM 1.5 illumination [24]

Using the monocarboxylated P3HT derivative poly-(3-thiopheneacetic acid) and the dicarboxylated derivative poly-(3-thiophenemalonic acid), Senadeera and colleagues obtained V_{oc} values of 0.3–0.4 V [163, 180], which were attributed to a shift in the position of the surface conduction band of TiO_2 due to protonation of the surface by the carboxylic groups. Also, a high density of carboxylic groups anchored to the surface of the n-type semiconductor was found to decrease the open circuit photovoltage due to the creation of interface dipoles at the TiO_2–polymer interface [181]. However, when Mwaura et al. used the same carboxylated P3HT derivative as used here, which consists of an alkyl chain of five CH_2- groups separating the π-conjugated structure from the carboxylate moiety, they observed relatively higher photovoltages of up to 0.54 V [178]. It was suggested that the carboxylic groups in close proximity to the site of charge generation in the molecule or monomeric unit of the polymer cause the reduction of the open circuit potential due either to high dark currents or shifts in band-offsets.

The carboxylated P3HT derivatives were reported to function poorly as hole transport materials [182]. Hence, instead of using a hole transporting material to deliver the holes to the contact, the hole collection contact was placed close to the point of hole generation by infiltrating the conducting polymer PEDOT/PSS into the nanotubes. However such solid state devices showed a poor fill factor which was attributed to poor electrical contact to the ITO coated glass slide contact, which is merely pressed onto the top of the polymer coated devices [24]; photocurrents were about 1.37 to 2 mA/cm^2 with a photovoltage of about 0.35 V [24].

Figure 5.51 shows the *J–V* characteristics of a device assembled from carboxylated P3HT sensitized TiO_2 nanotube array films spin coated with a P3HT-PCBM blend. This type of bulk heterojunction devices show a photocurrent density of 6.5 mA/cm^2, which is higher than that obtained with liquid junction cells using the same electrode and sensitizer. However, the efficiency of these devices is limited to

1% due to the lower open circuit potential and fill factor. A self-assembled polymeric sensitizer has a less than optimal surface coverage resulting in gaps on the surface where the hole transport material and the n-type TiO_2 are in direct contact, allowing for the recombination of photogenerated holes in the hole transport material with photogenerated electrons in the TiO_2 subsequent to charge separation [183]. The performance of solid state solar cells employing the all-organic non-polymeric sensitizer indoline is superior with reported efficiencies higher than 4% [179], probably due to the higher surface coverage of the dye and

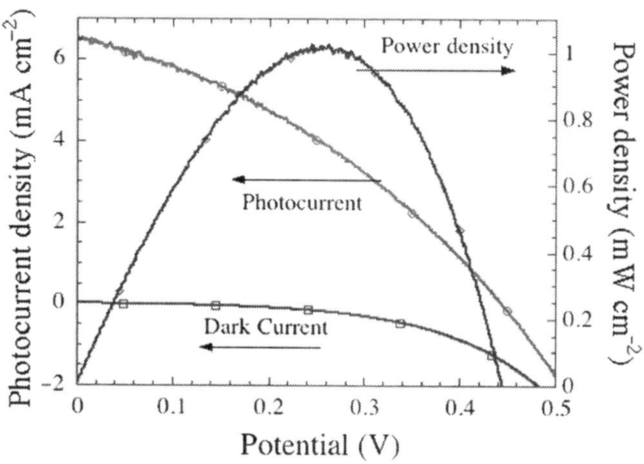

Fig. 5.51 *I–V* characteristics of backside illuminated double heterojunction solid state solar cell under 1 sun AM 1.5 illumination showing the dark current density, photocurrent density, and power density as a function of potential [24]

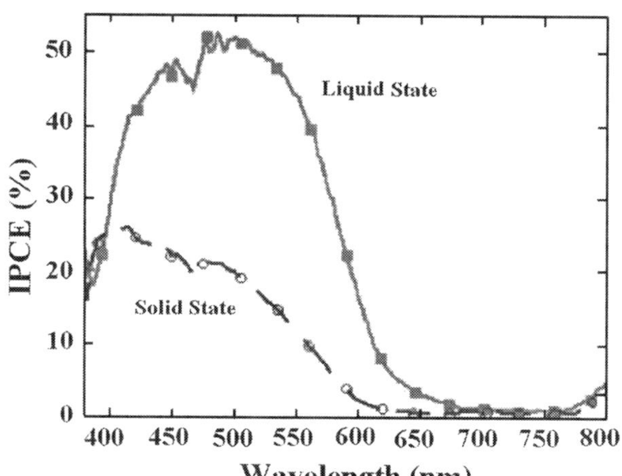

Fig. 5.52 IPCE spectra of backside illuminated liquid and solid state single heterojunction solar cells [24]

the use of spiro-OMETAD as the hole transport material, which shows excellent pore filling characteristics. A second sensitization step by small molecule dyes to coat areas where no polymeric sensitizer is present may ameliorate the recombination problem. Also, treatment of the devices with an ionic liquid containing imidazolium salts may improve device performance due to improvements in the electronic properties of the π-conjugated polymer [182].

The action spectra of the liquid junction and single heterojunction solid state solar cells are shown in Fig. 5.52. A maximum IPCE of 53% is obtained with the liquid junction solar cells, and 25% is obtained with the single heterojunction solid state solar cell [24].

References

1. O'Regan B, Gratzel M (1991) A low-cost, high-efficiency solar-cell based on dye-sensitized colloidal TiO_2 films. Nature 353:737–740
2. Ma WL, Yang CY, Gong X, Lee K, Heeger AJ (2005) Thermally stable, efficient polymer solar cells with nanoscale control of the interpenetrating network morphology. Adv Funct Mater 15:1617–1622
3. Shaheen SE, Ginley DS, Jabbour GE (2005) Organic-based photovoltaics: toward low-cost power generation. MRS Bull 30:10–15
4. Scharber MC, Mühlbacher D, Koppe M, Denk P, Waldauf C, Heeger AJ, Brabec CJ (2006) Design rules for donors in bulk-heterojunction solar cells – towards 10% energy-conversion efficiency. Adv Mater 18:789–794
5. Mayer AC, Scully SR, Hardin BE, Rowell MW, McGehee MD (2007) Polymer based solar cells. Mater Today 10:28–33
6. Forrest SR (2005) The limits to organic photovoltaic cell efficiency. MRS Bull 30:28–32
7. Janssen RAJ, Hummelen JC, Sariciftci NS (2005) Polymer–Fullerene bulk heterojunction solar cells. MRS Bull 30:33–36
8. Kim JY, Lee K, Coates NE, Moses D, Nguyen TQ, Dante M, Heeger AJ (2007) Efficient tandem polymer solar cells fabricated by all-solution processing. Science 317:222–225
9. Huynh WU, Dittmer JJ, Alivisatos AP (2002) Hybrid nanorod-polymer solar cells. Science 295:2425–2427
10. McDonald SA, Konstantatos G, Zhang SG, Cyr PW, Klem EJD, Levina L, Sargent EH (2005) Solution-processed PbS quantum dot infrared photodetectors and photovoltaics. Nat Mater 4:138–142
11. Sun BQ, Marx E, Greenham NC (2003) Photovoltaic devices using blends of branched CdSe nanoparticles and conjugated polymers. Nano Lett 3:961–963
12. Arango AC, Johnson LR, Bliznyuk VN, Schlesinger Z, Carter SA, Horhold HH (2000) Efficient titanium oxide/conjugated polymer photovoltaics for solar energy conversion. Adv Mater 12:1689–1692
13. Beek WJE, Wienk MM, Janssen RAJ (2004) Efficient hybrid solar cells from zinc oxide nanoparticles and a conjugated polymer. Adv Mater 16:1009–1013
14. Cui DH, Xu J, Zhu T, Paradee G, Ashok S, Gerhold M (2006) Harvest of near infrared light in PbSe nanocrystal-polymer hybrid photovoltaic cells. Appl Phys Lett 88:183111 (3pp)
15. Slooff LH, Kroon JM, Loos J, Koetse MM, Sweelssen J (2005) Influence of the relative humidity on the performance of polymer/TiO_2 photovoltaic cells. Adv Func Mater 15: 689–694

16. Kuo CY, Tang WC, Gau C, Guo TF, Jeng DZ (2008) Ordered bulk heterojunction solar cells with vertically aligned TiO_2 nanorods embedded in a conjugated polymer. Appl Phys Lett 93:033307
17. Shim HS, Na SI, Nam SH, Ahn HJ, Kim HJ, Kim DY, Kim WB (2008) Efficient photovoltaic device fashioned of highly aligned multilayers of electrospun TiO_2 nanowire array with conjugated polymer. Appl Phys Lett 92:183107
18. Snaith HJ, Arias AC, Morteani AC, Silva C, Friend RH (2002) Charge generation kinetics and transport mechanisms in blended polyfluorene photovoltaic devices. Nano Lett 2:1353–1357
19. Coakley KM, McGehee MD (2004) Conjugated polymer solar cells. Chem Mater 16:4533–4542
20. Kannan B, Castelino K, Majumdar A (2003) Design of nanostructured heterojunction polymer photovoltaic devices. Nano Lett 3:1729–1733
21. Coakley KM, McGehee MD (2003) Photovoltaic cells made from conjugated polymers infiltrated into mesoporous titania. Appl Phys Lett 83:3380–3382
22. Mor GK, Shankar K, Paulose M, Varghese OK, Grimes CA (2007) High efficiency double heterojunction polymer photovoltaic cells using highly ordered TiO_2 nanotube arrays. Appl Phys Lett 91:152111
23. Shankar K, Mor GK, Paulose M, Varghese OK, Grimes CA (2008) Effect of device geometry on the performance of TiO_2 nanotube array-organic semiconductor double heterojunction solar cells. J Non-Cryst Solids 354:2767–2771
24. Shankar K, Mor GK, Prakasam HE, Varghese OK, Grimes CA (2007) Self-assembled hybrid polymer–TiO_2 nanotube array heterojunction solar cells. Langmuir 23:12445–12449
25. Greene LE, Law M, Goldberger J, Kim F, Johnson JC, Zhang YF, Saykally RJ, Yang PD (2003) Low-temperature wafer-scale production of ZnO nanowire arrays. Angew Chem-Int Ed 42:3031–3034
26. Vayssieres L (2003) Growth of arrayed nanorods and nanowires of ZnO from aqueous solutions. Adv Mater 15:464–466
27. Vayssieres L, Gratzel M (2004) Highly ordered SnO_2 nanorod arrays from controlled aqueous growth. Angew Chem Int Ed 43:3666–3670
28. Olson DC, Piris J, Collins RT, Shaheen SE, Ginley DS (2006) Hybrid photovoltaic devices of polymer and ZnO nanofiber composites. Thin Solid Films 496:26–29
29. Peiro AM, Ravirajan P, Govender K, Boyle DS, O'Brien P, Bradley DDC, Nelson J, Durrant JR (2006) Hybrid polymer/metal oxide solar cells based on ZnO columnar structures. J Mater Chem 16:2088–2096
30. Ravirajan P, Peiro AM, Nazeeruddin MK, Gratzel M, Bradley DDC, Durrant JR, Nelson J (2006) Hybrid polymer/zinc oxide photovoltaic devices with vertically oriented ZnO nanorods and an amphiphilic molecular interface layer. J Phys Chem B 110:7635–7639
31. Takanezawa K, Hirota K, Wei QS, Tajima K, Hashimoto K (2007) Efficient charge collection with ZnO nanorod array in hybrid photovoltaic devices. J Phys Chem C 111:7218–7223
32. Novotny CJ, Yu ET, Yu PKL (2008) InP nanowire/polymer hybrid photodiode. Nano Lett 8:775–779
33. Tachibana Y, Moser JE, Gratzel M, Klug DR, Durrant JR (1996) Subpicosecond interfacial charge separation in dye-sensitized nanocrystalline titanium dioxide films. J Phys Chem 100:20056–20062
34. Kallioinen J, Benkö G, Sundström V, Korppi-Tommola JEI, Yartsev AP (2002) Electron transfer from the singlet and triplet excited states of Ru(dcbpy) 2(NCS) 2 into nanocrystalline TiO2 thin films. J Phys Chem B 106:4396–4404
35. Gratzel M (2005) Solar energy conversion by dye-sensitized photovoltaic cells. Inorg Chem 44:6841–6851
36. Redmond G, Fitzmaurice D (1993) Spectroscopic determination of flat-band potentials for polycrystalline TiO_2 electrodes in nonaqueous solvents. J Phys Chem 97:1426–1430

37. Pelet S, Moser JE, Gratzel M (2000) Cooperative effect of adsorbed cations and iodide on the interception of back electron transfer in the dye sensitization of nanocrystalline TiO_2. J Phys Chem B 104:1791–1795
38. Kay A, Humphry-Baker R, Gratzel M (1994) Artificial photosynthesis. 2. Investigations on the mechanism of photosensitization of nanocrystalline TiO2 solar-cells by chlorophyll derivatives. J Phys Chem 98:952–959
39. Howe RF, Gratzel M (1985) Electron-paramagnetic-res observation of trapped electrons in colloidal TiO_2. J Phys Chem 89:4495–4499
40. Gopel W, Anderson JA, Frankel D, Jaehnig M, Phillips K, Schafer JA, Rocker G (1984) Surface-defects of TiO_2 (110) – a combined XPS, XAES and ELS study. Surf Sci 139:333–346
41. Kumar A, Santangelo PG, Lewis NS (1992) Electrolysis of water at $SrTiO_3$ photoelectrodes – distinguishing between the statistical and stochastic formalisms for electron-transfer processes in fuel-forming photoelectrochemical systems. J Phys Chem 96:834–842
42. Schlichthorl G, Huang SY, Sprague J, Frank AJ (1997) Band edge movement and recombination kinetics in dye-sensitized nanocrystalline TiO_2 solar cells: a study by intensity modulated photovoltage spectroscopy. J Phys Chem B 101:8141–8155
43. Frank AJ, Kopidakis N, van de Lagemaat J (2004) Electrons in nanostructured TiO_2 solar cells: transport, recombination and photovoltaic properties. Coord Chem Rev 248:1165–1179
44. Wang P, Zakeeruddin SM, Comte P, Charvet R, Humphry-Baker R, Gratzel M (2003) Enhance the performance of dye-sensitized solar cells by co-grafting amphiphilic sensitizer and hexadecylmalonic acid on TiO_2 nanocrystals. J Phys Chem B 107:14336–14341
45. Wang P, Zakeeruddin SM, Humphry-Baker R, Moser JE, Gratzel M (2003) Molecular-scale interface engineering of TiO_2 nanocrystals: Improving the efficiency and stability of dye-sensitized solar cells. Adv Mater 15:2101–2104
46. Hara K, Dan-oh Y, Kasada C, Ohga Y, Shinpo A, Suga S, Sayama K, Arakawa H (2004) Effect of additives on the photovoltaic performance of coumarin-dye-sensitized nanocrystalline TiO_2 solar cells. Langmuir 20:4205–4210
47. He J, Benko G, Korodi F, Polvka T, Lomoth R, Akermark B, Sun L, Hagfeldt A, Sundstrom V (2002) Modified phthalocyanines for efficient near-IR sensitization of nanostructured TiO_2 electrode. J Am Chem Soc 124:4922–4932
48. Kay A, Gratzel M (1993) Artificial photosynthesis. 1. Photosensitization of TiO2 solar-cells with chlorophyll derivatives and related natural porphyrins. J Phys Chem 97:6272–6277
49. Huang SY, Schlichthorl G, Nozik AJ, Gratzel M, Frank AJ (1997) Charge recombination in dye-sensitized nanocrystalline TiO_2 solar cells. J Phys Chem B 101:2576–2582
50. Kusama H, Arakawa H (2004) Influence of aminothiazole additives in I^-/I_3^- redox electrolyte solution on Ru(II)-dye-sensitized nanocrystalline TiO_2 solar cell performance. Sol Energy Mater Sol Cells 82:457–465
51. Bauer C, Boschloo G, Mukhtar E, Hagfeldt A (2002) Interfacial electron-transfer dynamics in Ru(tcterpy)(NCS) 3-sensitized TiO2 nanocrystalline solar cells. J Phys Chem B 106:12693–12704
52. Nakade S, Saito Y, Kubo W, Kanzaki T, Kitamura T, Wada Y, Yanagida S (2003) Enhancement of electron transport in nano-porous TiO_2 electrodes by dye adsorption. Electrochem Commun 5:804–808
53. Zaban A, Ferrere S, Greg BA (1998) Relative energetics at the semiconductor sensitizing dye electrolyte interface. J Phys Chem B 102:452–460
54. Kang TS, Chun KH, Gong JS, Moon SH, Kim KJ (2000) Enhanced stability of photocurrent-voltage curves in Ru(II)-dye-sensitized nanocrystalline TiO_2 electrodes with carboxylic acids. J Electrochem Soc 147:3049–3053
55. Lindford MR, Fenter P, Eisenberger PM, Chidsey CED (1995) Alkyl monolayers on silicon prepared from 1-alkenes and hydrogen-terminated silicon. J Am Chem Soc 117:3145–3155

56. Lindford MR, Chidsey CED (1993) Alkyl monolayers covalently bonded to silicon surfaces. J Am Chem Soc 115:12631–12632
57. Bansal A, Li X, Lauermann I, Lewis NS (1996) Alkylation of Si surfaces using a two-step halogenation Grignard route. J Am Chem Soc 118:7225–7226
58. Grätzel M, Kalyanasundaram K (1994) Artificial photosynthesis – efficient dye-sensitized photoelectrochemical cells for direct conversion of visible-light to electricity. Curr Sci 66:706–714
59. Parkinson BA, Furtak TE, Canfield D, Kam K, Kline G (1980) Evaluation and reduction of efficiency losses at tungsten diselenide photoanodes. Discuss Faraday Soc 70:233–245
60. Kern R, van der Burg N, Chimel G, Ferber J, Hinsch A, Kinderman P, Kroon JM, Meyer A, Meyer T, Niepmann R, van Roosmalen J, Schill C, Sommeling P, Spath M, Uhlendorf I (2000) Long term stability of dye-sensitised solar cells for large area power applications. Opto-Electron Rev 8:284–288
61. Hagfeldt A, Gratzel M (1995) Light-Induced Redox Reactions in Nanocrystalline Systems. Chem Rev 95:49–68
62. Fisher AC, Peter LM, Ponomarev EA, Walker AB, Wijayantha KGU (2000) Intensity dependence of the back reaction and transport of electrons in dye-sensitized nanocrystalline TiO_2 solar cells. J Phys Chem B 104:949–958
63. Nazeeruddin MK, Pechy P, Renouard T, Zakeeruddin SM, Humphry-Baker R, Comte P, Liska P, Cevey L, Costa E, Shklover V, Spiccia L, Deacon GB, Bignozzi CA, Grätzel M (2001) Engineering of efficient panchromatic sensitizers for nanocrystalline TiO_2-based solar cells. J Am Chem Soc 123:1613–1624
64. Grätzel M (2004) Conversion of sunlight to electric power by nanocrystalline dye-sensitized solar cells. J Photochem Photobio A 164:3–14
65. Grätzel M (2005) Mesoscopic solar cells for electricity and hydrogen production from sunlight. Chem Lett 34:8–13
66. Grätzel M (2006) The advent of mesoscopic injection solar cells. Prog Photovolt 14:429–442
67. Solar Spectral Irradiance: Air Mass 1.5. http://rredc.nrel.gov/solar/spectra/am1.5/
68. Bucher K, Heidler K (1990) Photovoltaic measurements and calibration. Proc 5th Int Photovoltaic Sci Eng Conf Kyoto, pp 489–492
69. Emery K, Osterwald CR, Cannon TW, Myers DR, Burdick J, Glatfelter T, Czubatyj W, Yang J (1985) Proceeding of 18th IEEE Photovoltaic Specialist Conference. IEEE, New York, p 623
70. Osterwald CR (1986) Translation of device performance-measurements to reference conditions. Solar Cells 18:269–279
71. Shrotriya V, Li G, Yao Y, Moriarty T, Emery K, Yang Y (2006) Accurate measurement and characterization of organic solar cells. Adv Funct Mater 16:2016–2023
72. Field H, Emery K (1993) An uncertainty analysis of the spectral correction factor. Proceedings of the 23rd IEEE Photovoltaic Specialists Conference, Louisville, pp1180–1187
73. Emery K, Osterwald CR (1988) Current topics in photovoltaics, vol 3. Academic, London, Ch 4
74. Emery K (2003) Handbook of photovoltaic science and engineering. Luque A, Hegedus S (eds). Wiley, Chichester, UK, Ch 16
75. Ito S, Nazeeruddin MK, Liska P, Comte P, Charvet R, Pechy P, Jirousek M, Kay A, Zakeeruddin SM, Gratzel M (2005) Photovoltaic characterization of dye-sensitized solar cells: effect of device masking on conversion efficiency. Prog Photovolt 14:589–601
76. Koide N, Han L (2004) Measuring methods of cell performance of dye-sensitized solar cells. Rev Sci Instrum 75:2828–2831
77. Ito S, Matsui H, Okada K, Kusano S, Kitamura T, Wada Y, Yanagida S (2004) Calibration of solar simulator for evaluation of dye-sensitized solar cells. Sol Energy Mater Sol Cells 82:421–429

78. Peng TY, Hasegawa A, Qiu JR, Hirao K (2003) Fabrication of titania tubules with high surface area and well-developed mesostructural walls by surfactant-mediated templating method. Chem Mater 15:2011–2016
79. Cao F, Oskam G, Meyer GJ, Searson PC (1996) Electron transport in porous nanocrystalline TiO_2 photoelectrochemical cells. J Phys Chem 100:17021–17027
80. de Jongh PE, Vanmaekelbergh D (1996) Trap-limited electronic transport in assemblies of nanometer-size TiO_2 particles. Phys Rev Lett 77:3427–3430
81. Nelson J, Haque SA, Klug DR, Durrant JR (2001) Trap-limited recombination in dye-sensitized nanocrystalline metal oxide electrodes. Phys Rev B 63:205321
82. Dloczik L, Ileperuma O, Lauermann I, Peter LM, Ponomarev EA, Redmond G, Shaw NJ, Uhlendorf I (1997) Dynamic response of dye-sensitized nanocrystalline solar cells: characterization by intensity-modulated photocurrent spectroscopy. J Phys Chem B 101:10281–10289
83. Schlichthorl G, Park NG, Frank AJ (1999) Evaluation of the charge-collection efficiency of dye-sensitized nanocrystalline TiO_2 solar cells. J Phys Chem B 103:782–791
84. van de Lagemaat J, Frank AJ (2001) Nonthermalized electron transport in dye-sensitized nanocrystalline TiO_2 films: transient photocurrent and random-walk modeling studies. J Phys Chem B 105:11194–11205
85. Benkstein KD, Kopidakis N, van de Lagemaat J, Frank AJ (2003) Influence of the percolation network geometry on electron transport in dye-sensitized titanium dioxide solar cells. J Phys Chem B 107:7759–7767
86. Kopidakis N, Benkstein KD, van de Lagemaat J, Frank AJ, Yuan Q, Schiff EA (2006) Temperature dependence of the electron diffusion coefficient in electrolyte-filled TiO_2 nanoparticle films: evidence against multiple trapping in exponential conduction-band tails. Phys Rev B 73:045326
87. Cassiers K, Linssen T, Mathieu M, Bai YQ, Zhu HY, Cool P, Vansant EF (2004) Surfactant-directed synthesis of mesoporous titania with nanocrystalline anatase walls and remarkable thermal stability. J Phys Chem B 108:3713–3721
88. Law M, Greene LE, Johnson JC, Saykally R, Yang PD (2005) Nanowire dye-sensitized solar cells. Nat Mater 4:455–459
89. Tenne R, Rao CNR (2004) Inorganic nanotubes. Phil Trans R Soc A 362:2099–2125
90. Adachi M, Murata Y, Okada I, Yoshikawa S (2003) Formation of titania nanotubes and applications for dye-sensitized solar cells. J Electrochem Soc 150:G488–G493
91. Ong KG, Varghese OK, Mor GK, Shankar K, Grimes CA (2007) Application of finite-difference time domain to dye-sensitized solar cells: the effect of nanotube-array negative electrode dimensions on light absorption. Sol Energy Mater Sol Cells 91:250–257
92. Taflove A (1995) Computational electrodynamics: the finite-difference time-domain method. Artech House Inc, Boston
93. Ong KG, Varghese OK, Mor GK, Grimes CA (2005) Numerical simulation of light propagation through highly-ordered titania nanotube arrays: dimension optimization for improved photoabsorption. J Nanosci Nanotechnol 5:1801–1808
94. Roden JA, Gedney SD (1997) Efficient implementation of the uniaxial-based PML media in three-dimensional nonorthogonal coordinates with the use of the FDTD technique. Microwave Opt Tech Lett 14:71–75
95. Balanis CA (1989) Advanced engineering electromagnetics. Wiley, New York Ch 5
96. Nazeeruddin MK, Kay A, Rodicio I, Humphry-Baker R, Muller E, Liska P, Vlachopoulos N, Gratzel M (1993) Conversion of light to electricity by cis-x2bis(2, 2′-bipyridyl-4, 4′-dicarboxylate) ruthenium(II) charge-transfer sensitizers (x=Cl-, Br-, I-, CN-, and SCN-) on nanocrystalline TiO2 electrodes. J Am Chem Soc 115:6382–6390
97. Paulose M, Shankar K, Varghese OK, Mor GK, Grimes CA (2006) Application of highly-ordered TiO_2 nanotube-arrays in heterojunction dye-sensitized solar cells. J Phys D 39:2498–2503

98. Mor GK, Shankar K, Paulose M, Varghese OK, Grimes CA (2006) Use of highly-ordered TiO_2 nanotube arrays in dye-sensitized solar cells. Nano Lett 6:215–218
99. Mor GK, Varghese OK, Paulose M, Mukherjee N, Grimes CA (2003) Fabrication of tapered, conical-shaped titania nanotubes. J Mater Res 18:2588–2593
100. Yang DJ, Park H, Cho SJ, Kim HG, Choi WY (2008) TiO_2-nanotube-based dye-sensitized solar cells fabricated by an efficient anodic oxidation for high surface area. J Phys Chem Solids 69:1272–1275
101. Shankar K, Bandara J, Paulose M, Wietasch H, Varghese OK, Mor GK, LaTempa TJ, Thelakkat M, Grimes CA (2008) Highly efficient solar cells using TiO_2 nanotube arrays sensitized with a donor-antenna dye. Nano Lett 8:1654–1659
102. Karthikeyan CS, Wietasch H, Thelakkat M (2007) Highly efficient solid-state dye-sensitized TiO_2 solar cells using donor-antenna dyes capable of multistep charge-transfer cascades. Adv Mater 19:1091–1095
103. Handa S, Wietasch H, Thelakkat M, Durrant JR, Haque SA (2007) Reducing charge recombination losses in solid state dye sensitized solar cells: the use of donor-acceptor sensitizer dyes. Chem Commun 17:1725–1727
104. Haque SA, Handa S, Peter K, Palomares E, Thelakkat M, Durrant JR (2005) Supermolecular control of charge transfer in dye-sensitized nanocrystalline TiO_2 films: towards a quantitative structure-function relationship. Angew Chem Int Ed 44:5740–5744
105. Karthikeyan CS, Peter K, Wietasch H, Thelakkat M (2007) Highly efficient solid-state dye-sensitized TiO_2 solar cells via control of retardation of recombination using novel donor-antenna dyes. Sol Energy Mater Sol Cells 91:432–439
106. Sommeling PM, O'Regan BC, Haswell RR, Smit HJP, Bakker NJ, Smits JJT, Kroon JM, van Roosmalen JAM (2006) Influence of a $TiCl_4$ post-treatment on nanocrystalline TiO_2 films in dye-sensitized solar cells. J Phys Chem B 110:19191–19197
107. Park JH, Lee TW, Kang MG (2008) Growth, detachment and transfer of highly-ordered TiO_2 nanotube arrays: use in dye-sensitized solar cells. Chem Commun 25:2867–2869
108. Paulose M, Shankar K, Varghese OK, Mor GK, Hardin B, Grimes CA (2006) Backside illuminated dye-sensitized solar cells based on titania nanotube array electrodes. Nanotechnol 17:1446–1448
109. Varghese OK, Gong D, Paulose M, Ong KG, Dickey EC, Grimes CA (2003) Extreme changes in the electrical resistance of titania nanotubes with hydrogen exposure. Adv Mater 15:624–627
110. Balaur E, Macak JM, Tsuchiya H, Schmuki P (2005) Wetting behaviour of layers of TiO_2 nanotubes with different diameters. J Mater Chem 15:4488–4491
111. Shankar K, Mor GK, Prakasam HE, Yoriya S, Paulose M, Varghese OK, Grimes CA (2007) Highly-ordered TiO_2 nanotube-arrays up to 220 μm in length: use in water photoelectrolysis and dye-sensitized solar cells. Nanotechnol 18:065707
112. Kuang D, Brillet J, Chen P, Takata M, Uchida S, Miura H, Sumioka K, Zakeeruddin SM, Grätzel M (2008) Application of highly ordered TiO_2 nanotube arrays in flexible dye-sensitized solar cells. ACS Nano 2:1113–1116
113. Kuang DB, Ito S, Wenger B, Klein C, Moser JE, Humphry-Baker R, Zakeerudddin SM, Gratzel M (2006) High molar extinction coefficient heteroleptic Ruthenium complexes for thin film dye-sensitized solar cells. J Am Chem Soc 128:4146–4154
114. Kuang D, Uchida S, Humphry-Baker R, Zakeeruddin SM, Grätzel M (2008) Organic dye-sensitized ionic liquid based solar cells: remarkable enhancement in performance through molecular design of indoline sensitizers. Angew Chem Int Ed 47:1923–1927
115. Kuang DB, Klein C, Zhang ZP, Ito S, Moser JE, Zakeeruddin SM, Grätzel M (2007) Stable, high-efficiency ionic-liquid-based mesoscopic dye-sensitized solar cells. Small 3:2094–2102
116. Kuang DB, Wang P, Zakeeruddin SM, Grätzel M (2006) Stable mesoscopic dye-sensitized solar cells based on tetracyanoborate ionic liquid electrolyte. J Am Chem Soc 128:7732–7733

117. Yang Y, Wang X, Li L (2008) Synthesis and photovoltaic application of high aspect-ratio TiO_2 nanotube arrays by anodization. J Am Ceram Soc 91:3086–3089
118. Lin CJ, Yu WY, Chien SH (2008) Rough conical-shaped TiO_2-nanotube arrays for flexible back illuminated dye-sensitized solar cells. Appl Phys Lett 93:133107
119. Wang H, Yip CT, Cheung KY, Djurišić AB, Xie MH (2006) Titania-nanotube-array-based photovoltaic cells. Appl Phys Lett 89:023508
120. Han L, Koide N, Chiba Y, Islam A, Komiya R, Fuke N, Fukui A, Yamanaka R (2005) Improvement of efficiency of dye-sensitized solar cells by reduction of internal resistance. Appl Phys Lett 86:213501
121. Stergiopoulos T, Ghicov A, Likodimos V, Tsoukleris DS, Kunze J, Schmuki P, Falaras P (2008) Dye-sensitized solar cells based on thick highly ordered TiO_2 nanotubes produced by controlled anodic oxidation in non-aqueous electrolytic media. Nanotechnol 19:235602
122. Yip CT, Mak CSK, Djurišíc AB, Hsu YF, Chan WK (2008) Dye-sensitized solar cells based on TiO_2 nanotube/porous layer mixed morphology. Appl Phys A 92:589–593
123. Oey CC, Djurišic AB, Wang H, Man KKY, Chan WK, Xie MH, Leung YH, Pandey A, Nunzi JM, Chui PC (2006) Polymer-TiO2 solar cells: TiO2 interconnected network for improved cell performance. Nanotechnol 17:706–713
124. Wang H, Oey CC, Djurišíc AB, Xie MH, Leung YH, Man KKY, Chan WK, Pandey A, Nunzi JM, Chui PC (2005) Titania bicontinuous network structures for solar cell applications. Appl Phys Lett 87:023507
125. Xie ZB, Adams S, Blackwood DJ, Wang J (2008) The effects of anodization parameters on titania nanotube arrays and dye sensitized solar cells. Nanotechnol 19:405701
126. Zaban A, Greenshtein M, Bisquert J (2003) Determination of the electron lifetime in nanocrystalline dye solar cells by open-circuit voltage decay measurements. Chem Phys Chem 4:859–864
127. Fabregat-Santiago F, Garcia-Canadas J, Palomares E, Clifford JN, Haque SA, Durrant JR, Garcia-Belmonte G, Bisquert J (2004) The origin of slow electron recombination processes in dye-sensitized solar cells with alumina barrier coatings. J Appl Phys 96:6903–6907
128. Bisquert J, Zaban A, Greenshtein M, Mora-Sero I (2004) Determination of rate constants for charge transfer and the distribution of semiconductor and electrolyte electronic energy levels in dye-sensitized solar cells by open-circuit photovoltage decay method. J Amer Chem Soc 126:13550–13559
129. Niinobe D, Makari Y, Kitamura T, Wada Y, Yanagida S (2005) Origin of enhancement in open-circuit voltage by adding ZnO to nanocrystalline SnO_2 in dye-sensitized solar cells. J Phys Chem B 109:17892–17900
130. Mor GK, Shankar K, Paulose M, Varghese OK, Grimes CA (2005) Enhanced photocleavage of water using titania nanotube arrays. Nano Lett 5:191–195
131. Zhu K, Neale NR, Miedaner A, Frank AJ (2007) Enhanced charge-collection efficiencies and light scattering in dye-sensitized solar cells using oriented TiO_2 nanotubes arrays. Nano Lett 7:69–74
132. Macak JM, Tsuchiya H, Taveira L, Aldabergerova S, Schmuki P (2005) Smooth anodic TiO_2 nanotubes. Angew Chem Int Ed 44:7463–7465
133. Neale NR, Kopidakis N, van de Lagemaat J, Gratzel M, Frank AJ (2005) Effect of a coadsorbent on the performance of dye-sensitized TiO_2 solar cells: shielding versus band-edge movement. J Phys Chem B 109:23183–23189
134. van de Lagemaat J, Park NG, Frank AJ (2000) Influence of electrical potential distribution, charge transport, and recombination on the photopotential and photocurrent conversion efficiency of dye-sensitized nanocrystalline TiO_2 solar cells: a study by electrical impedance and optical modulation techniques. J Phys Chem B 104:2044–2052
135. Franco G, Gehring J, Peter LM, Ponomarev EA, Uhlendorf I (1999) Frequency-resolved optical detection of photoinjected electrons in dye-sensitized nanocrystalline photovoltaic cells. J Phys Chem B 103:692–698

136. Varghese OK, Gong DW, Paulose M, Grimes CA, Dickey EC (2003) Crystallization and high-temperature structural stability of titanium oxide nanotube arrays. J Mater Res 18:156–165
137. Zhu K, Vinzant TB, Neale NR, Frank AJ (2007) Removing structural disorder from oriented TiO_2 nanotube arrays: reducing the dimensionality of transport and recombination in dye-sensitized solar cells. Nano Lett 7:3739–3746
138. Jennings JR, Ghicov A, Peter LM, Schmuki P, Walker AB (2008) Dye-sensitized solar cells based on oriented TiO_2 nanotube arrays: transport, trapping, and transfer of electrons. J Am Chem Soc 130:13364–13372
139. Nunzi JM (2002) Organic photovoltaic materials and devices. C R Phys 3:523–542
140. Miranda PB, Moses D, Heeger AJ (2001) Ultrafast photogeneration of charged polarons in conjugated polymers. Phys Rev B 64:81201
141. Mozer A, Sariciftci NS (2006) Conjugated polymer photovoltaic devices and materials. C R Chim 9:568–577
142. Brabec C, Zerza G, Cerullo G, De Silvestri S, Luzatti S, Hummelen JC, Sariciftci NS (2001) Tracing photoinduced electron transfer process in conjugated polymer/fullerene bulk heterojunctions in real time. Chem Phys Lett 340:232–236
143. Parker ID (1994) Carrier tunneling and device characteristics in polymer light-emitting-diodes. J Appl Phys 75:1656–1666
144. Liu J, Shi YJ, Yang Y (2001) Solvation-induced morphology effects on the performance of polymer-based photovoltaic devices. Adv Funct Mater 11:420–424
145. Van Duren JJ, Loos J, Morissey F, Leewis CM, Kivits KP, Vanzendoorn LJ, Rispens MT, Hummelen JC, Janssen RAJ (2002) In-situ compositional and structural analysis of plastic solar cells. Adv Funct Mater 12:665–669
146. Bulle-Lieuwma CW, VanGenip WJH, Van Duren JKJ, Jankheim P, Janssen R, Niemantsverdiet JW (2003) Characterization of polymer solar cells by TOF-SIMS depth profiling. Appl Surf Sci 203:547–550
147. Van Duren J, Yang X, Loos J, Bulle-Lieuwma CWT, Sievel AB, Hummelen JC, Janssen RAJ (2004) Relating the morphology of poly(p-phenylene vinylene)/methanofullerene blends to solar-cell performance. Adv Funct Mater 14:425–434
148. Hoppe H, Glatzel T, Niggemann M, Schwinger W, Schaeffler F, Hinsch A, Lux-Steiner M, Sariciftci NS (2006) Efficiency limiting morphological factors of MDMO-PPV: PCBM plastic solar cells. Thin Solid Films 511–512:587–592
149. Hoppe H, Niggemann M, Winder C, Kraut J, Hiesgh R, Hinsch A, Meissner D, Sariciftci NS (2004) Nanoscale morphology of conjugated polymer/fullerene-based bulk-heterojunction solar cells. Adv Funct Mater 14:1005–1011
150. Gebeyehu D, Brabec CJ, Padinger F, Fromherz T, Hummelen JC, Badt D, Schindler H, Sariciftci NS (2001) The interplay of efficiency and morphology in photovoltaic devices based on interpenetrating networks of conjugated polymers with fullerenes. Synth Met 118:1–9
151. Martens T, Hoen JD, Munters T, Beelen Z, Goris L, Monca J, Oliesloeger MD, Vanderzende D, De Schopper L, Andriessen R (2003) Disclosure of the nanostructure of MDMO-PPV: PCBM bulk hetero-junction organic solar cells by a combination of SPM and TEM. Synth Met 138:243–247
152. Arias AC, MacKenzie JD, Stevenson R, Halls JM, Inbasekaran M, Woo EP, Richards D, Friend RH (2001) Photovoltaic performance and morphology of polyfluorene blends: a combined microscopic and photovoltaic investigation. Macromol 34:6005–6016
153. Gadisa A, Svensson M, Andersson MR, Inganas O (2004) Correlation between oxidation potential and open-circuit voltage of composite solar cells based on blends of polythiophenes/fullerene derivative. Appl Phys Lett 84:1609–1611
154. Schilinsky P, Waldauf C, Hauch J, Brabec CJ (2004) Simulation of light intensity dependent current characteristics of polymer solar cells. J Appl Phys 95:2816–2819

155. Riedel I, Dyakonov V (2004) Influence of electronic transport properties of polymer-fullerene blends on the performance of bulk heterojunction photovoltaic devices. Phys Status Solidi A 201:1332–1341
156. Padinger F, Rittberger R, Sariciftci NS (2003) Effects of postproduction treatment on plastic solar cells. Adv Funct Mater 13:85–88
157. Schilinsky P, Waldauf P, Brabec CJ (2002) Recombination and loss analysis in polythiophene based bulk heterojunction photodetectors. Appl Phys Lett 81:3885–3887
158. Kim J, Kim S, Lee H, Lee K, Ma WL, Huong X, Heeger AJ (2006) New architecture for high-efficiency polymer photovoltaic cells using solution-based titanium oxide as an optical spacer. Adv Mater 18:572–576
159. Prosa TJ, Winokur MJ, Moulton MM, Smith P, Heeger AJ (1992) X-ray structural studies of poly(3-alkylthiophenes) – an example of an inverse comb. Macromol 25:4364–4372
160. Schuller S, Schilinsky P, Hauch J, Brabec CJ (2004) Determination of the degradation constant of bulk heterojunction solar cells by accelerated lifetime measurements. Appl Phys A 79:37–40
161. Bao Z, Dodabalapur A, Lovinger A (1996) Soluble and processable regioregular poly (3-hexylthiophene) for thin film field-effect transistor applications with high mobility. Appl Phys Lett 69:4108–4110
162. Sirringhaus H, Tessler N, Friend RH (1998) Integrated optoelectronic devices based on conjugated polymers. Science 280:1741–1744
163. Mihailetchi VD, Van Duren JKJ, Blom PWM, Hummelen JC, Janssen RAJ, Kroon JM, Rispens MT, Verhees WJH, Wienk MM (2003) Electron transport in a methanofullerene. Adv Funct Mater 13:43–46
164. Günes S, Neugebauer H, Sariciftci NS (2007) Conjugated polymer-based organic solar cells. Chem Rev 107:1324–1338
165. Jenekhe S, Chen XL (1998) Self-assembled aggregates of rod-coil block copolymers and their solubilization and encapsulation of fullerenes. Science 279:1903–1907
166. Hadziioannou G, von Hutten F (2000) Semiconducting polymers. Wiley-VCH, Weinheim
167. Sariciftci NS, Sun SS (2005) Organic photovoltaics. Taylor & Francis, London
168. Tracz A, Jeszka JK, Watson MD, Pisula W, Mullen K, Pakula T (2003) Uniaxial alignment of the columnar super-structure of a hexa (alkyl) hexa-peri-hexabenzocoronene on untreated glass by simple solution processing. J Am Chem Soc 125:1682–1683
169. Olsa DC, Piris J, Collins RT, Shaheen S, Ginley D (2006) Hybrid photovoltaic devices of polymer and ZnO nanofiber composites. Thin Solid Films 496:26–29
170. Yu G, Gao J, Hummelen JC, Wudl F, Heeger AJ (1995) Polymer photovoltaic cells – enhanced efficiencies via a network of internal donor-acceptor heterojunctions. Science 270:1789–1791
171. Gur I, Fromer NA, Chen CP, Kanaras AG, Alivisatos AP (2007) Hybrid solar cells with prescribed nanoscale morphologies based on hyperbranched semiconductor nanocrystals. Nano Lett 7:409–414
172. Coakley KM, Liu YX, McGehee MD, Frindell KL, Stucky GD (2003) Infiltrating semiconducting polymers into self-assembled mesoporous titania films for photovoltaic applications. Adv Funct Mater 13:301–306
173. Peet J, Kim JY, Coates NE, Ma WL, Moses D, Heeger AJ, Bazan GC (2007) Efficiency enhancement in low-bandgap polymer solar cells by processing with alkane dithiols. Nat Mater 6:497–500
174. Coakley KM, Srinivasan BS, Ziebarth JM, Goh C, Liu Y, McGehee MD (2005) Enhanced hole mobility in regioregular polythiophene infiltrated in straight nanopores. Adv Funct Mater 15:1927–1932
175. Cadby AJ, Tolbert SH (2005) Controlling optical properties and interchain interactions in semiconducting polymers by encapsulation in periodic nanoporous silicas with different pore sizes. J Phys Chem B 109:17879–17886

176. Roux C, Leclerc M (1992) Rod-to-coil transition in alkoxy-substituted polythiophenes. Macromol 25:2141–2144
177. Yanagida S, Senadeera GKR, Nakamura K, Kitamura T, Wada Y (2004) Polythiophene-sensitized TiO_2 solar cells. J Photochem Photobiol A 166:75–80
178. Mwaura JK, Zhao XY, Jiang H, Schanze KS, Reynolds JR (2006) Spectral broadening in nanocrytalline TiO_2 solar cells based on poly(p-phenylene ethynylene) and polythiophene sensitizers. Chem Mater 18:6109–6111
179. Schmidt-Mende L, Bach U, Humphry-Baker R, Horiuchi T, Miura H, Ito S, Uchida S, Gratzel M (2005) Organic dye for highly efficient solid-state dye-sensitized solar cells. Adv Mater 17:813–815
180. Senadeera GKR, Kitamura T, Wada Y, Yanagida S (2005) Photosensitization of nanocrystalline TiO_2 films by a polymer with two carboxylic groups, poly(3-thiophenemalonic acid). Sol Energy Mater Sol Cells 88:315–322
181. Liu YX, Scully SR, McGehee MD, Liu JS, Luscombe CK, Frechet JMJ, Shaheen SE, Ginley DS (2006) Dependence of band offset and open-circuit voltage on the interfacial interaction between TiO_2 and carboxylated polythiophenes. J Phys Chem B 110:3257–3261
182. Senadeera GKR, Fukuri N, Saito Y, Kitamura T, Wada Y, Yanagida S (2005) Volatile solvent-free solid-state polymer-sensitized TiO2 solar cells with poly(3,4-ethylenedioxythiophene) as a hole-transporting medium. Chem Commun 17:2259–2261
183. Senadeera GKR, Pathirathne WMTC (2004) Utilization of conducting polymer as a sensitizer in solid-state photocells. Curr Sci 87:339–342

Chapter 6
Use of TiO_2 Nanotube Arrays for Biological Applications

6.1 Introduction

Protein immobilization on solid substrates underlies various experimental approaches in biology and biophysics [1–5]. Immobilized proteins are instrumental in identifying protein–protein, protein–DNA, and protein–molecule interactions for a variety of diagnostic and profiling purposes [5–9]. The support material must have active surface areas for protein binding, and good mechanical, thermal, and chemical stability. Bioelectrocatalytic systems allow the sensitive detection of affinity-based interactions between complementary molecule pairs [10] through electrical signals related to electrochemical reactions, while amperometric biosensors offer a convenient and potential application in the area of biomedical diagnosis as well as environmental analysis [11–13]. In this chapter, we consider the use of co-immobilized TiO_2 nanotube arrays as a biosensor platform for H_2O_2 and glucose detection.

Titanium and its alloys are widely used as implants due to its high strength, biocompatibility and high level of hemocompatibility [14, 15]. The high degree of Ti alloy biocompatibility is due to their ability to form stable and dense thin oxide layers in most environments. It is believed that thicker and more stable TiO_2-based oxide surfaces are generally favorable for surface bioactivity [16, 17]. Spark anodization is commonly used to increase the biocompatibility of titanium and its alloys, with the process leading to the formation of a disordered oxide structure several hundred nanometers thick [18, 19]. In contrast to this approach, the electrochemical formation of highly ordered TiO_2 nanotube arrays offer a unique surface for biomedical implants that offers both biocompatibility as well as drug eluting properties. We review the use of TiO_2 nanotube arrays to enhance apatite formation, cell activity, drug elution, and the application of TiO_2 nanotubular membranes for protein separation and drug delivery.

C.A. Grimes and G.K. Mor, *TiO_2 Nanotube Arrays*,
DOI 10.1007/978-1-4419-0068-5_6, © Springer Science+Business Media, LLC 2009

6.2 Biosensors

6.2.1 H_2O_2 Detection: Nanotubes Co-immobilized with HRP and Thionine

Liu et al. reported the use of horseradish peroxidase (HRP), which can be reoxidized chemically by hydrogen peroxide (H_2O_2) [20–22], and thionine coadsorbed on TiO_2 nanotube arrays for the development of a novel H_2O_2 electrochemical biosensor [23]. Nanostructured TiO_2 shows excellent biocompatibility and retention of biological activity for protein binding [24, 25]. Further, the TiO_2 nanotube array substrates possess large surface areas together with good uniformity and conformability over large areas, traits desirable for electrochemical biosensor design [26, 27]. Liu [23] prepared TiO_2 nanotube arrays in 1:8 acetic acid–water solution containing 1.0 vol% hydrofluoric acid at 20 V for 45 min. The substrates were then co-immobilized with HRP and thionine by immersing the TiO_2 nanotube electrodes in 10 μM thionine solution in 5 mM phosphate buffer (PB) at pH 7.0 containing 2 mU/mL of HRP for over 5 h [23]. For comparison, TiO_2 nanotube electrodes were immersed in (a) 5 mM PB at pH 7.0, (b) 2 mU/mL of HRP in 5 mM PB solution at pH 7.0, and (c) 10 μM thionine solution in 5 mM PB at pH 7.0 for 5 h to generate the bare, HRP, and thionine-modified TiO_2 nanotube electrodes (TiO_2, HRP/TiO_2, and Th/TiO_2), respectively. The resulting electrodes were stored in 5 mM PB at pH 7.0 and 4°C.

Cyclic voltammograms (CV) of Th/TiO_2 electrode were measured in 0.1 M PB at pH 6.8 with a scan rate of 50 mV/s. A well-defined reduction peak appeared at −0.477 V for the Th/TiO_2 electrode during the first cycle, which was attributed to the reduction of the adsorbed thionine on the electrode surface [23]. No peak is observed for both the bare Ti foil and TiO_2 nanotube electrode in the same potential range. No oxidative peaks are seen in the reverse scans, and no peaks in the following scans can be clearly resolved for the Th/TiO_2 electrode, indicating that the immobilized thionine on the TiO_2 nanotube electrode is reduced during the first scan. Increasing the electrolyte pH (from 5.0 to 9.0) resulted in a linear negative shift of the reduction peak potential [28]. Similar to the Th/TiO_2, a reduction peak was observed for the Th/HRP/TiO_2 electrode only during the first cycle in 0.1 M PB at pH 6.8, with no peak seen at the HRP/TiO_2 electrode in the same potential range. The response of Th/HRP/TiO_2 electrode was, therefore, attributed to reduction of the adsorbed thionine. However, the peak current was smaller and the peak potential had a positive shift compared to the Th/TiO_2 reduction peak. The results indicate that HRP and thionine were co-immobilized on the TiO_2 nanotube array surface, and that adsorption of HRP decreased the thionine load on the electrode surface [23]. The absorbance of the immobilized HRP (403 nm) was very close to the absorbance of the free HRP in solution (405 nm), indicating that the immobilized HRP retained its native structure [29] as desired for electrochemical biosensor applications [23].

Figure 6.1 shows CV recorded for the Th/HRP/TiO_2 electrode in 0.1 M PB at pH 7.0, 50 mV/s in: (a) the absence of H_2O_2, and (b) the presence of 0.2 mM H_2O_2 [23]. The CV curves show a dramatic increase of the reduction current in curve b compared to curve a. In contrast, the reduction current of both Th/TiO_2 and HRP/TiO_2 electrodes (curves c and d in the inset of Fig. 6.1, respectively) only slightly increases after injecting 0.2 mM H_2O_2 into the electrochemical cell. Thus, the large increase of reduction current for the Th/HRP/TiO_2 electrode shows a high electrocatalytic activity of the immobilized HRP for thionine oxidation in the presence of H_2O_2. Both immobilized HRP and thionine are necessary to maintain the cycle of the electrocatalytic reactions. The reduction current of Th/HRP/TiO_2 increases with increasing H_2O_2 concentration. Figure 6.2 presents the chronoamperometric response of the: (a) Th/HRP/TiO_2, and (b) Th/TiO_2 electrodes with successive additions of H_2O_2 into 0.1 M PB at pH 6.8, recorded under a stirred system at the potential of -0.45 V [23]. The detection limit was estimated as 1.2×10^{-6}M. The Th/HRP/TiO_2 electrode stored in PB at 4°C retained 86% of its initial current response after 2 weeks of storage. The relative standard deviation was 8.4% for six

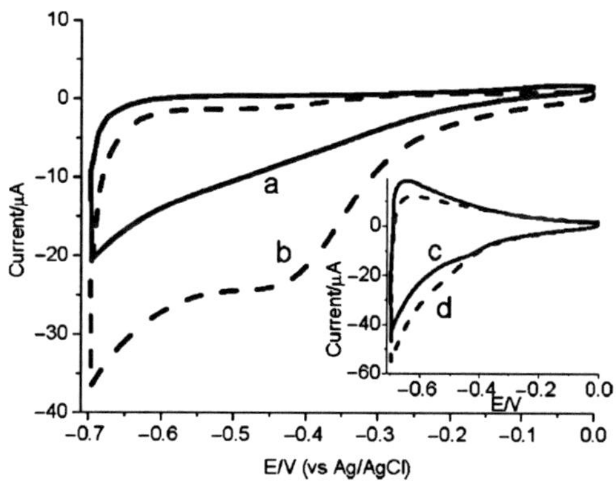

Fig. 6.1 Cyclic voltammograms of Th/HRP/TiO_2 and Th/TiO_2 (*inset*) in 0.1 M PB at pH 6.8 (**a** and **c**) before and (**b** and **d**) after the addition of 0.2 mM H_2O_2. (Reprinted with permission from [23] copyright 2005 American Chemical Society)

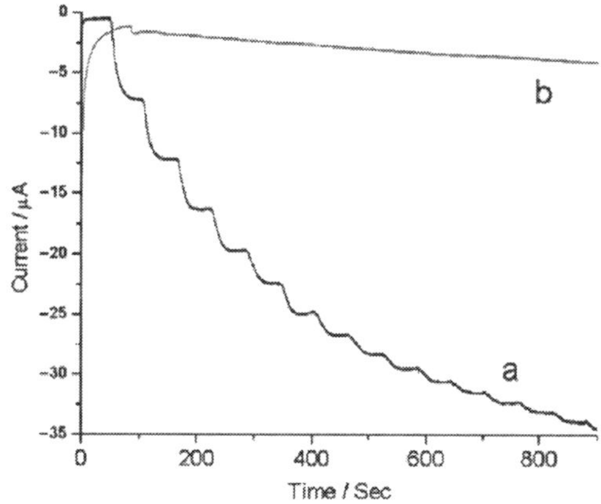

Fig. 6.2 Amperometric response of (**a**) Th/HRP/TiO_2 and (**b**) Th/TiO_2 at –0.45 V upon successive additions of 0.134 mM H_2O_2 into 0.1 M PB at pH 6.8. (Reprinted with permission from [23] copyright 2005 American Chemical Society)

successive determinations at an H_2O_2 concentration of 0.2 mM. The fabrication of six electrodes showed reproducibility with a relative standard deviation of 12.2% for the current determined at 0.2 mM H_2O_2 [23].

Xiao et al. co-adsorbed horseradish peroxidase and thionine chloride in KF prepared TiO_2 nanotubes and studied their sensitivity to hydrogen peroxide by means of cyclic voltammetric and galvanostatic measurements [30]. These biosensors demonstrated a hydrogen peroxide sensitivity in the range of 10^{-5} to 3×10^{-3} M at pH 6.7 and at a potential of −0.6 V (vs. Ag/AgCl).

6.2.2 Co-Immobilized with Cytochrome c

Zhao et al. coated TiO_2 nanotube array films with gold nanocrystals for examining the immobilization and direct electrochemistry of cytochrome *c* (Cyt *c*). The gold nanocrystals were found to significantly enhance the electron transfer rate of Cyt *c* [31]. A pair of redox peaks due to the direct electron transfer of Cyt *c* was obtained both in Cyt *c* solution and for the immobilized Cyt *c*. The Cyt *c*/Au–TiO_2 nanotubes/Ti electrode was used to electrocatalyze the reduction of H_2O_2 as a biosensor with a detection range from 2×10^{-6} to 3.49×10^{-4} mol/L in the pH range 3–8 [31].

6.2.3 Detection of H_2O_2 and Glucose

Xie et al. employed a TiO_2 nanotube array substrate in developing a biosensor based on a bioelectrocatalytic redox system for amperometeric detection [32]. Glucose oxidase (GO) can catalyze glucose with concomitant release of H_2O_2, which enables an electrochemical redox on the TiO_2/Ti electrode for amperometric detection. The current intensity primarily reflects the amount of interfacial electron transfer on the enzyme-functionalized TiO_2/Ti nanotube array electrode. TiO_2/Ti electrodes were immersed in 50 g/L GO, 0.1 M phosphate buffer solution for 12 h at 4°C, with surface immobilization achieved with a 2.0 mM pyrrole conductive monomer at 0.8 V for 20 min [32]. Current intensity versus H_2O_2 concentration was initially conducted in phosphate buffered saline (PBS) under a constant potential to establish a calibration curve. Quantitative determination of glucose was achieved by amperometric measurement of H_2O_2 based on electrochemical reduction. To keep the original bioactivity of GO the electrochemical process was carried out in 50 mL 0.1 M PBS electrolyte aerated with 25 mL/min oxygen [32].

The electrochemical behavior of a pure TiO_2 nanotube array electrode in the presence of H_2O_2 in PBS electrolyte showed a strong reduction peak below −0.20 V vs. SCE [32]; the reduction current amplitude depended on H_2O_2 concentration as well as the applied potential. In the presence of glucose (PBS electrolyte) the GO–TiO_2 nanotube array exhibited a strong reduction peak when the negative potential was below −0.21 V, a potential less negative than in the absence of glucose. GO catalyzes the aerobic oxidation of glucose to form H_2O_2 in oxygen-saturated PBS. The quantitative determination of glucose as well as H_2O_2 becomes feasible when the amperometric response of the GO–TiO_2/Ti electrode is initially calibrated. The reduction current was found to increase with cathodic potential as did the response time; for sensing applications it was decided to set the working electrode potential at −0.4 V vs. SCE which had showed a responsive time 2.1 s. A standard calibration curve was obtained for quantitative measurement of H_2O_2 amount using this TiO_2/Ti electrode, which demonstrates a linear relationship between steady current ($I_{H_2O_2}$, μA) and H_2O_2 concentration ($C_{H_2O_2}$, mM) in the full range of 0.1–2.0 mM, as shown in Fig. 6.3. According to this calibration plot, the detection limit of H_2O_2 was determined as 2.0×10^{-4}M for pure TiO_2/Ti nanotube array at a signal-to-noise of 3.

Amperometric detection of glucose was performed based on enzymic catalysis and electrochemical reduction reaction on GO–TiO_2/Ti nanotube array electrode at −0.40 V vs. SCE [32]. When the glucose concentration in PBS increased from 0.01 to 1.0 mM, the steady current jumped from −27.2 to −136.2 μA while the responsive time increased from 1.1 to 5.6 s. Figure 6.4 reveals that a nonlinear dependent relationship was obtained in the full concentration range of 0.01–1.0 mM between

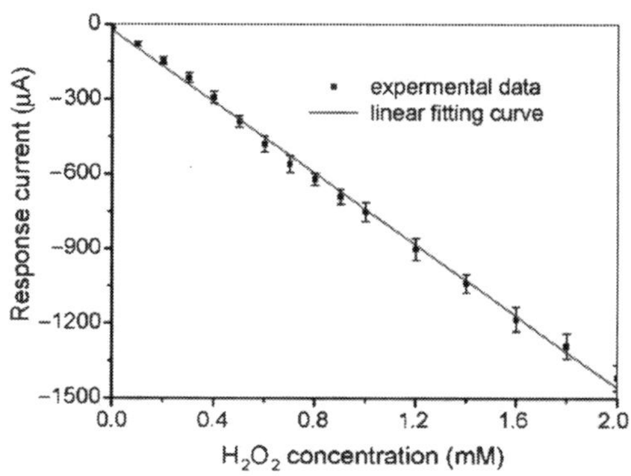

Fig. 6.3 Standard calibration plot of pure TiO_2/Ti nanotube electrode for amperometric detection of H_2O_2. (Reprinted from [32]. Copyright 2007, with permission from Elsevier)

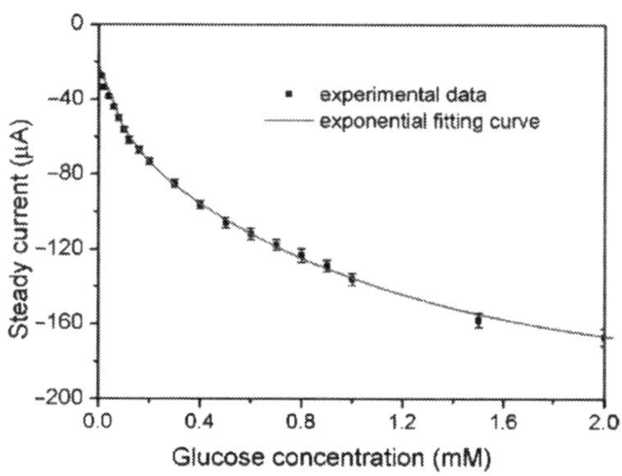

Fig. 6.4 Standard calibration plot of GO–TiO_2/Ti nanotube electrode for amperometric detection of glucose. (Reprinted from [32]. Copyright 2007, with permission from Elsevier)

the steady current ($I_{glucose}$, μA) and the glucose concentration ($C_{glucose}$, mM). According to this calibration plot, the detection limit of glucose was determined as 2.0×10^{-3} M for GO–TiO_2 /Ti biosensor at a signal-to-noise of 3. The reproducible current response of GO–TiO_2 /Ti nanotube electrode with relative standard deviation of 2.3% was observed for the constant 1.0 mM glucose with detection sensitivity of 45.5 μA/mM/cm^2 [32].

Pang et al. recently fabricated a glucose biosensor by using GO on a modified TiO_2 nanotube/Carbon nanotube/Pt coated electrode and demonstrated a sensitivity of 0.24 μA/mM/cm^2 to glucose in the range of 0.006–1.5 mM with a response time of less than 3 s and a detection limit of 5.7 μM at a signal/noise ratio of 3 [33].

6.3 Enhanced Blood Clotting

The main biological purpose of blood coagulation is formation of an obstacle to prevent blood loss of hydraulic strength sufficient to withstand the blood pressure. An enhancement of coagulation kinetics is particularly useful in trauma patients suffering from severe hemorrhage. Different methods have been adopted for activating the clotting process, which include administering activated coagulation factor VII (rFVIIa) [34], fresh frozen plasma (FFP), platelet concentrates, antifibrinolytic agents [35] and drugs of the Liquemin family [36]. Recently, Roy et al. reported the use of biocompatible TiO_2 nanotubes on the clotting kinetics of whole blood [37]. They demonstrated that blood containing dispersed TiO_2 nanotubes, and blood in contact with gauze pads surface-decorated with nanotubes showed significantly stronger clot formation at reduced clotting times compared to pure blood [37].

In this particular application, DMSO synthesized TiO_2 nanotube arrays were ultrasonically agitated in a water-ethanol solution to separate and subsequently disperse the nanotubes. The water-ethanol solution was centrifuged to separate nanotubes, which were then dried in an oven. A stock solution was formed from 10 mg of nanotube powder dispersed in 2 mL saline solution. Magnetoelastic sensors were used to quantify the blood-clotting kinetics [38–44]. For a magnetoelastic sensor in contact with blood, clot kinetics (clot rate, clot strength) are characterized by tracking the time-dependent shifts in the resonance amplitude. Bovine blood was used in the experiments; to avoid unwanted coagulation, the blood was mixed with citrate salt during collection. A mixture of calcium chloride and sodium chloride was used to reverse the citrate anticoagulant. The reversed blood [42, 43] was injected into a small glass vial containing the magnetoelastic sensor. After addition of blood to the vial it was sealed with a Teflon cap, then inserted into the sensor-reader electronics for measurement; all measurements were done at body temperature ~37.5°C. Blood samples containing dispersed TiO_2 nanotubes were analyzed, with an appropriate amount (between 50 and 200 mL) of the saline nanotube stock solution added to the blood mixture. Gauze pads (Johnson & Johnson, NJ, USA) were immersed in a saline solution of variable nanotube concentration (0.1, 0.5, or 1 mg/mL) then allowed to dry at 120°C for 1 h.

After drying a nanotube, surface-decorated gauze bandage was adhered to the metal, ribbon-like sensor surface using a spray-on adhesive.

Figure 6.5 shows the time-dependent change in the resonance amplitude [42, 43, 45] of blood immersed magnetoelastic sensors to which varying concentrations of nanotubes were added; comparative amplitude–time data for a pure blood sample, and a blood sample containing TiO_2 nanoparticles are also shown [37]. The amplitude curves are characterized by three distinct regions that mirror the clotting cascade: a slow initial decrease, followed by a rapid fall, and then finally, a steady decrease till it becomes almost constant with time. These three regions represent different parts of the coagulation cascade. A slow initial reduction of amplitude is caused by the beginning of the coagulation cascade, where different enzymes and proteins involved in the clotting process are activated leading to a steady increase in the blood viscosity. In the second stage the actual clotting is initiated; here, thrombin acts on fibrinogen, a high molecular weight protein, converting it into fibrin monomers that are rapidly converted into thread-like polymeric structures that form the clot matrix [46]. This cascade results in a rapid increase in the blood viscosity resulting in a sharp decrease in the amplitude of the sensor vibrations. Once the clot formation takes place the coagulation process gradually comes to a halt, marked by a slow but steady amplitude decrease similar to the first region.

The second stage of the clot profile, characterized by a rapid decrease in the signal amplitude, contains the needed clot information. The minimum of the first derivative of the amplitude-time response in this region defines the clotting time, while the magnitude of the derivative minimum (maximum clot rate) correlates to clot strength [42, 45, 47]. The amplitude-time derivatives of the data seen in Fig. 6.5 are plotted in Fig. 6.6 [37]. The activated clotting time of pure blood was estimated to be 285 s, reducing to about 260 s for the nanotube containing blood samples. For whole blood, the elasticity was 2.21 kPa. Larger elasticity values, that are stronger clot strengths, were found for increasing nanotube concentrations, reaching 3.87 kPa for blood containing 1 mg/mL of nanotubes. In comparison, the clotting

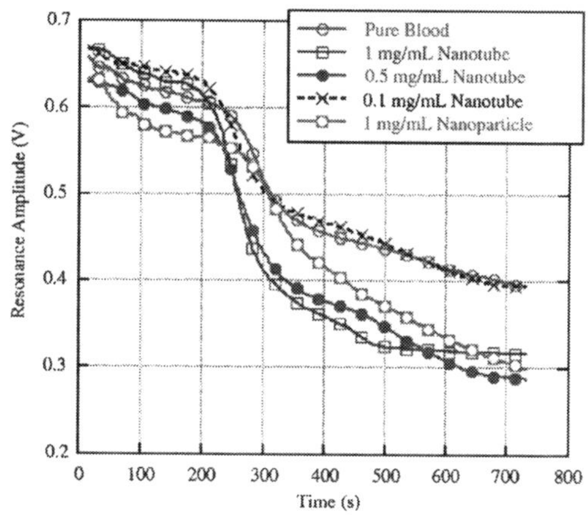

Fig. 6.5 Time variation in sensor resonance amplitude when immersed in pure blood, nanoparticle containing blood, and blood containing nanotubes of varying concentrations. 5% of the data points are shown. Similar results are obtained for blood in contact with a nanotube/particle decorated gauze bandage [37]

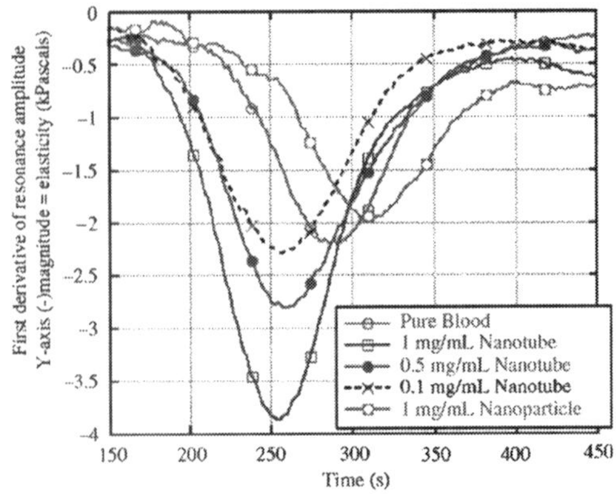

Fig. 6.6 The first derivative of the resonance amplitude–time data of Fig. 6.5, from which the clotting time (time of minimum) and clot strength (elasticity, from slope amplitude) are determined. 5% of the data points are shown [37]

time increased to 310 s for blood in combination with the TiO_2 nanoparticles, and the clot strength decreased to 1.97 kPa. The experimental study suggested that gauze bandages surface decorated with TiO_2 nanotubes could be used to improve the rate of blood clot formation and the strength of the resulting clot.

6.4 Cell Adhesion and Osteoblast Growth

Ti and its alloys are biocompatible metals, suitable for osseointegration, hence their widespread use as orthopaedic and dental implants [48, 49]. To enhance Ti bioactivity various surface treatments have been performed such as roughening the surface by sand blasting [50], covering the Ti surface with a hydroxyapatite (HAp) layer, or chemical treatment [51–55]. As suggested by Webster and co-workers, implant bioactivity and osteoblast adhesion can be significantly enhanced using nanostructured oxide materials [56, 57]. For bio-implant applications, the adhesion and mechanical integrity of the TiO_2 layer is of critical importance. Ti anodization in fluoride-based aqueous or non-aqueous solvents provides strongly adherent vertically oriented titania nanotube arrays. Oh and co-workers have used this structure for bone growth studies [58]. In another study, to facilitate the preferential nucleation of bioactive calcium phosphate compounds, anodic TiO_2 layers for bio-medical applications were electrochemically fabricated in electrolytes with sodium silicate solution; ions incorporated on the anodic TiO_2 surface acted as preferential nucleation sites for calcium phosphate by interaction with Ca^{2+} ion in the biological fluid [59]. Kim found that F^- ion-enriched surface oxides favor cell attachment, whereas, PO_4^{3-} ion enriched oxides encourage osteoblast growth [60, 61].

Bone is a calcium phosphate-based mineral, which contains 70% HAp-like material with the remainder consisting mostly of collagen. In order to evaluate the hydroxyapatite formation (i.e., calcium phosphate), the effects of nanostructure formation are investigated by exposing the nanotube arrays to a simulated body solution. For that, the TiO_2 nanotube array films were dipped in 20 mL of a simulated body fluid (SBF) solution at 36.5°C for 1, 2, 3 and 5 days. Ion concentration in the SBF solution was nearly the same to those of human blood plasma [62] with respect to Na^+, K^+, Ca^{2+}, Mg^{2+}, Cl^-, HCO_3^-, HPO_4^{2-}, and SO_4^{2-} ions. After soaking, the specimens were gently rinsed with distilled water, and then dried at 60°C for 24 h. Morphological changes on the nanotube surface were investigated via SEM. For cell adhesion studies, MC3T3-E1 osteoblast cells (mice cells of the type CRL-2593, sub-clone 4, ATCC, Rockville, MD) were used [63]. Each 1 mL of cells was mixed with 10 mL of alpha minimum essential medium (α-MEM) in the presence of 10% fetal bovine serum and 1% penicillin–streptomycin. The cell suspension was plated in a cell culture dish and the cell was incubated under 37°C, 5% CO_2 environment. When the concentration of the MC3T3-E1 osteoblastic cells reached ~3×10^5 cells/mL, they were seeded onto the nanotube array films, which were then placed on a 12-well polystyrene plate, and stored in a CO_2 incubator to observe cell morphology and count viable attached cells as a function of incubation time. The concentration of the cells seeded onto the specimen substrate was ~5×10^4 cells/mL. After incubation, the specimens were washed with 0.1 M PB and distilled water, respectively, and fixed with 2.5% glutaraldehyde in 0.1 M PBS for 1 h. After fixing, they were rinsed three times with 0.1 M PBS for 10 min. The samples were then dehydrated in a graded series of alcohol, dried by supercritical CO_2.

It is known that anatase phase TiO_2 is much more efficient for nucleation and growth of HAp (bone growth) than rutile [50]. In crystallized TiO_2 nanotubes, the walls are anatase while rutile exists at the base of nanotubes above the Ti substrate. Hence, the nanotube structure would be the excellent material for the growth of HAp. NaOH treated nanotubes showed the preferential growth of nanofibers on top of the nanotubes. Compositional analysis by EDX indicated the presence of Na, Ti and O supporting the formation of sodium titanate, which is known for accelerating HAp formation in SBF solution as well as in-vivo [64–66]. TiO_2 nanotube array substrates covered with sodium titanate nanofibers, ≈25 nm in diameter, were immersed in SBF solution for 1 day. The resulting microstructure exhibits a HAp-like structure, with EDX, see Fig. 6.7, showing the presence of Ca, P, and O (the ingredients of HAp). SBF soaking for 2–5 days resulted in a complete Hap surface coverage. Compared to unmodified TiO_2 nanotube array substrates, where it took 7 days for formation of detectable amount of HAp, HAp formation was accelerated with the sodium titanate nanofiber modified TiO_2 nanotube array substrates.

Images of MC3T3-E1 cells cultured on pure Ti and TiO_2 nanotube array substrates indicated that cell growth and filopodia propagation were much faster on the TiO_2 nanotube substrates [63]. The nanotube pore, and the gap between the nanotubes serve as a useful pathway for continuous supply of ions, nutrients, and proteins, which is likely to positively contribute to the health of the growing cells. Moreover, it should be noted that these nanotubes have strong mechanical adherence to the underlying Ti substrate. Figure 6.8 is a SEM-view showing the adhesion of the osteoblast cells (after 2 h) on vertically oriented TiO_2 nanotube arrays. The micrograph clearly shows that the filopodia of propagating osteoblast cells go into the nanotubes pores. The observed rapid adherence and spread of osteoblastic cells

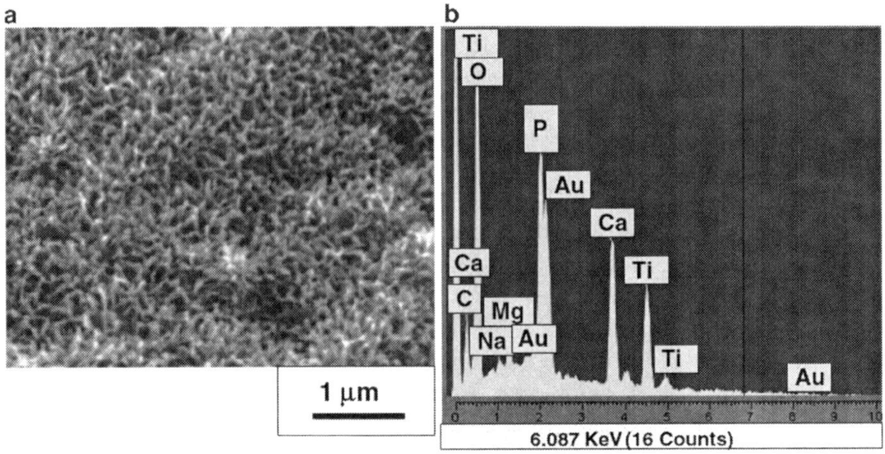

Fig. 6.7 (**a**) SEM microstructure of the nanoscale HAp phase formed on chemically treated TiO_2 nanotubes, and (**b**) EDX spectrum showing the presence of Ca and P in the Hap phase. (Reprinted from [63]. Copyright 2006, with permission from Elsevier)

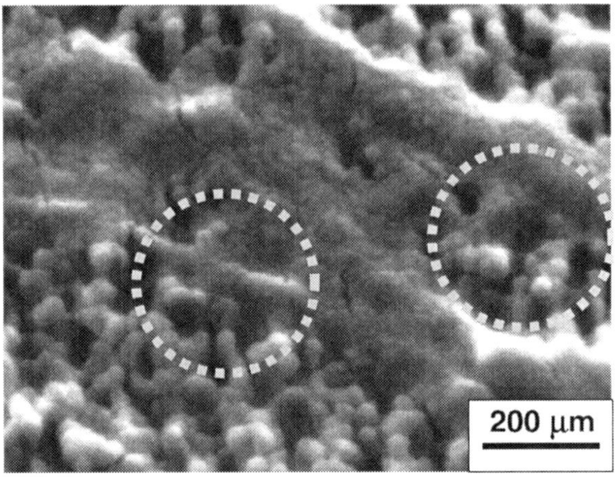

Fig. 6.8 Micrograph showing the growth and adhesion of the osteoblast cells. The circles indicate the TiO_2 nanotube and the cell growth into nanotube pores. (Reprinted from [63]. Copyright 2006, with permission from Elsevier)

cultured on the TiO_2 nanotube array films could be caused by the enormously larger surface area, the pronounced up-and-down vertical topology contributing to the locked-in cell configuration, and the presence of the fluid pathways.

Kunze et al. report rapid apatite formation on anatase TiO_2 nanotube arrays [67]. In the initial stages of apatite growth more nuclei are formed on the nanotube arrays than on compact TiO_2. While the crystallographic structure of the substrate was observed to play a less important role than the morphology in the initial nucleation stages, it was found to be of great importance in the later stages of apatite crystal growth. The nanotubular morphology combined with an anatase structure lead to the formation of apatite layers with a thickness of greater than 6 nm in less than 2 days. No stable apatite layer was observed on amorphous TiO_2 films. XPS, FT-IR and XRD measurements revealed that carbonated hydroxyapatite (CHA) of low crystallinity formed on annealed TiO_2 nanotube arrays, which should thus facilitate their use as osseointegrative implants. In other report, Li et al. used 500°C annealed long (12 μm) TiO_2 nanotubes anodically prepared in fluorinated DMSO solution for soaking in SBF solution and detected a thick apatite layer (13 μm) over the whole surface of nanotube arrays after 14 days of soaking; indicating their excellent in vitro bioactivity, which is mainly attributed to their high specific surface area and the anatase phase [68].

To accelerate hydroxyapatite (HA) formation, Ma et al. tried precalcification of nanotube array surfaces [69]. Nanotube array samples were annealed at 500°C for 2 h for crystallization; typical morphology of the annealed samples was outer diameter 100 nm, inner diameter 70 nm and length of about 200 nm. In the precalcification process, the crystallized nanotubes were soaked in 0.5 M Na_2HPO_4 solution for 24 h and then in saturated $Ca(OH)_2$ solution for 5 h [70]. After a DI water rinse, the precalcified and non-precalcified nanotube samples were immersed in a supersaturated calcium phosphate solution (SCP) for Ca–P deposition at 37.5°C. The SCP solution was prepared by dissolving 8 g NaCl, 0.42 g $CaCl_2$, 0.27 g $Na_2HPO_4 \cdot 2H_2O$ in 1 l ultra-pure water and was buffered at pH=7.4 with tris-hydroxymethyl aminomethane (TRIS) and HCl, in which the calcium and phosphor ions concentrations were 1.5 times higher than those in human blood plasma. The SCP for Ca–P deposition was refreshed every 2 days. After 7 days immersion in SCP, the surface of the precalcified nanotube was completely covered by a flake-like layer, as shown in Fig. 6.9a. However, no such precipitation was observed on the non-precalcified sample exposed to the same conditions of SCP immersion. After a 14-day immersion, only a discontinuous loose layer was formed on the non-precalcified sample, with hollow nanotubes still observed in some regions, as

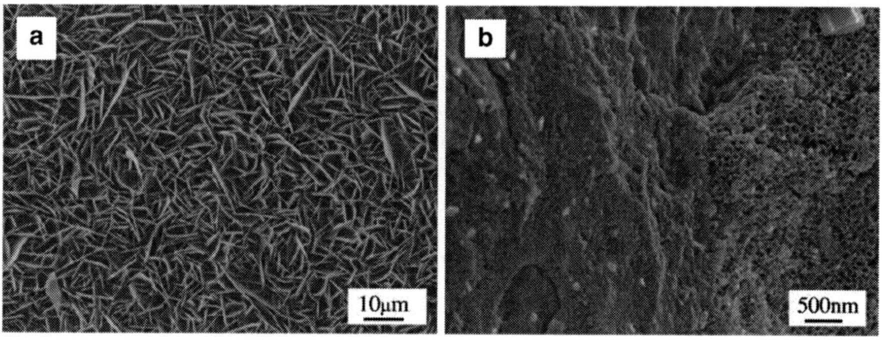

Fig. 6.9 The surface morphologies for: (**a**) the precalcified TiO_2 nanotube array sample after immersion in SCP for 7 days, and (**b**) the non-precalcified nanotube array sample after immersion in SCP for 14 days. (Reprinted from [69]. Copyright 2008, with permission from Elsevier)

seen in Fig. 6.9b. EDS spectrums of the precalcified sample showed neither Ca nor P peaks. After immersion in SCP for 7 days, the EDS detected the presence of Ca, P and O, the essential ingredients of hydroxyapatite.

6.5 Drug Elution from TiO_2 Nanotubes

The success of orthopedic implant surgery depends not only on the bone-implant integration, but also on the presence of a sterile environment around the implant, which will prevent bacterial infection. Acute infection or chronic osteomyelitis develops in as many as 5–33% of implant surgeries despite the use of strict antiseptic operative procedures employed [71, 72]. When a bone becomes infected the soft, inner part (bone marrow) often swells. As the swollen tissue presses against the rigid outer wall of the bone, the blood vessels in the bone marrow may become compressed, which reduces the blood supply to the bone. Without an adequate blood supply, parts of the bone may die and thus drastically impair the patient's recovery process. To reduce the chance of the infection, patients are subjected to antibiotic drug therapy upon initial surgery where the antibiotics are systemically delivered intravenously, intramuscularly or topically. Systemic antibiotic delivery entails certain drawbacks such as systemic toxicity and limited bioavailability. Further, for the drug to be effective at the site of implantation, high doses are required, which can result in undesired side effects. Thus, local antibiotic therapy is the preferred way of administering drugs. Popat and co-workers [73–75] report TiO_2 nanotube array surfaces for local delivery of antibiotics at the site of implantation, with demonstrated prevention of bacterial adhesion while maintaining the osseointegrative properties of the nanostructured surface [73]. Precise control of the nanotube length and diameter enable different amounts of drugs at different rates to be eluted at the implant site.

Gentamicin is a commonly used aminoglycoside antibiotic able to treat many types of bacterial infections, particularly Gram-negative infection. It works by binding the 30S subunit of the bacterial ribosome interrupting protein synthesis, and is one of the few heat-stable antibiotics that remain active even after autoclaving. Films comprised of TiO_2 nanotube arrays, nanotube pore diameter 80 nm and 400 nm length, were loaded with different amounts of gentamicin using a lyophilization method, with approximately 70–85% of the drug retained inside the nanotubes after an initial wash [73]. To facilitate gentamicin release from the nanotubes, the surfaces were immersed in 500 μL of PBS in a 24-well plate at room temperature with orbital shaking. Samples were collected periodically and the concentration determined using a colorimetric assay. Figure 6.10 shows the release data obtained from the gentamicin loaded nanotubes. There is slower and sustained release from the nanotubes loaded with a higher amount of drug compared to those loaded with lower amounts. For the given tube dimensions, the 200 μg gentamicin loading eluted within 45 min. Alternately, for the nanotubes loaded with 400 μg of gentamicin all the drug was eluted within 90 min. However, it took almost 150 min for all the drug to elute out of nanotubes that were loaded with 600 μg of gentamicin.

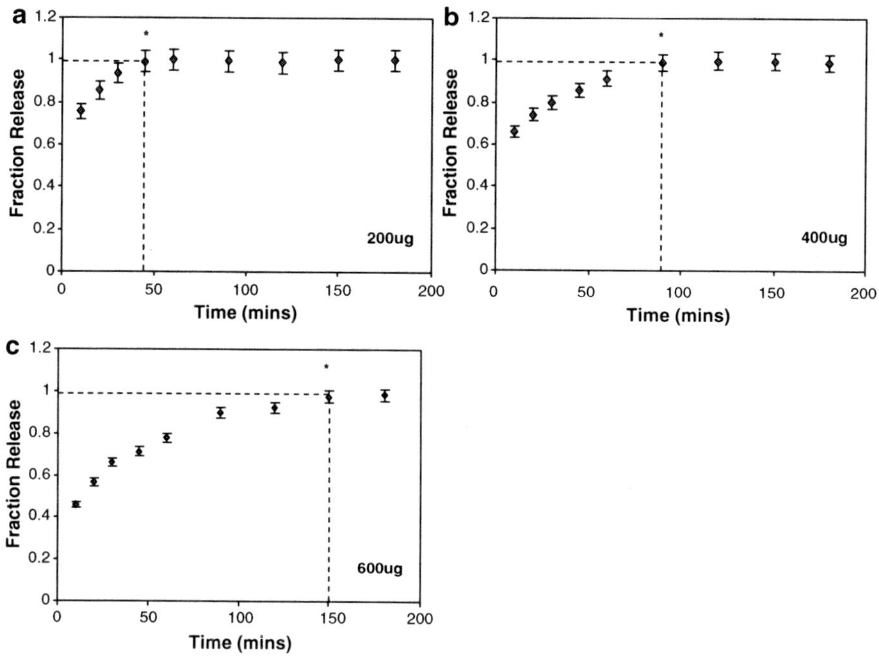

Fig. 6.10 Fraction released from nanotubes filled with 200, 400 and 600 μg of gentamicin, versus time [73]

Bacterial adhesion to a material surface can be described as a two-phase process including an initial, instantaneous and reversible physical phase, followed by a time-dependent and irreversible molecular and cellular phase. After initial adhesion, the surface growth of adhering microorganisms leads to the formation of a biofilm, which is significant in the pathogenesis of biomaterials-related infections. Growth of these organisms in direct contact with a surface play a pivotal role in biofilm formation as they link the entire biofilm to the surface. Hence, we investigated bacterial adhesion on nanotubes loaded with 600 μg gentamicin (NT-G) to evaluate the efficacy of eluting drug in preventing bacteria adhesion and subsequent colony and biofilm formation. *S. epidermis* bacteria strain was cultured on different surfaces and stained with Syto 9 (Molecular Probes). Bacteria cell counts were then completed using a fluorescence microscope in a field of view of 200 μm × 200 μm. Figure 6.11 shows the fluorescence microscopy images of bacteria colonies formed on Ti, NT and NT-G after 1 h of culture. These images suggest that NT-G has a significantly reduced bacteria colony present on the surface compared to Ti and NT. The bacteria colony count indicated an ≈70% decrease in bacteria colonies on NT-G compared to Ti and NT after 4 h of culture. This suggests that gentamicin elution from the nanotubes is significantly able to reduce *S. epidermis* colony forming units. It is expected that bacterial adhesion can be completely eliminated by increasing the amount of drug loaded in the nanotubes. SEM images of bacterial

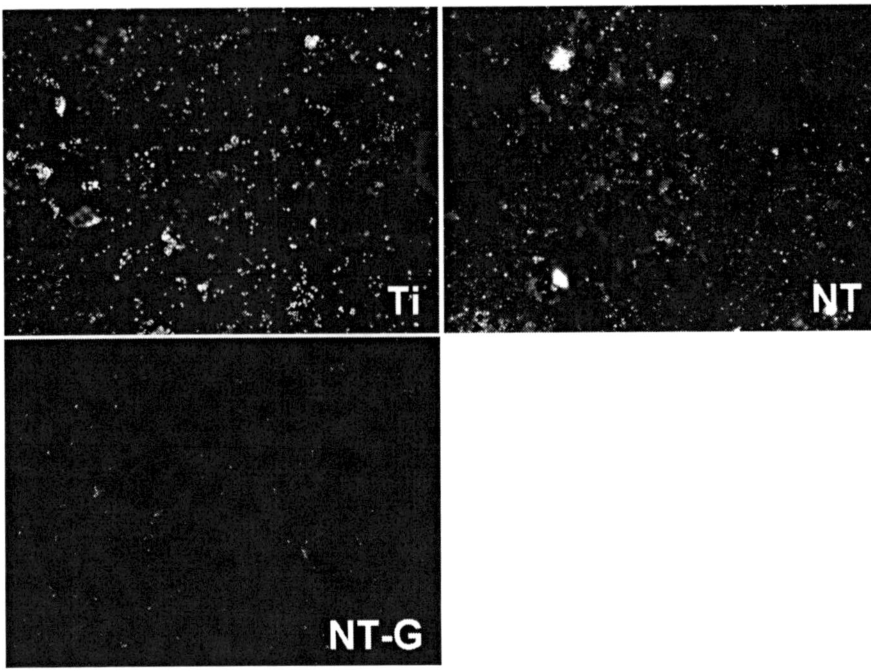

Fig. 6.11 Fluorescence microscope images of bacteria stained with Syto 9 after 1 h of culture on titanium (Ti), nanotubes (NT) and nanotubes filled with gentamicin (NT-G) [73]

colony formation after 1 h of culture on: (a) Ti, (b) NT and (c) NT-G show that bacterial colonies on Ti and NT are much larger then those on NT-G, with the relative difference growing with time [73]. Gentamicin is known to have prolonged effects on bacterial growth even after the drug is completely utilized or absorbed.

The release rates of bovine serum albumin (BSA), lysozyme (LYS), and gentamicin can be controlled by varying the amount loaded into the nanotubes and the polarity of the loaded molecule [74]. By changing the nanotube diameter, wall thickness and length, the release kinetics can be altered to achieve either rapid or sustained release. Hence these nanotubular surfaces have potential applications in orthopedics implants, dental implants and stents where not only controlled release of drugs such as antibiotics or growth factors but appropriate bio-integration is desired. BSA and LYS were used as model proteins to investigate loading and release efficiencies from nanotube platforms [74]. BSA is a relatively larger molecule (67 kDa) with a net negative charge at neutral pH (−18), LYS (14 kDa) has a net positive charge at neutral pH (+7). The nanotubes, 80 nm pore diameter and 400 nm length, were filled via a simplified lyophilization method [74]. Before the release studies were performed the efficiency of protein loading in the nanotubes was evaluated by comparing the concentrations of the original and the rinse solutions via a commercially available assay kit. Loading efficiencies, expressed as a percentage of loaded protein before and after washing, for nanotubes loaded with

200, 400 and 800 μg of both BSA and LYS were always higher than 70%. The results indicate that the majority of the protein is retained in the nanotubes after washing regardless of initial loading.

The surfaces were immersed in 500 μL of PBS in a 24-well plate at room temperature with orbital shaking, with samples periodically collected. The samples were analyzed for protein content using a commercially available Micro-BCA assay kit and the concentration adjusted for dilutions due to replacement of fresh PBS. Figure 6.12 shows the release data obtained from nanotubes loaded with 200, 400 and 800 μg of BSA, while Fig. 6.13 shows the release data obtained from nanotubes loaded with 200, 400 and 800 μg of LYS. The amount of protein eluted is

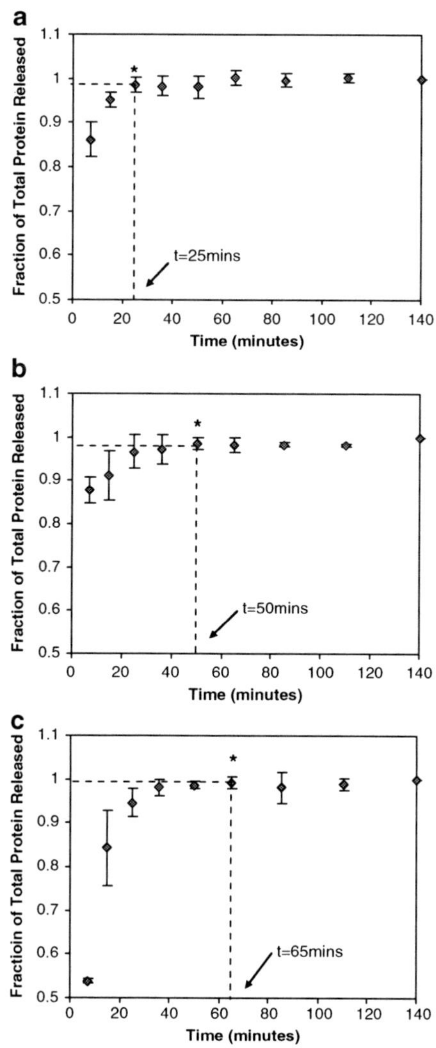

Fig. 6.12 Fraction of total bovine serum albumin released from nanotubes filled with 200, 400 and 800μg of BSA (**a–c**), respectively (*top to bottom*). The time point at which all the protein is released is indicated by the dotted line; concentrations at these time points are significantly different then those for time points before, however not significantly different then the time points after, $p < 0.05$, $n = 3$ [74]

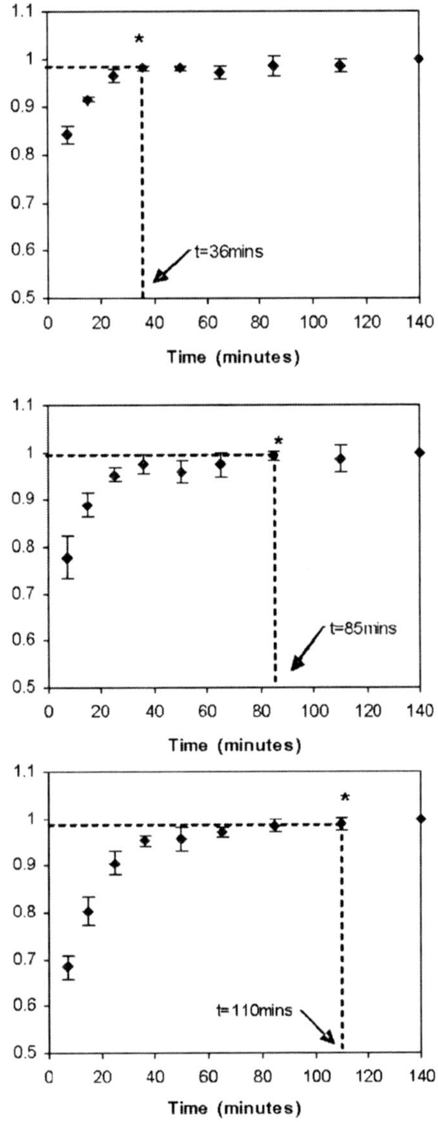

Fig. 6.13 Fraction of total lysozyme released from nanotubes filled with: (*Top*) 200 μg, (*Center*) 400 μg, and (*Bottom*) 800 μg of LYS. The time point at which all the protein is released is indicated by the dotted line; concentrations at these time points are significantly different then those for time points before, however not significantly different then the time points after, $p < 0.05$, $n = 3$ [74]

expressed in terms of fraction of total protein released. The results show that two different proteins, the relatively larger negative molecule BSA, and the relatively smaller positive molecule LYS, can be easily released from the nanotubes. As expected, there is slower and sustained release from the nanotubes loaded with a higher amount of protein compared to those loaded with lower amounts of protein. However, although LYS is a smaller molecule than BSA, data indicate that release of positively-charged LYS from the nanotubes is much slower compared to that of negatively-charged BSA [74]. This behavior is believed due to the difference between the negatively and positively charged proteins interacting with surface charge of the nanotube interface. The surfaces of most metal oxide films are inherently charged as a consequence of the equilibration of charged crystalline lattice defects within the surface. Depending on the net concentration of lattice defects, the surface may be positively or negatively charged. The titania nanotube surface consists of terminal hydroxyl groups, which results in a mild negative charge on the surface. Thus, the fact that the release of LYS which is positively charged is much slower compared to that of BSA, which is negatively charged, maybe, due to a stronger electrostatic interaction between the LYS and the titania surface. There was no release detected from surfaces that were *adsorbed* with BSA or LYS.

6.6 Hydrophobic Nanotubes: SAMs on Surface on Hydrophilic Nanotubes

Cell adhesion on biomedical devices is a critical factor for the integration process of a biomaterial after implantation, with surface characteristics such as chemistry, topography, surface energy, morphology and wettability viewed as critical factors [76–81]. To illustrate the complexities involved, it was found that cells tend to attach better to hydrophilic surfaces than to hydrophobic surfaces by several authors [82–84], with opposite results reported by others [85–87]. Reports have indicated the ability to achieve super-hydrophobic behavior with the nanotube array surfaces [88, 89]. Bauer et al. [90] reported on the response of mesenchymal stem cells to strongly hydrophobic and super-hydrophobic surfaces obtained via organic modification of TiO_2 nanotubes with self-assembled monolayers (SAMs) of octadecylphosphonic acid (ODPA). Before surface modification with the SAMs, the samples were washed in acetone, ethanol and ultrapure water each for 10 min, followed by exposure to ultraviolet (UV) light. Attachment of SAMs to the TiO_2 nanotube array surfaces was achieved by soaking the samples for 24 h in tetrahydrofuran containing 150 μM of ODPA ($C_{18}H_{37}PO(OH)_2$). The samples were then removed, subsequently washed with tetrahydrofuran and ultrapure water, and dried under nitrogen gas flow and then at 70°C overnight. Rat mesenchymal stem cells were isolated from fresh bone marrows from femurs of 4-week-old Wistar rats [91]. Cell isolation and expansion were performed [92], with the expanded cells showing the potential to differentiate into multiple mesenchymal lineages, including osteoblasts, chondroblasts, adipocytes, myoblasts, myofibroblasts and endothelial cells, in differentiation experiments in vitro.

In order to alter the wettability of TiO_2 nanotube array substrates, SAMs of ODPA were attached to the TiO_2 surfaces. As a result, the wetting behavior changed remarkably for all tube diameters. The originally hydrophilic surfaces became hydrophobic with water contact angles of about 108° for 15 nm diameter tubes up to superhydrophobic 167° for 100 nm tubes. X-ray photoelectron spectroscopy analyses confirmed the presence of phosphonic acid-SAMs on TiO_2 nanotube. It should be noted that, by washing with acetone, NaOH (1 M) and water, followed by drying at 70°C overnight, the original contact angles of these surfaces could be restored to the original values before exposure to culture medium. In order to investigate whether cell behavior responds critically to different contact angles in the super-hydrophobic range, green fluorescent protein (GFP)-labeled rat mesenchymal stem cells were seeded in culture medium containing 2% FCS at the same cell density on the different nanotube layers. As shown earlier, cell attachment on unmodified TiO_2 nanotubes depends on the nanotube diameter [92]. Unmodified TiO_2 nanotube surfaces of pore sizes 30 nm or less have shown strongly enhanced cellular activities compared to smooth TiO_2 surfaces [93]. For a tube diameter larger than 50 nm, cell adhesion and spreading were severely affected, resulting in reduced cellular activity and a high extent of programmed cell death. However, when coated with a monolayer of ODPA, the tube diameter had no significant influence on attachment. However, cell adhesion was found to be independent of nanotube diameter with the surface super-hydrophobic modification [92].

6.7 Biological Fluids Filtration and Drug Delivery Using TiO_2 Nanotubular Membrane

The majority of membranes, currently used for separation of sub-micron particles in biochemical applications, are made from polymers such as polysulfone, polyacrylonitrile, polyamides, etc., or ceramics such as alumina (Millipore isopore or Whatman filters). There are several incompatibilities associated with such membranes, such as sensitization to sterilization techniques; wide pore size distribution; adhesion of various proteins and biomolecules to membranes thus leading to biofouling and leaching of contaminants from polymers [94]. Most common polymeric membranes are made through a solvent-casting process, which results in a pore size distribution with variations as large as 30%. The use of ion track etching to form membranes (e.g., Millipore Isopore) produces a much tighter pore size distribution (±10%). However, these membranes have low porosities ($<10^9$ pores/cm^2), limited pore sizes, and the pores are randomly distributed across the surface. Porous alumina membranes (e.g., Whatman filters) offer high pore density ($>10^{10}$ pores/cm^2); however, pore sizes are generally limited to $\geq$40 nm. Thus, development of well-controlled, stable and uniform membranes capable of complete separation of viruses, proteins or peptides is an important consideration for biofiltration applications. To that end, Grimes and co-workers have developed fabrication strategies for TiO_2 nanotubular/porous membranes with well-controlled and uniform pores capable of separating nano-scale particles and drug delivery [95–97]. Of particular interest is that TiO_2 is an active photocatalytic material, giving rise to the possibility of self-cleaning, bioactive filters when used in combination UV light.

Glucose (MW: 180 Da), and Immunoglobulin G (IgG; MW: 150 kDa) were chosen as model molecules to study the diffusive transport through TiO_2 nanoporous membranes. The molecular weight of glucose is similar to that of the metabolic wastes (MW < 12 kDa) associated with kidney failure (e.g., urea MW: 60 Da, uric acid MW: 168 Da, etc.) that is targeted for removal by dialysis and haemofiltration, and hence was used as model molecule for diffusion studies. Further, IgG is a serum antibody secreted by b-cells and plasma cells that accounts for 75% of serum antibodies. Unlike the other types of immunoglobulins, IgG is the smallest class of immunoglobulin and circulates in the body as a monomer. It aids the immune system in protecting the body against potentially harmful antigens such as bacteria, viruses, and cancer cells by binding to the antigens and eliminating them through neutralization, opsonization, and complement activation. It also plays a role in transplant rejection since they are designed to recognize any foreign particles that might be harmful. Furthermore, the specificity of IgG makes them ideal candidates for targeted therapeutic applications. Currently, therapeutic antibodies are used to treat a variety of illnesses such as cancer, autoimmune disease, and cardiovascular diseases. Controlled diffusion of IgG through membranes is important for successful immunoisolation type devices. Phenol red (MW: 354.4 Da) and BSA (MW: 66 kDa) were also used as model molecules for diffusion studies since their molecular weights are between glucose and IgG.

Titania nanotubular membranes were fabricated by anodizing titanium foils of 0.25 mm thickness in an electrolyte composition of 0.3 wt% ammonium fluoride and 2 vol% water in ethylene glycol; the applied potential, e.g., 60 V, controlled the pore size, while length was controlled through anodization duration. The anodized samples were dipped in ethyl alcohol and subjected to ultrasonic agitation to separate the nanotube array film from the underlying Ti substrate. The compressive stress at the barrier layer–metal interface facilitates detachment from the substrate. Figure 6.14a shows the nanotube array membrane immersed in ethyl alcohol after it was separated from the underlying Ti substrate by ultrasonication and Fig. 6.14b

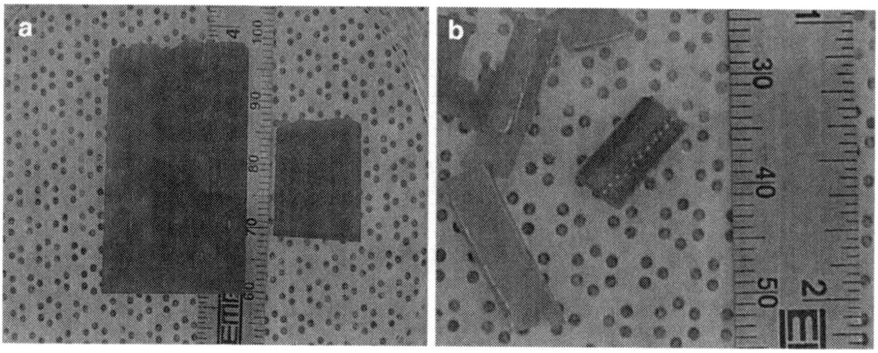

Fig. 6.14 Digital photographs of (**a**) flat membranes kept in ethyl alcohol after separation from Ti foil (**b**) membranes taken directly from water/ethanol and dried in air, showing the initial curling of the membrane as it dries [95]

membranes taken directly out of solution and then dried in air, note the extensive curling of the originally flat membranes. As clearly demonstrated by Fig. 6.14, the planar membranes, perfectly flat while wet, curled dramatically after they were removed from the liquid and dried in air due to surface tensional forces of the solution acting on the membrane.

To overcome membrane curling, low surface tension liquids such as hexamethyldisilizane can be used to wash the membranes. Also, critical point drying can be successfully used to dry the membrane after removal from solution. CO_2 critical point drying is a well-known technique used to prepare biological specimens for SEM studies. Figure 6.15 shows a photograph of a flat nanotube array membrane after critical point drying, the membrane remains flat indefinitely. Membranes of area 10 cm by 10 cm have been fabricated, to date. Membranes of 40 μm thickness or more were found robust enough for easy handling. The surface of the membrane after critical point drying occasionally showed a nanofiber surface, which could be removed by subjecting the membrane to ultrasonic agitation or a rapid acid rinse. It was believed that TiO_2 nanoparticles occasionally detach from nanotubes forming wires in the high pressure and temperature conditions used for critical point drying; similar to the situation where TiO_2 nanowires are grown by hydrothermal treatment of TiO_2 particles [98].

Figure 6.16a shows the backside, i.e., the barrier layer side of an as-fabricated nanotube array film. To open the closed ends, a dilute hydrofluoric acid/sulfuric acid solution was applied to the barrier layer side of the membrane to etch away the oxide and the membranes were then rinsed with ethyl alcohol. The acid rinse is repeated until the pores are completely opened as seen in Fig. 6.16b. Without disturbing their flatness, the samples were crystallized in anatase phase by annealing in an oxygen environment at 280°C for 1 h [99–101]. Diffusive transport through TiO_2 nanotubular membranes were studied using the apparatus shown in Fig. 6.17. The membrane was adhered with a cyanoacrylate adhesive to an aluminum frame, and then sealed between the two diffusion chambers, A and B. Chamber A was filled with 2 mL of 1 mg/mL solute solution (either glucose, phenol red, BSA

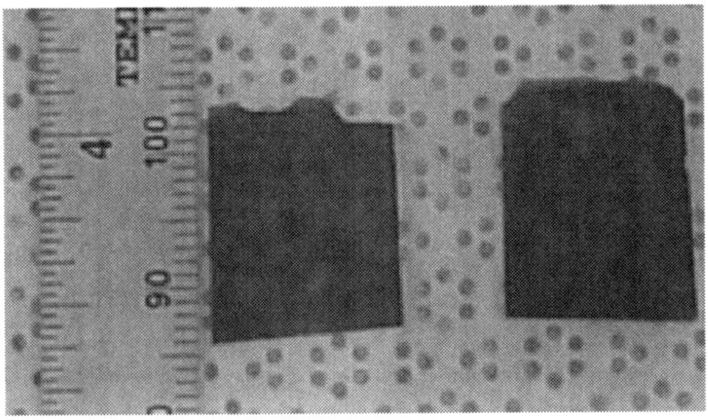

Fig. 6.15 Digital photograph of titania nanotube array membranes after critical point drying; the membranes are perfectly flat [95]

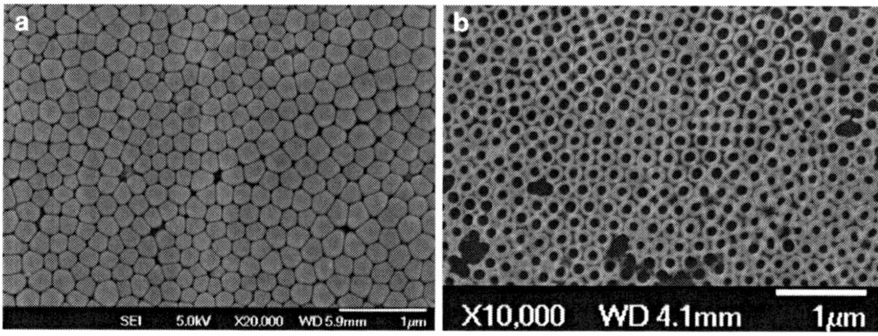

Fig. 6.16 FESEM images of: (**a**) back side (barrier layer) of as-fabricated nanotube array film; (**b**) etched backside of self-standing membrane at different degrees of magnification [95]

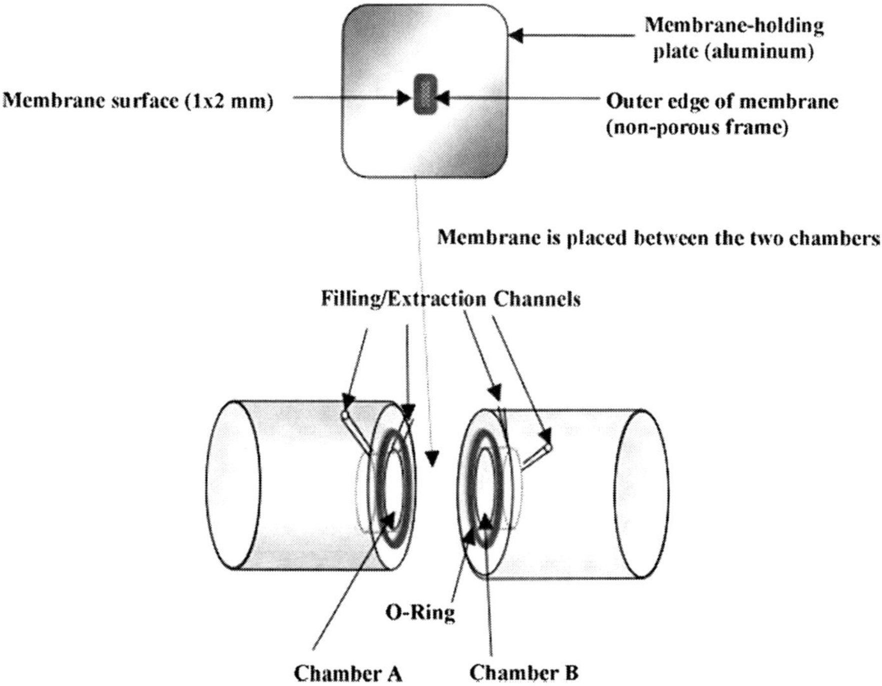

Fig. 6.17 Experimental setup for testing diffusion characteristics of TiO_2 membranes [95]

or IgG) and chamber B was filled with 2 mL of DI water (for glucose diffusion) or PBS (for phenol red, BSA and IgG diffusion). The assembled setup was rotated at 4 rpm throughout the experiment to eliminate any boundary layer effects. Samples were collected from chamber B at specific intervals of time. Glucose diffusion studies were carried out for 3 h, whereas IgG diffusion studies were carried out for 20 days. The concentration was measured by means of a quantitative enzymatic assay and colorimetric reading was determined using a spectrophotometer.

The ratio of measured concentration (C) with original concentration (C_0) was plotted against time to determine the diffusive transport through the membranes.

Diffusion across a membrane separating two well-stirred compartments A and B can be described by Fick's first law of diffusion as follows:

$$J = D_{\text{eff}} A_{\text{eff}} \left(\frac{C_{\text{A}} - C_{\text{B}}}{L} \right) \quad (6.1)$$

where J is mass flux, D_{eff} is the effective diffusion coefficient, A_{eff} is the cross-sectional pore area, L the membrane thickness, and C_{A} and C_{B} the measured concentrations of chamber A (donor) and B (recipient), respectively. The flux can be considered steady state since over the course of the experiment, chamber A acts as an infinite source of solute with a negligible change in its concentration.

Figure 6.18a, b show the measured B-side concentration versus time for glucose and IgG, respectively. There is a high degree of linearity indicating a zero-order diffusion system. By coupling this with the mass balance equation, the diffusion coefficient can be calculated using the following expression [102]:

$$\frac{1}{2} \ln \left(\frac{C_{\text{A0}} - 2C_{\text{B}}}{C_{\text{A0}}} \right) = \frac{A_{\text{eff}} D_{\text{eff}} t}{\Delta L \cdot V} \quad (6.2)$$

where C_{A0} is the initial concentration in chamber A, C_{B} the measured concentration in chamber B, ΔL the membrane thickness, V the total volume in chamber (same for A and B) and t is time. The diffusion coefficients were then compared with diffusion coefficient in water, calculated according to Stokes–Einstein equation:

$$D_{\text{H}_2\text{0}} = \frac{kT}{6\pi\eta R_{\text{d}}} \quad (6.3)$$

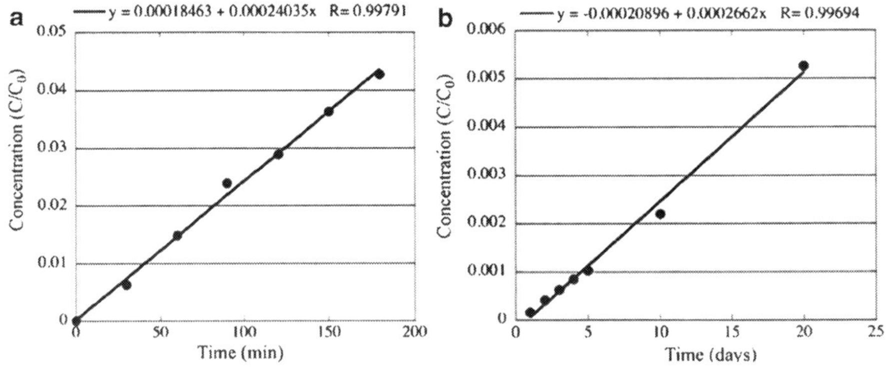

Fig. 6.18 (**a**) Time-dependent diffusion of glucose through a TiO_2 membrane 200 μm thick having 125 nm pore size. (**b**) Time-dependent diffusion of IgG through a TiO_2 membrane 200 μm thick having 125 nm pore size [95]

where k is Boltzmanns' constant, T is temperature, η the solvent viscosity, and R_d Stokes' radius. Using the above equation, D_{H2O} was found to be 6.14×10^{-6}. The membrane (200 μm thick, 125 nm pore size) diffusion coefficients were calculated as: $D_{glucose} = 2.42 \times 10^{-4}$ min^{-1}, $D_{IgG} = 1.73 \times 10^{-7}$ min^{-1}, $D_{Phenol\ Red} = 6.64 \times 10^{-4}$ min^{-1}, and $D_{BSA} = 3.70 \times 10^{-5}$ min^{-1}. Figure 6.19 shows the plot of molecular weight versus natural log of the diffusion coefficient for the four test molecules.

Figure 6.20 shows release profile of phenol red diffusion through a TiO_2 nanotubular membrane; the phenol red diffusion is constrained with approximately

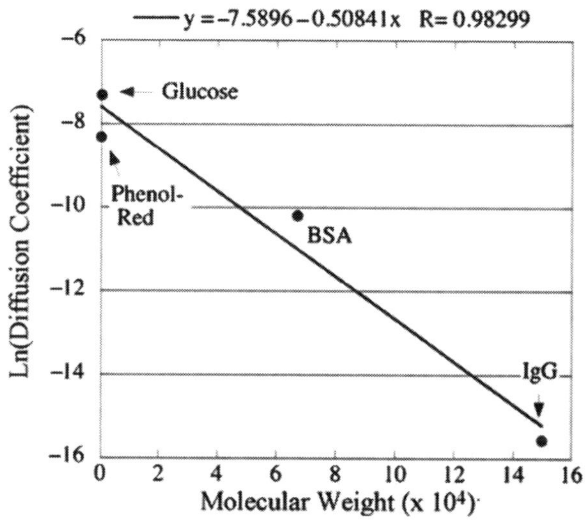

Fig. 6.19 Molecular weight versus natural log of the diffusion coefficient for the four test molecules [95]

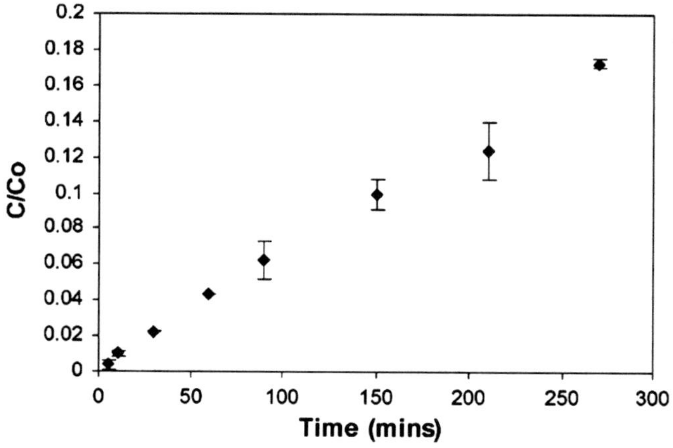

Fig. 6.20 Phenol red diffusion through a TiO_2 nanotubular membrane of 120 nm pore size and 200 μm thickness [96]

20% of phenol red diffusing in 5 h [96]. The origin of the constrained diffusion through nanotubular membrane may be related to the phenomenon of single-file diffusion, involving particles that cannot pass each other due to constrained pathways. Single-file diffusion phenomena have been recently reported in one-dimensional zeolites and also in biological ion exchange [103–105]. Another possible explanation of constrained diffusion could be the drag effects. When a colloidal particle suspended in a quiescent fluid is close to a flat wall, Stokes drag force acting on it is increased relative to that when far from the wall; therefore, the diffusion coefficient is smaller than that when from the wall. The increase of the drag force is attributed to the alteration of the hydrodynamic interaction between the particle and the fluid generated by the boundary condition imposed by the nearby wall. Using Fick's law of diffusion, the diffusion coefficient for phenol red through TiO_2 nanotubular membranes was calculated to be $9 \times 10^{-7} cm^2/s$. The Stokes–Einstein diffusivity for phenol red in aqueous solution is $2.84 \times 10^{-6} cm^2/s$. This difference in the two values is due to the presence of the nanotubular membrane and its interaction with the diffusant.

6.8 Application of Photocatalytic TiO_2 Nanotube Properties

TiO_2 is a well-known photocatalyst [106–108]. The photogenerated charge carriers form reactive oxygen species including O_2^- and $OH^\bullet$ radicals. Matsunaga et al. showed the ability of UV-irradiated platinized TiO_2 powder to kill bacterial cells in an aqueous environment [109]. Since then, the photoinduced devitalization of various cells using TiO_2 was reported [110–114]. For example, Kalbacova and co-workers reported the use of TiO_2 nanotube arrays for enhanced photo-induced killing of cancer cells. For this purpose, nanotube arrays were plated with HeLa G cancer cells [115]. The incubated cells were exposed to UV, 5 mW/cm^2 intensity, with changes in the cell morphology and viability observed [115]. Cell morphology cultured on TiO_2 nanotube array surfaces compared to that on glass revealed a slight variation in size. After UV exposure, cells on the TiO_2 nanotube array layers showed significant size reduction along with changes in shape due to the photocatalytic activity of the nanotube arrays.

Ionic silver demonstrates known anti-microbial activity, with silver impregnated polymers, and silver-impregnated hydroxyapatite films on alumina exhibiting excellent antibacterial properties against gram-positive and gram-negative bacteria [116, 117]. Das and co-workers investigated the influence of surface modified TiO_2 nanotube array surfaces on the in-vitro colonization of bacteria as an index of antibacterial activity as well as evaluation of the human osteoblast cells material interactions [118–120]. Human osteoblast bone cell attachment and growth behavior were studied on the nanotubes, using an osteo-precursor cell line, for durations of up to 11 days to evaluate cell-materials interaction. Silver was electrodeposited on the nanotube surfaces, which were then evaluated for antibacterial activity using the gram-negative bacterial strain Pseudomonas aeruginosa [121]. Nanotube surfaces with and without the presence of silver showed good cell-to-cell attachment, high cell proliferation and enhanced bone cell-material interactions in comparison to Ti-control [118]. The Ag deposited nanotube surface had an antibacterial activity over 99% against the growth of Pseudomonas aeruginosa colonies, whereas the Ti control and TiO_2 nanotube surfaces had no influence on the growth of Pseudomonas aeruginosa [118].

References

1. DeGrado WF (2003) Computational biology – biosensor design. Nature 423:132–133
2. MacBeath G, Schreiber SL (2000) Printing proteins as microarrays for high-throughput function determination. Science 289:1760–1763
3. Zhu H, Bilgin M, Bangham R, Hall D, Casamayor A, Bertone P, Lan N, Jansen R, Bidlingmaier S, Houfek T, Mitchell T, Miller P, Dean RA, Gerstein M, Snyder M (2001) Global analysis of protein activities using proteome chips. Science 293:2101–2105
4. Hultschig C, Kreutzberger J, Seitz H, Konthur Z, Bussow K, Lehrach H (2006) Recent advances of protein microarrays. Curr Opin Chem Biol 10:4–10
5. Phizicky E, Bastiaens PI, Zhu H, Snyder M, Fields S (2003) Protein analysis on a proteomic scale. Nature 422:208–215
6. Hall DA, Zhu H, Zhu X, Royce T, Gerstein M, Snyder M (2004) Regulation of gene expression by a metabolic enzyme. Science 306:482–484
7. Huang J, Zhu H, Haggarty SJ, Spring DR, Hwang H, Jin F, Snyder M, Schreiber SL (2004) Finding new components of the target of rapamycin (TOR) signaling network through chemical genetics and proteome chips. Proc Natl Acad Sci USA 101:16594–16599
8. Haab BB (2006) Applications of antibody array platforms. Curr Opin Biotechnol 17:415–421
9. Robinson WH (2006) Antigen arrays for antibody profiling. Curr Opin Chem Biol 10:67–72
10. Lojou E, Bianco P (2006) Application of the electrochemical concepts and techniques to amperometric biosensor devices. J Electroceram 16:79–91
11. Akyilmaz E, Sezginturk MK, Dinckaya E (2003) A biosensor based on urate oxidase-peroxidase coupled enzyme system for uric acid determination in urine. Talanta 61:73–79
12. Dzyadevych SV, Anh TM, Soldatkin AP, Chien ND, Jaffrezic-Renault N, Chovelon JM (2002) Development of enzyme biosensor based on pH-sensitive field-effect transistors for detection of phenolic compounds. Bioelectrochem 55:79–81
13. Yemini M, Reches M, Gazit E, Rishpon J (2005) Peptide nanotube-modified electrodes for enzyme-biosensor applications. Anal Chem 77:5155–5159
14. Liu XY, Chu PK, Ding CX (2004) Surface modification of titanium, titanium alloys, and related materials for biomedical applications. Mater Sci Eng R 47:49–121
15. Long M, Rack HJ (1998) Titanium alloys in total joint replacement – a materials science perspective. Biomater 19:1621–1639
16. Sul YT, Johansson CB, Petronis S, Krozer A, Jeong Y, Wennerberg A (2002) Characteristics of the surface oxides on turned and electrochemically oxidized pure titanium implants up to dielectric breakdown: the oxide thickness, micropore configurations, surface roughness, crystal structure and chemical composition. Biomater 23:491–501
17. Yang BC, Uchida M, Kim HM, Zhang XD, Kokubo T (2004) Preparation of bioactive titanium metal via anodic oxidation treatment. Biomater 25:1003–1010
18. Dyer CK, Leach JSL (1978) Breakdown and efficiency of anodic oxide growth on titanium. J Electrochem Soc 125:1032–1038

19. Marchenoir JJC, Loup JP, Masson JE (1980) Study of porous layers formed by anodic-oxidation of titanium at high voltages. Thin Solid Films 66:357–369
20. Ruan C, Yang F, Lei C, Deng J (1998) Thionine covalently tethered to multilayer horseradish peroxidase in a self-assembled monolayer as an electron-transfer mediator. Anal Chem 70:1721–1725
21. Ruan C, Yang R, Chen X, Deng J (1998) A reagentless amperometric hydrogen peroxide biosensor based on covalently binding horseradish peroxidase and thionine using a thiol-modified gold electrode. J Electroanal Chem 455:121–125
22. Wollenberger U, Drungiliene A, Stocklein W, Kulys JJ, Scheller FW (1996) Direct electrocatalytic determination of dissolved peroxidases. Anal Chim Acta 329:231–237
23. Liu S, Chen A (2005) Coadsorption of horseradish peroxidase with thionine on TiO_2 nanotubes for biosensing. Langmuir 21:8409–8413
24. Topoglidis E, Cass AEG, Gilardi G, Sadeghi S, Beaumont N, Durrant JR (1998) Protein adsorption on nanocrystalline TiO_2 films: an immobilization strategy for bioanalytical devices. Anal Chem 70:5111–5113
25. Zhang JK, Cass AEG (2001) A study of his-tagged alkaline phosphatase immobilization on a nanoporous nickel-titanium dioxide film. Anal Biochem 292:307–310
26. Gong D, Grimes CA, Varghese OK, Hu W, Singh RS, Chen Z, Dickey EC (2001) Titanium oxide nanotube arrays prepared by anodic oxidation. J Mater Res 16:3331–3334
27. Holt RE, Cotton TM (1989) Surface-enhanced resonance raman and electrochemical investigation of glucose-oxidase catalysis at a silver electrode. J Am Chem Soc 111:2815–2821
28. Clavilier J, Svetlicic V, Zutic V (1995) Thionine self-assembly on polyoriented gold and sulfur-modified gold electrodes. J Electroanal Chem 386:157–163
29. Nassar AEF, Willis WS, Rusling JF (1995) Electron-transfer from electrodes to myoglobin – facilitated in surfactant films and blocked by adsorbed biomacromolecules. Anal Chem 67:2386–2392
30. Xiao P, Garcia BB, Guo Q, Liu D, Cao G (2007) TiO_2 nanotube arrays fabricated by anodization in different electrolytes for biosensing. Electrochem Commun 9:2441–2447
31. Zhao G, Lei Y, Zhang Y, Li H, Liu M (2008) Growth and favorable bioelectrocatalysis of multishaped nanocrystal Au in vertically aligned TiO_2 nanotubes for hemoprotein. J Phys Chem C 112:14786–14795
32. Xie Y, Zhoua L, Huang H (2007) Bioelectrocatalytic application of titania nanotube array for molecule detection. Biosens Bioelectron 22:2812–2818
33. Pang X, He D, Luo S, Cai Q (2009) An amperometric glucose biosensor fabricated with Pt nanoparticle-decorated carbon nanotubes/TiO2 nanotube arrays composite. Sens Actuators B 137:134–138
34. Rizoli SB, Nascimento B, Osman F, Netto FS, Kiss A, Callum J, Brenneman FD, Tremblay L, Tien HC (2006) Recombinant activated coagulation factor VII and bleeding trauma patients. J Trauma-Inj Infect Crit Care 61:1419–1425
35. Fries D, Haas T, Velik-Salchner C, Lindner K, Innerhofer R (2005) Management of coagulation after multiple trauma. Anaesthesist 54:137–154
36. Wirz S, Knuefermann P, Baumgarten G, Potzsch B, Schaller C, Nadstawek J (2003) Head trauma and blood coagulation disorders. Anaesthaseol Intensiv 44:478–490
37. Roy SC, Paulose M, Grimes CA (2007) The effect of TiO_2 nanotubes in the enhancement of blood clotting for the control of hemorrhage. Biomater 28:4667–4672
38. Grimes CA, Ong KG, Loiselle K, Stoyanov PG, Kouzoudis D, Liu Y, Tong C, Tefiku F (1999) Magnetoelastic sensors for remote query environmental monitoring. J Smart Mater Struct 8:639–646
39. Stoyanov PG, Grimes CA (2000) A remote query magnetostrictive viscosity sensor. Sens Actuators A 80:8–14
40. Puckett LG, Barrett G, Kouzoudis D, Grimes CA, Bachas LG (2003) Monitoring blood coagulation with magnetoelastic sensors. Biosens Bioelectron 18:675–681

41. Puckett LG, Lewis JK, Urbas A, Cui X, Gao D, Bachas LG (2005) Magnetoelastic transducers for monitoring coagulation, clot inhibition, and fibrinolysis. Biosens Bioelectron 20:1737–1743
42. Zeng K, Roy SC, Grimes CA (2007) Quantification of blood clotting kinetics, Part I: determination of activated clotting times as a function of heparin concentration using magnetoelastic sensors. Sens Lett 5:439–445
43. Roy SC, Ong KG, Zeng K, Grimes CA (2007) Quantification of blood clotting kinetics, Part II: thromboelastograph analysis and measurement of erythrocyte sedimentation rate using magnetoelastic sensors. Sens Lett 5:446–454
44. de Lacheisserie E duT (1993) Magnetostriction: theory and applications of magnetoelasticity. In: Handbook of chemistry and physics, CRC Press, New York
45. Ong KG, Leland JM, Zeng K, Barrett G, Zourob M, Grimes CA (2006) A rapid highly-sensitive endotoxin detection system. Biosens Bioelectron 21:2270–2274
46. Guyton AC, Hall JE (2000) Hemostasis and blood coagulation In: A textbook of medical physiology. Saunders WB Company, USA, pp. 390–399
47. Carr ME, Krischnaswami A, Martin E (2007) Method of using platelet contractile force and whole blood clot elastic modulus as clinical markers. US patent 7192726
48. Black J, Hasting G (eds) (1998) Handbook of biomaterial properties. Chapman & Hall, London, pp 179–200
49. Ratner BD, Hoffman AS, Schoen FJ, Lemons JE (1996) Biomaterials science. Academic Press, San Diego, CA
50. Uchida M, Kim HM, Kokubo T, Fujibayashi S, Nakamura T (2003) Structural dependence of apatite formation on titania gels in a simulated body fluid. J Biomed Mater Res 64:164–170
51. Lugscheider E, Weber T, Knepper M, Vizethum F (1991) Production of biocompatible coatings by atmospheric plasma spraying. Mater Sci Eng A 139:45–48
52. Ducheyne P, van Raemdonck W, Heughebaert JC, Heughebaert M (1986) Structural analysis of hydroxyapatite coatings on titanium. Biomater 7:97–103
53. Cooley DR, Van Dellen AF, Burgess JO, Windeler AS (1992) The advantages of coated titanium implants prepared by radio frequency sputtering from hydroxyapatite. J Prosthet Dent 67:93–100
54. de Andrade MC, Sader MS, Filgueiras MRT, Ogasawara T (2000) Microstructure of ceramic coating on titanium surface as a result of hydrothermal treatment. J Mater Sci: Mater Med 11:751–755
55. Kim HM, Miyaji F, Kokubo T, Nakamura T (1996) Preparation of bioactive Ti and its alloys via simple chemical surface treatment. J Biomed Mater Res 32:409–417
56. Webster TJ, Ergun C, Doremus RH, Siegel RW, Bizios R (2000) Enhanced functions of osteoblast on nanophase ceramics. Biomater 21:1803–1810
57. Webster TJ, Schandler LS, Siegel RW, Bizios R (2001) Mechanisms of enhanced osteoblast adhesion on nanophase alumina involve vitronectin. Tissue Eng 7:291–301
58. Oh SH, Finones RR, Daraio C, Chen LH, Jin S (2005) Growth of nano-scale hydroxyapatite using chemically treated titanium oxide nanotubes. Biomater 26:4938–4943
59. Oh HJ, Lee JH, Kim YJ, Suh SJ, Lee JH, Chi CS (2008) Surface characteristics of porous anodic TiO_2 layer for biomedical applications. Mater Chem Phys 109:10–14
60. Kim SE, Lim JH, Lee SC, Nam SC, Kang HG, Choi J (2008) Anodically nanostructured titanium oxides for implant applications. Electrochim Acta 53:4846–4851
61. Xiao XF, Liu RF, Tian T (2008) Preparation of bioactive titania nanotube arrays in HF/Na_2HPO_4 electrolyte. J Alloys Compd 466:356–362
62. Kokubo T, Kushitani H, Sakka S, Kitsugi T, Yamamuro T (1990) Solution able to reproduce in-vivo surface-structure changes in bioactive glass-ceramic. J Biomed Mater Res 24:721–734
63. Oh S, Jin S (2006) Titanium oxide nanotubes with controlled morphology for enhanced bone growth. Mater Sci Eng C 26:1301–1306

64. Chen M, Yang X, He F, Liu Y, Zhu S, Cui Z (2003) Effect of NaOH concentration on formation of bone-like apatite layer on NiTi shape memory alloy. Jinshu Xuebao 39:859–864
65. Wang XX, Hayakawa S, Tsuru K, Osaka A (2001) A comparative study of in vitro apatite deposition on heat-, H_2O_2-, and NaOH-treated titanium surfaces. J Biomed Mater Res 54:172–178
66. Uchida M, Kim HM, Kokubo T, Fujibayashi S, Nakamura T (2002) Effect of water treatment on the apatite-forming ability of NaOH treated titanium metal. J Biomed Mater Res 63:522–530
67. Kunze J, Muller L, Macak JM, Greil P, Schmuki P, Muller FA (2008) Time-dependent growth of biomimetic apatite on anodic TiO_2 nanotubes. Electrochim Acta 53:6995–7003
68. Li MO, Xiao X, Liu R (2008) Synthesis and bioactivity of highly ordered TiO_2 nanotube arrays. Appl Surf Sci 255:365–367
69. Ma Q, Li M, Hu Z, Chen Q, Hu W (2008) Enhancement of the bioactivity of titanium oxide nanotubes by precalcification. Mater Lett 62:3035–3038
70. Wen HB, Wolke JGC, de Wijin JR, Liu Q, Cui FZ, de Groot K (1997) Fast precipitation of calcium phosphate layers on titanium induced by simple chemical treatments. Biomater 18:1471–1478
71. Ostermann PA, Henry SL, Seligson D (1993) The role of local antibiotic therapy in the management of compound fractures. Clin Orthop Relat Res 295:102–111
72. Hendriks JG, van Horn JR, van der Mei HC, Busscher H (2004) Backgrounds of antibiotic-loaded bone cement and prosthesis-related infection. Biomater 25:545–556
73. Popat KC, Eltgroth M, LaTempa TJ, Grimes CA, Desai TA (2007) Decreased Staphylococcus epidermis adhesion and increased osteoblast functionality on antibiotic-loaded titania nanotubes. Biomater 28:4880–4888
74. Popat KC, Eltgroth M, LaTempa TJ, Grimes CA, Desai TA (2007) Titania nanotubes: a novel platform for drug-eluting coatings for medical implants? Small 11:1878–1881
75. Peng L, Mendelsohn AD, LaTempa TJ, Yoriya S, Grimes CA, Desai TA (2009) Long-term small molecule and protein elution from TiO_2 nanotubes. Nano Letters 9:1932–1936
76. Dalby MJ, McCloy D, Robertson M, Agheli H, Sutherland D, Affrossman S (2006) Osteoprogenitor response to semiordered and random nanotopographies. Biomater 27:2980–2987
77. Dalby MJ, McCloy D, Robertson M, Wilkinson CD, Oreffo RO (2006) Osteoprogenitor response to defined topographies with nanoscale depths. Biomater 27:1306–1315
78. Popat KC, Daniels RH, Dubrow RS, Hardev V, Desai TA (2006) Nanostructured surfaces for bone biotemplating applications. J Orthop Res 24:619–627
79. Ratner BD (2004) Correlation, surfaces and biomaterials science. In: Ratner BD, Hoffman AS, Schoen FJ, Lemons JE (eds) Biomaterials science: an introduction to materials in medicine. Elsevier/Academic Press, Amsterdam, p 765
80. Boyan BD, Hummert TW, Dean DD, Schwartz Z (1996) Role of material surfaces in regulating bone and cartilage cell response. Biomater 17:137–146
81. Balasundarama G, Webster TJ (2006) A perspective on nanophase materials for orthopedic implant applications. J Mater Chem 16:3737–3745
82. Altankov G, Grinnell F, Groth T (1996) Studies on the biocompatibility of materials: fibroblast reorganization of substratum bound fibronectin on surfaces varying in wettability. J Biomed Mater Res 30:385–391
83. Groth T, Altankov G (1996) Studies on cell–biomaterial interaction: role of tyrosine phosphorylation during fibroblast spreading on surfaces varying in wettability. Biomater 17:1227–1234
84. Chen M, Zamora PO, Som P, Pena LA, Osaki S (2003) Cell attachment and biocompatibility of polytetrafluoroethylene (PTFE) treated with glow-discharge plasma of mixed ammonia and oxygen. J Biomater Sci Polym Edit 14:917–935
85. Faucheux N, Schweiss R, Lutzow K, Werner C, Groth T (2004) Self assembled monolayers with different terminating groups as model substrates for cell adhesion studies. Biomater 25:2721–2730

86. Ikada Y (1994) Surface modification of polymers for medical applications. Biomater 15:725–736
87. Lee JH, Khang G, Lee JW, Lee HB (1998) Interaction of different types of cells on polymer surfaces with wettability gradient. J Colloid Interf Sci 205:323–330
88. Balaur E, Macak JM, Taveira L, Schmuki P (2005) Tailoring the wettability of TiO_2 nanotube layers. Electrochem Commun 7:1066–1070
89. Balaur E, Macak JM, Tsuchiya H, Schmuki P (2005) Wetting behaviour of layers of TiO_2 nanotubes with different diameters. J Mater Chem 15:4488–4491
90. Bauer S, Park J, von der Mark K, Schmuki P (2008) Improved attachment of mesenchymal stem cells on super-hydrophobic TiO_2 nanotubes. Acta Biomater 4:1576–1582
91. Jiang YH, Jahagirdar BN, Reinhardt RL, Schwartz RE, Keene CD, Ortiz-Gonzalez XR, Reyes M, Lenvik T, Lund T, Blackstad M, Du JB, Aldrich S, Lisberg A, Low WC, Largaespada DA, Verfaillie CM (2002) Pluripotency of mesenchymal stem cells derived from adult marrow. Nature 418:41–49
92. Park J, Bauer S, von der Mark K, Schmuki P (2007) Nanosize and vitality: TiO_2 nanotube diameter directs cell fate. Nano Lett 7:1686–1691
93. Popat KC, Leoni L, Grimes CA, Desai TA (2007) Influence of engineered titania nanotubular surfaces on bone cells. Biomaterials 28:3188–3197
94. Zeman LJ, Zydney A (1996) Microfiltration and Ultrafiltration. CRC Press, New York
95. Paulose M, Peng L, Popat KC, Varghese OK, LaTempa TJ, Bao N, Desai TA, Grimes CA (2008) Fabrication of mechanically robust, large area, polycrystalline nanotubular/porous TiO_2 membranes. J Membrane Sci 319:199–205
96. Paulose M, Prakasam HE, Varghese OK, Peng L, Popat KC, Mor GK, Desai TA, Grimes CA (2007) TiO_2 nanotube arrays of 1,000 μm length by anodization of titanium foil: phenol red diffusion. J Phys Chem C 111:14992–14997
97. Popat KC, Eltgroth M, LaTempa TJ, Grimes CA, Desai TA (2007) Titania nanotubes: a novel platform for drug-eluting coatings for medical implants? Small 11:1878–1881
98. Daoud WA, Pang GK (2006) Direct synthesis of nanowires with anatase and TiO_2-B structures at near ambient conditions. J Phys Chem B 110:25746–25750
99. Mor GK, Carvalho MA, Varghese OK, Pishko MV, Grimes CA (2004) A room temperature TiO_2-nanotube hydrogen sensor able to self clean photoactively from environmental contamination. J Mater Res 19:628–634
100. Mor GK, Shankar K, Paulose M, Varghese OK, Grimes CA (2005) Enhanced photocleavage of water using titania nanotube arrays. Nano Lett 5:191–195
101. Mor GK, Shankar K, Paulose M, Varghese OK, Grimes CA (2006) Use of highly-ordered TiO_2 nanotube arrays in dye-sensitized solar cells. Nano Lett 6:215–218
102. Sharkawy AA, Klitzman B, Truskey GA, Reichert WM (1998) Engineering the tissue, which encapsulates subcutaneous implants. II. Plasma–tissue exchange properties. J Biomed Mater Res 40:586–597
103. Aityan SK, Portnov VI (1986) Computer-simulation of single-file transport. Gen Physiol Biophys 5:351–364
104. Aityan SK, Portnov VI (1988) Analysis of models of single-file diffusion. Gen Physiol Biophys 7:591–611
105. Hahn K, Karger J, Kukla VV (1996) Single-file diffusion observation. Phys Rev Lett 76:2762–2765
106. Serpone N, Pelizzetti E (1989) Photocatalysis – fundamentals and applications. Wiley, New York
107. Hoffmann MR, Martin ST, Choi W, Bahnemann DW (1995) Environmental applications of semiconductor photocatalysis. Chem Rev 95:69–96
108. Mills A, LeHunte S (1997) An overview of semiconductor photocatalysis. J Photochem Photobiol A 108:1–35
109. Matsunaga T, Tomoda R, Nakajima T, Wake H (1985) Photoelectrochemical sterilization of microbial-cells by semiconductor powders. FEMS Microbiol Lett 29:211–214

110. Saito T, Iwase T, Morioka T (1992) Mode of photocatalytic bactericidal action of powdered semiconductor TiO_2 on streptococci. J Photochem Photobiol B 14:369–379
111. Cai RX, Kubota Y, Shuin T, Sakai H, Hashimoto K, Fujishima A (1992) Induction of cytotoxicity by photoexcited TiO_2 particles. Cancer Res 52:2346–2348
112. Maness PC, Samolinski S, Blake DM, Huang Z, Wolfrum EJ, Jacoby WA (1999) Bactericidal activity of photocatalytic TiO_2 reaction: toward an understanding of its killing mechanism. Appl Environ Microbiol 65:4094–4098
113. Lu ZX, Zhou L, Zhang ZL, Shi WL, Xie ZX, Xie HY, Pang DW, Shen P (2003) Cell damage induced by photocatalysis of TiO_2 thin films. Langmuir 19:8765–8768
114. Seo JW, Chung H, Kim MY, Lee J, Choi IH, Cheon J (2007) Development of water-soluble single-crystalline TiO_2 nanoparticles for photocatalytic cancer-cell treatment. Small 3:850–853
115. Kalbacova M, Macak JM, Schmidt-Stein F, Mierke CT, Schmuki P (2008) TiO_2 nanotubes: photocatalyst for cancer cell killing. Phys Stat Sol 2:194–196
116. Feng QL, Kim TN, Wu J, Park ES, Kim JO, Lim DY, Cai FZ (1998) Antibacterial effects of Ag-HAp thin films on alumina substrates. Thin Solid Films 335:214–219
117. Kawahara K, Tsuruda K, Morishita M, Uchida M (2000) Antibacterial effect of silver-zeolite on oral bacteria under anaerobic condition. Dent Mater 16:452–455
118. Das K, Bose S, Bandyopadhyay A, Karandikar B, Gibbins BL (2008) Surface coatings for improvement of bone cell materials and antimicrobial activities of Ti implants. J Biomed Mater Res 87:455–460
119. Das K, Bandyopadhyay A, Bose S (2008) Biocompatibility and in situ growth of TiO_2 nanotubes on Ti using different electrolyte chemistry. J Am Ceram Soc 91:2808–2814
120. Das K, Balla VK, Bandyopadhyay A, Bose S (2008) Surface modification of laser-processed porous titanium for load-bearing implants. Scr Mater 59:822–825
121. Kunin CM (1987) Detection, prevention and management of urinary tract infection, 4th edn. Lea & Febiger, Philadelphia, PA, pp 245–288

Chapter 7
Conclusions and New Directions

7.1 Conclusions

We considered four synthesis generations of TiO_2 nanotube arrays [1–3]. The first being aqueous solutions using HF as reported by Gong and co-workers [4], the second as reported by Cai et al. [5] using buffered KF or NaF aqueous solutions to achieve nanotube array lengths of several microns. The third synthesis generation, as initially reported by Paulose and co-workers [6–10], dwelt on the use of organic electrolytes, which extended the achievable nanotube array growth up to 1 mm. A fourth synthesis generation involves replacement of the fluorine ions, used in the first three synthesis generations with, most notably, chlorine as reported by Allam et al. [11, 12]. Each synthesis generation has provided new material properties, which in turn have enabled new applications.

First generation nanotube arrays, obtained in HF aqueous electrolytes, resulted in nanotubes with a pore size from 22 to 76 nm and a maximum achievable thickness of about 500 nm. A pore size of 120 nm and length of 1.1 μm were grown in acidic HF (i.e., 0.3 wt% HF+1.0 M H_3PO_4) based electrolytes, while a pore size of about 90–140 nm and length 0.69–2.47 μm was achieved using acidic NH_4F (NH_4F/H_3PO_4) based electrolytes with non-Pt based transition elements (Fe, Co, Cu, Ta, and W) as counter electrodes [13]. An enabling discovery was that addition of acetic acid to the anodization electrolyte transformed the nanotubes from something quite fragile, prone to breaking during handling, to something mechanically robust. Conical shape nanotubes can be obtained by linearly varying the anodization voltage [14], while nanotube wall thickness can be varied through temperature for aqueous electrolytes [15], or acid content for organic electrolytes [1].

Second-generation nanotubes [3, 5] saw an increase in length to approximately 7 μm through the use of buffered solutions having higher pH values while remaining acidic that reduced chemical dissolution of TiO_2 during anodization. On varying the pH of KF-based electrolyte (containing citrate acid and $NaHSO_4$ solution) from 1 to 5.5, the length of nanotubes (anodized at 25 V) increased from 0.56 to 6.4 μm, respectively.

C.A. Grimes and G.K. Mor, *TiO_2 Nanotube Arrays*,
DOI 10.1007/978-1-4419-0068-5_7, © Springer Science+Business Media, LLC 2009

Third generation nanotubes [6–10, 16–18] were grown in organic electrolytes containing less than 5% water, key to achieving ultra long nanotube arrays. For example, nanotube arrays of 70 μm length, pore size about 132 nm and wall thickness ~24 nm were grown using an electrolyte of 0.27 M NH_4F in formamide (FA) containing 5% water by anodization at 35 V for 48 h [6, 18]. On replacing FA with *N*-methyl formamide (NMF), nanotubes with a smaller pore diameter and about 10% greater length were obtained. On increasing the anodization voltage, the length and the pore size of nanotube increase, whereas the nanotube wall thickness decreases. The range of applied anodization potentials over which nanotube arrays are obtained is significantly extended in organic electrolytes, with nanotubes formed from 10 to 100 V, with nanotube length largely found to depend on anodization duration. While studying the effect of different cationic species on the formation of TiO_2 nanotube arrays in formamide-water mixtures, it was found that the cation choice is a *key parameter* influencing both the nanotube growth rate and resulting nanotube length [18]. Under similar conditions in FA electrolytes, those containing the tetrabutylammonium cation resulted in the longest nanotubes (~94 μm), while the shortest nanotubes (~3 μm) were obtained when H^+ ions were the sole cationic species in the anodization electrolyte. This was attributed to the inhibitory effect of the quarternary ammonium ions that restrict the thickness of the interfacial (barrier) oxide layer; a thinner interfacial oxide layer facilitates ionic transport thus enhancing the nanotube growth.

In DMSO/HF-based electrolytes [6, 10, 19] the maximum nanotube length obtained was 101 μm with a pore size of 150 nm, 2.0% HF-DMSO at 60 V for 70 h. Effects of electrolyte aging are clearly seen using DMSO electrolytes, with tubes grown from used electrolytes, that is electrolytes previously used for an anodization, having relatively thicker walls and smaller pore sizes that those obtained with fresh, i.e., previously unused, electrolytes. The addition of water to the DMSO electrolyte was found to improve the nanotube array adhesion to the underlying Ti substrate, and also reduce the presence of debris.

In ethylene glycol (EG) electrolytes [6–9, 20], TiO_2 nanotube arrays were observed for anodization voltages 20–65 V in the NH_4F concentration range (0.1–0.5 wt%) and H_2O (1–4%). Ti foil anodized at 60 V in a fresh solution of 0.3 wt% NH_4F and 2 vol% H_2O in ethylene glycol for 4, 17, 21, 48 and 96 h exhibited a length of ~58, 160, 188, 289 and 360 μm, respectively. In the given range of NH_4F and H_2O concentrations, the anodic dissolution due to the increased wt% of NH_4F was compensated by the increase in H_2O concentration resulting in greater growth rates and hence, longer nanotube lengths. From 0.1 to 0.5 wt% NH_4F with 2% water, 0.25 mm foil samples were completely anodized resulting in two ≈360 μm nanotube arrays across a thin barrier layer. A maximum individual nanotube array length of 1,005 μm was obtained upon anodizing a 2.0 mm thick Ti foil sample at 60 V for 216 h (9 days) in 0.5 wt% NH_4F and 3.0% water in EG. Use of EG electrolytes, and the resulting close-packed strongly adhered nanotube arrays several hundred microns in length enabled the successful fabrication of mechanically robust nano-filtration membranes. After removal of the nanotube arrays from the underlying Ti substrate [7, 20], the back-side plugged end of the nanotube array was opened with an acid rinse resulting in, after drying, a stable mechanically robust nanoporous titania membrane.

Nanotubes formed in EG electrolytes are close packed, generally approaching hexagonal packing, while there is clear separation between adjacent nanotubes fabricated in FA/NMF. Nanotubes grown from DMSO electrolytes generally have a slight conical nature, with the top layers well separated and the bottom layers densely packed. Nanotube separation is increased dramatically with the use of di-ethylene glycol (DEG) electrolytes containing HF as the fluoride source [16]. Tube-to-tube spacing of several microns could be obtained, with the nanotube length ranging from 3 to 7.5 μm, pore size from ~140 nm (40 V) to ~190 nm (100 V).

In combination with field emission scanning electron microscope (FESEM) images taken at different stages of the anodization process, observation of current density versus time during potentiostatic anodization in fluoride-based solution allowed an often repeated TiO_2 nanotube array formation mechanism to be put forth by Mor and co-workers in 2003 [3, 14]. In organic electrolytes, nanotube growth underlies a several competing reactions including water content, cation size, acid content, electrolyte viscosity and conductivity [21].

For crystallization the as-anodized amorphous nanotubes are typically annealed in oxygen ambient, at elevated temperatures, for several hours [22]. The temperature at which samples crystallize without disturbance of the nanotubular structure is found to be a function of the nanotube length, e.g., for short nanotubes it is 480°C, and 580°C for nanotubes of several microns length. These nanotubes begin to crystallize in anatase phase at about 280°C, with the crystallite size and crystallized volume increasing with annealing temperature. The nanotube walls remain anatase irrespective of annealing temperature provided the nanotubular structure remains intact, whereas the metal-nanotube region starts transforming anatase to rutile at 430°C for short nanotubes and at 530°C for medium length (6–7 μm) nanotubes. Glancing angle X-ray diffraction of 14 μm long nanotubes annealed at 580°C for 4 h exhibit only anatase peaks, as the underlying rutile barrier layer is not seen by the GAXRD scan. The crystallinity of the annealed nanotubes was demonstrated to be enhanced upon solvothermal treatment in ethanol vapor [23]. In an effort towards low temperature crystallization, anodizing oxidized Ti foil, treated with H_2O_2 at 80°C or $(NH_4)_2S_2O_8$ at 120°C, resulted in the direct fabrication of crystallized nanotubes up to 1.4 μm in length. In another report, nanotubes grown in DEG/NH_4F/2% H_2O electrolyte at 80 V were found to be partially crystalline [16].

Each synthesis generation has enabled new applications of the TiO_2 nanotube arrays. As of today, applications of the TiO_2 nanotube array architecture include use as hydrogen gas sensors, dye sensitized solar cells (DSCs), polymer-TiO_2 nanotube based bulk heterojunction photovoltaics, drug eluting implant surfaces, biofiltration membranes, biosensors, photocatalysts, hydrophilic surfaces, electro-chromic devices, super capacitors, etc. The remarkable properties that the TiO_2 nanotube arrays possess is illustrated in their use as hydrogen sensors. At room temperature, nanotubes of 1 μm length (prepared in KF electrolyte, pH ~4.0, 10 V) exhibit a resistance change of 8.7 orders of magnitude in response to 1,000 ppm hydrogen; this represents the largest known sensitivity of any material, to any gas, at any temperature [24–26]. While such testing chambers are not readily available, linear extrapolation of the sensitivity indicates the ability of the room temperature nanotube-array to detect hydrogen at 20 ppt (parts per trillion). Combined with their outstanding photocatalytic properties, the hydrogen sensors exhibit excellent self-cleaning capabilities and are hence able to recover from environmental contamination [27, 28].

The nanotube array architecture, achieved by anodization of the base metal, has now been extended to a variety of other materials, including tantala [29] and zirconia [30], and more recently iron oxide, specifically α-Fe_2O_3 (hematite) [31] as shown in Fig. 7.1. The novel properties offered by the nanotube array architecture can be expected to be extended to other materials. For example, nanotubular tantala has already proven itself to be a most interesting substrate for osseointegration and drug elution applications [29].

The nanotube array structure has proven itself to be an, if not the, ideal architecture for water photoelectrolysis [6, 15, 32, 33]. Nanotube arrays approximately 20μm in length, 205 nm outer diameter, prepared using a formamide electrolyte and then annealed at 580°C for 1 h for crystallization, demonstrated a maximum efficiency of 16.25% (320–400 nm illumination, 100 mW/cm^2) [6, 21, 33] with an IPCE of approximately 80%. Hence, thin wall hematite nanotube arrays offer an exciting opportunity in application to water splitting, with the $\approx$2.2 eV bandgap well suited for capturing solar spectrum energy. The short minority carrier diffusion length of hematitie, $\approx$5 nm, demands thinner nanotube walls than those seen in Fig. 7.1 of approximately 25 nm [1, 31]. This should be a tractable problem as, we note, nanotube arrays with 5 nm thick walls have been achieved using a formamide electrolyte [18]. Introduction of controlled amounts of lower band gap oxide semiconductors into the TiO_2 nanotube walls offers a path to combine the excellent material strength and photocorrosion stability of TiO_2 with the superior light-harvesting abilities of smaller band gap oxides in the nanotube array architecture [34, 35]. For example, visible light responding Fe–Ti–O ternary oxide nanotube arrays were synthesized by anodization of Fe:Ti films. A maximum photocurrent of 1.5–2 mA/cm^2 was observed in translucent Ti–Fe–O nanotube films ($Ti_{93.4}Fe_{6.6}$ film before anodization) at 0.5 V vs. Ag/AgCl in 1 M NaOH solution under AM 1.5 light. Various efforts have been focused on either doping the TiO_2 nanotube arrays, with various anionic and cationic dopants such as N, C, S, B and Si, to achieve useful visible light properties, without great or sometimes even repeatable success, or combining the nanotube arrays with a visible light responsive low bandgap semiconductors such as WO_3 [36], CdS [37] or CdTe [38].

DSCs require facile electronic transport of the photogenerated electrons to the back contact. Superior electronic transport allows for improved light harvesting, and hence device efficiency, since thicker films can be used to increase the optical density enabling absorption of low energy photons in the red and infrared regions without losing the additionally harvested charge carriers to recombination. The high surface area TiO_2 nanotube-array geometry, with tubes oriented perpendicular to the surface, permits directed charge transfer along the length of the photoelectrode. Open circuit voltage decay measurements on N719 dye sensitized TiO_2

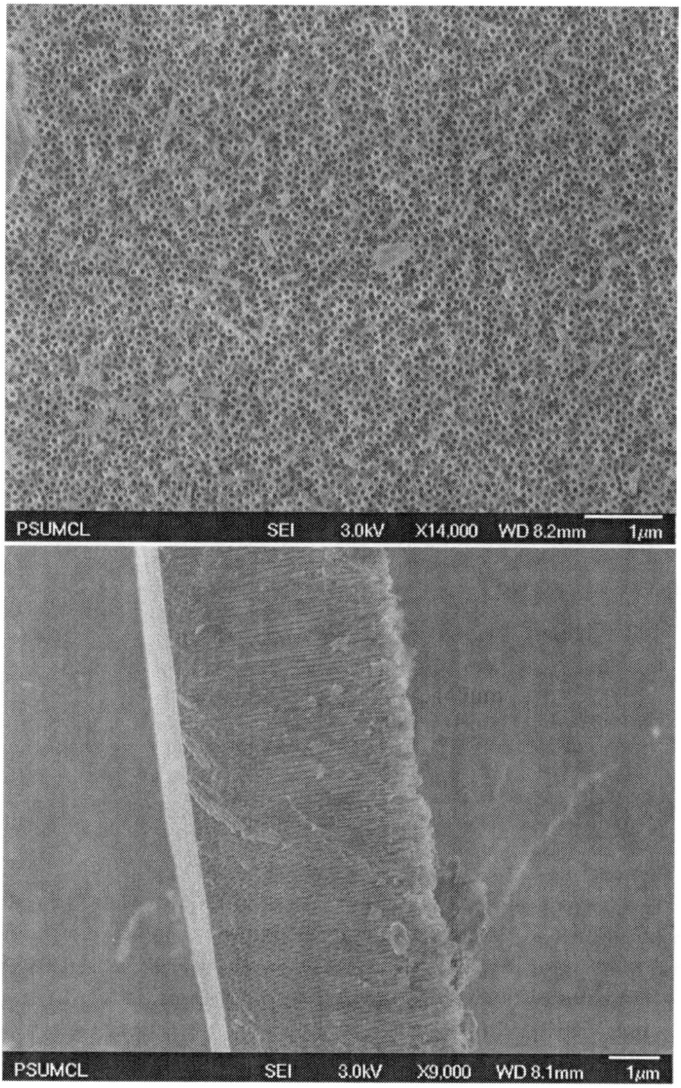

Fig. 7.1 Top and cross-sectional views of hematite nanotube arrays, 4.4 μm long, achieved by anodization of iron foil [1, 31]

nanotubes indicate comparatively slow recombination of photogenerated electrons, indicating that fewer potential surface recombination sites exist in nanotube arrays than in the commonly used nanoparticle films [33, 39]. Using impedance-modulated photocurrent/photovoltage spectroscopy it was shown that the recombination time constants for nanotube array DSCs are an order of magnitude larger than those for nanoparticle films, with the collection efficiency of the nanotube array films 25%

larger than that of the nanoparticle films [40]. With respect to DSC application, a $TiCl_4$ treated 3,600 nm thick transparent nanotube array on FTO glass (front illumination geometry), infiltrated with N-719 in acetonitrile, exhibited an overall conversion efficiency of 4.7%, while a backside illuminated DSC, comprised of a 20 μm long nanotube array on Ti foil, showed an overall conversion efficiency of 6.89% [33]. TiO_2 nanotube arrays have shown great use in construction of polymer-based solid state solar cells, with a FTO/TiO_2 nanotube array/active layer/PEDOT:PSS/Au general structure [41–43]. An overall conversion efficiency of 4.07% was obtained with 270 nm long transparent nanotube array films, pore size of ~50 nm, using a blend of P3HT polymer and $PC_{70}BM$ molecules [41].

TiO_2 nanotube arrays have been investigated for a variety of biological applications. TiO_2 nanotube arrays offer an exciting prospect for achieving drug eluting surfaces for biotemplating and osseointegration applications [44–47], while mechanically robust nanotube array membranes have shown great utility in biofiltration applications [7, 20]. Biosensors using nanotube arrays co-immobilized with horseradish peroxidase and thionine have been shown quite useful for H_2O_2 detection [48], while nanotubes immobilized with glucose oxidase enabled quantification of glucose [49]. Gauze bandages sensitized with dispersed nanotubes were shown to reduce the clotting time of blood while increasing clot strength [50].

7.2 Some Future Directions

7.2.1 Intercalation and Supercapacitors

Fabregat-Santiago and co-workers [51] investigated the electronic charging and conducting properties of TiO_2 nanotube arrays formed by anodization of Ti foil samples. Depending on the electrochemical conditions two rather different electronic behaviours were observed. Nanotube array samples in basic medium show behaviour analogous to that of nanoparticulate TiO_2 films commonly used in dye-sensitized solar cells: a chemical capacitance and electronic conductivity that increases exponentially with bias potential indicating a displacement of the Fermi level. Nanotube array samples in acidic medium, or samples in a basic medium submitted to a strong negative bias exhibit a large increase in capacitance, to super capacitor levels, and conductivity indicating Fermi level pinning. The contrasting behaviours are ascribed to proton intercalation of the TiO_2 suggesting a route for controlling the electronic properties of the ordered metal-oxide nanostructures for their use in applications including supercapacitors, DSCs, and gas sensing.

A crucial property determining the photovoltaic performance of nanostructured semiconducting films is the electronic carrier density, which, in turn, is dependent on the dopant level; carrier density plays a key role in determining electronic conductivity. The carrier density and dopant level in nanostructured films may vary widely depending on the materials, method of preparation, architecture, and electrochemical or thermal treatments. As discussed in a recent review, one can generally distinguish two basic limiting situations [52]. First, standard colloidal TiO_2 films for DSCs made of small, $\approx$10 nm diameter, nanoparticles usually show very low (unmeasurable) charge carrier densities since the dopant levels are minimal and the equilibrium Fermi level is far from the conduction band. When the Fermi level is raised by a bias potential the films respond with a chemical capacitance [53] that normally displays an exponential bias dependence reflecting the density of localized states (DOS) in the bandgap [54]. Over prolonged operation the DOS can be modified by cation intercalation in the TiO_2 nanoparticle bulk [55].

The nanotube array geometry can be viewed as a combination of the characteristic DSC nanoparticle size in the radial direction, with a macroscopic continuous pathway in the axial direction. Hence the characteristic carrier density and electrical field distribution are not *a priori* clear, i.e., whether the film should respond with a chemical capacitance, or with some other behavior. Fabregat-Santiago and coworkers [51] used impedance spectroscopy to analyze a range of nanotube array samples, with the simultaneous detection of transport resistance and capacitance allowing characterization of the carrier density dependence on potential.

TiO_2 nanotube array films were made by anodic oxidation of titanium foil, 20 V in 0.27 M tetrabutyl ammonium fluoride in formamide containing 3.0% de-ionized water [18, 51], with nanotube outer diameter, inner diameter and wall thickness of, respectively, 100, 56 and 22 nm. A typical top and cross-sectional view of an approximately 8 μm long TiO_2 nanotube array film, achieved using the formamide electrolyte, is shown in Fig. 7.2. All samples were crystallized by annealing in oxygen gas at 525ºC for 2 h with a ramp-up and ramp-down rate of 1°C/min [51]. DSCs were prepared by nanotube array immersion in dye solution (N719 0.3 mM in ethanol) overnight. The nanotube array electrode was clamped between two glass sheets, one of them thermally platinized TCO glass (Pilkington TEC15) with a 30 μm suryline tape spacer. The cell was then filled with a drop of electrolyte (0.5 M of LiI, 0.05 M of I_2 and 0.5 M of MBI in Acetonitrile). Voltammetric and impedance spectroscopy measurements were performed using TiO_2 nanotube

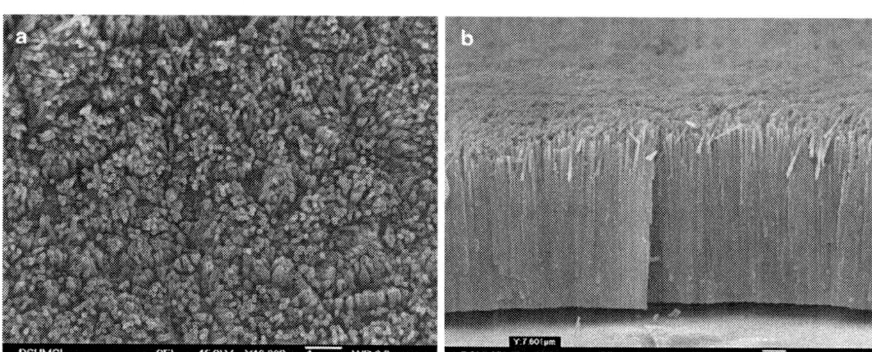

Fig. 7.2 FESEM views of the formamide TiO_2 nanotubes: (**a**) top view, (**b**) cross section [51]

samples of variable length: 1.5, 5, 8, 10 and 20 μm [51] in two potential ranges: between 0.1 and –0.75 V for both acidic and basic media, aqueous solutions adjusted to pH 11 with KOH and pH 2 with H_2SO_4, and from 0.1 to -1.5 V for basic media. Potentials were reported versus an Ag/AgCl (3 M KCl) reference electrode, hence there is a -0.31 V offset between DSC and electrochemical potentials (I_3^-/I^- redox potential vs. Ag/AgCl reference electrode).

For an array of TiO_2 column-like tubes of length (height) L immersed in an electrolyte, the impedance behavior of a typical porous electrode is described by a transmission line model shown in Fig. 7.3a [56, 57]:

$$Z(\omega) = \left(\frac{R_{tr}R_{ct}}{1 + i\omega/\omega_{rec}}\right)^{1/2} \coth\left[(R_t/R_{ct})^{1/2}\,(1 + i\omega/\omega_{rec})^{1/2}\right] \tag{7.1}$$

Here, R_{tr} is the total transport resistance and R_{ct} is the charge transfer resistance at the solid/electrolyte interface along the resistive transport channel. ω_{rec} is the radian frequency of the charge transfer process, $\omega_{rec}=(R_{ct}C_3)^{-1}$, related to electron lifetime by $\tau_n = \omega_{rec}^{-1}$. C_3 is the total capacitance in the solid phase and the solid-electrolyte

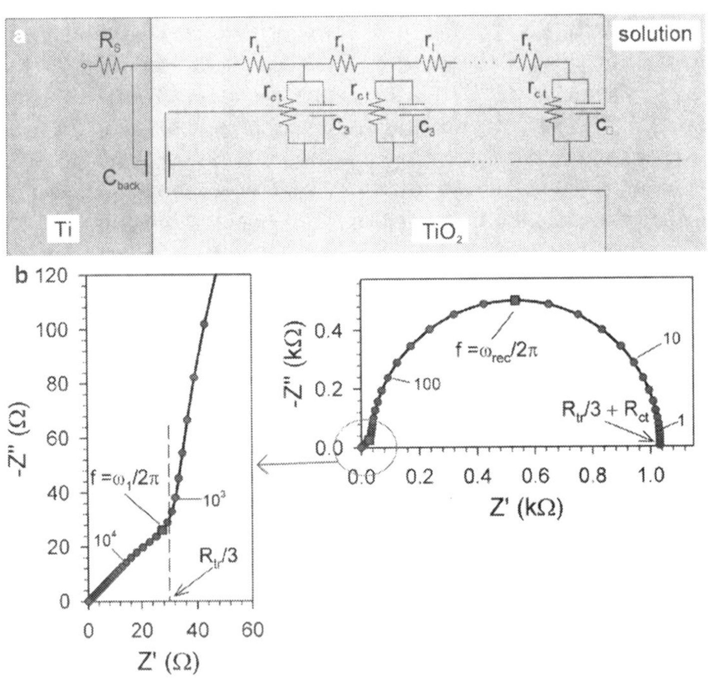

Fig. 7.3 (**a**) Transmission line equivalent circuit model applied to the TiO_2 nanotube array films. (**b**) Impedance response of a nanotube electrode. The large arc is the parallel combination of R_{ct} and C_3. The left panel shows in expansion the high frequency 45° like behavior due to transport of electrons in the semiconductor; frequencies ω_1 and ω_{rec} are defined in text [51]

interface. C_3 has several possible contributions [53]: (1) the chemical capacitance of the semiconductor, C_μ; (2) the depletion capacitance in the semiconductor surface, C_{dpl}; and (3) the Helmholtz capacitance of the surface, C_H. Total film capacitance has an additional contribution from the exposed surface of the barrier layer at the bottom of the columns, C_{back}, see Fig. 7.3a.

In the case of low reactivity with respect to electron transport, as given by the condition $R_{ct}>R_t$ [56], for frequencies higher than $\omega_1=(R_tC_3)^{-1}$ (7.1) reduces to:

$$Z = \left(\frac{R_{tr}}{C_3}\right)^{1/2} (i\omega)^{-1/2} \tag{7.2}$$

Equation 7.2 is often termed the "Warburg" impedance, appearing as a 1-slope line in the complex impedance plot shown in the left of Fig. 7.3b. It should be emphasized that, in contrast to semi-infinite diffusion, in the transmission line model the 45° characteristic appears only in the high frequency part of the impedance spectrum. As the frequency decreases ($\omega<\omega_1$) the charge-transfer resistance gives rise to a semicircle at low frequency, consisting of the parallel combination of R_{ct} and C_3 as shown in the right of Fig. 7.3b.

The nanotube array presents the characteristic behavior of nanoparticulate TiO_2 networks usually found in DSCs [52, 58] with little intrinsic doping by foreign atoms and consequently a homogeneous conduction band level throughout the semiconductor. The bias potential V determines the Fermi level position with respect to the conduction band, while the transport resistance in the TiO_2 displays the potential dependence [59]:

$$R_{tr} = R_{0tr}\ \exp\left[\frac{qV}{kT}\right] \tag{7.3}$$

with q the elementary positive charge. R_{tr} is large at low potentials due to the absence of free carriers, therefore no contribution from the TiO_2 nanotubes to capacitance and charge transfer resistance is expected. In this regime the majority of the cell capacitance will come from the exposed barrier layer, with a value of the order of 10 μF cm^{-2}. The TiO_2 capacitance increases with applied potential by [54]:

$$C_\mu = C_{0\mu} \exp\left[-\alpha \frac{qV}{kT}\right] \tag{7.4}$$

with $\alpha < 1$, until it reaches the value of the C_H given by the total surface of the tubes.

Electrochemical measurement of the nanotube array samples provided a variety of behaviors. First, while keeping the samples in a basic (pH 11) electrolyte medium at moderate applied potentials (< 1 V), they presented the general behavior described by (7.1), shown in Fig. 7.4. The parameter values obtained from impedance (C_3, R_{tr}, R_{ct}) follow quite accurately the trends stated in (7.3) and (7.4). The resulting parameters for the 10 μm sample are shown in Fig 7.5, with a chemical

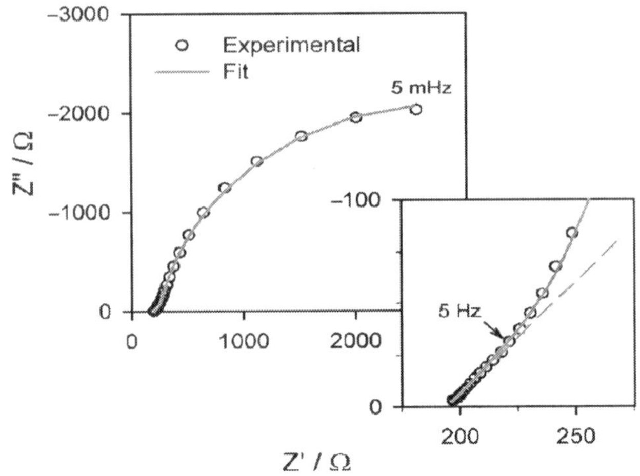

Fig. 7.4 Impedance spectra of a TiO_2 nanotube array film initially held at –1 V for 20 min, i.e., "aged," pH 11, showing typical transmission line behavior with low frequency recombination arc and high frequency 45° line. Impedance was performed at –0.7 V [51]

capacitance value of $\alpha=0.21$, charge transfer resistance coefficient of 0.40, and transport resistance slope of 84 mV/decade, the values of which are comparable to those obtained for nanocolloid-based films [52, 58]. The clear experimental observation of transmission line behavior (Fig. 7.3) indicates that the capacitance and charge-transport resistance are due to the electronic response of the TiO_2 nanotubes, while the impedance behavior can be unambiguously interpreted in terms of the Fig. 7.3a model providing information on the carrier distribution within the tubes.

As seen in Fig. 7.5, when the applied bias potential was brought to large negative values (< -1.5 V) or kept at moderate potentials (≈ -1 V for more than 20 min) in basic media, the impedance results changed dramatically. The transport resistance is reduced and maintains a nearly constant value, indicating a large charge carrier density insensitive to applied voltage. This is a strong indication of highly doped behavior. In addition, the chemical capacity of TiO_2 increases enormously, with values ranging from 1 to 6 mF cm^{-2} at low potentials, reaching close to 80 mF cm^{-2} at the highest applied potential. These values approach those of supercapacitors ($\approx$0.3–1 mF cm^{-2}) [60], see Fig. 7.5a. Larger values of capacitance are obtained with increasing nanotube length. The observed modifications under large negative bias in basic medium are attributed to proton uptake in the nanotube walls, which also changed the clear nanotube walls to a blue color. Proton uptake in nanocrystalline TiO_2 was demonstrated by Hupp and co-workers using a photoelectrochemical quartz crystal microbalance technique [61]. Cation intercalation within the TiO_2 nanotubes results in free electrons that, in turn, decrease the electrical resistance and results in a much larger capacitance that is most likely due to formation of a depletion region at the semiconductor surface [62–64].

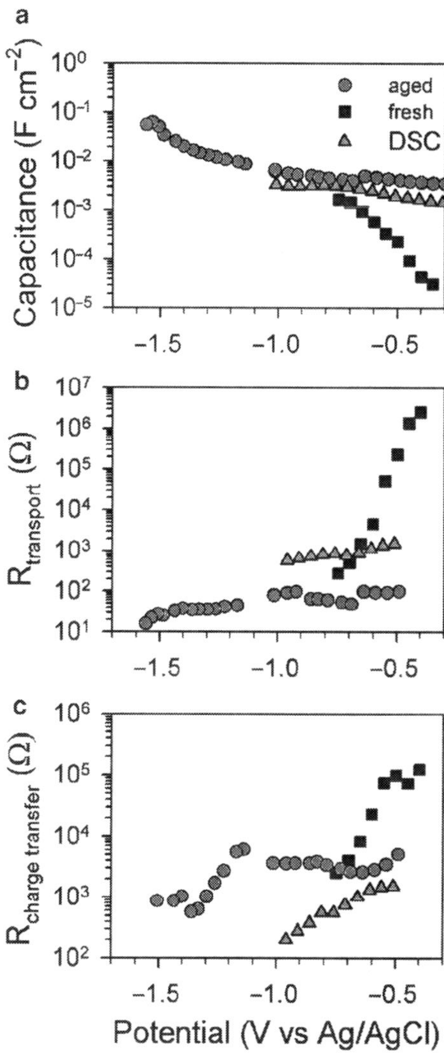

Fig. 7.5 Comparison of parameters from impedance data, from the equivalent circuit model of Fig. 7.3a, as a function of electrode potential, for a 10 μm sample before (fresh) and after being held at –1 V for 20 min in basic solution (aged), and in a DSC. (**a**) Capacitance per unit geometric area, C_3, (**b**) transport resistance, and (**c**) charge-transfer resistance. Note that for comparison DSC values have been shifted –0.31 V (I_3^-/I^- redox potential vs. Ag/AgCl reference electrode) [51]

Figure 7.6a shows cyclic voltammetric measurements of a 10μm long nanotube sample prior to and after cation intercalation. The capacitive current [65] is much higher for the cation intercalated films indicating a great increase in the capacitance at low potentials. Figure 7.6b shows the dynamic intercalation process in an 8μm sample. During the first cathodic scan, starting at 0 V, at low negative potentials the capacity is small, while in the return anodic scan the capacitance is much larger. In the second cathodic scan the capacitance is large even at low potentials, and with respect to the first scan further increases in the anodic return.

Measurements were also done in acidic media (pH 2) and in this case high capacitance values were observed in all samples, see Fig. 7.7a. The transport

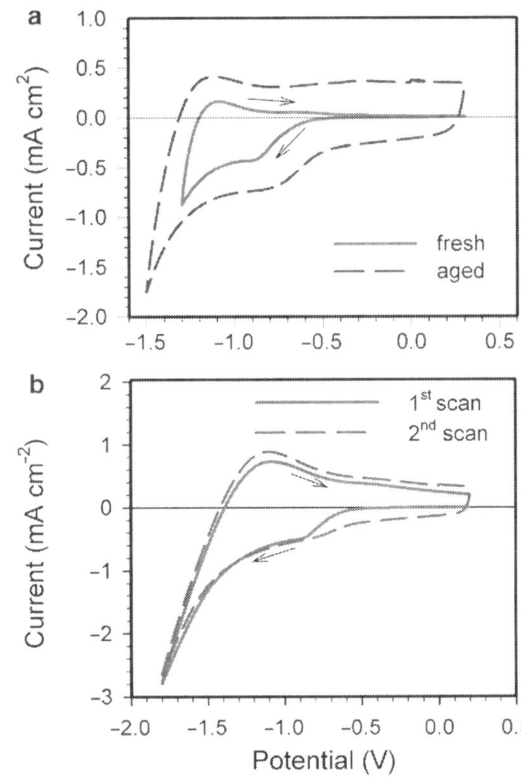

Fig. 7.6 (**a**) Cyclic voltammetries of a 10 μm sample before (fresh) and after application of a strong bias potential (aged) while measuring impedance. (**b**) The intercalation process during a voltammetry of sufficient potential in a 8 μm long nanotube array sample [51]

resistance is so low that it cannot be reliably measured. The capacitance increase at the highest negative potentials appears due to the unpinning of the Fermi level that occurs at potentials more negative than −0.8 V allowing the Fermi level to approach the conduction band [65]. The results indicate that the nanotube array samples are proton intercalated even in the absence of the negative bias that was required in basic media. Figure 7.7 enables comparison of the capacitance behavior of the nanotube array films as a function of nanotube array length in both basic and acid electrolytes. The capacitance scales with nanotube length. The small dispersion of the rescaled values in Fig. 7.7b are attributed to minor differences in wall thickness of the different samples.

Thus, while "fresh" nanotube array samples in basic medium display the standard behavior of crystalline TiO_2 nanoparticles, consisting of initially low transport resistance and chemical capacitance that exponentially increase with bias potential [58, 59], these properties change dramatically upon negative bias whereupon the sample displays the behavior characteristic of a highly doped semiconductor, i.e., a high capacitance and a low charge transport resistance nearly independent of bias. This change is attributed to the intercalation of protons from water into the nanotubes. In acidic media the high proton concentration of the electrolyte

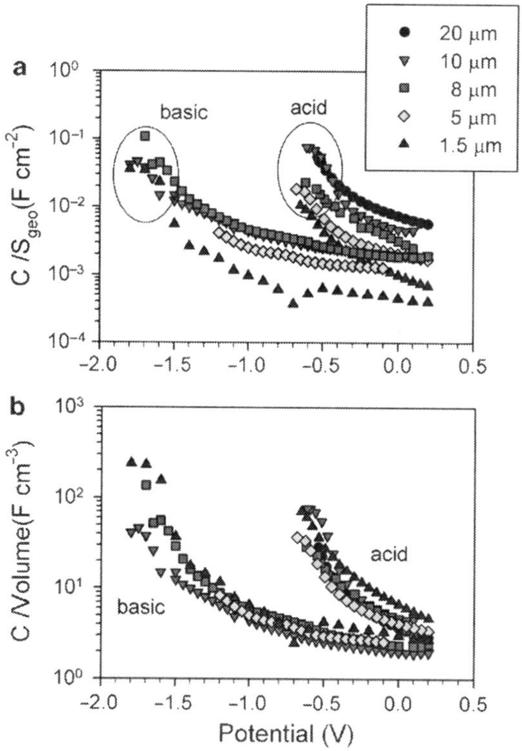

Fig. 7.7 (**a**) TiO_2 nanotube array capacitance, per unit geometric area of the film, in basic and acidic media. (**b**) Same capacitance data rescaled to volume density, obtained by dividing by nanotube length [51]

facilitates intercalation, while in basic solution it requires application of a negative potential.

Back-side illuminated DSCs were fabricated using 10μm long TiO_2 nanotube array films on a Ti foil substrate. Prior to DSC fabrication the nanotube array film was held at −1 V for 20 min in acidic solution where the performance of the electron transporting nanotube arrays are expected to be altered by proton intercalation. From the current density-potential (j–V) curve, Fig. 7.8, an efficiency of 3.1% is obtained, short circuit current of 3.1 mA cm^{-2}, photopotential 0.63 and fill factor of 0.44. Although un-optimized, this DSC is significantly less efficient than previous reports for the same device geometry [3, 33]. The impedance measurements of Fig. 7.5 allow identification of the central factors determining the relatively poor shape of the j–V curve [66, 67]: (i) The high capacitance obtained for the DSC, see Fig. 7.5a, indicates that the TiO_2 nanotubes have been cation intercalated. This property may limit the movement of the Fermi level in the semiconductor, thus limiting V_{oc}. (ii) The recombination resistance, see Fig. 7.5c, is comparatively low [58, 66] indicating that cation intercalation, or doping, enhances recombination. It appears enhanced recombination is responsible for the pronounced slope at low potentials which contributes to a reduction in the fill factor and, additionally, limits V_{oc} [66, 67]. (iii) TiO_2 intercalation, observed through

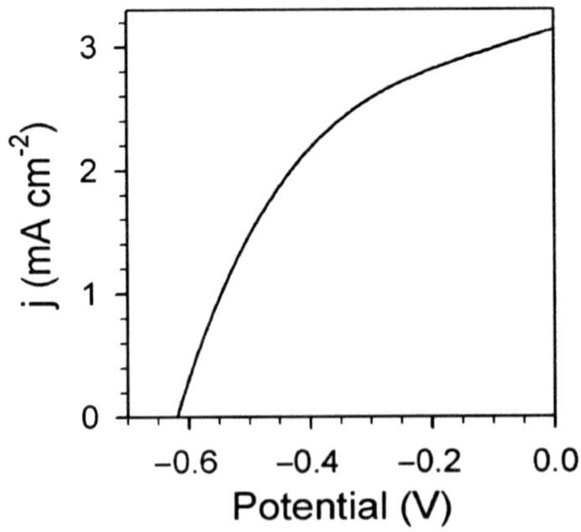

Fig. 7.8 Current density-potential curve of a DSC made from a 10 μm nanotube film "aged" by being held at −1 V for 20 min [51]

capacitance measurements, is confirmed by the nearly constant value obtained for the transport resistance. The values obtained for R_{tr} in the DSC samples are higher than expected. These transport resistance values accurately describe the slope of the $j-V$ curve at high negative potentials, related to a high overall series resistance and poor fill factor [66, 67]. The contributions to the overall resistance found for these DSC samples were: 20 Ω from the conducting glass and titanium foil, 7 Ω from the Pt counter-electrode, from 20 to 100 Ω (depending on the potential) from the diffusion of the redox in the electrolyte, and values above 200 Ω (=$R_{tr}/3$ [56, 57, 68]) from electron transport in the TiO_2.

From these results it is inferred that *unintentional* doping, i.e., intercalation, may have detrimental effects on DSC performance due mainly to the pinning of the Fermi level and increase of recombination. Strategies for improving DSC performance by electrochemical doping include the formation of fast transport states [69] and the formation of core-shell or graded doping of the metal-oxide including a fast transport channel [70]. Such approaches could decrease the electron transport resistance to improve the photocurrent while allowing for Fermi level displacement and consequently a larger photovoltage. The possibility of unwanted TiO_2 intercalation suggests long-term DSC stability may be enhanced by substitution of the lithium ions by larger ones, such as tetrabutylammonium or guanidinium.

In summary, in basic media, the films display the standard behavior of nanoparticulate TiO_2 films with a chemical capacitance exponentially dependent on the bias potential. This can be modified to demonstrate a very high capacitance, most likely due to the intercalation of protons into the TiO_2, offering the possibility for application of nanotube array films in supercapacitors. Proton intercalation, or doping of the nanotubes in acidic media is readily achieved while that in basic media requires application of strong negative potentials. DSCs prepared with the nanotubes previously intercalated with protons from solution showed limited performance. In this case charge transport within the titanium dioxide was found to be the main contribution to the DSC series resistance.

7.2.2 CO_2 Reduction Using Visible Light

Varghese and co-workers have recently reported the use of TiO_2 nanotube arrays, in combination with copper oxide, for high rate conversion of CO_2 and water vapor into hydrocarbon fuels [71]. The rapid increase in the level of anthropogenic carbon dioxide is a matter of great concern. While discussions have begun on means to reduce carbon dioxide emissions, it is apparent that atmospheric carbon dioxide concentrations will continue to monotonically increase for the foreseeable future due to fossil fuel consumption. For example, each 12 days the world consumes approximately 1 billion barrels of oil representing atmospheric release of almost 1 trillion pounds of CO_2. Suggestions have been made to sequester carbon dioxide [72], although the ability to store several billion tons of carbon emitted in the form of carbon dioxide every year is questionable as are the environmental consequences. Recycling of carbon dioxide via conversion into a high energy-content fuel, suitable for use in the existing hydrocarbon-based energy infrastructure, is an attractive option however the process is energy intense and useful only if a renewable energy source can be used for the purpose.

A possible avenue for sustainable development is to use photocatalysts for the conversion of CO_2 into hydrocarbons with the help of solar energy. A process involving only water and carbon dioxide is very promising as it can form a useful carbon cycle:

$$CO_2 + 2H_2O \Leftrightarrow CH_4 + 2O_2$$

The photocatalytic methane forming reaction requires eight photons, with additional photons required for other hydrocarbons. Hence, achieving significant hydrocarbon reaction yields requires an efficient photocatalyst that utilizes a maximum amount of solar energy.

In 1979, Honda and co-workers demonstrated photocatalytic reduction of carbon dioxide to organic compounds, i.e., formic acid, formaldehyde, methanol and methane, by suspending both oxide and nonoxide semiconductor particles in water [73]. Additional research efforts on photocatalytic conversion of carbon dioxide dissolved in liquids followed [74–77]. With respect to gas phase CO_2, Graetzel and co-workers reported the production of methane from a mixture of carbon dioxide and hydrogen in argon using Ru/RuO_x sensitized titania nanoparticles [78]. An initial methane production rate of about 116 μl/h per 100 mg of nanoparticles was obtained under simulated light, which was believed due to the contributions from thermal as well as ultraviolet (UV) light-induced effects in the Ru–RuO_x/titania system. However, after a comprehensive study Melsheimer et al. [79] concluded that it was solely a thermal effect.

Titania has been considered the most appropriate candidate for photocatalytic processes due to its powerful oxidation properties, superior charge transport properties as well as corrosion resistance. Yet, it has so far yielded only low carbon dioxide conversion rates despite using ultraviolet illumination for bandgap excitations. Adachi and co-workers achieved a total hydrocarbon (methane, ethylene and ethane) generation rate of about 1.7 μl/h g under xenon lamp illumination when copper loaded titania nanoparticles were dispersed in CO_2-pressurized water [76]. Tan et al. used titania pellets (100 g) to obtain a maximum rate of about 0.25 μmol/h of methane from moist carbon dioxide under monochromatic ultraviolet (253.7 nm wavelength) illumination [80, 81]. Using a hydrogen (90%), water, carbon dioxide combination, a rate of 4.1 μmol/h g was obtained by Lo and co-workers using UV illumination at wavelengths 254 and 365 nm [82]. Nishimura et al. used sol–gel derived titania multilayer films inside a copper tube to obtain a total product (CO, CH_4, C_2H_4 and C_2H_6) formation rate of about 3.2 nmol/cm^{2}h under UV illumination of 2.4 mW/cm^2 [83]. Thus, while the intrinsic idea of photocatalytic conversion of carbon dioxide and water (liquid/vapor) into hydrocarbon fuels is appealing, the process has historically suffered from low conversion rates.

As depicted in Fig. 7.9, Varghese and co-workers [71] sought to enhance photocatalytic carbon dioxide conversion rates by using the following strategies: (i) employ high surface area titania nanotube arrays, with a wall thickness low enough to facilitate efficient transfer of photogenerated charge carriers to the surface species; (ii) modify the titania bandgap to absorb and utilize the visible portion of the solar spectrum where the bulk of the solar energy lies; (iii) distribute co-catalyst nanoparticles on the nanotube array surface to adsorb the reactants and help the redox process.

TiO_2 nanotube arrays formed by anodic oxidation [1–4] of titanium have very high surface areas [23], comparable to porous titania nanoparticle films, and proven photocatalytic properties [28]. While the TiO_2 band gap, 3.0 and 3.2 eV, respectively for rutile and anatase, restricts excitation wavelengths to less than ≈400 nm Varghese et al. succeeded in incorporating nitrogen into the nanotubes In situ during anodization, with a subsequent heat treatment resulting in crystallized nanotubes with N 2p states formed above the titania valence band shifting the absorption edge of titania to ~540 nm. These nitrogen doped titania nanotube

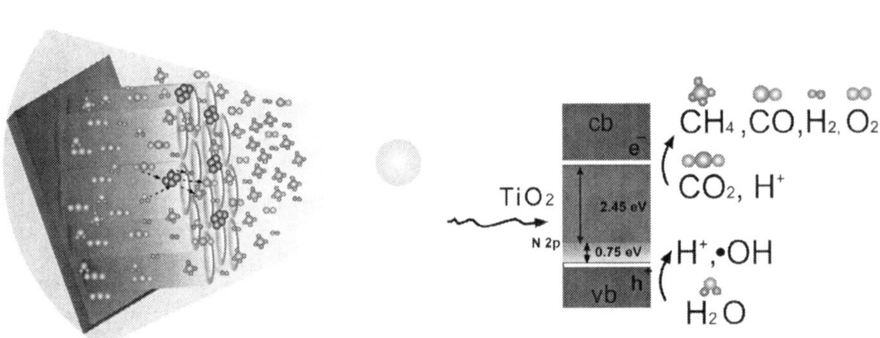

Fig. 7.9 Depiction of sunlight driven photocatalytic carbon dioxide conversion to hydrocarbon fuels using nitrogen doped titania nanotube arrays surface-loaded with Cu and/or Pt co-catalyst nanoparticles [71]

arrays, with platinum and/or copper nanoparticles dispersed upon the top of the nanotube array surface, were used to convert carbon dioxide and water vapor into hydrocarbons under natural (outdoor) sunlight. The nitrogen doped titania nanotube arrays were synthesized by anodizing titanium foil in an electrolyte consisting of 0.3 M ammonium fluoride (NH_4F) in 2 vol% water containing ethylene glycol at 55 V [6]. Nanotube array samples up to $\approx 130\,\mu$m in length were prepared by changing the anodization duration. The debris from the anodization bath re-depositing on top of the original nanotubes was removed by ultrasonic agitation and/or critical point drying (CPD) [7]. The average pore diameter of all the samples estimated from FESEM images was about 95 nm (standard deviation 13 nm) and wall thickness about 20 nm (standard deviation 5 nm). While the nitrogen incorporated within the nanotubes during their anodization is primarily supplied by the NH_4F, experiments also showed that to a lesser extent atmospheric nitrogen dissolved into the electrolyte from where it was incorporated into the nanotubes. This observation was verified by fabricating the nanotubes using an electrolyte of 0.7 vol% hydrofluoric acid in ethylene glycol at 120 V [6, 8], which still resulted in residual nitrogen incorporation into the nanotubes. Ti anodization in fluorine containing electrolytes yields amorphous nanotube arrays [2], hence elevated temperature oxygen annealing is needed to maintain titania stoichiometry and reduce lattice defects. However, annealing the nitrogen incorporated samples at high temperatures in an oxygen rich environment generally removes the nitrogen by oxidation. The incorporated nitrogen within the nanotube samples was retained by limiting the oxygen flow across the sample during annealing in a tube furnace. The samples were annealed at 460°C or 600°C for 3 h.

A digital image of a 32 μm long as-prepared sample is shown in Fig. 7.10a; a 460°C annealed nanotube array sample of thickness ~25 μm is shown in Fig. 7.10b and ~130 μm long in Fig. 7.10c; a 600°C annealed ≈40 μm long nanotube array sample is shown in Fig. 7.10d. For the 460°C heat-treated nanotube array samples the color changed from dark green to yellowish green as the thickness increased from a few tens of microns to 100 μm, while the samples became pale yellow after annealing at 600°C. The change of color to yellowish green for the longer nanotubes is due to the greater absorption in the green region of the visible spectrum, where the absorption coefficient of nitrogen doped titania is very low, as the light traverses to greater depths. Lindgren and co-workers [84] estimated an absorption coefficient in the range of $10^2\,cm^{-1}$ in the 500–550 nm wavelength region for their nitrogen doped porous titania films.

The diffuse reflectance spectra of nanotube array samples of nearly the same thickness (~130 μm) annealed at 460 and 600°C, expressed in terms of the Kubelka-Munk function F(R) are shown in Fig. 7.10e. Nitrogen doping in the samples is indicated by the hump above the intrinsic titania absorption edge at ≈400 nm, which extends up to ≈540 nm. The 600°C annealed samples exhibit lower light absorption in the visible region indicating a lower level of nitrogen doping, a result confirmed via X-ray photoelectron spectroscopy, see Fig. 7.10f.

Since estimating a doping concentration from porous samples could lead to significant errors, the nanotube array membrane was separated from the foil [6, 7]

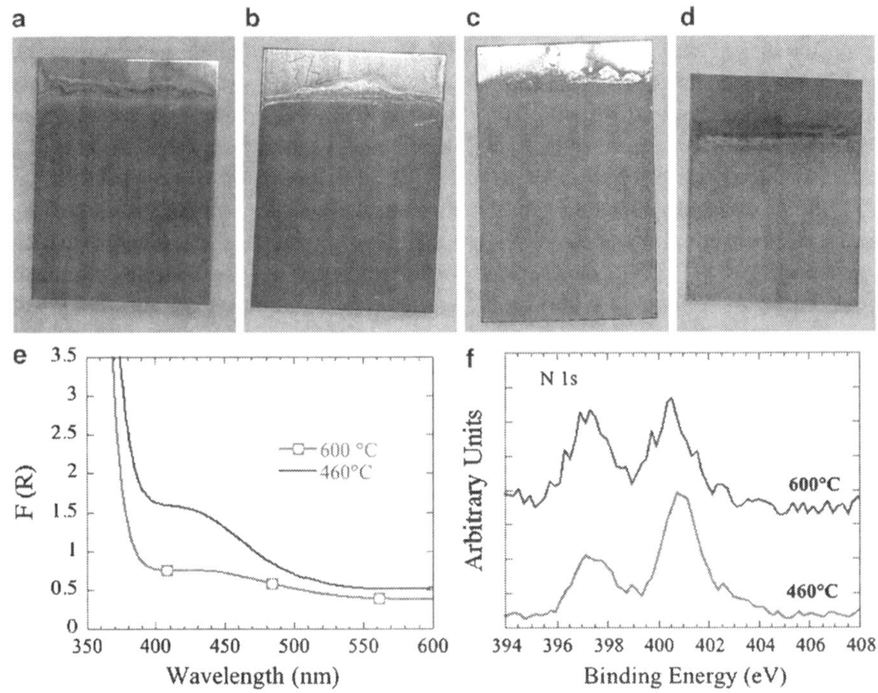

Fig. 7.10 (a-d) Digital images of the nitrogen doped nanotube array samples: (**a**) 32 μm long as prepared; 460°C annealed sample; (**b**) 23 μm long; (**c**) 130 μm long; (**d**) 600°C annealed sample 40 μm long. Plots **(e)** and **(f)** show, respectively, the diffuse reflectance spectra plotted as K–M function F(R) and the X-ray photoelectron spectra of samples annealed at 460°C and 600°C [71]

and the "bottom-up" geometry used for XPS analysis. The bottom ends of the nanotubes are closed and hence present the appearance of a dense film having nanometer-scale roughness. The N 1s core level spectra given in Fig. 7.10f were collected after initial "cleaning" by sputter-removal of approximately 30–40 nm of the film. The nitrogen peaks appear at binding energies of about 397 and 400.5 eV. Out of several samples studied most samples, irrespective of nanotube length, yielded only the 400.5 eV peak when the data was collected from the bare surface without the sputter etching. While there exists disagreement among researchers regarding the assignment of peaks arising from nitrogen doping [85], there appears a consensus in the assignment of the peaks in the 396–397 eV region as arising from nitrogen occupying anionic sites in titania as assigned by Asahi and co-workers [86]. Varghese et al. suggest that substitutional nitrogen (β-N) in the nanotubes is giving rise to the peak at 397 eV, while the peak at 400.5 eV cannot be from molecularly chemisorbed nitrogen (γ-N_2) [86] but from oxidized nitrogen (NO) in the lattice [85, 87, 88]. The total nitrogen concentration as estimated from the high resolution scans were, respectively, approximately 0.75 atomic percent (at%) and 0.4 at% for the 460 and 600°C annealed nanotube array samples.

Prior to the photocatalytic carbon dioxide reduction experiments, the annealed nanotube array samples were coated with an ultrathin layer of platinum or copper nanoparticles using dc sputtering (dc power 1.8 W/cm^2, pressure 20 mTorr, 5 cm distance between the target and sample, deposition duration 13 s for Pt and 40 s for Cu). High resolution transmission electron microscope (HRTEM) studies on platinum coated samples showed nanoscale platinum islands/particles (spread of the island less than ~40 nm) attached to the nanotube walls, primarily near the open end of the nanotube. No pore clogging was observed through either HRTEM or FESEM imaging. Samples of ~2×2 cm size were loaded into a stainless steel chamber having valves for evacuation and gas feeding, a side port with a septum attached to it for gas sampling and an O-ring sealed quartz window at the top for admitting solar radiation. Two chambers having identical construction and total volumes 7.5 and 8.6 cm^3 were used allowing for simultaneous exposure of two samples. Anodization of titanium foil creates nanotube arrays on both sides of the foil; to avoid any contribution from the unexposed side, this side was masked using opaque adhesive tape. The samples were fixed inside the cells using copper tape with the nanotube array surface facing the quartz window. After sample loading the chamber was evacuated to about 10 mTorr using a mechanical pump and then sealed. Carbon dioxide (99.99% pure), via a mass flow controller (MKS instruments), was passed through a bubbler containing de-ionized water before entering the reaction chamber. Both inlet and outlet valves were kept open with the chamber flushed for more than 10 min, and then closed maintaining a nominal excess pressure of less than 1.0 PSI.

All photocatalytic carbon dioxide conversion experiments were done under outdoor sunlight. The results reported in this article were based on our experiments conducted under clear skies or sky with a few clouds at University Park, Pennsylvania (latitude 40°49′W and longitude 77°51′N) during July to September, 2008, although preliminary outdoor experiments were started in April 2008. Figure 7.11a

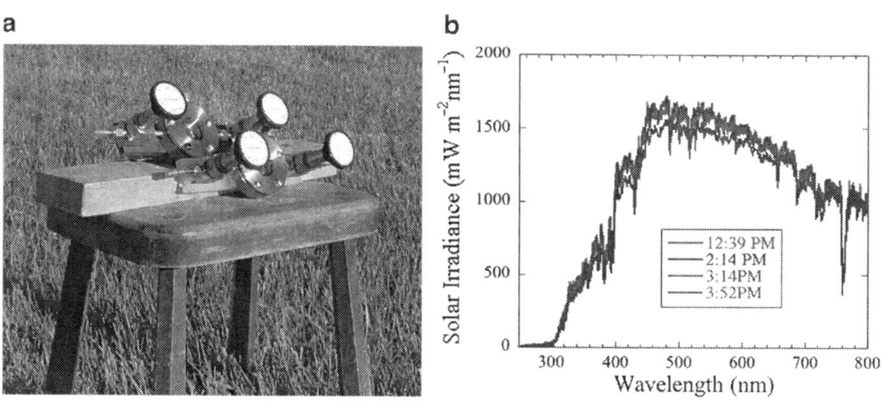

Fig. 7.11 **(a)** Digital photograph of the reaction chambers kept under natural sunlight for photocatalytic CO_2 conversion. **(b)** Spectral irradiance recorded from 12:39 PM to 3:52 PM for an experiment conducted on September 1, 2008, University Park, PA [71]

shows a digital photograph of both reaction chambers loaded with nanotube array samples kept under sunlight for CO_2 conversion. The solar power density was measured at constant intervals (from 1 to 30 min depending upon atmospheric conditions) using a calibrated thermopile detector (Newport Corporation). The solar spectra were recorded as needed using an optical spectrometer equipped with an integrating sphere (model OSM 2-400DUV-U, Newport Corporation). As an example, the solar spectra recorded during experiments on September 1, 2008 are given in Fig. 7.11b.

The average power density P_s of exposure was determined from the total energy incident over the duration T of the sunlight exposure using the relation:

$$P_s = \frac{\int_0^T P(t)\mathrm{d}t}{T}$$

Since the average power density varied slightly day to day, product formation rates were normalized to a global AM 1.5 value of 100 mW/cm^2. Quantitative exposures were made for a duration of about 2.5 h, up to a maximum of 3.5 h, only when the sunlight power density was between about 102 and 75 mW/cm^2. The equilibrium temperature of the samples was about 44°C (±4°C). Although experiments were conducted at lower temperatures (~35°C), no definite influence of temperature on product formation rates was seen.

Reaction products were analyzed using Shimadzu (GC-17A) gas chromatograph equipped with flame ionization (FI) and thermal conductivity (TC) detectors. The FI detector enabled detection of most hydrocarbons, and the TC detector was used for other products. Gas sample analysis of the reaction products predominately showed methane, while ethane, propane, butane, pentane and hexane as well as olefins and branched paraffins were also found in low concentrations. Figure 7.12a shows the hydrocarbon generation rates for 600°C annealed samples surface-loaded with Pt catalyst (labeled NT/Pt-600, ≈35 μm long) or Cu (NT/Cu-600, ≈50 μm

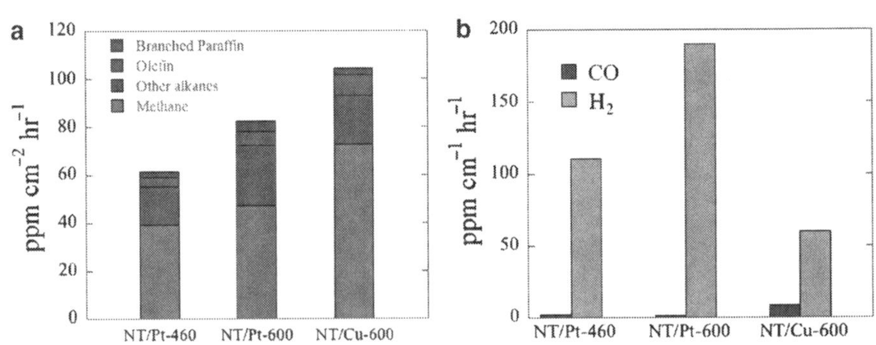

Fig. 7.12 Nitrogen doped nanotube array films with Pt (NT/Pt) and Cu (NT/Pt) co-catalysts, 460 or 600°C annealed as indicated, under sunlight illumination: (**a**) Hydrocarbon generation rate; (**b**) H_2 and CO generation rates [71]

long), and 460°C annealed samples supporting Pt catalyst (NT/Pt-460, 70 μm). A hydrocarbon production rate of about 104 ppm/cm^{2}h (0.78 μl/cm^{2}h or 31.9 nmol/cm^{2}h) was obtained with the 600°C annealed copper surface-loaded sample. A direct comparison of this value with those in the literature is difficult since most previous work was done using nanoparticles (except the work of Nishimura et al. [83]) primarily under UV illumination [80–82]. However, since the mass of a 1 cm^2 nanotube array membrane, i.e., the nanotube array removed from the metal foil substrate, is a few milligrams (~5 mg measured for a 1 cm^2 membrane of 35 μm length) it can be seen that this rate of production obtained under outdoor sunlight (≈150 μl/g h) is at least 20× higher than those reported under laboratory conditions using UV illumination [76, 77, 80–82].

Experiments done with NT/Pt samples of different nanotube lengths showed an increase in hydrocarbon production rates up to a nanotube length of about 25 μm; no definite correlation between the reaction rates and nanotube length was obtained for lengths between 25 and 70 μm, and beyond this nanotube array length generally a decrease in production rate was observed. HRTEM studies showed the Pt nanoparticles concentrated near the tube openings. The distribution of the sputtered co-catalyst nanoparticles, which remain at the top surface of the nanotubes rather than inside, appears responsible for the observed behavior.

As seen in Fig. 7.12b, in addition to methane and carbon dioxide the TC detector showed the presence of hydrogen and carbon monoxide. Although oxygen was also detected, its actual origin and quantification was difficult due to the trace amounts. A striking difference between the NT/Pt and NT/Cu sample behaviors is the amount of H_2 and CO intermediates. With the NT/Pt samples the H_2 generation rate exceeded the hydrocarbon rate, whereas with the NT/Cu samples this rate was low. In contrast, the NT/Cu samples generated ≈5× more CO concentration than the NT/Pt samples. Considering all products, the total production rate using NT/Pt samples is about 273 ppm/cm^{2}h.

Since co-catalysts Pt and Cu influence the product formation in different ways, with a view towards enhancing hydrocarbon production rates both co-catalysts were used together on a single 600°C annealed 40 μm long sample but covering different areas, i.e., different regions of the sample. A NT array sample with ~52% of the surface area decorated by Cu nanoparticles, and the remaining surface area by Pt nanoparticles, showed an enhanced hydrocarbon generation rate of about 111 ppm/cm^{2}h (≈160 μl/g h, or 0.83 μl/cm^{2}h) with no (or below GC detection limits) carbon monoxide generation, see Fig. 7.13. It appears that a homogenous distribution of co-catalyst nanoparticles over the entire nanotube array surface area would further improve utilization of the intermediates for hydrocarbon generation.

Although the nitrogen concentration was higher in the 460°C annealed samples (0.75 vs. 0.4 at%), the hydrocarbon production rates, and those of other products, were lower (Fig. 7.12). The GAXRD patterns of two 60 μm long NT array samples annealed at 460 and 600°C, recorded under identical instrument conditions, showed a higher degree of crystallinity for the 600°C annealed NT array as indicated by the increase in the reflection intensities, see Fig. 7.14. Therefore, a lower degree of crystallization in the 460°C annealed samples is a factor responsible for its

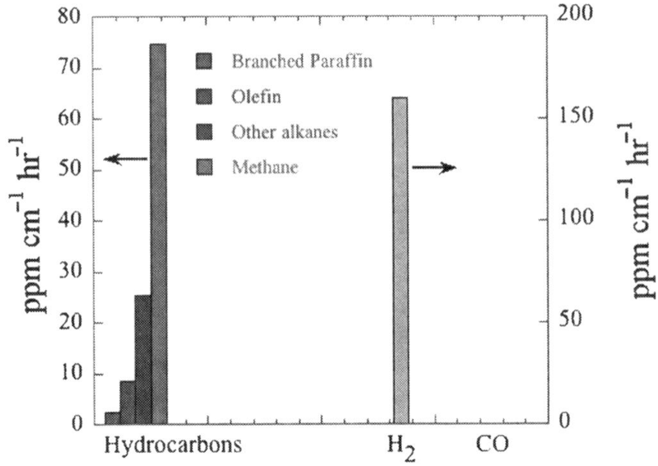

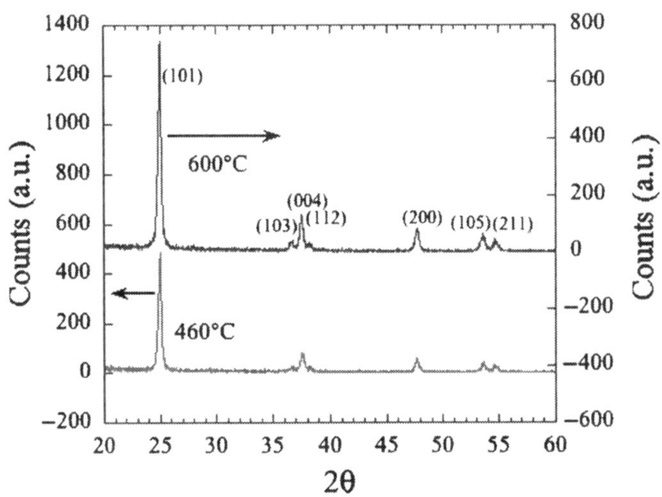

Fig. 7.14 Glancing angle X-ray diffraction patterns of two otherwise identical nanotube array films annealed at 460 and 600°C for 3 h in an oxygen ambient. All reflections are from anatase phase [71]

relatively inferior performance. Only anatase phase is observed in the nanotube walls of the annealed samples.

To understand the role of nitrogen doping the carbon dioxide conversion experiment was performed using a high-pass filter with a "cut on" wavelength of 400 nm to remove the UV component of sunlight. Figure 7.15a shows the total yield

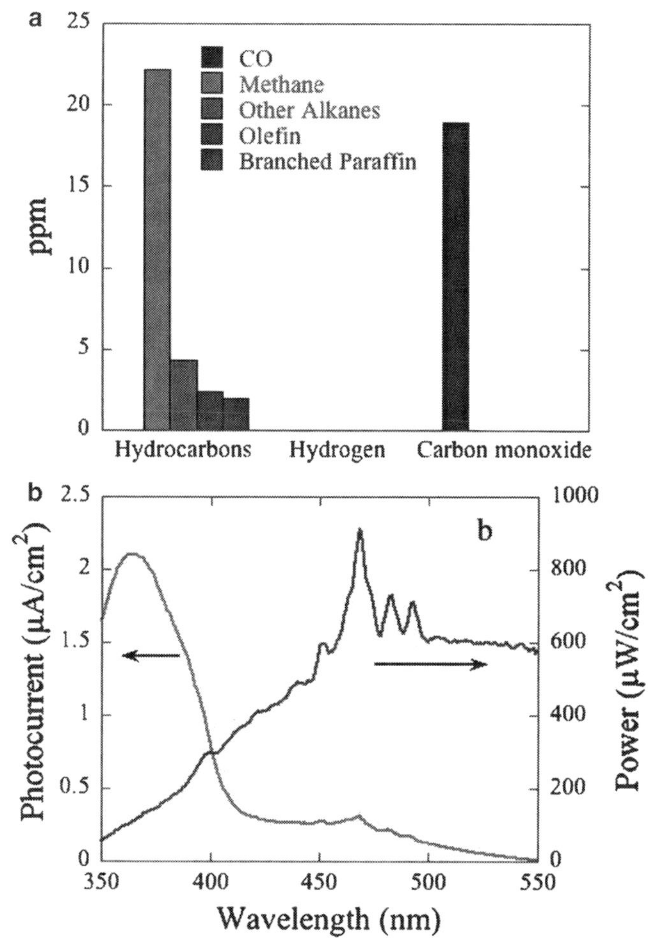

Fig. 7.15 **(a)** Product yield from a nitrogen doped nanotube array sample when a 400 nm wavelength high-pass filter was used to remove the UV component from the solar radiation (sample area 3.6 cm^2, measured filter output 78.5 W/cm^2 sunlight, 3 h exposure duration). **(b)** Photocurrent spectrum of the nanotubes in Na_2SO_4/water electrolyte (*left*), with the incident spectrum (*right*) [71]

obtained from a NT/Cu array sample (600°C annealed) of area ≈3.6 cm^2 exposed to 78.5 W/cm^2 sunlight (measured filter output) for 3 h. The rate of hydrocarbon production is about 3% of the rate shown in Fig. 7.12a for the whole solar spectrum. Lindgren et al. [84] observed that only six to seven electrons are produced per hundred photons (internal quantum efficiency 6.8%) within the maximum visible light absorption region (between 400 and 450 nm) in a 850 nm thick nitrogen doped nanocrystalline porous titania film. Therefore, for a CO_2-to-hydrocarbon reaction process that involves a minimum of eight electrons the obtained visible-light hydrocarbon yield using high surface area nanotube array films is reasonable.

To help verify the contribution of individual wavelengths above 400 nm towards hydrocarbon production, the photocurrent spectrum of 600°C annealed nanotubes (no co-catalysts) kept in an aqueous 0.1 M Na_2SO_4 solution, platinum counter electrode, were collected using a monochromator and a 300 W Xe arc lamp. A dc bias of 0.5 V was applied across the electrode. The photocurrent and the excitation spectrum are given in Fig. 7.15b. Although the samples did generate charge carriers in the 400–500 nm wavelength range, their contribution was very low compared to that from the UV region. It has been shown that nitrogen doping creates localized states favoring carrier recombination [85, 89]. Ti^{3+} may be formed as a result of lattice reduction during nitrogen doping creating carrier trap states in the titania bandgap [90, 91]. The fact that the nanotubes failed to show a significant improvement in visible light utilization despite annealing them in oxygen ambient and thereby reducing the probability of Ti^{3+} -induced trap states shows that there are other trap states inherent to nitrogen doping, which may fundamentally prevent the samples from yielding the desired high visible light utilization for photocatalytic conversions.

In order to obtain a basic understanding of the CO_2 conversion process the following experiments were performed: (i) In the dark with a Cu/NT sample loaded in the reaction chamber. (ii) In sunlight using a reaction chamber filled with CO_2 and water vapor with either no sample inside, or (iii) a nanotube array sample with no co-catalysts. Analysis of the chamber gases after keeping a Cu/NT sample (area 3.5 cm^2, annealed at 600°C) at about 45°C in the dark for 3 h yielded, in addition to CO_2, 2.1 ppm of methane, <1 ppm of ethane and traces of oxygen and nitrogen. The dark reaction produced no significant hydrocarbon amounts, indicating the CO_2 conversion process is photocatalytically driven. Nearly the same gas concentrations, 1.7 ppm CH_4 and <1 ppm C_2H_6, were obtained when the reactant (CO_2, water vapor) filled chamber with no nanotube array sample inside was exposed to sunlight (90.6 mW/cm^2) for 3 h.

The reaction performed with a nanotube array sample (area 3.9 cm^2, 88 mW/cm^2 average sunlight illumination for 3.5 h) without co-catalyst surface-loading yielded a nominal hydrocarbon concentration of 10.5 ppm, indicating that without an appropriate co-catalyst titania cannot efficiently drive the CO_2 conversion reaction. It is well known that the titania valence band potential is much more positive (Normal Hydrogen Electrode, NHE, scale) than the oxidation potential of water, while the conduction band potential is close to the water reduction potential [92]. Therefore, titania readily oxidizes H_2O forming H^+ and the hydroxyl radical. A recent study shows the flat band potentials of titania nanotube arrays of lengths 0.5 and 1.0μm are -0.49 and -0.32 V at the isoelectric point of anatase titania (pH 5.8) [93]. These are close to the reduction potential of water, which is -0.34 V (vs. NHE) at this pH. The reduction potential of CO_2 is about 0.11 V more negative than that of water [94]. Therefore, titania cannot effectively reduce water nor CO_2 unless there is a high electron degeneracy, explaining the low hydrocarbon yield for nanotube array samples without any co-catalysts.

Temperature programmed desorption studies conducted on hydrothermally prepared titania nanotubes by Yu and co-workers [95] indicated that the ability of the nanotubes to adsorb CO_2 was considerably enhanced by Pt nanoparticle loading. In the case of Pt loaded titania nanoparticles used for photocatalytic applications, it is generally believed that the photo-excited electrons are trapped by the Pt islands, which then transfer to the surface species for reduction [96]. Hence it is reasonable to believe that Cu and Pt co-catalysts on the surface of the nanotube array help with water and carbon dioxide reduction. However, the fact that the CO concentration was higher for the case of Cu surface-loaded samples, and H_2 concentration was higher in the case of Pt loaded samples indicate that Cu is more active in reducing carbon dioxide while Pt prefers water reduction. Hence, an appropriate dispersion of Cu–Pt co-catalysts on the nanotube array surface has the advantage of carrying out both reactions at enhanced rates. The hole diffusion length in titania is ≈10 nm [97], the electron diffusion length is ≈10 μm [98], while half the nanotube wall thickness is approximately 10 nm. The high rate of carbon dioxide conversion can be attributed to the high surface area and nano-scale wall thickness of the nanotubes, enabling the surface species to readily receive both charge carriers generated near the surface due to the wave function overlap and those generated deep inside the wall via diffusion.

Tanaka et al. based on their XPS and Auger electron spectroscopy studies on TiO_2 and Pt/TiO_2 samples, demonstrated that CO_2 is reduced to CO and O_2 on Pt/TiO_2 [99]. Thus, a likely process in the photocatalytic reduction of CO_2 by Cu or Pt loaded samples is the reduction of CO_2 via the reaction $CO_2 + 2\,e^- \rightarrow CO + 1/2\,O_2$, which involves a free energy change of about 257 KJ/mol (1.33 eV per electron). The CO, thus formed, would react with atomic hydrogen to form hydrocarbons. Based on this hypothesis, the possible reactions involved in the photocatalytic formation of methane suggested by Varghese and co-workers are [71]:

$$H_2O + h^+ \rightarrow \bullet OH + H^+$$
$$H^+ + e^- \rightarrow H^\bullet$$
$$H^\bullet + H^\bullet \rightarrow H_2$$
$$2CO_2 + 4e^- \rightarrow 2CO + O_2$$
$$CO + 6e^- + 6H^+ \rightarrow CH_4 + H_2O$$

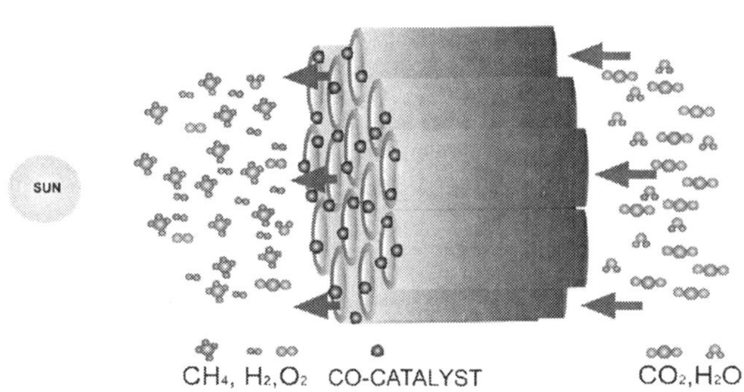

Fig. 7.16 Depiction of co-catalyst loaded flow-through nanotube array membrane for high rate photocatalytic conversion of CO_2 and water vapor into hydrocarbon fuels [71]

A possible back reaction is oxidation of the formed hydrocarbons by the TiO_2. However, CO_2 reduction experiments were conducted at about 85% relative humidity, a humidity level at which the titania surface must be covered with chemisorbed hydroxyls followed by physisorbed water molecules [100]. Further, since the hydrocarbon concentration is low compared to the reactant concentration the hydrocarbons will have limited access to the titania. Varghese and co-workers [71] suggest a stronger titania-hydrocarbon interaction would be expected when the hydrocarbon concentration increases to high values, hence rather than a batch-reactor as used for the work herein a flow-through photocatalytic membrane design would be desirable as depicted in Fig. 7.16 opening a new avenue for carbon recycling using renewable sources. It is an exciting prospect for the future.

References

1. Shankar K, Basham JI, Allam NK, Varghese OK, Mor GK, Feng X, Paulose M, Seabold JA, Choi KS, Grimes CA (2009). A review of recent advances in the use of TiO_2 nanotube and nanowire arrays for oxidative photoelectrochemistry. J Phys Chem C 113:6327–6359
2. Grimes CA (2007) Synthesis and application of highly-ordered arrays of TiO_2 nanotubes. J Mater Chem 17:1451–1457
3. Mor GK, Varghese OK, Paulose M, Shankar K, Grimes CA (2006) A review on highly ordered, vertically oriented TiO_2 nanotube arrays: Fabrication, material properties, and solar energy applications. Sol Energy Mater Sol Cells 90:2011–2075
4. Gong D, Grimes CA, Varghese OK, Hu W, Singh RS, Chen Z, Dickey EC (2001) Titanium oxide nanotube arrays prepared by anodic oxidation. J Mater Res 16:3331–3334
5. Cai Q, Paulose M, Varghese OK, Grimes CA (2005) The effect of electrolyte composition on the fabrication of self-organized titanium oxide nanotube arrays by anodic oxidation. J Mater Res 20:230–236
6. Paulose M, Shankar K, Yoriya S, Prakasam HE, Varghese OK, Mor GK, Latempa TA, Fitzgerald A, Grimes CA (2006) Anodic growth of highly ordered TiO_2 nanotube arrays to 134 μm in length. J Phys Chem B 110:16179–16184
7. Paulose M, Prakasam HE, Varghese OK, Peng L, Popat KC, Mor GK, Desai TA, Grimes CA (2007) TiO_2 Nanotube arrays of 1,000 μm in length by anodization of titanium foil: Phenol red diffusion. J Phys Chem C 111:14992–14997
8. Albu SP, Ghicov A, Macak JM, Schmuki P (2007) 250 μm long anodic TiO_2 nanotubes with hexagonal self-ordering. Physica Status Solidi 1:R65–R67
9. Prakasam HE, Shankar K, Paulose M, Grimes CA (2007) A new benchmark for TiO_2 nanotube array growth by anodization. J Phys Chem C 111:7235–7241
10. Yoriya S, Paulose M, Varghese OK, Mor GK, Grimes CA (2007) Fabrication of vertically oriented TiO_2 nanotube arrays using dimethyl sulfoxide electrolytes. J Phys Chem C 111:13770–13776
11. Allam NK, Grimes CA (2007) Formation of vertically oriented TiO_2 nanotube arrays using a fluoride free HCl aqueous electrolyte. J Phys Chem C 111:13028–13032
12. Allam NK, Shankar K, Grimes CA (2008) Photoelectrochemical and water photoelectrolysis properties of ordered TiO_2 nanotubes fabricated by Ti anodization in fluoride-free HCl electrolytes. J Mater Chem 18:2341–2348
13. Allam NK, Grimes CA (2008) Effect of cathode material on the morphology and photoelectrochemical properties of vertically oriented TiO_2 nanotube arrays. Sol Energy Mater Sol Cells 92:1468–1475

14. Mor GK, Varghese OK, Paulose M, Mukherjee N, Grimes CA (2003) Fabrication of tapered, conical-shaped titania nanotubes. J Mater Res 18:2588–2593
15. Mor GK, Shankar K, Paulose M, Varghese OK, Grimes CA (2005) Enhanced photocleavage of water using titania nanotube arrays. Nano Lett 5:191–195
16. Yoriya S, Mor GK, Sharma S, Grimes CA (2008) Synthesis of ordered arrays of discrete, partially crystalline titania nanotubes by Ti anodization using diethylene glycol electrolytes. J Mater Chem 18:3332–3336
17. Shankar K, Mor GK, Prakasam HE, Yoriya S, Paulose M, Varghese OK, Grimes CA (2007) Highly-ordered TiO_2 nanotube-arrays up to 220 μm in length: Use in water photoelectrolysis and dye-sensitized solar cells. Nanotechnology 18:065707
18. Shankar K, Mor GK, Fitzgerald A, Grimes CA (2007) Cation effect on the electrochemical formation of very high aspect ratio TiO_2 nanotube arrays in formamide-water mixtures. J Phys Chem C 111:21–26
19. Ruan C, Paulose M, Varghese OK, Mor GK, Grimes CA (2005) Fabrication of highly ordered TiO_2 nanotube arrays using an organic electrolyte. J Phys Chem B 109: 15754–15759
20. Paulose M, Peng L, Popat KC, Varghese OK, LaTempa TJ, Bao N, Desai TA, Grimes CA (2008) Fabrication of mechanically robust, large area, polycrystalline nanotubular/porous TiO_2 membranes. J Membrane Sci 319:199–205
21. Grimes CA, Varghese OK, Ranjan S (2008) Light, water, hydrogen: The solar generation of hydrogen by water photoelectrolysis. Springer, New York
22. Varghese OK, Gong D, Paulose M, Ong KG, Grimes CA, Dickey EC (2003) Crystallization and high-temperature structural stability of titanium oxide nanotube arrays. J Mater Res 18:156–165
23. Sharma S, Varghese OK, LaTempa TJ, Mor GK, Allam NK, Grimes, CA (2009) Enhanced crystallization and photoelectrochemical performance of titania nanotube array photoanodes by a post-thermal anneal vapor treatment. J Mater Chem 19:3895–3898
24. Varghese OK, Gong D, Paulose M, Ong KG, Grimes CA (2003) Hydrogen sensing using titania nanotubes, Sens Actuators B 93:338–344; Varghese OK, Mor GK, Grimes CA, Paulose M, Mukherjee N (2004) A titania nanotube-array room-temperature sensor for selective detection of hydrogen at low concentrations. J Nanosci Nanotechnol 4:733–737; Varghese OK, Gong D, Paulose M, Ong KG, Dickey EC, Grimes CA (2003) Extreme changes in the electrical resistance of titania nanotubes with hydrogen exposure. Adv Mater 15:624–662; Mor GK, Varghese OK, Paulose M, Ong KG, Grimes CA (2006) Fabrication of hydrogen sensors with transparent titanium oxide nanotube-array thin films as sensing elements. Thin Solid Films 496:42–48
25. Paulose M, Varghese OK, Mor GK, Grimes CA, Ong KG (2006) Unprecedented ultra-high hydrogen gas sensitivity in undoped titania nanotubes. Nanotechnology 17:398–402
26. Varghese OK, Yang X, Kendig J, Paulose M, Zeng K, Palmer C, Ong KG, Grimes CA (2006) A transcutaneous hydrogen sensor: From design to application. Sens Lett 4:120–128
27. Mor GK, Varghese OK, Paulose M, Grimes CA (2003) A self-cleaning room temperature titania-nanotube hydrogen gas sensor. Sens Lett 1:42–46
28. Mor GK, Carvalho MA, Varghese OK, Pishko MV, Grimes CA (2004) A room temperature TiO_2 nanotube hydrogen sensor able to self-clean photoactively from environmental contamination. J Mater Res 19:628–634
29. Allam NK, Feng X, Grimes CA (2008) Self-assembled fabrication of vertically oriented Ta_2O_5 nanotube arrays, and membranes thereof, by one-step Tantalum anodization. Chem Mater 20:6477–6481. Ruckh T, Porter JR, Allam NK, Feng X, Grimes CA, Popat KC (2009) Nanostructured tantala as a template for enhanced osseointegration. Nanotechnology 20:045102
30. Shin Y, Lee S (2009) A freestanding membrane of highly ordered anodic ZrO_2 nanotube arrays. Nanotechnology 20:Art. No. 105301
31. LaTempa TJ, Grimes CA (2009) Hematite nanotube arrays by potentiostatic anodization of Ti foil: Photoelectrochemical properties. J Phys Chem C 113 (in press)

32. Varghese OK, Paulose M, Shankar K, Mor GK, Grimes CA (2005) Water-photolysis properties of micron-length highly-ordered titania nanotube-arrays. J Nanosci Nanotechnol 5:1158–1165
33. Shankar K, Mor GK, Prakasam HE, Yoriya S, Paulose M, Varghese OK, Grimes CA (2007) Highly-ordered TiO_2 nanotube arrays up to 220 μm in length: Use in water photoelectrolysis and dye-sensitized solar cells. Nanotechnology 18:065707–065711
34. Mor GK, Prakasam HE, Varghese OK, Shankar K, Grimes CA (2007) Vertically oriented Ti–Fe–O nanotube array films: Toward a useful material architecture for solar spectrum water photoelectrolysis. Nano Lett 7:2356–2364
35. Mor GK, Varghese OK, Wilke RHT, Sharma S, Shankar K, Latempa TJ, Choi KS, Grimes CA (2008) p-Type Cu–Ti–O nanotube arrays and their use in self-biased heterojunction photoelectrochemical diodes for hydrogen generation. Nano Lett 8:1906–1911
36. Zhao J, Wang X, Kang Y, Xu X, Li Y (2008) Photoelectrochemical activities of W-doped titania nanotube arrays fabricated by anodization. IEEE Photonics Technol Lett 20:1213–1215
37. Chen SG, Paulose M, Ruan C, Mor GK, Varghese OK, Grimes CA (2006) Electrochemically synthesized CdS nanoparticle-modified TiO_2 nanotube-array photoelectrodes: Preparation, characterization, and application to photoelectrochemical cells. J Photochem Photobiol A 177:177–184
38. Seabold JA, Shankar K, Wilke RHT, Paulose M, Varghese OK, Grimes CA, Choi KS (2008) Photoelectrochemical properties of heterojunction $CdTe/TiO_2$ electrodes constructed using highly ordered TiO_2 nanotube arrays. Chem Mater 20:5266–5273
39. Mor GK, Shankar K, Paulose M, Varghese OK, Grimes CA (2006) Use of highly-ordered TiO_2 nanotube arrays in dye-sensitized solar cells. Nano Lett 6:215–218
40. Zhu K, Neale NR, Miedaner A, Frank AJ (2006) Enhanced charge-collection efficiencies and light scattering in dye-sensitized solar cells using oriented TiO_2 nanotubes arrays. Nano Lett 7:69–74
41. Mor GK, Shankar K, Paulose M, Varghese OK, Grimes CA (2007) High efficiency double heterojunction polymer photovoltaic cells using highly ordered TiO_2 nanotube arrays. Appl Phys Lett 91:152111
42. Shankar K, Mor GK, Paulose M, Varghese OK, Grimes CA (2008) Effect of device geometry on the performance of TiO_2 nanotube array-organic semiconductor double heterojunction solar cells. J Non-Cryst Solids 354:2767–2771
43. Shankar K, Mor GK, Prakasam HE, Varghese OK, Grimes CA (2007) Self-assembled hybrid polymer-TiO_2 nanotube array heterojunction solar cells. Langmuir 23:12445–12449
44. Popat KC, Eltgroth M, LaTempa TJ, Grimes CA, Desai TA (2007) Decreased Staphylococcus epidermis adhesion and increased osteoblast functionality on antibiotic-loaded titania nanotubes. Biomaterials 28:4880–4888
45. Popat KC, Eltgroth M, LaTempa TJ, Grimes CA, Desai TA (2007) Titania nanotubes: A novel platform for drug-eluting coatings for medical implants? Small 11:1878–1881
46. Popat KC, Leoni L, Grimes CA, Desai TA (2007) Influence of engineered titania nanotubular surfaces on bone cells. Biomaterials 28:3188–3197
47. Peng L, Mendelsohn AD, LaTempa TJ, Yoriya S, Grimes CA, Desai TA (2009) Long-term small molecule and protein elution from TiO_2 nanotubes. Nano Lett 9:1932–1936
48. Liu S, Chen A (2005) Coadsorption of horseradish peroxidase with thionine on TiO_2 nanotubes for biosensing. Langmuir 21:8409–8413
49. Xie Y, Zhoua L, Huang H (2007) Bioelectrocatalytic application of titania nanotube array for molecule detection. Biosens Bioelectron 22:2812–2818
50. Roy SC, Paulose M, Grimes CA (2007) The effect of TiO_2 nanotubes in the enhancement of blood clotting for the control of hemorrhage. Biomaterials 28:4667–4672
51. Fabregat-Santiago F, Barea EM, Bisquert J, Mor GK, Shankar K, Grimes CA (2008) High carrier density and capacitance in TiO_2 nanotube arrays induced by electrochemical doping. J Am Chem Soc 130:11312–11316

52. Bisquert J (2008) Physical electrochemistry of nanostructured devices. Phys Chem Chem Phys 10:49–72
53. Bisquert J (2003) Chemical capacitance of nanostructured semiconductors: Its origin and significance for nanocomposite solar cells. Phys Chem Chem Phys 5:5360–5364
54. Bisquert J, Fabregat-Santiago F, Mora-Seró I, Garcia-Belmonte G, Barea EM, Palomares E (2007) A review of recent results on electrochemical determination of the density of electronic states of nanostructured metal-oxide semiconductors and organic hole conductors. Inorganica Chim Acta 361:684–698
55. Kopidakis N, Benkstein KD, van de Lagemaat J, Frank AJ (2003) Transport-limited recombination of photocarriers in dye-sensitized nanocrystalline TiO_2 solar cells. J Phys Chem B 107:11307–11315
56. Bisquert J, Garcia-Belmonte G, Fabregat-Santiago F, Ferriols NS, Bogdanoff P, Pereira EC (2000) Doubling exponent models for the analysis of porous film electrodes by impedance. Relaxation of TiO_2 nanoporous in aqueous solution. J Phys Chem B 104:2287–2298
57. Bisquert J (2002) Theory of the impedance of electron diffusion and recombination in a thin layer. J Phys Chem B 106:325–333
58. Fabregat-Santiago F, Bisquert J, Garcia-Belmonte G, Boschloo G, Hagfeldt A (2005) Influence of electrolyte in transport and recombination in dye-sensitized solar cells studied by impedance spectroscopy. Solar Energy Mater Solar Cells 87:1–4
59. Wang Q, Ito S, Grätzel M, Fabregat-Santiago F, Mora-Seró I, Bisquert J, Bessho T, Imai H (2006) Characteristics of high efficiency dye-sensitized solar cells. J Phys Chem B 50:25210–25221
60. Winter M, Brodd RJ (2004) What are batteries, fuel cells, and supercapacitors? Chem Rev 104:4245–4269
61. Lemon BI, Hupp JT (1996) Photochemical quartz crystal microbalance study of the nanocrystalline titanium dioxide semiconductor electrode/water interface: Simultaneous photoaccumulation of electrons and protons. J Phys Chem 100:14578–14580
62. Patermarakis G, Moussoutzanis K (1995) Mathematical models for the anodization conditions and structural features of porous anodic Al_2O_3 films on aluminum. J Electrochem Soc 142:737–743
63. Muñoz AG (2007) Semiconducting properties of self-organized TiO_2 nanotubes. Electrochim Acta 52:4167–4176
64. Ghicov A, Tsuchiya H, Hahn R, Macak JM, Muñoz AG, Schmuki P (2006) TiO_2 nanotubes: H^+ insertion and strong electrochromic effects. Electrochem Commun 8:528–532
65. Fabregat-Santiago F, Mora-Seró I, Garcia-Belmonte G, Bisquert J (2003) Cyclic voltammetry studies of nanoporous semiconductors. Capacitive and reactive properties of nanocrystalline TiO_2 electrodes in aqueous electrolyte. J Phys Chem B 107:758–768
66. Fabregat-Santiago F, Bisquert J, Palomares E, Otero L, Kuang D, Zakeeruddin SM, Gratzel M (2007) Correlation between photovoltaic performance and impedance spectroscopy of dye-sensitized solar cells based on ionic liquids. J Phys Chem C 111:6550–6560
67. Sze SM (1981) Physics of semiconductor devices, 2nd edn. John Wiley and Sons, New York
68. Yan SG, Hupp JT (1996) Semiconductor-based interfacial electron-transfer reactivity: Decoupling kinetics from pH-dependent band energetics in a dye-sensitized titanium dioxide aqueous solution system. J Phys Chem 100:6867–6870
69. Wang Q, Zhang Z, Zakeeruddin SM, Gratzel M (2008) Enhancement of the performance of dye-sensitized solar cell by formation of shallow transport levels under visible light illumination. J Phys Chem C 112:7084–7092
70. Chappel S, Grinis L, Ofir A, Zaban A (2005) Extending the current collector into the nanoporous matrix of dye sensitized electrodes. J Phys Chem B 109:1643–1647
71. Varghese OK, Paulose M, LaTempa TJ, Grimes CA (2009) High-rate solar photocatalytic conversion of CO_2 and water vapor to hydrocarbon fuels. Nano Lett 9:731–737
72. Buesseler KO, Doney SC, Karl DM, Boyd PW, Fei C, Coale KH, De Baar JWH, Falkowski PG, Johnson KS, Lampitt RS, Michaels AF, Nagvi SWA, Smetacek V, Takeda S, Watson AJ

(2008) Ocean iron fertilization- moving forward in a sea of uncertainty. Science 319:162–163
73. Inoue T, Fujishima A, Konishi S, Honda K (1979) Photoelectrochemical reduction of carbon dioxide in aqueous suspensions of semiconductor powder. Nature 277:637–638
74. Halmann M, Ulman M, Blajeni BA (1983) Photochemcial solar collector for the photo-assisted reduction of aqueous carbon dioxide. Solar Energy 31:429–431
75. Cook RL, MacDuff RC, Sammells AF (1988) Photoelectrochemical carbon dioxide reduction to hydrocarbons at ambient temperature and pressure. J Electrochem Soc 135:3069–3070
76. Adachi K, Ohta K, Mizuna T (1994) Photocatalytic reduction of carbon dioxide to hydrocarbon using copper-loaded titanium dioxide. Solar Energy 53:187–190
77. Anpo M, Yamashita H, Ichihashi Y, Ehara S (1995) Photocatalytic reduction of CO_2 with H_2O on various titanium oxide catalysts. J Electroanal Chem 396:21–26
78. Thampi KR, Kiwi J, Gratzel M (1987) Methanation and photomethanation of carbon dioxide at room temperature and atmospheric pressure. Nature 327:506–508
79. Melsheimer J, Guo W, Ziegler D, Wesemann M, Schlogl R (1991) Methanation of carbon dioxide over Ru/titania at room temperature: explorations for a photoassisted catalytic reaction. Catal Lett 11:157–168
80. Tan SS, Zou L, Hu E (2007) Photosynthesis of hydrogen and methane as key components for clean energy system. Sci Technol Adv Mater 9:89–92
81. Tan SS, Zou L, Hu E (2006) Photocatalytic reduction of carbon dioxide into gaseous hydrocarbon using TiO_2 pellets. Catal Today 115:269–273
82. Lo C-C, Hung C-H, Yuan C-S, Wu J-F (2007) Photoreduction of carbon dioxide with H_2 and H_2O over TiO_2 and ZrO_2 in a circulated photocatalytic reactor. Solar Energy Mater Solar Cells 91:1765–1774
83. Nishimura A, Sugiura N, Kato S, Maruyama N (2004) High yield CO_2 conversion into CH_4 by photocatalyst multiplayer film, 2nd International Energy Conversion Engineering Conference, 16–19 August 2004. Rhode Island, USA
84. Lindgren T, Mwabora JM, Avendano E, Jonsson J, Hoel A, Granqvist CG, Lindquist SE (2007) Photoelectrochemical and optical properties of nitrogen doped titanium dioxide films prepared by reactive dc magnetron sputtering. J Phys Chem B 107:5709–5716
85. Emeline AV, Kuznetsov VN, Rybchuk VK, Serpone N (2008) Visibe-light-active titania photocatalysts: The case of N-doped TiO_2S- Properties and some fundamental issues. Inter J Photoenergy 258394:1–19
86. Asahi R, Morikawa T, Ohwaki T, Aoki K, Taga Y (2001) Visible-light photocatalysts in nitrogen-doped titanium oxides. Science 293:269–271
87. Chen X, Lou Y, Samia ACS, Burda C, Gole JL (2005) Formation of oxynitride as the photocatalytic enhancing site in nitrogen-doped titania nanocatalysts: Comparison to a commercial nanopowder. Adv Funct Mater 15:41–49
88. Liu G, Li F, Wang D-W, Tang D-M, Liu C, Ma X, Lu GQ, Cheng H-M (2008) Electron field emission of a nitrogen-doped TiO_2 nanotube array, Nanotechnology 19 025606:1–6
89. Torres GR, Lindgren T, Lu J, Granqvist CG, Lindquest SE (2004) Photoelectrochemical study of nitrogen doped titanium dioxide for water oxidation. J Phys Chem B 108:5995–6003
90. Batzill M, Morales EH, Diebold U (2006) Influence of nitrogen doping on the defect formation and surface properties of TiO_2 rutile and anatase, Phys Rev Lett 96 Art. No. 026103
91. Henderson MA, White JM, Uetsuka H, Onishi H (2003) Photochemical charge transfer and trapphing at the interface between an organic adlayer and an oxide semiconductor. J Am Chem Soc 125:14974–14975
92. Gerischer H (1979) Solar photoelectrolysis with semiconductor electrodes. In: Seraphin BO (ed) Solar energy conversion: Solid-state physics aspects, vol 31, Topics in applied physics. Springer, NY, pp 115–172

93. Munoz AG (2007) Semiconducting properties of self-organized TiO_2 nanotubes. Electrochim Acta 52:4167–4176
94. Sayama K, Arakawa H (1993) Photocatalytic decomposition of water and photocatalytic reduction of carbon dioxide over ZrO_2 catalyst. J Phys Chem 97:531–533
95. Yu KP, Yu WY, Kuo MC, Liou YC, Chien SH (2008) Pt/titania nanotube: A potential catalyst for CO_2 adsorption and hydrogenation. Appl Catal B 84:112–118
96. Linsebigler AL, Lu G, Yates JT (1995) Photocatalysis on TiO_2 surfaces: Principles, mechanisms and selected results. Chem Rev 95:735–758
97. Salvador P (1984) Hole diffusion length in n-TiO_2 single crystals and sintered electrodes: Photoelectrochemical determination and comparative analysis. J. Appl Phys 55:2979–2983
98. Dloczik L, Ileperuma O, Lauermann I, Peter LM, Ponomarev EA, Redmond G, Shaw NJ, Uhlendorf I (1997) Dynamic response of dye-sensitized nanocrystalline solar cells: Characterization by intensity-modulated photocurrent spectroscopy. J Phys Chem B 101:10281–10286
99. Tanaka K, Miyahara K, Toyoshima I (1984) Adsorption of CO_2 on TiO_2 and Pt/TiO_2 studied by X-ray photoelectron spectroscopy and Auger electron spectroscopy. J Phys Chem 88:3504–3508
100. Varghese OK, Grimes CA (2003) Metal oxide nanoarchitectures for environmental sensing. J Nanosci Nanotechnol 3:277–293

Index

A

D

G

H

I

Q

R